MW01531331

ACS SYMPOSIUM SERIES **974**

Vanadium: The Versatile Metal

Kenneth Kustin, Editor
Brandeis University, Emeritus

João Costa Pessoa, Editor
IST–Technical University of Lisboa

Debbie C. Crans, Editor
Colorado State University

Sponsored by the
Division of Inorganic Chemistry, Inc.

American Chemical Society, Washington, DC

Library of Congress Cataloging-in-Publication Data

Vanadium : the versatile metal / Kenneth Kustin, editor, João Costa Pessoa, editor, Debbie C. Crans, editor ; sponsored by the Division of Inorganic Chemistry, Inc.

p. cm.—(ACS symposium series ; 974)

Includes bibliographical references and index.

ISBN 978–0–8412–7446–4 (alk. paper)

1. Vanadium. 2. Vanadium compounds.

I. Kustin, Kenneth, 1934- II. Pessoa, João Costa. III. Crans, Debbie Catharina, IV. American Chemical Society. Division of inorganic Chemistry.

QD181.V2V36   2007
546'.522—dc22

2007060812

The paper used in this publication meets the minimum requirements of American National Standard for Information Sciences—Permanence of Paper for Printed Library Materials, ANSI Z39.48–1984.

Copyright © 2007 American Chemical Society

Distributed by Oxford University Press

All Rights Reserved. Reprographic copying beyond that permitted by Sections 107 or 108 of the U.S. Copyright Act is allowed for internal use only, provided that a per-chapter fee of $36.50 plus $0.75 per page is paid to the Copyright Clearance Center, Inc., 222 Rosewood Drive, Danvers, MA 01923, USA. Republication or reproduction for sale of pages in this book is permitted only under license from ACS. Direct these and other permission requests to ACS Copyright Office, Publications Division, 1155 16th Street, N.W., Washington, DC 20036.

The citation of trade names and/or names of manufacturers in this publication is not to be construed as an endorsement or as approval by ACS of the commercial products or services referenced herein; nor should the mere reference herein to any drawing, specification, chemical process, or other data be regarded as a license or as a conveyance of any right or permission to the holder, reader, or any other person or corporation, to manufacture, reproduce, use, or sell any patented invention or copyrighted work that may in any way be related thereto. Registered names, trademarks, etc., used in this publication, even without specific indication thereof, are not to be considered unprotected by law.

PRINTED IN THE UNITED STATES OF AMERICA

Foreword

The ACS Symposium Series was first published in 1974 to provide a mechanism for publishing symposia quickly in book form. The purpose of the series is to publish timely, comprehensive books developed from ACS sponsored symposia based on current scientific research. Occasionally, books are developed from symposia sponsored by other organizations when the topic is of keen interest to the chemistry audience.

Before agreeing to publish a book, the proposed table of contents is reviewed for appropriate and comprehensive coverage and for interest to the audience. Some papers may be excluded to better focus the book; others may be added to provide comprehensiveness. When appropriate, overview or introductory chapters are added. Drafts of chapters are peer-reviewed prior to final acceptance or rejection, and manuscripts are prepared in camera-ready format.

As a rule, only original research papers and original review papers are included in the volumes. Verbatim reproductions of previously published papers are not accepted.

ACS Books Department

Contents

Haloperoxidases: Mechanism and Model Studies

Enzymology, Toxicology, and Transport

Coordination Chemistry: Speciation and Structure

New Materials and Processes

Indexes

Preface: Recent Advances and Highlights of the 5th International Vanadium Symposium

João Costa Pessoa [1], Debbie C. Crans[2], and Kenneth Kustin[3]

[1]Centro Química Estrutural, Instituto Superior Técnico, TU Lisbon,
Av. Rovisco Pais, 1049–001 Lisboa, Portugal
[2]Department of Chemistry, Colorado State University,
Fort Collins, CO 80523–1872
[3]Department of Chemistry, Emeritus, Brandeis University, MS015,
Waltham, MA 02454–9110

Knowledge of the chemistry and biochemistry of vanadium has increased enormously since the early 1980s, particularly that of vanadium(III), (IV) and (V). Part of this interest arose through the synthetic utility of several vanadium complexes in catalysis and as new materials, but particularly after the new understanding that vanadium compounds have insulin-enhancing properties (*1*), the discovery of several vanadium haloperoxidases in marine macroalgae (*2-4*) and vanadium-containing nitrogenases (*5*). The presence of vanadium in several accumulators, Amanita toadstools (*6*), ascidians (*7*), peroxidases of fungi and lichens, and fan worms, (*8-9*) has also contributed to interest in vanadium chemistry and biochemistry. The catalytic effects of vanadium on phosphoryl transfer and other enzymes (*10-12*) has also been the subject of intensive research. Recently, a report on the existence of vanadate-respiring bacteria, suggests that anaerobic metal respiration (including vanadium) may be associated with bacteria in hydrothermal vents (*13*). The possibility that these processes involve ATP production was proposed, (*14*) and if confirmed is likely to represent a new area of research in vanadium science.

These and many other advances in vanadium science were presented at the 5th International Symposium on the Chemistry and Biological Chemistry of Vanadium held September 10-14, 2006 at San Francisco, CA as part of the National ACS Meeting. This was the first International Vanadium Symposium to take place on U. S. soil, the previous meetings being held at Cancun, Mexico (1), Berlin, Germany (2), Osaka, Japan (3), and Szeged, Hungary (4). Below, we summarize highlights of the 5th Vanadium Symposium, and show how each chapter in the book relates to mainstream chemistry and biological chemistry.

Catalysis

The design of novel reactions that proceed with high atom economy and that enable multiple transformations through a shorter reaction sequence is a fundamental task of modern chemistry. Over the past two decades vanadium chemistry has developed into an important component of organic synthesis addressing these goals. Namely vanadium complexes exhibit a rich redox chemistry providing potential tools in organic synthesis (15). The use of oxovanadium compounds in C-C bond formation is well demonstrated in many organic transformations already reported. For example, oxovanadium compounds such as $VOCl_3$ and VOF_3 promoted oxidative biaryl coupling (16); low valent V species such as $CpV(CO)_4$ and Cp_2VCl_2, in combination with Zn and R_3SiCl are used for efficient pinacolic coupling of aliphatic and aromatic aldehydes, ketones and aldimines in a highly selective manner (17); and this methodology has also been used for the synthesis of several bioactive molecules (18). Vanadium salen-catalyzed asymmetric silylcyanation of aldehydes has also become a novel route to the synthesis of chiral cyanohydrins (19). The compound $VO(OEt)Cl_2$ has been extensively explored by the group of Hirao in the oxidation of main group examples of organometallics such as organoboranes (20), organolithium and organomagnesium (21), organoaluminum (22), organozinc (23-25) and organozirconate (26) compounds.

Aldol-type condensations have been carried out in the presence of vanadium (27), as well as Michael-type base-catalyzed additions (28) and asymmetric hetero Diels-Alder reactions (29). Several other vanadium mediated C-C bond formation reactions have also been reported (30), e.g., oxidative coupling of phenols, naphthols and arenes, asymmetric synthesis of BINOLs, Mannich-type (31) and modified Mannich-type reactions. The sulfoxidation of alkanes has also been reported (32).

Vanadium exists in oxidation states ranging from -3 to +5 and generally converts between oxidation states via one-electron redox processes. The inclusion of vanadium in enzymes such as haloperoxidases and nitrogenases reveals the importance of its redox chemistry. Vanadium complexes, including organometallic compounds, exist in a variety of configurations depending on their oxidation states and coordination numbers. Their properties have permitted the development of a wide range of organic reactions by controlling the redox processes of the vanadium catalysts or activators.

Radical species are useful intermediates in organic synthesis and as emphasized by Hirao in Chapter 1 one-electron reduction or oxidation of organic compounds using the redox processes of metals, particularly those of early transition metals including vanadium, titanium, and manganese provide useful and practical routes to generate anion radicals or cation radicals, and have been extensively employed for this purpose (33-34). The redox function may be tuned by a ligand or a solvent, allowing an efficient interaction through the orbitals of metals and substrates for facile electron transfer.

The use of stoichiometric or excess amounts of metallic reducing or oxidizing agents to complete the reactions may be a limitation of the procedures. Therefore, to avoid the use of expensive and/or toxic metallic reagents, catalytic systems should be developed. For this purpose the selection of stoichiometric co-reductants or co-oxidants is essential for the reversible cycle of a catalyst (33-34), and an alternative method for recycling a catalyst is electrolysis (35). A metallic co-reductant is ultimately converted to the corresponding metal salt in a higher oxidation state, which may work as a Lewis acid to facilitate the reduction reaction or, on the contrary, to impede the reaction. Taking these interactions into account, the requisite catalytic system for efficient one-electron reduction is envisioned to be formed by multi-component interaction. As described by Hirao in Chapter 1 and in previous reports (33-34), the active catalyst is smoothly regenerated by redox interaction with the co-reductant. Steric control by means of coordination may permit stereoselective and/or stereospecific transformations. Additive promoters, which are considered to contribute to the redox cycle, possibly facilitate the electron transfer and liberate the catalyst from the reaction adduct.

Redox interaction between a metal and a metal or a ligand (or an organic group) through inner- or outer-sphere electron transfer is considered to change the oxidation state of these components. Precise regulation of the redox process based on the electronic interaction between the HOMO and LUMO orbitals is required for this purpose, and this interaction has allowed the development of novel and selective methods for reductive or oxidative transformations of organic compounds under redox potential control (33). Chapter 1 by Hirao describes the synthetically useful redox reactions promoted or catalyzed by vanadium compounds, namely, the low-valent vanadium-catalyzed reduction reactions including dehalogenation, pinacol coupling, radical cyclization (33) and the related radical reaction have been developed by constructing multi-component redox systems in combination with a co-reductant and an additive promoter. High stereoselectivity is reported to be attained in these catalytic transformations. This kind of catalytic reaction is very important biologically and investigation of nitrogen fixation models is challenging from this point of view (36-37).

Oxovanadium(V) compounds, which are Lewis acids with oxidation capability, may induce one-electron oxidation reactions and a variety of oxidative transformations have been developed and have been reviewed by Hirao in Chapter 1 and refs. (15, 33). The oxidation capability and redox potential are effectively controlled by the ligand of the oxovanadium(V) compounds. Hirao in Chapter 1 (36) also summarizes its synthetic utility, which permits versatile dehydrogenative aromatization, allylic oxidation, ring-opening oxygenation, oxidative decarboxylation, ring-opening oxidation, and so on (38-39) Oxidative ligand coupling of organoaluminums, organoborons and organozincs, and the ate complexes are also achieved by treatment with oxovanadium(V) compounds, and a catalytic reaction of organoborates is reported to proceed under molecular oxygen.

As a consequence of their low radius/charge ratio, vanadium(V) centres are usually strong Lewis acids, and this property makes them suitable for the activation of peroxidic reagents. The ability of oxovanadium(V) compounds to activate hydrogen peroxide has been used for promoting several oxidation reactions (brominations, epoxidations of alkanes and allylic alcohols, oxidation of sulfides to sulfoxides and sulfones, hydroxylations of alkanes and arenes, oxidations of primary and secondary alcohols, etc.). The active species has been identified in stoichiometric reactions as mononuclear oxoperoxovanadium(V) complexes.

The functionalization of alkanes, under mild conditions, toward the synthesis of organic products with commercial value (e.g., alcohols, ketones or carboxylic acids) is a challenge to modern chemistry (40). Currently, alkanes are mainly used as fuels, but other applications are being looked for, which has been hampered by their unreactivity. This can be promoted by the use of a suitable metal centre which can activate (i) the alkane itself or (ii) a reagent (like O_2 or H_2O_2) that can thus generate a highly reactive species (e.g. a metal-oxo radical, a metal-peroxo or a metal-hydroperoxo species or an hydroxo radical) capable of reacting with the alkane. During the last few years, a number of vanadium complexes, in particular with biological or pharmacological significance, have been applied successfully as catalysts or catalyst precursors for the oxidation and carboxylation of alkanes.

In Chapter 4 Pombeiro and coworkers discuss the use of a number of vanadium compounds, including Amavadine and its models, as catalysts for the hydroxylation, oxygenation, halogenation or carboxylation of alkanes. The substrates are gaseous (in particular methane) and liquid (both linear and cyclic), and afford, under mild or moderate conditions, the corresponding alcohols, ketones, organohalides or carboxylic acids. Some of the systems operate in peroxidative liquid biphasic media, whereas others involve dioxygen as the oxidant and V-catalysts supported on modified silica gel. Scheme 1 summarizes the reactions reported by Pombeiro and coworkers (41) in Chapter 4.

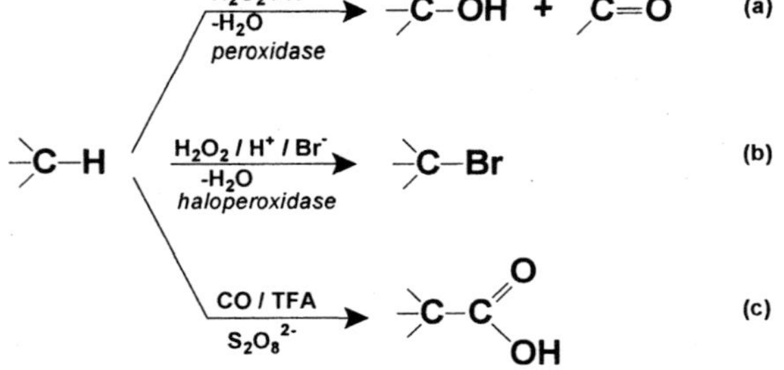

Scheme 1 - Peroxidative oxidation (a), haloperoxidation (b) and carboxylation (c) of alkanes catalysed by vanadium complexes. Reproduced from Chapter 4 of this volume.

This work was initiated by searching for peroxidase and haloperoxidase behavior (toward alkanes) of Amavadine, and its model $[V(HIDA)_2]^{2-}$ [HIDA = 2,2'-(hydroxyimino)diacetate]. These vanadium complexes act as (i) an electron-transfer mediator in the electrocatalytic oxidation of thiols according to a Michaelis-Menten mechanism (41), and (ii) a peroxidase toward thiols by catalysing their oxidation by H_2O_2 (42).

Vanadium complexes with N,O- or O,O-ligands act as remarkably active catalysts or catalyst precursors for a variety of alkane functionalization reactions under mild conditions, including their (i) hydroxylation or oxygenation with H_2O_2 or O_2 to give alcohols and ketones, (ii) peroxidative halogenation with H_2O_2 to afford organohalides, and (iii) carboxylation, not only with CO but also in the absence of this gas when the trifluoroacetic acid solvent behaves as the carbonyl source.

Various vanadium(IV or V) complexes with N,O- or O,O-donor groups other than Amavadine [e.g. $VO(CF_3COO)_2$, $VO(CF_3SO_3)_2$ and $VOSO_4$] were also tested and shown to exhibit peroxidase and haloperoxidase behaviour toward alkanes, and to catalyze their carboxylations. Amavadine and its models normally appear as the most active catalyst precursors and, although amavadine's biological role still remains unknown, the studies suggest it can act as a peroxidase and/or a haloperoxidase eventually involved in the protection system of the Amanita fungi.

From the perspective of developing sustainable reaction media, and considering the improvement obtained by reducing the amount of volatile organic solvents used in an industrial process, numerous research groups have either developed catalysts with better performances, or significantly decreased the environmental impact of solvents. For example, in Chapter 2, a simple and effective procedure for oxybromination of unsaturated substrates in a biphasic system H_2O/CH_2Cl_2 has been proposed by Conte's group (43-44), where the V-catalyst, oxidant, and bromide are dissolved in the aqueous phase, while the substrate is dissolved in the organic phase. In this synthetic method mild conditions are used, namely a sustainable oxidant, H_2O_2, and a sustainable source of positive bromine such as KBr. Moreover, the by-products are bromide and water.

Ionic liquids (ILs), because of their peculiar properties, have been proposed as novel and environmentally benign reaction media for several organic syntheses, and the use of some hydrophilic and hydrophobic ILs in V(V)-catalyzed oxidations with peroxides has been addressed. Chlorinated solvents were substituted by ILs and a remarkable increase of both yield and selectivity were achieved in the oxidation of styrene and ethynylbenzene, obtaining synthetically interesting results (45). More recently, a modification of the two-phase procedure for oxybromination of double and triple bonds with $M/H_2O_2/Br^-$ (M=V(V) or Mo(VI)) has been presented, in which the chlorinated solvent is replaced by hydrophobic ILs (45-46). Evidence was given that such variation gives rise to even more sustainable processes, characterized by higher rates and better selectivity.

It is well known that vanadate-dependent haloperoxidases from marine macro-algae (such as *Ascophyllum nodosum* and *Corallina officinalis*, and the fungus *Curvularia inaequalis*) contain vanadate(V).

Vanadium is the center of a trigonal bipyramid (see below), and the enzyme catalyzes the oxidation, by peroxide, of halide to a species, presumably hypohalous acid, which can halogenate non-enzymatically a large variety of organic substrates. In the course of the catalytic reaction, a peroxo intermediate is formed (*47-48*) and the peroxo group is subject to protonation (*48*), thus providing a center for nucleophilic attack of the substrate, e.g. bromide, leading to the generation of a hypobromito intermediate (*49*) which is then released in the form of hypobromous acid.

Many vanadium complexes are functional models of V-haloperoxidases and show catalytic activity towards oxidative halogenations, along with other oxidation reactions (epoxidation, hydroxylation, alcohol oxidation, sulfoxidation), including asymmetric processes (*50-64*). Immobilization of some of these materials on polymer supports (*65*), and encapsulation of complexes in zeolite-type materials has also been reported by Maurya (*66-67*) and others (*68*) as approaches to avoid/decrease the decomposition of the complexes during the catalytic reactions. Protonation of these compounds is essential for reactivity and is supported by quantum chemical studies showing a decrease in the reaction barrier for the oxidation of bromine and thioethers (*64, 69*). The haloperoxidases also exhibit a sulfideperoxidase activity (*70*) in so far as (prochiral) sulfides are oxygenated to (chiral) sulfoxides plus some sulfone. This latter reaction is of particular interest, because of the potential applications of chiral sulfoxides (*70-71*).

Consequently, several groups have been working on vanadium-based model systems of the haloperoxidases during the past decade (*72-77*), aiming at the synthesis of chiral enantio-pure sulfoxides. Several approaches have been developed namely the use of salen-type ligands (*78*), complexone derivatives (*62*), and di- to tetradentate aminoalcohols (*75*). In Chapter 5 the Rehder group reports on the usefulness of aminoalcohols containing an allyl substituent attached to the nitrogen. These complexes of composition [VO(OAl)L], where OAl derives from alcohols and L is an aminoalcohol, are *functional* models of the sulphide-peroxidase activity of the enzymes since they mimic the oxygenation, by peroxide, of prochiral sulfides to chiral sulfoxides. In the first part of Chapter 3 Hartung discusses the use of tridentate Schiff-base vanadium(V) complexes in oxidation for stereoselective tetrahydrofuran formation from alkenols and *tert*-butyl hydroperoxide. The second part this chapter deals with the synthesis of vanadium(V) complexes starting from 2,6-dihydroxymethyl-substituted piperidines and their use in oxidation catalysis with emphasis on reactions in protic media.

Since the early stages of his studies Natta established the ability if simple di- and tetravalent vanadium salts to perform random and super-random olefin copolymerization (*79*), and the ability of vanadium to act as a catalyst for olefin homo- and co-polymerization was reviewed by Gambarotta (*80*). Although the catalytic activity is a few orders of magnitude smaller than that displayed by Group IV based catalysts, the unique quality of the polymers produced by vanadium catalysts has made them irreplaceable for the manufacture of synthetic rubber and elastomers (EPDM). The stability of V-C bonds towards reductive elimination plays a critical role in the behavior of V complexes as Ziegler-Natta catalysts, and several authors have ascribed the poor activity of V catalysts to the tendency to perform reductive elimination towards inactive divalent species. For this reason the use of mild oxidizing agents capable of restoring the trivalent state is a crucial point in the V catalyzed EPDM technology (*81*). Gambarotta (*80*) examined several ligand features and their ability to affect the vanadium oxidation state has been related to the stability of the V-C bond and catalytic activity, but no systematic features for the design of efficient vanadium polymerization catalysts have yet been revealed.

The communication of the Rosenthal group in Chapter 6 contains new results on the synthesis and application of oxovanadium alkoxides with ether alcohol ligands. For a better understanding of the recent results a summary of earlier work by the authors (*82-84*) is also included in this chapter, and completed by an extended discussion. The ether alcohol ligands offer the possibility of stabilizing vanadium compounds in their high oxidation states through chelation by an additional oxygen donor. With this approach Rosenthal and coworkers are able to synthesize compound classes which were previously inaccessible. The presence of remaining chloride ligands in mixed alkoxo chloro complexes makes the metal atom functionalizable; *e.g.* to introduce alkyl groups, and are also essential for olefin polymerization (*85*).The catalytic epoxidation of the complexes with *tert*-butyl hydroperoxide is also discussed. Catalytic polymerizations in combination with different co-catalysts and methyl trichloro acetate as promoter generate different reaction products than without promoter.

Insulin-Enhancement

To the non-specialist, the term diabetes signifies a disease in which the body is unable to metabolize ingested carbohydrates properly. Although many forms of this disease exist, two limiting classifications are commonly used to describe two fundamentally different causes of diabetes. One, and the more devastating, is the self-destruction of insulin-producing cells, usually early in the life of an individual thus afflicted, referred to as Type 1 diabetes. A second, less acute form of the disease, is the inability of the body's insulin specifically to effect transport of glucose from the blood stream into the cell, referred to as Type 2 diabetes. This type of diabetes usually occurs later in life and is more widespread than Type 1 diabetes. Type 1 diabetes can only be treated with injected or inhaled insulin. Type 2 diabetes is amenable to treatment with orally or transdermally administered medication other than insulin, although combination therapies are often used.

In the past twenty or so years, research such as that led by Shechter (*86*), Kahn (*87*), Posner (*88*), Brichard (*89*), Rosetti (*90*), Sakurai (*91*), Orvig (*92*), Willsky (*93*), and Rehder (*94*) has shown that much work is needed to find a viable vanadium-based therapy. Studies have focused on development of new and better insulin-enhancing compounds, evaluation of the compounds currently available, reducing the toxicity of the compounds available as well as understanding the mechanism of action of the compounds. There have been several recent reviews on this topic (*95-98*). Design strategies for developing different types of vanadium compounds that have the potential to accomplish therapeutic goals are being investigated by many research groups. One representative contribution by Sakurai (Chapter 9) discusses how structure-activity relationships can improve these compounds. The timeliness of this work is perhaps best illustrated by the fact that one vanadium compound, bismaltolatooxovanadium(IV), has just entered Phase II clinical trials underlining the importance of these types of compounds.

In addition to structure-activity approaches, this monograph contains three contributions that explore how vanadium compounds affect biological systems. In Chapter 7 the Makinen group provides new insights by studying how *bis*(acetylacetonato)oxovanadium(IV) [VO(acac)$_2$] interacts with a protein tyrosine kinase. Their steady-state kinetics studies of the catalytic activity of a tyrosine phosphatase in the presence of this complex show that [VO(acac)$_2$] functions synergistically with insulin to potentiate tyrosine phosphorylation of the insulin receptor. Willsky and coworkers (Chapter 8) focus on a different complex: vanadium dipicolinic acid in three different oxidation states. They assess the status of these vanadium complexes in the body through a novel experimental protocol. Blood glucose-lowering results following oral administration of the vanadium dipicolinic acid complexes have been published (*99*); this administrative route is termed "chronic." In the results reported herein, the Willsky group injects the vanadium complex into muscle, or into tissue so that intraperitoneal injection can be modelled. This administration route is termed "acute." Important differences in the effectiveness of vanadium dipicolinic acid complexes were measured, depending on the route of administration. Their results provide new clues as to how much oxidation state, serum residence time, and lipid content at the target site affect the treatment of hyperglycemia.

In past years a controversy has existed whether insulin-enhancing properties of vanadium compounds in tissue culture studies accurately represented the effects of these compounds in animal system studies. Researchers were divided between those that believed tissue culture studies modelled animal studies and those that did not (*95-98*). Although such assumptions are readily made in studies with other drugs, vanadium compounds seem more sensitive to methods of administration and types of biological systems. The Willsky contribution in this volume provides experimental evidence to lend support that a correlation exists between results obtained in tissue culture and animal model systems.

The contribution by the Roess/Crans group in Chapter 10 explores how vanadium compounds may affect membrane organization and mediation of signal transduction. Most emphasis on insulin-enhancing abilities of vanadium complexes has been associated with downstream effects in the insulin cascade. From this point of view, membrane effects are judged unlikely to explain all or many of the insulin-enhancing effects exhibited by vanadium compounds. However, effects at the site of the membrane could take place, and the magnitude and nature of such effects warrant investigation. This study is a first step in evaluating whether vanadium compounds play a role in translocating insulin in the membrane.

Haloperoxidases

Vanadium-dependent haloperoxidases (VHPO) are metalloenzymes containing a mononuclear vanadium(V) co-factor that catalyzes the oxidation of halides in the presence of peroxides such as hydrogen peroxide. The oxidized species (HOX, X_3^-, or X_2) subsequently adds to organic molecules, yielding halogenated products. VHPOs can also catalyze the stereospecific two-electron oxidation of organic sulfides to chiral sulfoxides. Vanadium bromoperoxidases (VBrPO) are universally present in marine macroalgae; vanadium chloroperoxidases (VClPO) have been found in terrestrial fungi and in lichens (100-103). The enzyme nomenclature is based on the most electronegative halide that the enzyme is able to oxidize (i.e., chloroperoxidases oxidize Cl^-, Br^-, and I^-).

In the brown macroalgae Ascophyllum nodosum, one vanadate is required per subunit for enzyme activity. The redox state of vanadium does not change during the turnover of the enzyme and the main function of the metal centre is to bind H_2O_2 to yield the activated peroxo intermediate, which is able to react with bromide to produce HOBr. The role of VBrPO in the algae has been suggested to be involvement in the polymerization of polyphenols in order to hold the zygotes to the membranes during the reproductive cycle of the cell (104). In the fungi and lichens a role in the degradation of plant material or in the organism's defence mechanism has been suggested (105-106).

When using high concentrations of phosphate in crystallisation conditions, in some cases it was found that phosphate will prevent vanadate binding. It was also observed that if phosphate binds to vanadium haloperoxidases, phosphatase activity can be observed (107). Interestingly, the enzyme active sites of the VHPOs have been shown to have some conservation with acid phosphatases, which constitute a totally different group of enzymes, suggesting that nature has developed the same binding site for vanadium and phosphate. This similarity has been confirmed by the structural determination of an acid phosphatase from Escherichia blattae (108). Acid phosphatases that are related to the VHPOs enzymes are considered to be histidine phosphatases (108, 109).

The kinetics and mechanism of haloperoxidases have been the subject of multiple studies (110, 111) in part due to their potential use as industrial antimicrobial agents and disinfectants (112, 113) as well as halogenation catalysts (114, 115). A few thermostability studies have also been done (e.g. 116).

A number of X-ray crystal structures are now available for VHPO's (47, 109, 117, 118, and Chapter 11 of this book), including peroxide bound (47) and unbound (47, 118) forms of a VClPO from the fungus *Curvularia inaequalis*. Representations of X-ray crystal structure determinations for the VHPO of the red seaweed *Corallina officinalis* are shown and discussed by Littlechild and coworkers in Chapter 11. Vanadate is the prosthetic group of VHPOs and is fixed in the active site cavity by just one coordinative bond to a histidine residue and embedded in an environment of extensive hydrogen bonds. The coordination geometry is trigonal bipyramidal, with an oxygen atom in the axial position trans to an N-atom of histidine, three other oxygen atoms complete the coordination sphere. The peroxide-bound form is best described as having a distorted tetragonal geometry when treating the peroxo moiety as a single η^2 donor. The structure of a novel acid phosphatase from *Escherichia blattae* has been published (108), and despite having relatively low amino acid sequence identity (18%), it does have the same overall structural folding and conservation of most residues at the active site around the phosphate binding site.

The crystallographic studies have revealed details of the structure of the enzymes and increased our understanding of halide specificity, stability, substrate binding and enzymatic mechanism. An efficient process to produce the enzyme in recombinant form after refolding of inclusion bodies has been developed and a novel truncated mutant dimeric form of the enzyme has been constructed for use in commercial biotransformation experiments (Chapter 11). However, several structural uncertainties concerning the prosthetic group still remain. This uncertainty is related to the general resolution problem of protein crystallographic data (2.03 Å for the native VCPO and 2.24 Å for the peroxo form), which leads to an estimated mean positional error of the atoms of about ±0.23 Å (47). In fact, the presence and positions of hydrogen atoms could not be unambiguously determined from the crystal structure; therefore, the nature of the vanadium first coordination sphere ligands, namely, whether a particular group is oxo- or hydroxo- still remains the subject of debate.

The active site structure of vanadium haloperoxidase enzymes is the same for VClPO and VBrPOs and, moreover, is observed to be rather rigid as the structure of the apo protein is compared with the vanadate and tungstate derivatives of the VClPO (119). Again we emphasize the observation that this active site is also observed for a series of acid phosphatases, indicating consequences of the chemical similarities of vanadate and phosphate (120). Enzymological (121), spectroscopic (122), and both synthetic (57, 43) and computational models (48, 123-125) studies have elucidated key steps in the major catalytic steps in these systems (121). There is no evidence for redox cycling of the vanadate co-factor during catalysis (*121, 122*). It is commonly believed that vanadium is acting as a Lewis acid to activate the terminal oxidant hydrogen peroxide by polarizing the peroxo bond.

Synthetic models have shown that protonation of the peroxo-vanadium complex is critical for oxidation of halides and organic sulfides (*126, 127*). Hydrogen peroxide was shown to be the source of oxygen transferred to the thioether during sulfide oxidation via isotopic labeling studies using the enzyme (*76*).

Coordination complexes of oxovanadium(V) and their peroxo analogues have been shown by a number of groups to be functional mimics (*43, 75a, 127-132*) for VHPOs. In fact, many peroxo-oxovanadium complexes are excellent oxidation catalysts and have been used in a number of oxidative transformations (*50*) relevant to organic synthesis, including alcohol oxidation, halide oxidation, alkene oxidation, and sulfoxidation. Of these reactions, the oxidation of achiral sulfides to chiral sulfoxides has sparked most attention in recent years because of the potential applications of chiral sulfoxides (*133*). A number of catalysts have been shown to be competent with respect to sulfide oxidation and in some cases to be stereoselective (*50, 75a, 78, 127, 134, 135*).

As Pecoraro and Plass and coworkers point out (Chapters 12 and 13) despite the many studies reported by multiple laboratories addressing various aspects of chemistry and biology of vanadium haloperoxidases, questions are still being raised about both the enzyme and model systems, namely: (i) what is the protonation state of vanadium species (both resting and active forms), (ii) which oxygen of the peroxo ligand is transferred to the substrate, (iii) what is the geometry of the transition state, (iv) what is the protonation state and orientation of the vanadate cofactor in the resting state and the peroxo form, (v) what is the mechanistic pathway from the resting state to the peroxo form, (vi) what are the specific roles of several residues around the active site, (vii) which are the relevant supramolecular effects of the protein matrix on structure and mode of action of the vanadate cofactor.

Polenova and coworkers also emphasize (Chapter 14) the factors determining the substrate specificity, *i.e.* whether a particular enzyme will or will not display chlorinating activity is not well understood (*136-138*). A delicate balance of multiple interactions between the active site residues has been proposed to be responsible for the chlorinating activity of the haloperoxidases (*137*).

Quantum chemical methods have been useful tools to investigate structural, electronic and reactivity properties of transition metal complexes and models of the active site of metallo-enzymes (*139-141*). In fact, computational investigations of vanadium(V) complexes have been reported (*48, 123-125, 142*).

As mentioned above, the architectures of the Cl- and Br-peroxidase active sites were found to be very similar (*117, 118, 143*), and a delicate balance of multiple interactions between the active site residues has been proposed to be responsible for chlorinating activity of the haloperoxidases (137), i.e. the electronic structure of the vanadate cofactor is modulated by the protein. Apparently the high catalytic activity of the vanadate cofactor in these enzymes is caused by the interaction with the hydrogen bonding network with a specifically tailored active site pocket which shows a rather high degree of rigidity. This allows for a reasonably sized approximation of the direct protein environment and thereby for DFT (density functional theory) investigations on a comparatively high level of theory, in particular as the employed basis sets are concerned. Nevertheless, subtle changes in this system are expected to bring about variations concerning the catalytic properties, as becomes apparent from the comparison of the structure-function relationships of vanadium haloperoxidases and their active site homologous acid phosphatases (119). Therefore any kind of investigation on the mechanism of this system is a difficult task, be it theoretical or experimental.

Hybrid QM/MM calculations were conducted independently by two groups of investigators addressing extended active site models of VCPO treated quantum mechanically and significant portions of the protein treated with classical mechanics (125, 144). In parallel, DFT calculations were carried out on large models of VClPO active site (145). The above calculations suggest that in the resting state, at least one equatorial oxygen needs to be protonated to stabilize the metal cofactor (123, 125, 144, 145). According to the DFT results, the equatorial hydroxo group is likely to be bound to Ser-402 (145). QM/MM calculations indicate that the protein environment is crucial for creating the long-range electrostatic field necessary for stabilization of the resting state and suggest that the protonated equatorial oxygen accepts two hydrogen bonds from Arg-360 and Arg-390 residues (125). The roles of the individual amino acid residues and of the oxo atoms of the cofactor have been re-examined, and a revised mechanism for VCPO catalyzed halide oxidation reaction has been proposed (125). In another QM/MM study, Raugei and Carloni concluded that one of the equatorial oxygens is protonated, in agreement with the above studies, but the equatorial hydroxy group is hydrogen bonded to either Ser-402 or to Lys-353 residue (144). Thus, the coordination environment of the vanadate cofactor including the protonation states of the individual oxygen atoms still remains an open question.

As emphasized by Polenova and coworkers in Chapter 14, the ^{51}V solid-state spectra are typically dominated by the anisotropic quadrupolar and chemical shielding interactions. These interactions in turn report on the geometric and electronic structure of the vanadium site (146-148). As demonstrated for the model bioinorganic oxovanadium complexes mimicking the active site of haloperoxidases, the NMR parameters extracted from numerical simulations of the ^{51}V MAS spectra (MAS: magic angle spinning NMR spectroscopy) are sensitive to the coordination environment of the vanadium atom beyond the first coordination sphere. The NMR fine structure constants calculated for the crystallographically characterized complexes using

DFT, are in good agreement with the experimental results, illustrating that a combination of solid-state NMR experiments and quantum mechanical calculations presents a powerful approach for deriving coordination geometry in these systems.

Polenova and coworkers in Chapter 14 also discusses ^{51}V MAS NMR spectra of vanadium chloroperoxidase from *C. inaequalis* and of VBrPO from *A. nodosum*. The spectra reveal different electronic environments of the vanadate cofactor in each species. In both enzymes, the spectra are dominated by a large quadrupolar interaction, providing the first direct experimental evidence of the asymmetric electronic charge distribution at the vanadium site. The isotropic chemical shifts in VClPO and VBrPO are quite different, suggesting that the vanadate oxygens are likely not to have the same protonation states. In VClPO, the NMR observables were extracted from the spectra, and the DFT calculations of these observables in the extensive series of the active site models whose electronic structure and energetics were previously addressed by De Gioia, Carlson, Pecoraro and co-workers (*123*) indicate that one equatorial and one axial oxygen are protonated resulting in an overall anionic vanadate. These experimental results are in remarkable agreement with the quantum mechanical calculations discussed (*123, 125, 144, 145*). With the protocol used in Chapter 14 the detailed coordination environment of the vanadate cofactor in vanadium chloroperoxidase was obtained. This strategy is expected to be generally applicable to studies of diamagnetic V(V) sites in other vanadium containing proteins.

Pecoraro and coworkers in Chapter 12 also emphasize that the proton dependence of peroxide binding is very complex. There is a first order dependence on protons at stoichiometric proton concentrations, but the presence of excess protons had no effect on the rate of peroxide coordination Additionally, the coordination of peroxide occurs in the absence of additional protons at a slower rate. This observation is consistent with the fact that H_2O_2 can act as a proton donor to the complex, followed by rapid coordination of peroxide, alluding to the importance of acid/base catalysis within the active site of the enzyme.

Of the complexes discussed by the group of Pecoraro in Chapter 12, Hheida was shown to be most efficient functional model. In the presence of one acid equivalent a protoned peroxo-vanadium complex is generated as demonstrated by shifts in the UV-Vis and ^{51}V NMR spectra. Upon protonation, this complex is capable of oxidizing bromide, iodide (*126*) and thioethers (*127*). It was proposed that protonation of the vanadium complex led to a hydroperoxo intermediate that is responsible for catalysis.

These small molecule studies have provided important insights into the mechanism of VHPOs for both halide and thioether oxidation, allowing a detailed investigation of each catalytic step not available to enzymatic studies. From these studies demonstrating the importance of acid base catalysis in the active site of the enzyme Pecoraro and coworkers (Chapter 12) establish the importance of protons in both peroxide binding and the substrate oxidation.

Based on the results presented in his contribution, including the relevant cited data from literature, a reaction scheme is also proposed by Plass and coworkers for the catalytic cycle of the halide oxidation by vanadium haloperoxidases), and several important features are summarized (see Figure 12 in Chapter 13).

Enzymology, Toxicology and Transport

Several striking occurrences of vanadium in living systems have long been known, yet, the precise role of the element within these systems remains unknown or only partially elucidated (*149a,b 150*). A primary example is the concentration of vanadium in the +3 oxidation state in certain blood cells of the marine protochordates commonly known as sea squirts (tunicates). What is the function of these blood cells, referred to as vanadocytes? How does the animal concentrate vanadium against an approximately 10^7-fold unfavorable concentration gradient? How does a biological system achieve sufficient electrochemical potential to reduce V(V) to V(III)? What is the function of the vanadium in vanadocytes? The answer to each of these questions is "We do not know." (*151*) To a lesser degree, this lack of understanding is also encountered for vanadium toxicity and enzymology. The contributions on these subjects in this monograph provide new data, insights, and perspectives on these issues.

Michibata and coworkers have found an interesting series of proteins, which they call vanabins, involved in the binding of vanadate from sea water. In Chapter 19 they present evidence for the further participation of related proteins in the transport of vanadium to its target cell, the vanadocyte, and in vanadium reduction. The problem of vanadium reduction is further addressed by Frank and coworkers in Chapter 20. They present a detailed analysis of the chelation and EMF requirements of an enzymatic redox system capable of reducing vanadium(V) all the way to vanadium(III).

The metallurgy of vanadium led to its earliest important uses, and the concomitant mining and machining of the element led to its first known toxicological effects (*152-155*). The deleterious effect of vanadium in the lungs arises from the two facets of vanadium chemistry that surface in many other contexts. These are its interference in phosphate metabolism (*156, 157*) and the stress induced by its labile oxidation-reduction behavior. As Ghio and Samet show in Chapter 17, damage occurs to lung tissue and as Cohen shows in Chapter 16, it also ramifies into the immunological system as well.

Important biological effectors are usually present in a single oxidation state (calcium(II)) or perhapsin two such states (iron(II) and ironIII)). In contrast, vanadium occurs in three different oxidation states. Because two of these states have relatively high oxidation numbers, different hydrolytic forms within one such state can accumulate in cells and tissues, and can have different effects, or have single effects that differ in the degree of their effectiveness. Such is the case for vanadium(V) and one of its forms, decavanadate (*158-160*). In Chapter 18 Aureliano and coworkers show how oxidative stress and phosphate mimicry interact to exact profound effects on signaling, muscle contraction and lipid peroxidation for different vanadium(V) species.

In considering these biological effects, it should be remembered that vanadium has been shown to be an essential element in animal studies (*149a,b*). Few mechanistic studies exist, however, in which the effect of vanadium on development has been probed. In Chapter 15 Etcheverry and coworkers present compelling new experimental evidence that vanadium can have a tremendous and positive effect on the formation of bone tissue. The dark side of normal growth and development is the unrestricted cellular proliferation that occurs in tumorigenesis and metastasis. In some tissues, such as bone tissue, vanadium exerts anti-tumor activity (*161-163*). In Chapter 21 Wilker and coworkers address this vanadium anti-cancer activity, and propose a mechanism to explain the element's therapeutic effects.

Coordination Chemistry

Several reviews have been published on vanadium coordination chemistry, but here we emphasize those in Comprehensive Coordination Chemistry I, 1987 (*164*) and *II*, 2003 (*165*). More specific recent reviews were included in volume 237 (2003) of *Coord. Chem. Rev.*, and volume 77 (2005) of *Pure & Appl. Chem.*, both dedicated to vanadium chemistry.

Since 2003 fundamental coordination chemistry continued to be developed in several fronts and theoretical studies have also improved significantly the last few years to model the structure of vanadium compounds, as well as to calculate relevant spectroscopic parameters, namely, for EPR (*166, 167*) and NMR spectroscopies (*168-170*). In Chapter 22 Bühl discusses the application of DFT and QM/MM studies to models of peroxidases; already in Chapter 13 Plass and coworkers have applied DFT methods to investigate the structure of the resting state of the prosthetic group in the enzyme pocket and to elucidate the mechanism of the formation of its peroxo complex. Bühl also presents molecular dynamics simulations for the calculation of ^{51}V NMR chemical shifts.

Peroxovadium complexes continue to be the center of much interest, partly due to their reactivities and their application in catalysis. Coordination complexes and some of their applications continue to be reported by many groups, and include salen-type complexes (e.g. *19, 66, 171-176*), pyrone, pyridinone and pyrimidinone complexes (*177-179*), several types of Schiff bases and related complexes (*66, 180-182*), picolinato (e.g. *183*), quinoline derivatives (*184*) dipicolinato complexes (*185*), compounds involving ligands such as amino acids or derivatives (*28, 186-197*), hydroxycarboxylates (*198, 199*), but novel classes of coordination compounds and other new reports on previously known compounds are too many to name.

The nature of the oxovanadate(V) species that form in aqueous solutions is relatively well known, particularly after the use of combined potentiometric and NMR studies, and Crans (*165, 200*) recently reviewed this subject. Many polyoxovanadates have also been characterized in the solid state, and a donor-acceptor intermolecular survey was also reported, four different main motifs having been found: monomers, dimers, 1D and 2D arrays (*201*).

The same is not true for the oxovanadium(IV) system: since the review of Vilas-Boas and Costa Pessoa in 1987 (*164*), no significant advance was made in the knowledge of the hydrolytic products of oxovanadium(IV). For vanadium(III) the amount of equilibrium data is also rather limited in the literature, partly because of the strong tendency of V(III) to become oxidized and to hydrolyse in aqueous solution even below pH 2. Buglyó recently revised the procedures to prepare and store V(III) stock solutions (*99, 202*).

The coordination chemistry of vanadium(II) was reviewed by Crans and Smee (*165*); besides the short mention to this oxidation state in the catalytic systems described by Hirao in Chapter 1, or Rosenthal and coworkers in Chapter 6, and the polynuclear vanadium complexes presented by Armstrong and coworkers (*203*), no other V(II) complexes were described in the V5 Symposium. The chemistry of vanadium in oxidation states lower than +2 is comprised primarily of organometallic compounds, but no communication was presented on the chemistry or coordination chemistry of these low-valent compounds. The review by Crans and Smee (*165*) and the volumes of *Comprehensive Organometallic Chemistry* should be consulted for information about these complexes.

The coordination chemistry of vanadium in oxidation state III was reviewed by Crans and Smee (165), and besides the catalytic systems described by Hirao in Chapter 1 of this book, some other reports were presented in the V5 Symposium or in the literature. Phenoxide and other related complexes have been of interest in catalysis because of their role as procatalysts for ethylene polymerization (*80, 204, 205*) and because these complexes are found in tunicates (*7, 206-209*).

Some reports of the formation of V(III) complexes with biologically relevant ligands have been published within the last few years, *e.g.* solution studies of complex formation with the amino acids L-alanine and L-aspartic acid (*210*) and with L-cysteine and S-methylcysteine (*211*), and some other systems (202). Globally the stability of the vanadium complexes with amino acids follow the order V(V) < V(IV) < V(III) (*211*). The molecular structure of Na[V^{III}(L-cys)$_2$] is known since 1993 (*212*), and interesting *in vitro* cytotoxic effects towards tumor cells were reported for a distinct V(III)-cysteine compound (*213*) as well as upon using V(III)-cysteine solutions (*211*). The preparation of several V(III) complexes containing dipeptides and *o*-phenanthroline in the coordination sphere have also been reported (*192*).

Although as mentioned above V(III) complexes are normally not stable enough against hydrolysis or against oxidation, some show reasonable air stability (e.g. V(maltolato)$_3$) and several V(III) complexes with hydroxypyrones and pyridinones have been reported to have insulin-enhancing properties (214).

Only one communication dealing mainly with V(III) compounds was made at the V5 Symposium. Namely Krystek and coworkers (215) discussed complexes in which the V(III) ion is coordinated by three thiolates, and various other donor atoms (phosphorus, nitrogen, chloride) yielding a total coordination number ranging from 5 to 7. Often the ground spin state of V(III) complexes is a non-Kramers triplet (S = 1) with typically large zero-field splitting, thus frequently "silent" in conventional EPR. Krystek and coworkers (215) reported on a high-frequency and -field EPR investigation (HFEPR, frequency 95-700 GHz, magnetic field up to 25 Tesla) of his series of complexes, and signals are readily observable. Spin Hamiltonian parameters obtained from HFEPR and from parallel magnetometric experiments, in conjunction with optical absorption spectra and X-ray diffraction structural information, give insight into the electronic structure of the novel complexes. HFEPR will certainly be useful and used for other V(III) compounds in the future.

The coordination chemistry of vanadium in oxidation states IV and V was reviewed in 2003 by Crans and Smee (165), and since then several relevant reports have appeared developing new ligands and/or new applications (e.g. catalysis, insulin-enhancing, anti-tumor) of already known compounds. For example: Schiff bases of the salen-type (66), reduced Schiff bases of salan-type (171, 172, 173), sal-amino acidate and other Schiff bases, including the reduced-type derivatives, (28, 78, 193, 195, 197, 216, 217), amino acids or peptides (186-189, 196), hydroxycarboxylates (198, 218, 219, 221), dipic (185), picolinato (e.g. 183), pyrones, piridinones, pyrimidinones (177-179), hydroxamate (220), acac-type (221-223), or other ligands (e.g. 65, 158, 180, 184, 224-234).

At the V5 Symposium several communications dealing with V(IV) and V(V) compounds were presented. For example, in Chapter 25 Drouza and Keramidas re-examined co-ordination compounds containing *p*-dioxolene ligands in the form of hydroquinone, semiquinone or quinone, and Gambino and coworkers (235) new V^{IV}O(L)$_2$ complexes with quinoxaline N,N'-dioxide derivatives bearing *in vitro* biological activity. Rehder and coworkers in Chapter 5 also report the syntheses of multi-chiral, multi-dentate amino alcohols, containing varying substituents at the nitrogen (such as vinyl, glycyl, imidazolyl). When reacted with oxovanadium(V), these ligands yield penta-coordinate vanadium complexes [VO(OR)L] which structurally model the active center of the VHPOs. Reaction of these complexes with organic hydroperoxides produces the peroxo complexes [VO(O$_2$R)L]. The compounds model functionally the active center of vanadate-dependent haloperoxidases as well as the sulfideperoxidase activity of these enzymes.

Sivák and coworkers (236), based on crystal structures of monoperoxovanadium(V) complexes with the VO(O$_2$)$^+$ core, discussed the stereochemical rules for occupation of the "free" positions of heteroligand complexes by N or O donor atoms of neutral or anionic bidentate ligands, and Schwendt and coworkers (237) reported peroxovanadium complexes containing, besides vanadium, other transition metals (M=Zn, Ni, Cu), characterized by x-ray diffraction. In Chapter 27, Salifoglou through meticulous pH-synthetic studies, characterized binary $[V_2O_4(L)_2]^{n-}$, $[V_2O_2(L)_2]^{p-}$ and $[V_2O_2(O_2)_2(L)_2]^{q-}$ (n,p,q=2-6) species with hydroxycarboxylate ligands (citrate and malate). These studies may contribute significantly to the delineation of the aqueous structural speciation of the relevant binary and ternary systems related to that metal's potential insulin mimetic properties. S. Baruah (238) reported the preparation and characterization by various spectroscopic methods of several vanadium dipicolinate complexes. With the chlorodipicolinate ligand vanadium(III, IV and V) complexes were prepared in enough amounts for their evaluation in STZ induced rats. Electron spin lattice relaxation rates for four vanadium(IV) complexes were studied by a team led by Gareth and Sandra Eaton. In Chapter 26 they report that as ligand coordination number increases, electron spin lattice relaxation rates decrease, indicating an increase in the complex's rigidity.

The group of Mahroof-Tahir (239) presented the synthesis, characterization and some theoretical calculations of vanadium flavonoid complexes in solid and solution states and Moriuchi and coworkers (240) presented structural characterization of novel organoimido vanadium(V) complexes,

Pessoa and coworkers (241a,b) reported the synthesis of a new ligand derived from pyrimidinone heterocycles and studies of its complexation with $V^{IV}O^{2+}$ and $V^{V}O_2^+$. This ligand binds strongly to V(IV) and V(V), particularly in the pH range 3-7. The same group (242) also reported the synthesis of salDPA, a salan type ligand derived from the condensation of salicylaldehyde with D,L- or L-diaminopropionic acid, the preparation of its V(IV) and V(V) complexes and the characterization of the V(IV)- and V(V)-salDPA systems in aqueous solutions in the pH range 2-10. The group of Correia and Costa Pessoa (61) also reported the synthesis and characterization of several new V(V) and V(IV) chiral salen- and salan-type complexes, some of them being water-soluble. These complexes might prove adequate for more sustainable asymmetric oxidation reactions.

Vanadium compounds exert preventive effects against chemical carcinogenesis, but furthermore vanadium compounds may induce cell-cycle arrest and/or cytotoxic effects through DNA cleavage and fragmentation, and plasma membrane lipoperoxidation. These subjects have been reviewed (243). The compounds more often considered in this respect are either bis(cyclopentadienyl)vanadium(IV) compounds, or complexes containing bipy or o-phenanthroline in the coordination sphere, but several other compounds have been studied (212, 243-247). Bisperoxovanadium(V) have been widely investigated as anti-cancer agents. In the V5 Symposium Abu-Omar (248) discussed the mechanism of DNA cleavage by bisperoxovanadium(V) phenanthroline and bipyridine complexes. The studies presented identify the hydroxyl radical, produced from the photooxidation of the peroxo ligand on vanadium, as the active species in DNA cleavage. In Chapter 24 and (*249*) Costa Pessoa, Cavaco and coworkers present studies of the action of several vanadium complexes on plasmid DNA, namely salen- and salan-type ligands. Some of the ligands were shown to be able to cleave plasmid DNA in the absence of activators or light; this was also shown for solutions containing $VO(acac)_2$ and $VO(phen)_2(SO_4)$. Costa Pessoa and coworkers (*250*) emphasized that all *in vitro* studies should be careful in understanding which species is in fact exerting the biological effect being measured. Examples were given where more than one vanadium species were proven to be present in the experimental conditions used for the *in vitro* measurements reported previously.

In Chapter 23 and (*251*), Kiss and coworkers discuss the interactions of insulin-mimetic oxovanadium(IV) complexes with cell constituents. From among the important cell constituents, glutathione will possibly take part in the reduction of V(V) to V(IV), and will help keep V(IV) in this oxidation state. As a strong V(IV) binder, ATP will bind the metal ion, forming binary and/or ternary complexes. This strongly suggests that ATP binds relevant V(IV) species under cell conditions, and thus might somehow be involved in the insulin-enhancing action of V(IV) compounds (*251*). In the same chapter, Kiss and coworkers also discuss the transport processes of the insulin-enhancing vanadium complexes in the blood (in most cases transferrin completely displaces the carrier ligand).

New Materials and Processes

An important aspect of vanadium-based materials such as vanadium-oxide and polyoxometalate chemistry is closely linked to structural characterization of these systems (*252*). This line of research favors less systematic planning and more open-ended observation and response. As Pasteur said, "In the fields of observation, chance favors only the mind that is prepared." The final three contributions of this Symposium are in line with this statement and represent examples of structure-guided and exploratory studies in solid state and solution chemistry.

The oxovanadium organodiphosphonate materials are prototypical composite materials, which can exhibit a range of structures, including molecular clusters, chains, layers and three-dimensional frameworks (*253*). The Zubieta group has been studying these materials and others using hydrothermal reactions (*254*), and varying conditions such as fill volume, pH, temperature and the presence of mineralizers such as HF. The structural determinants may include the presence or absence of cations, the cation identity, and the tether length and additional functionality of the anion under investigation. The contribution in Chapter 28 by Ouellette and Zubieta focuses on investigations into the diphosphonate family of materials (*255*).

During the last 15 years a new type of metal-oxide based host-guest chemistry has developed (*256*). The Müller group has been especially interested in creations where aesthetical beauty and function merge in harmony. Although most of the supermolecular structures have been based on polymolybdate chemistry, vanadium can replace molybdenum and generate new structures containing vanadium (*257*). In addition, some polyvanadates have been found to form, which generates a multitude of polyoxovanadates. This line of investigation is presented in Chapter 29 by the Hayashi group, in which they describe a new class of vanadium-oxide structures. Unlike the chemistry of a cyclic metaphosphate or silicate, metavanadate species in acetonitrile form tri- or tetra-cyclic species. The Hayashi group discovered an expansion of cyclic tri- and tetrameric condensed metavanadates obtained by the reaction of metavanadate with various transition metal cations (*257*). The aggregates have new and potentially useful structures and topologies. The transition metal cation can serve as a Lewis acid and shift the equilibrium toward larger metavanadate rings. Examples that will be described include the tetravanadate-cobalt complex, hexavanadate-palladium complex, octavanadate-di-copper complex, decavanadate-µ-hydroxo-tetra-nickel complex. These disk-shaped com-plexes can be utilized as a scaffold to synthesize a spherical molecule in a rational manner.

In Chapter 30, Prof. Kan Kanamori and one of his graduate students, Yuya Shirosaka, describe a serendipitous observation that occurred when storing solutions of what they thought were new vanadium(III) complexes to be retrieved later for spectroscopic studies. After a few days had passed, they went to fetch one of the green-colored solutions they had stored, and encountered an orange-colored solution in its place. Careful follow-up investigations determined the correct nature of the complex in the flask, its properties, and the fact that the color change was actually part of a more complex series of reactions leading to the discovery of a new homogeneous chemical oscillator. Several dozen chemical oscillators are known, and most of these are based on oxidation-reduction reactions of halogen oxoanions (*258*). Not many oscillators have as a core a metal ion that has multiple accessible oxidation states. One such metal ion is manganese, and now vanadium joins manganese in an exclusive club of metal-based chemical oscillators.

Acknowledgment

João Costa Pessoa thanks the FEDER and Fundação para a Ciência e Tecnologia, project POCI/QUI/56949/2004 for financial support.

References

1. a. Orvig, C.; Thompson, K. H.; Battell, M.; McNeill, J. H. in *Metal Ions in Biological Systems*, Sigel, H.; Sigel, A. Eds. Marcel Dekker, Inc.: New York, **1995**; Vol. 31, pp 575-594, b. Sakurai, H.; Kojima, Y.; Yoshikawa, Y.; Kawabe, K.; Yasui, H. *Coord. Chem. Rev.* **2002**, *226*, 187-198. c. Thompson, K. H.; Orvig, C. *J. Chem. Soc. Dalton Trans.* **2000**, 2885-2892. d. Sakurai, H.; Katoh, A.; Yoshikawa, Y. *Bull. Chem. Soc. Jpn.* **2006**, *79*, 1645-1664. e. Thompson, K. H.; Orvig, C. *Dalton Trans.* **2006**, 1925-1935.
2. a. Wever, R.; Hemrika, W. in *Vanadium in the Environment* in *Adv. Environmental Science and Technology*, Nriagu, J. O. Ed. John Wiley & Sons, Inc.: New York, **1998**; Vol. 30, pp 285-305, b. Butler, A.; Walker, J. V. *Chem. Rev.* **1993**, *93*, 1937-1944.
3. Vilter, H. in *Metal Ions in Biological Systems*, Sigel, H.; Sigel, A. Eds. Marcel Dekker, Inc.: New York, **1995**; Vol. 31, pp 325-362.
4. Eady, R. R. in *Metal Ions in Biological Systems*, Sigel, H.; Sigel, A. Eds. Marcel Dekker, Inc.: New York, **1995**; Vol. 31, pp 363-405.
5. a. Rehder, D.; Jantzen, S. in *Vanadium in the Environment* in *Adv. Environmental Science and Technology*, Nriagu, J. O. Ed. John Wiley & Sons, Inc.: New York, **1998**; Vol. 30, pp 251-284. b. Rehder, D. *Angew. Chem. Int. Ed. Engl.* **1991**, *30*, 148-167.
6. Bayer, E. in *Metal Ions in Biological Systems*, Sigel, H.; Sigel, A. Eds. Marcel Dekker, Inc.: New York, **1995**; Vol. 31, pp 407-421.
7. a. Michibata, H.; Kanamori, K. in *Vanadium in the Environment* in *Adv. Environmental Science and Technology*, Nriagu, J. O. Ed. John Wiley & Sons, Inc.: New York, **1998**; Vol. 30, pp 217-249. b. Kustin, K.; McLeod, G. C.; Gilbert, T. R.; Briggs, L. B. R., 4th *Structure and Bonding* **1983**, *53*, 139-161.
8. Aureliano, M.; Madeira, V. M. C. in *Vanadium in the Environment* in *Adv. Environmental Science and Technology*, Nriagu, J. O. Ed. John Wiley & Sons, Inc.: New York, **1998**; Vol. 30, pp 333-357.
9. Ishii, T.; Nakai, I.; Okoshi, K. in *Metal Ions in Biological Systems*, Sigel, H.; Sigel, A. Eds. Marcel Dekker, Inc.: New York, **1995**; Vol. 31, pp 491-509.
10. Mendz, G. L. in *Vanadium in the Environment* in *Adv. Environmental Science and Technology*, Nriagu, J. O. Ed. John Wiley & Sons, Inc.: New York, **1998**; Vol. 30, pp 307-332.
11. Stankiewicz, P. J.; Tracey, A. S. in *Metal Ions in Biological Systems*, Sigel, H.; Sigel, A. Eds. Marcel Dekker, Inc.: New York, **1995**; Vol. 31, pp 249-285.
12. Stankiewicz, P. J.; Tracey, A. S.; Crans, D. C. in *Metal Ions in Biological Systems*, Sigel, H.; Sigel, A. Eds. Marcel Dekker, Inc.: New York, **1995**; Vol. 31, pp 287-324.
13. Csotonyi, J. T:; Stackebrandt, E.; Yurkov, V. *Appl. and Environ Microbiol.* **2006**, *72*, 4950-4956.

14. Carpentier, W.; De Smet, L.; Van Beeumen, J. *J. Bacteriology* **2005**, *187*, 3293-3301
15. Hirao, T. *Chem. Rev.* **1997**, *97*, 2707 and references therein.
16. Evans, D. A.; Dinsmore, C. J.; Evrard, D. A.; DeVries, K. M. *J. Am. Chem. Soc.* **1993**, *115*, 6426.
17. a. Hirao, T.; Takeuchi, H.; Ogawa, A.; Sakurai, H. *Synlett* **2000**, *11*, 1658. b. Hirao, T.; Hatano, B.; Imamoto, Y.; Ogawa, A. *J. Org. Chem.* **1999**, *64*, 7665. c. Hirao, T.; Asahara, M.; Muguruma, Y.; Ogawa, A. *J. Org. Chem.* **1998**, *63*, 2812. d. Hatano, B.; Ogawa, A.; Hirao, T. *J. Org. Chem.* **1998**, *63*, 9421.
18. a. Banfi, L.; Guanti, G.; Basso, A. *Eur. J. Org. Chem.* **2000**, 939. b. Nazaré, M.; Waldmann, H. *Chem. Eur. J.* **2001**, *7*, 3363.
19. a. Belokon, Y. N.; North, M.; Parsons, T. *Org. Lett.* **2000**, *2*, 1617-1619; b. Belokon, Y.N.; Green, B.; Ikonnikov, N.S.; North, M.; Parsons, T.; Tararov, V.I. *Tetrahedron* **2001**, *57*, 771.
20. Ishikawa, T.; Nonaka, S.; Ogawa, A.; Hirao, T. *J. Chem. Soc. Chem. Commun.* **1998**, 1209.
21. Ishikawa, T.; Ogawa, A.; Hirao, T. *Organometallics* **1998**, *17*, 5713.
22. Ishikawa, T.; Ogawa, A.; Hirao, T. *J. Am. Chem. Soc.* **1998**, *120*, 5124.
23. Takada, T.; Sakurai, H.; Hirao, T. *J. Org. Chem.* **2001**, *66*, 300.
24. Hirao, T.; Takada, T.; Ogawa, A. *J. Org. Chem.* **2000**, *65*, 1511.
25. Hirao, T.; Takada, T.; Sakurai, H. *Org. Lett.* **2000**, *2*, 3659.
26. Ishikawa, T.; Ogawa, A.; Hirao, T. *J. Organomet. Chem.* **1999**, *575*, 76.
27. a. Trost, B. M.; Oi, S. *J. Am. Chem. Soc.* **2001**, *123*, 1230. b. Chen, C.T.; Hon, S.W.; Weng, S.S. *Synlett* **1999**, 816.
28. Costa Pessoa, J; Calhorda, M. C.; Cavaco, I.; Costa, P. J.; Correia, I.; Costa, D.; Vilas-Boas, L.; Felix, V.; Gillard, R. D.; Henriques, R. T.; Wiggins, R. *Dalton Trans.* **2004**, 2855-2866.
29. Togni, A. *Organometallics* **1990**, *9*, 3106.
30. Reddy, P. P.; Chu, C. Y.; Hwang, D. R.; Wang, S. K.; Uang, B. J., *Coord. Chem. Rev.* **2003**, *237*, 257-269.
31. Hwang, D. R.; Uang, B. J. *Org. Lett.* **2002**, *4*, 463-466.
32. Ishii, Y.; Matsunaka, K.; Sakaguchi, S. *J. Am. Chem. Soc.* **2000**, *122*, 7390-7393.
33. Hirao, T. *Coord. Chem. Rev.* **2003**, *237*, 271-279.
34. Hirao, T. *Pure Appl. Chem.* **2005**, *77*, 1539-1557.
35. a. Scheffold, R.; Busato, S. *Helv. Chim. Acta* **1994**, *77*, 92. b. Ozaki, S.; Horiguchi, I.; Matsushita, H.; Ohmori, H. *Tetrahedron Lett.* **1994**, *35*, 725. c. Ozaki, S.; Matsui, E.; Waku, J.; Ohmori, H. *Tetrahedron Lett.* **1997**, *38*, 2705. d. Rollin, Y.; Derien, S.; Duñach, E.; Gebehenne, C.; Perichon, J. *Tetrahedron* **1993**, *49*, 7723. e. Mubarak, M. S.; Pagel, M.; Marcus, L. M.; Peters, D. G. *J. Org. Chem.* **1998**, *63*, 1319.
36. Davies, S. C.; Hughes, D. L.; Janas, Z.; Jerzykiewicz, L. B.; Richards, R. L.; Sanders, J. R.; Silverston, J. E.; Sobota, P. *Inorg. Chem.* **2000**, *39*, 3485.
37. Chu. W.C.; Wu, C.C.; Hsu, H.F. *Inorg. Chem.* **2006**, *45*, 3164.

38. a. Hirao, T.; Mori, M.; Ohshiro, Y. *Bull. Chem. Soc. Jpn.* **1989**, *62*, 2399. b. Hirao, T.; Mori, M.; Ohshiro, Y. *J. Org. Chem.* **1990**, *55*, 358. c. Hirao, T.; Mori, M.; Ohshiro, Y. *Chem. Lett.* **1991**, 783.
39. a. Hirao, T.; Fujii, T.; Tanaka, T.; Ohshiro, Y. *J. Chem. Soc., Perkin Trans.1* **1994**, 3. b. Hirao, T.; Fujii, T.; Tanaka, T.; Ohshiro, Y. *Synlett* **1994**, 845. c. Hirao, T.; Sakaguchi, M.; Ishikawa, T.; Ikeda, I. *Synth. Commun.* **1995**, *25*, 2579.
40. Periana, R. A.; Mironov, O.; Taube, D.; Bhalla, G.; Jones, C. J. *Science* **2003**, *301*, 814.
41. Guedes da Silva, M. F. C.; Silva, J. A. L.; Fraústo da Silva, J. J. R.; Pombeiro, A. J. L.; Amatore, C.; Verpeaux, J.-N. *J. Am. Chem. Soc.* **1996**, *118*, 7568.
42. Matoso, C. M. M.; Pombeiro, A. J. L.; Fraústo da Silva, J. J. R.; Guedes da Silva, M. F. C.; Silva, J. A. L.; Baptista-Ferreira, J. L.; Pinho-Almeida, F. in *Vanadium Compounds. Chemistry, Biochemistry, and Therapeutic Applications ACS Symposium Series* Tracey, A. S.; Crans, D. C. Eds. American Chemical Society: Washington, DC, **1998**, Vol.711, pp 241-247.
43. a. Conte, V.; Bortolini, O.; Carrano, M.; Moro, S. *J. Inorg. Biochem.* **2000**, *80*, 41-49; b. Bortolini, O.; Carrano, M.; Conte, V.; Moro, S. *Eur. J. Inorg. Chem.* **2003**, 699-704.
44. Welton, T. *Chem. Rev.* **1999**, *99*, 2071-2083.
45. Conte, V.; Floris, B.; Galloni, P.; Silvagni, A. *Pure & Appl. Chem.* **2005**, *77*, 1575-1581.
46. Conte, V.; Floris, B.; Galloni, P.; Silvagni, A. *Adv. Syn.& Catal.* **2005**, *347*, 1341.
47. Messerschmidt, A.; Prade, L; Wever, R. *J. Biol. Chem.* **1997**, *378*, 309-315.
48. Zampella, G.; Fantucci, P.; Pecoraro, V. L.; De Gioia, L. *J. Am. Chem. Soc.* **2005**, *127*, 953-960.
49. Conte, V.; Di Furia, F.; Moro, S. *J. Phys. Org. Chem.* **1996**, *9*, 329-336.
50. Bolm, C. *Coord. Chem. Rev.* **2003**, *237*, 245-256.
51. Clague, M.J.; Keder N.N.; Butler, A. *Inorg. Chem.* **1993**, *32*, 4754.
52. Conte, V.; Di Furia, F.; Licini, G. *Appl. Catal. A* **1997**, *157*, 335
53. Ligtenbarg, A. G. J.; Hage, R.; Feringa, B. L. *Coord. Chem. Rev.* **2003**, *237*, 89-101.
54. Rehder, D.; Santoni, G.; Licini, G. M.; Schulzke, C.; Meier, B. *Coord. Chem. Rev.* **2003**, *237*, 89-101.
55. Conte, V.; Di Furia, F.; Moro, S. *Tetrahedron Lett.* **1994**, *35*, 7429-7432.
56. Andersson, M.; Conte, V.; Di Furia, F.; Moro, S. *Tetrahedron Lett.* **1995**, *36*, 2675-2678.
57. Conte, V.; Di Furia, F.; Moro, S.; Rabbolini, S. *J. Mol. Cat. A: Chem.* **1996**, *113*, 175-184.
58. Yamamoto, H. *5th International Symposium on the Chemistry and Biological Chemistry of Vanadium*, ACS National Meeting, San Francisco, California, USA, September, **2006**, INOR 856.

59. Chen, C.T. *5th International Symposium on the Chemistry and Biological Chemistry of Vanadium*, ACS National Meeting, San Francisco, California, USA, September, **2006**, INOR 858.

60. Maurya, M. R.; Amit Kumar, A.; Ebel, M.; Rehder, D. *5th International Symposium on the Chemistry and Biological Chemistry of Vanadium*, ACS National Meeting, San Francisco, California, USA, September, **2006**, INOR 992.

61. Correia, I.; Adão, P.; Costa Pessoa, J. *5th International Symposium on the Chemistry and Biological Chemistry of Vanadium*, ACS National Meeting, San Francisco, California, USA, September, **2006**, INOR 385.

62. Schneider, C. J.; Kampf, J. W.; Pecoraro, V. L. *5th International Symposium on the Chemistry and Biological Chemistry of Vanadium*, ACS National Meeting, San Francisco, California, USA, September, **2006**, INOR 387

63. Lippold, I.; Nica, S.; Mancka, M.; Plass, W. *5th International Symposium on the Chemistry and Biological Chemistry of Vanadium*, ACS National Meeting, San Francisco, California, USA, September, **2006**, INOR 829.

64. Jakusch, T.; Pecoraro, V. L. *5th International Symposium on the Chemistry and Biological Chemistry of Vanadium*, ACS National Meeting, San Francisco, California, USA, September, **2006**, INOR 830.

65. Maurya, M. R.; Kumar, U.; Manikandan, P. *Dalton Trans.* **2006**, 3561-3575.

66. Maurya, M. R.; Kumar, M.; Tutuinchi, S. J. J.; Abbo, H. S.; Chand, S. *Catal. Lett.* **2003**, *86*, 97-105.

67. Ulagappan, N.; Krishnasamy, V. *Ind. J. Chem.* **1996**, *35*, 787.

68. Balkus, K. J.; Khanmamedova, A. K.; Dixon, K. M.; Bedioui, F. *Appl. Catal. A: General* **1996**, *143*, 159.

69. Pecoraro, V. L.; C. J.; Schneider, V. L.; Jakusch, T.; Zampella, G.; DeGioia, L.; Kampf, J. W.; Penner-Hahn, J. E. *5th International Symposium on the Chemistry and Biological Chemistry of Vanadium* ACS National Meeting, San Francisco, California, USA, September, **2006**, INOR 409.

70. ten Brink, H. B.; Holland, H. L.; Schoemaker, H. E.; van Lingen, H.; Wever, R. *Tetrahedron: Asymmetry.* **1999**, *10*, 4563-4572.

71. a. Andersen, K. K. in: Patai, S.; Rappoport, Z.; Stirling, C. J. M. Eds. *The Chemistry of Sulfones and Sulfoxides*, John Wiley & Sons: Chichester, **1998**, ch. 3 and 16. b. Carreño, M. C. *Chem. Rev.* **1995**, *95*, 1717.

72. Nakajima, K.; Kojima, M.; Kojima, K.; Fujita, J. *Bull. Chem. Soc. Jpn.* **1990**, *63*, 2620.

73. Bolm, C.; Bienewald, F. *Angew. Chem. Int. Ed.* **1995**, *34*, 2640.

74. Kagan, H. B. in *Asymmetric Oxidation of Sulfides,* Ojima, I. Ed., Wiley-VCH: New York, **2000**, ch. 6c.

75. a. Santoni, G.; Licini, G.; Rehder, D. *Chem. Eur. J.* **2003**, *9*, 4700-4708, b. Wikete, C.; Wu, P.; Zampella, G.; De Gioia, L.; Licini, G.; Rehder, D. *Inorg. Chem.* 2007, 46, 196-207.

76. ten Brink, H. B.; Schoemaker, H. E.; Wever, R. *Eur. J. Biochem.* **2001**, *268*, 132-138.

77. Santoni, G.; Licini, G.; Rehder, D. *Chem. Eur. J.* **2003**, *9*, 4700-4708.
78. Sun, J.; Zhu, C.; Dai, Z.; Yang, M.; Pan, Y.; Hu, H. *J. Org. Chem.* **2004**, *69*, 8500-8503.
79. Natta, G.; Pino, P.; Corradine, P.; Danusso, F.; Mantica, E.; Mazzanti, G.; Moraglio, G. *J. Am. Chem. Soc.* **1955**, *77*, 1708.
80. Gambarotta, S. *Coord. Chem. Rev.* **2005**, *237*, 229-243 and references therein.
81. Cucinella, S.; Mazzei, A. US Patent 3,711,455, CI. 260-85.3, **1973**.
82. Rosenthal, E. C. E.; Girgsdies, F. *Z. Anorg. Allg. Chem.* **2002**, *628*, 1917.
83. Rosenthal, E. C. E.; Cui, H.; Lange, K. C. H.; Dechert, S. *Eur. J. Inorg. Chem.* **2004**, 4681.
84. Rosenthal, E. C. E.; Cui, H.; Koch, J.; Escarpa Gaede, P.; Hummert, M.; Dechert, S. *Dalton Trans.* **2005**, 3108.
85. Hagen, H.; Boersma, J.; van Koten, G. *Chem. Soc. Rev.* **2002**, *31*, 357.
86. Shechter, Y.; Karlish, S. J. D. *Nature* **1980**, *284*, 556-558.
87. Goldfine, A. B.; Simonson, D. C.; Folli, F.; Patti, M.-E.; Kahn, C. R. *Mol. and Cell. Biochem.* **1995**, *153*, 217-231.
88. Fantus, I. G.; Kadota, S.; Deragon, G.; Foster, B.; Posner, B. I. *Biochemistry* **1989**, *28*, 8864-8871.
89. Brichard, S. M.; Pottier, A. M.; Henquin, J. C. *Endocrinology* **1989**, *125*, 2510-2516.
90. Cohen, N.; Halberstam, M.; Shlimovich, P.; Chang, C. J.; Shamoon, H.; Rossetti, L. *J. Clin Invest.* **1995**, *95*, 2501-2509.
91. Sakurai, H.; Tsuchiya, K.; Nukatsuka, M.; Kawada, J.; Ishikawa, S.; Yoshida, H.; Komatsu, M. *J. Clin. Biochem. Nutr.* **1990**, *8*, 193-200.
92. McNeill, J. H.; Yuen, V. G.; Hoveyda, H. R.; Orvig, C. *J. Med. Chem.* **1992**, *35*, 1489-1491.
93. Willsky, G. R.; Goldfine, A. B.; Kostyniak, P. J. in *Vanadium Compounds. Chemistry, Biochemistry, and Therapeutic Applications ACS Symposium Series* Tracey, A. S.; Crans, D. C. Eds. American Chemical Society: Washington, DC, **1998**, Vol.711, pp 278-296.
94. Rehder, D. *Inorg. Chem. Commun.* **2003**, *6*, 604-617.
95. Sekar, N.; Li, J.; Shechter, Y. *Am J Physiol* **1997**, *272*, E30-5.
96. Thompson, K. H.; McNeill, J. H.; Orvig, C. *Chem. Rev.* **1999**, 2561-2571.
97. Saltiel, A. R.; Kahn, C. R. *Nature* **2001**, *414*, 799-806.
98. Crans, D. C. *J. Inorg. Biochem.* **2000**, *80*, 123-131.
99. Buglyó, P.; Crans, D. C.; Nagy, E. M.; Lindo, R. L.; Yang, L.; Smee, J.; Chi, L.-H.; Godzala, M. E.; Willsky, G. R. *Inorg. Chem.* **2005**, *44*, 5416-5427.
100. Vilter, H. *Phytochem.* **1984**, *23*, 1387-1390.
101. Vollenbroek, E. G. M.; Simons, L. H.; Van Schijndel, J. W. P. M.; Barnett, P.; Balzar, M.; Dekker, H.; Van Der Linden, C.; Wever, R. *Biochem. Soc. T.* **1995**, *23*, 267-271.
102. Butler, A. *Coord. Chem. Rev.* **1999**, *187*, 17-35.

103. Butler, A. *Vanadium-Dependent Redox Enzymes: Vanadium Haloperoxidase* in *Comprehensive Biological Catalysis*, Sinnott, M., Ed. Academic Press:New York, **1998**; pp. 1-12.

104. Vreeland, V.; Waite, J. H.; Epstein, L. *J. Phycol.* **1998**, 1-8.

105. Neidlleman, S. L.; Geigert, J. *Biohalogenation: Principles, Basic Roles and Applications*, Ellis Horwood: Chichester, **1986**; pp. 13-44.

106. Simons, B. H.; Barnett, P.; Vollenbroek, E. G. M.; Dekker, H. L.; Muijsers, A. O.; Messerschmidt, A.; Wever, R. *Eur. J. Biochem.* **1995**, 566-574.

107. Hemrika, W.; Renirie, R.; Dekker, H.; Barnett, P.; Wever, R. *Proc. Nat. Acad. of Sci.* **1997**, *94*, 2145-2149.

108. Ishikawa, Y.; Mihara, Y.; Gondoh, K.; Suzuku, E.I.; Asano, Y. *EMBO J.* **2000**, *19*, 2412-2423.

109. Littlechild, J.; Garcia-Rodriguez, E.; Dalby, A.; Isupov, M. *J. Mol. Recognit.* **2002**, *15*, 291-296.

110. Van Schijndel, J. W. P. M.; Barnett, P.; Roelse, J.; Vollenbroek, E. G. M.; Wever, R. *Eur. J. Biochem.* **1994**, *225*, 151-157.

111. Macedo-Ribeiro, S.; Hemrika, W.; Renirie, R.; Wever, R.; Messerschmidt, A. *J. Biol. Inorg. Chem.* **1999**, *4*, 209-219.

112. Svendsen, A.; Jorgensen, L. *Haloperoxidases with Altered Ph Profiles.* US Patent 6,221,821 B1, 04/24/1002, **2001**.

113. Hansen, E. H.; Albertsen, L.; Schafer, T.; Johansen, C.; Frisvad, J. C.; Molin, S.; Gram, L. *Appl. Environ. Microb.* **2003**, *69*, 4611-4617.

114. Munir, I. Z.; Hu, S. H.; Dordick, J. S. *Adv. Synth. Catal.* **2002**, *344*, 1097-1102.

115. Littlechild, J. *Curr. Opin. Chem. Bio.* **1999**, *3*, 28-34.

116. Garcia-Rogriguez, E.; Ohshiro, T.; Aibara, T.; Izumi,Y.; Littlechild, J. *J. Biol. Inorg. Chem.* **2005**, *10*, 275-282.

117. Weyand, M.; Hecht, H. J.; Vilter, H.; Schomburg, D. *Acta. Crystallogr., Sec. D: Biol. Crystallogr.* **1996**, *D52*, 864-865.

118. Messerschmidt, A.; Wever, R. *Proc. Natl. Acad. Sci. U. S. A.* **1996**, *93*, 392-396.

119. Messerschmidt, A.; Wever, R. *Inorg. Chim. Acta* **1998**, *273*, 160-166.

120. Plass, W. *Angew. Chem. Int. Ed.* **1999**, *38*, 909-912.

121. De Boer, E.; Wever, R. *J. Biol. Chem.* **1988**, *263*, 12326-12332.

122. Arber, J. M.; De Boer, E.; Garner, C. D.; Hasnain, S. S.; Wever, R. *Biochemistry* **1989**, *28*, 7968-7973.

123. Zampella, G.; Kravitz, J. Y.; Webster, C. E.; Fantucci, P.; Hall, M. B.; Carlson, H. A.; Pecoraro, V. L.; De Gioia, L. *Inorg. Chem.* **2004**, *43*, 4127-4136.

124. Kravitz, J. Y.; Pecoraro, V. L. *Pure Appl. Chem.* **2005**, *77*, 1595-1605.

125. Kravitz, J. Y.; Pecoraro, V. L.; Carlson, H. A. *J. Chem. Theor. Comp.* **2005**, *1*, 1265-1274.

126. Colpas, G. J.; Hamstra, B. J.; Kampf, J. W.; Pecoraro, V. L. *J. Am. Chem. Soc.* **1996**, *118*, 3469-3478.

127. Smith II, T. S.; Pecoraro, V.L. *Inorg. Chem.* **2002**, *41*, 6754-6760.
128. Kimblin, C.; Bu, X.; Butler, A. *Inorg. Chem.* **2002**, *41*, 161-163.
129. Maurya, M. R.; Agarwal, S.; Bader, C.; Rehder, D. *Eur. J. Inorg. Chem.* **2005**, 147-157.
130. Plass, W. *Coord. Chem. Rev.* **2003**, *237*, 205-212.
131. Slebodnick, C.; Hamstra, B. J.; Pecoraro, V. L. *Struct. Bonding (Berlin)* **1997**, *89*, 51-108.
132. Butler, A.; Clague, M. J.; Meister, G. E. *Chem. Rev.* **1994**, *94*, 625-638.
133. Fernandez, I.; Khiar, N. *Chem. Rev.* **2003**, *103*, 3651-3705.
134. Du, G.; Espenson, J. H. *Inorg. Chem.* **2005**, *44*, 2465-2471.
135. Blum, S. A.; Bergman, R. G.; Ellman, J. A. *J. Org. Chem.* **2003**, *68*, 153-155.
136. Almeida, M.; Filipe, S.; Humanes, M.; Maia, M. F.; Melo, R.; Severino, N.; da Silva, J. A. L.; da Silva, J. J. R. F.; Wever, R. *Phytochem.* **2001**, *57*, 633-642.
137. Renirie, R.; Hemrika, W.; Wever, R. *J. Biol. Chem.* **2000**, *275*, 11650-11657.
138. Murphy, C. D. *J. Appl. Microbiol.* **2003**, *94*, 539-548.
139. Friesner, R. A.; Baik, M.-H.; Gherman, B. F.; Guallar, V.; Wirstam, M.; Murphy, R. B.; Lippard, S. J. *Coord. Chem. Rev.* **2003**, *238-239*, 267-290.
140. Lovell, T.; Himo, F.; Han, W.-G.; Noodleman, L. *Coord. Chem. Rev.* **2003**, *238-239*, 211-232.
141. Siegbahn, P. E. M.; Blomberg, M. R. A. *Chem. Rev.* **2000**, *100*, 421-437.
142. Buhl, M. *J. Comput. Chem.* **1999**, *20*, 1254-1261.
143. Weyand, M.; Hecht, H. J.; Kiess, M.; Liaud, M. F.; Vilter, H.; Schomburg, D. *J. Mol. Biol.* **1999**, *293*, 595-611.
144. Raugei, S.; Carloni, P. *J. Phys. Chem. B.* **2006**, *110*, 3747-3758.
145. Bangesh, M.; Plass, W. *J. Mol. Struc-Theochem.* **2005**, *725*, 163-175.
146. Smith, M. E.; van Eck, E. R. H. *Prog. NMR Spec.* **1999**, *34*, 159-201.
147. (HJCP-69). Pooransingh, N.; Pomerantseva, E.; Ebel, M.; Jantzen, S.; Rehder, D.; Polenova, T. *Inorg. Chem.* **2003**, *42*, 1256-1266.
148. Lapina, O. B.; Shubin, A. A.; Khabibulin, D. F.; Terskikh, V. V.; Bodart, P. R.; Amoureux, J. P. *Catal. Today.* **2003**, *78*, 91-104.
149. a. Nielsen, F. H. in *Metal Ions in Biological Systems*, Sigel, H.; Sigel, A. Eds. Marcel Dekker, Inc.: New York, **1995**; Vol. 31, pp 43-573. b. Nielsen, F. H. in *Vanadium Compounds. Chemistry, Biochemistry, and Therapeutic Applications ACS Symposium Series* Tracey, A. S.; Crans, D. C. Eds. American Chemical Society: Washington, DC, **1998**, Vol.711, pp 297-307.
150. Kaplan, D. I.; Adriano, D. C.; Carlson, C. L.; Sajwan, K. S. *Water Air Soil Pollut.* **1990**, *49*, 81-91.
151. Michibata, H. *Zool. Sci.* **1996**, 13, 489-502.
152. Waters, M. D. in *Advances in Modern Toxicology*, Goyer, R. A.; Mehlman, M. A., Eds. John Wiley & Sons: New York, **1977**; Vol. 2, pp 147-189.

153. Cohen, M. D.; Sisco, M.; Prophete, C.; Chen, L.-C.; Zelikoff, J. T.; Ghio, A. J.; Stonehuerner, J. D.; Smee, J. J.; Holder, A. A.; Crans, D. C. *J. Immunotoxicology* **2007**, *4*, 49-60.

154. Cohen, M. D. *J. Immunotoxicology*, **2004**, *1*, 39-69.

155. Kadiiska, M. B.; Ghio, A. J.; Mason, R. P. *Spectroch. Acta. Part A. Mol. Biomol. Spec.* **2004**, *60*, 1371-1377.

156. Kustin, K. in *Vanadium Compounds. Chemistry, Biochemistry, and Therapeutic Applications ACS Symposium Series* Tracey, A. S.; Crans, D. C. Eds. American Chemical Society: Washington, DC, **1998**, Vol.711, pp 170-183.

157. Dikanov, S. A.; Liboiron, B. D.; Orvig, C. *J. Am. Chem. Soc.* **2002**, 124, 2969-2978.

158. Aureliano, M.; Tiago, T.; Gandara, R. M. C.; Sousa, A.; Moderno, A.; Kaliva, M.; Salifoglou, A.; Duarte, R. O.; Moura, J. J. *J. Inorg. Biochem.* **2005**, *99*, 2355-2361.

159. Aureliano, M. *J. Inorg. Biochem.* **2007**, *101*, 80-88.

160. Tiago, T.; Aureliano, M.; Gutierrez-Merino, C. *Biochemistry* **2004**, *43*, 5551-5561.

161. Djordjevic, C.; Wampler, G. L. *J. Inorg. Biochem.* **1985**, *25*, 51-55.

162. Sakurai, H.; Tamura, H.; Okatani, K. *Biochem. Biophys. Res. Comm.* **1995**, *206*, 133-137.

163. Hamilton, E. E.; Fanwick, P. E.; Wilker, J. J. *J. Am. Chem. Soc.* **2006**, *128*, 3388-3395.

164. Vilas Boas, L. F.; Costa Pessoa, J. in *Comprehensive Coordination Chemistry*, Wilkinson, G.; Gillard, R.D.; McCleverty, J.A. Eds., Pergamon Press: Oxford, **1987**; Vol. 3, pp 453-584.

165. Crans, D.C.; Smee, J. J. in *Comprehensive Coordination Chemistry II*, McCleverty, J. A.; Meyer, T. J.; Wedd, A. G. Eds., Elsevier: Amsterdam, **2004**; Vol. 4, pp 175-239.

166. aladino, A. C.; Larsen, S. C. *J. Phys. Chem. A* **2003**, *107*, 1872-1878.

167. Saladino, A. C.; Larsen, S. C. *J. Phys. Chem. A* **2003**, *107*, 4735-4740.

168. Bühl, M.; Schurhammer, R.; Imhof, P. *J. Am. Chem. Soc.* **2004**, *126*, 3310.

169. Bühl, M.; Mauschick, F. T.; Terstegen, F.; Wrackmeyer, B. *Angew. Chem. Int. Ed.* **2002**, *41*, 2312.

170. Bühl, M.; Grigoleit, S.; Kabrede, H.; Mauschick, F. T. *Chem. Eur. J.* **2006**, *12*, 477.

171. Correia, I.; Costa Pessoa, J.; Duarte, M. T.; Henriques, R. T.; Piedade, M. F. M.; Veiros, L. F.; Jackush, T.; Dornyei, A.; Kiss, T.; Castro, M. M. C. A.; Geraldes, C. F. G. C.; Avecilla, F. *Eur. J. Chem.* **2004**, *10*, 2301-2317.

172. Correia, I.; Costa Pessoa, J.; Duarte, M. T.; Piedade, M. F. M.; Jakusch, T.; Kiss, T.; Castro, M. M. C. A.; Geraldes, C. F. G. C.; Avecilla, F. *Eur. J. Inorg. Chem.* **2005**, 732-744.

173. Costa Pessoa, J.; Marcão, S.; Correia, I.; Gonçalves, G.; Dornyei, A.; Kiss, T.; Jakusch, T.; Tomaz, I.; Castro, M. M. C. A.; Geraldes, C. F. G. C.; Avecilla, F. *Eur. J. Inorg. Chem.* **2006**, 3595-3606.

174. Smith, K. I.; Borer, L.; Olmstead, M. M. *Inorg. Chem.* **2003**, *42*, 7410-7415.
175. Neves, A.; Tamanini, M.; Rosa Correia, V.; Vencato, I. *J. Braz. Chem. Soc.* **1997**, *8*, 519-522.
176. Neves, A.; Romanowski, S. M. M.; Vencato, I.; Mangrich, A. S. *J. Braz. Chem. Soc.* **1998**, *9*, 426-429.
177. Katoh, A.; Taguchi, K.; Saito, R.; Fujisawa, Y.; Takino, T.; Sakurai, H. *Heterocycles* **2003**, *60*, 1147-1159.
178. Rangel, M.; Leite, A.; Amorim, M. J.; Garribba, E.; Micera, G.; Lodyga-Chruscinska, E. *Inorg. Chem.* **2006**, *45*, 8086-8097.
179. Yamaguchi, M.; Wakasugi, K.; Saito, R.; Adachi, Y.; Yoshikawa, Y.; Sakurai, H.; Katoh, A. *J. Inorg. Biochem.* **2006**, *100*, 260-269.
180. Maurya, M. R.; Agarwal, S.; Abid, M.; Azam, A..; Bader, C.; Ebel, M.; Rehder, D. *Dalton Trans.* **2006**, 937-947.
181. Weng, S. S.; Shen, M. W.; Kao, J. Q.; Munot, Y. S.; Chen, C. T. *Proc. Natl. Acad. Sci. U.S.A.* **2006**, *103*, 3522-3527.
182. Wang, D.; Ebel, M.; Schulzke, C.; Grüning, C.; Hazari, S. K. S.; Rehder, D., *Eur. J. Inorg. Chem.* **2001**, 1935-942.
183. Sakurai, H.; Katoh, A.; Yoshikawa, Y. *Bull. Chem. Soc. Jpn.* **2006**, *79*, 1645-1664.
184. Garribba, E.; Micera, G.; Sanna, D.; Lodyga-Chruscinska, E. *Inorg. Chim. Acta* **2003**, *348*, 97-106.
185. a. Yang, L.; La Cour, A.; Anderson, O. P.; Crans, D. C. *Inorg. Chem.* **2002**, *41*, 6322-6331. b. Crans, D. C.; Yang, L. *Inorg. Chem.* **2000**, *39*, 4409-4416. c. Jakusch, T.; Jin, W.; Yang, L.; Kiss, T.; Crans, D. C. *J. Inorg. Biochem.* **2003**, *95*, 1-13.
186. Yue, H.; Zhang, D.; Chen, Y.; Shi, Z.; Feng, S. *Inorg. Chem. Commun.* **2006**, *9*, 959-961.
187. Garribba, E.; Lodyga-Chruscinska, E.; Micera, G.; Panzanelli, A.; Sanna, D. *Eur. J. Inorg. Chem.* **2005**, 1369-1382.
188. Baruah, B.; Das, S.; Chakravorty, A. *Inorg. Chem.* **2002**, *41*, 4502-4508.
189. Durupthy, O.; Coupé, A.; Tache, L.; Rager, M. N.; Maquet, J.; Coradin, T.; Steunou, N.; Livage, J. *Inorg. Chem.* **2004**, *43*, 2021-2030.
190. (CJCP-227). Schmidt, H.; Andersson, I.; Rehder, D.; Petersson, L. *Eur. J. Chem.* **2001**, *7*, 251-257.
191. Costa Pessoa, J.; Tomaz, I.; Kiss, T.; Kiss, E.; Buglyó, P., *J. Biol. Inorg. Chem.* **2002**, *7*, 225-240
192. Tasiopoulos, A. J.; Tolis, E. J.; Tsangaris, J. M.; Evangelou, A.; Woollins, J. D.; Slawin, A. M.; Costa Pessoa, J.; Correia, I.; Kabanos, T. A. *J. Biol. Inorg. Chem.* **2002**, *7*, 363-374.
193. Costa Pessoa, J.; Calhorda, M. J.; Cavaco, I.; Correia, I.; Duarte, M. T.; Felix, V.; Henriques, R. T.; Piedade, M. F. M.; Tomaz, I., *J. Chem. Soc. Dalton Trans.* **2002**, 4407-4415.
194. Kiss, T.; Jakusch, T.; Costa Pessoa, J.; Tomaz, I., *Coord. Chem. Rev.* **2003**, *237*, 123-133.

195. Jakusch, T.; Dörnyei, Á.; Correia, I.; Rodrigues, L.; Tóth, G. K.; Kiss, T.; Costa Pessoa, J.; Marcão, S. *Eur. J. Inorg. Chem.* **2003**, 2113-2122.

196. Maurya, M. R.; Khurana, S.; Rehder, D. *Trans. Met. Chem.* **2003**, *28*, 511-517.

197. Yue, H.; Zhang, D.; Shi, Z.; Feng, S. *Solid State Sci.* **2006**, *8*, 1368-1372.

198. Kaliva, M.; Raptopoulou, C. P.; Terzis, A.; Salifoglou, A. *Inorg Chem.* **2004**, *43*, 2895-2905

199. Kaliva, M.; Raptopoulou, C. P.; Terzis, A.; Salifoglou, A. *J. Inorg. Biochem.* **2003**, *93*, 161-173

200. Crans, D. C. *Pure & Appl.Chem.* **2005**, *77*, 1497-1527.

201. Ferreira da Silva, J. L.; M. F. M. Piedade; Duarte, M. T. *Inorg. Chim. Acta* **2003**, *356*, 222-242

202. Buglyó, P.; Nagy, E. M.; Sóvágó, I. *Pure & Appl. Chem.* **2005**, *77*, 1583-1594.

99. Buglyó, P.; Crans, D. C.; Nagy, E. M.;Lindo, R. L.; Yang, L.; Smee, J. J.; Chi, L.-H.; Godzala, M. E.; Willsky, G. R. *Inorg. Chem.* **2005**, *44*, 5416-5427.

203. Armstrong, W. H.; Abu-Sbeih, K. *5th International Symposium on the Chemistry and Biological Chemistry of Vanadium*, ACS National Meeting, San Francisco, California, USA, September, **2006**, INOR 72.

204. Karol, F. J.; Kao, S.-C. *Stud. Surf. Sci. Catal.* **1994**, *89*, 389-403.

205. Mazzanti, M.; Floriani, C.; Chiesi-Villa; A.; Guastatini, C. *J. Chem. Soc. Dalton Trans.* **1989**, 1793-1798.

206. Wever, R.; Kustin, K. *Adv. Inorg. Chem.* **1990**, *35*, 81-115.

207. Smith, M. J.; Ryan, D. E.; Nakanishi, K.; Frank, P.; Hodgson, K. O. *Met. Ions Biol. Syst.* **1995**, *31*, 423-490.

208. Smith, M. J. *Experientia* **1989**, *45*, 452-457.

209. Meier, R.; Boddin, M.; Mitzenhein, S. *Bioinorg. Chem.* **1997**, 69-97.

210. Bukietynska, K.; Podsiadly, H.; Karwecka, Z., *J. Inorg. Biochem.* **2003**, *94*, 317-325.

211. Osinnska-Krolicka, I.; Podsiadly, H.; Bukietynska, K.; Zemanek-Zboch, M.; Nowak, D.; Suchoszek-Lukaniuk, K.; Malicka-Blaszkiewicz, M., *J. Inorg. Biochem.* **2004**, *98*, 2087-2098.

212. Maeda, H.; Kanamori, K.; Michibata, H.; Konno, T.; Okamoto, K.I.; Hidaka, J. *Bull. Chem. Soc. Jpn.* **1993**, *66*, 790-796.

213. Papaioannou, A.; Manos, M.; Karkabounas, S.; Liasko, R.; Kalfakakou, V.; Correia, I.; Enangelou, A.; Costa Pessoa, J.; Kabanos, T. A. *J. Inorg. Biochem.* **2004**, *98*, 959-968.

214. Melchior, M.; Rettig, S. J.; Liboiron, B. D.; Thompson, K. H.; Yuen, V. G.; McNeil, J. H.; Orvig, C. *Inorg. Chem.* **2001**, *40*, 4686-4690.

215. Krzystek, J.; Ozarowski, A.; Chu, W.-C.; Wang, Z.-C.; Tsai, Y.-F.; Chen, K.-Y.; Wu, C.-C.; Hsu H.-F.; Telser; J. *5th International Symposium on the Chemistry and Biological Chemistry of Vanadium*, ACS National Meeting, San Francisco, California, USA, September, **2006**, INOR 71.

216. Costa Pessoa, J.; Correia, I.; Kiss, T.; Jakusch, T.; Castro, M. M. C. A.; Geraldes, C. F. G. C. *J. Chem. Soc., Dalton Trans.* **2002**, 4440-4450.

217. Jakusch, T.; Marcão, S.; Rodrigues, L. M.; Correia, I.; Costa Pessoa, J.; Kiss, T. *J. Chem. Soc., Dalton Trans.* **2005**, 3072-3078.

218. Schwendt, P.; Ahmed, M.; Marek, J. *Inorg. Chim. Acta* **2005**, *358*, 3572-3580.

219. Kaliva, M.; Kyriakakis, E.; Salifoglou, A. *Inorg. Chem.* **2002**, *41*, 7015-7023.

220. Kawabe, K.; Sasagawa, T.; Yoshikawa, Y.; Ichimura, A.; Kumekawa, K.; Yanagihara, N.; Takino, T.; Sakurai, H.; Kojima, Y. *J. Biol. Inorg. Chem.* **2003**, *8*, 893-906.

221. Mahroof-Tahir, M.; Brezina, D.; Fatima, N.; Choudhary, M. I.; Rahman, A. *J. Inorg. Biochem.* **2005**, *99*, 589-599.

222. Makinen, M. W.; Brady, M. J. *J. Biol. Chem.* **2002**, *277*, 12215-12220.

223. Garribba, E.; Micera, G.; Sanna, D. *Inorg. Chim. Acta* **2006**, *359*, 4470-4476.

224. Storr, T.; Mitchell, D.; Buglyó, P.; Thompson, K. H.; Yuen, V.G.; McNeill, J. H.; Orvig, C. *Bioconjugate Chem.* **2003**, *14*, 212-221.

225. Nakai, M.; Obata, M.; Sekigushi, F.; Kato, M.; Shiro, M.; Ichimura, A.; Kinoshita, I.; Mikuriya, M.; Inohara, T.; Kawabe, K.; Sakurai, H.; Orvig, C.; Yano, S. *J. Inorg. Biochem.* **2004**, *98*, 105-112.

226. Okazaki, K.; Saito, K. *Bull. Chem. Soc. Jpn.* **1982**, *55*, 785-791.

227. Nekola, H.; Wang, D.; Grüning, C.; Gatjjens, J.; Behrens, A.; Rehder, D. *Inorg. Chem.* **2002**, *41*, 2379-2374.

228. Davies, S. C.; Hughes, D. L.; Richards, R. L.; Sanders, J. R. *J. Chem Soc., Dalton Trans.* **2002**, 1442-1447.

229. Wolff, F.; Lorber, C.; Choukroun, R.; Donnadieu, B. *Inorg. Chem.* **2003**, *42*, 7839-7845.

230. Thompson, K. H.; Bohmerle, K.; Polishcuk, E.; Martins, C.; Toleikis, P.; Tse, J.; Yuen, V.; McNeil, J. H.; Orvig, C. *J. Inorg. Biochem.* **2004**, *98*, 2063-2070.

231. Garribba, E.; Micera, G.; Lodyga-Chruscinska, E.; Sanna, D. *Eur. J. Inorg. Chem.* **2006**, 2690.

232. Chilas, G. I.; Miras, H. N.; Manos, M. J.; Woollins, J. D.; Slawin, A. M. Z.; Stylianou, M.; Keramidas, A. D.; Kabanos, T. A. *Pure & Appl. Chem.* **2005**, 1529-1538.

233. Cocco, M. T.; Onnis, V.; Ponticelli, G.; Meier, B.; Rehder, D.; Garribba, E.; Micera, G. *J. Inorg. Biochem.* **2007**, *101*, 19-29.

234. Xing, Y.; Zhang, Y.; Sun, Z.; Ye, L.; Xu, Y.; Ge, M.; Zhang, B.; Niu, S. *J. Inorg. Biochem.* **2007**, *101*, 36-43.

235. Gambino, D.; Urquiola, C.; Noblía, P.; Vieites, M.; Torre, M. H.; Cerecetto, H.; Lavaggi, M.; Aguirre, G.; González, M.; Costa-Filho, A.; Azqueta, A.; López, A.; Monge, A.; Parajón-Costa, B. *5th International Symposium on the Chemistry and Biological Chemistry of Vanadium*, ACS

National Meeting, San Francisco, California, USA, September, **2006**, INOR 76.

236. Sivák, M.; Schwendt, P.; Pacigová, S.; Tatiersky, J. *5th International Symposium on the Chemistry and Biological Chemistry of Vanadium*, ACS National Meeting, San Francisco, California, USA, September, **2006**, INOR 384.

237. Schwendt, P.; Sivák, M.; Dudášová, D.; Marek, J. *5th International Symposium on the Chemistry and Biological Chemistry of Vanadium*, ACS National Meeting, San Francisco, California, USA, September, **2006**, INOR 831.

238. Baruah, S.; Smee, J. J.; Epps, J. A.; Ding, W. Crans, D.C. *5th International Symposium on the Chemistry and Biological Chemistry of Vanadium*, ACS National Meeting, San Francisco, California, USA, September, **2006**, INOR 822.

239. Mahroof-Tahir, M.; Gregory, D. D.; Bushkofsky, J. R.; Swingley, L. E.; Eannelli, M. A. *5th International Symposium on the Chemistry and Biological Chemistry of Vanadium*, ACS National Meeting, San Francisco, California, USA, September, **2006**, INOR 926.

240. Moriuchi, T.; Ishino, K.; Beppu, T.; Hirao, T. *5th International Symposium on the Chemistry and Biological Chemistry of Vanadium*, ACS National Meeting, San Francisco, California, USA, September, **2006**, INOR 990.

241. a. Avecilla, F.; Palacio, L.; Figueiredo, A.; Faneca, H.; Costa Pessoa, J.; Gonçalves, G.; Geraldes, C. F. G. C.; Lima M. C. P.; Castro, M. M. C. A. *5th International Symposium on the Chemistry and Biological Chemistry of Vanadium*, ACS National Meeting, San Francisco, California, USA, September, **2006**, INOR 376. b. (CJCP-287). Avecilla, F.; Palacio, L.; Maestro, M.; Gonçalves, G.; Tomaz, I.; Costa Pessoa, J.; Castro, M. M. C. A.; Geraldes, C. F. G. C. *5th International Symposium on the Chemistry and Biological Chemistry of Vanadium*, ACS National Meeting, San Francisco, California, USA, September, **2006**, INOR 377.

242. Costa Pessoa, J.; Marcão, S.; Correia, I.; Gonçalves, G.; Dörnyei, Á.; Kiss, T.; Jakusch, T.; Tomaz, I.; Castro, M. M. C. A.; Geraldes, C. F. G. C.; Avecilla, F. *5th International Symposium on the Chemistry and Biological Chemistry of Vanadium*, ACS National Meeting, San Francisco, California, USA, September, **2006**, INOR 827.

243. Evangelou, A. M. *Crit. Rev. Oncol. Hematol.* **2002**, *42*, 249-265.

244. D'Cruz, O. J.; Uckun, F. M. *Expert Opin. Investig. Drugs* **2002**, *11*, 1829-1836.

245. Hwang. J. H.; Larson, R. K.; Abu-Omar, M. M. *Inorg. Chem.* **2003**, *42*, 7967-7977.

246. Noblía, P.; Vieites, M.; Parajón-Costa, B. S.; Baran, E. J.; Cerecetto, H.; Draper, P.; Gonzalez, M.; Piro, O. E.; Castellano, E. E.; Azqueta, A.; Ceráin, A. L.; Monge-Vega, A.; Gambino, D. *J. Inorg. Biochem.* **2005**, *99*, 443-451.

247. Chakraborty, T.; Chatterjee; A.; Saralaya, M. G.; Chaterjee, M. *J. Biol. Inorg. Chem.* **2006**, *11*, 855-866.

248. Abu-Omar, M. M. *5th International Symposium on the Chemistry and Biological Chemistry of Vanadium*, ACS National Meeting, San Francisco, California, USA, September, **2006**, INOR 411.

249. Cavaco, I.; Ribeiro, V.; Correia, I.; Tomaz, I.; Brotas, G.; Vale, I.; Marcão, S.; Costa Pessoa, J. *5th International Symposium on the Chemistry and Biological Chemistry of Vanadium*, ACS National Meeting, San Francisco, California, USA, September, **2006**, INOR 381.

250. Costa Pessoa, J.; Correia, I.; Tomaz, I.; Vale, I.; Cavaco, I.; Ribeiro, V.; Castro, M. M. C. A.; Geraldes C. C. F. G. *5th International Symposium on the Chemistry and Biological Chemistry of Vanadium*, ACS National Meeting, San Francisco, California, USA, September, **2006**, INOR 79.

251. Dornyei, A.; Marcão, S.; Costa Pessoa, J.; Jakusch, T.; Kiss, T. *Eur. J. Inorg. Chem.* **2006**, 3614-3621.

252. a. Pope, M. T. in *Comprehensive Coordination Chemistry*, Wilkinson, G.; Gillard, R.D.; McCleverty, J.A. Eds., Pergamon Press: Oxford, **1987**; Vol. 3, pp 1023-1058. b. Cronin, L. in *Comprehensive Coordination Chemistry II*, McCleverty, J. A.; Meyer, T. J.; Wedd, A. G. Eds., Elsevier: Amsterdam, **2004**; Vol. 7, pp 1-56. c. Hill, C. L. in *Comprehensive Coordination Chemistry II*, McCleverty, J. A.; Meyer, T. J.; Wedd, A. G. Eds., Elsevier: Amsterdam, **2004**; Vol. 4, pp 679-759.

253. Khan, I.; Zubieta, J. *Prog. Org. Chem.* **1995**, *43*, 29-50.

254. Lisnard, L.; Dolbecq, A.; Mialane, P.; Marrot, J.; Riviere, E.; Borshch, S. A.; Petit, S.; Robert, V.; Duboc, C.; McCormac, T.; Secheresse, F. *Dalton Trans.* **2006**, *43*, 5141-5148.

255. a. Müller, A.; Kuhlmann, C.; Bogge, H.; Schmidtmann, M.; Baumann, M.; Krickemeyer, E. *Eur. J. Inorg. Chem.* **2001**, *9*, 2271-2277, b. Müller, A.; Peters, F.; Pope, M. T.; Gatteschi, D. *Chem. Rev.* **1998**, *98*, 239-272.

256. a. Tsukerblat, B.; Müller, A., Tarantul, A. *J. Chem. Phys.* **2006**, *125*, 054714/1-054714/10. b. Gatteschi, D.; Pardi, L.; Barra, A. L.; Müller, A. in *Polyoxometalates: From Platonic Solids to Anti-Retroviral Activity* Pope, M. T.; Müller, A. Eds.; Kluwer Academic Publishers: Dordrecht, **1994**; pp 219-231.

257. Kurata, T.; Uehara, A.; Hayashi, Y.; Isobe, K. *Inorg. Chem.* **2005**, *44*, 2524-2530.

258. Epstein, I. R.; Pojman, J. A. *An Introduction to Nonlinear Chemical Dynamics*; Oxford University Press: New York, **1998**; pp 78-82.

Vanadium Catalysis of Synthesis: Organic Compounds and Polymers

Chapter 1

Synthetic Transformations via Vanadium-Induced Redox Reactions

Toshikazu Hirao

Department of Applied Chemistry, Graduate School of Engineering, Osaka University, Yamada-oka, Suita, Osaka 565–0871, Japan

Low-valent vanadium-catalyzed reduction reactions including dehalogenation, pinacol coupling, and the related radical reaction have been developed by constructing a multi-component redox system in combination with a co-reductant and an additive promoter. High stereoselectivity is attained in these catalytic transformations. Oxovanadium(V) compounds, which are evaluated as Lewis acids with oxidation capability, induce one-electron oxidative desilylation of organosilicon compounds. Oxidative ligand coupling of organoaluminums, organoborons and organozincs, and the ate complexes is achieved by treatment with oxovanadium(V) compounds. A catalytic reaction of organoborates proceeds under molecular oxygen.

© 2007 American Chemical Society

Vanadium is a biologically essential element. Its inclusion in enzymes such as haloperoxidase and nitrogenase reveals the importance of its redox chemistry. Vanadium complexes, including organovanadium compounds, exist in a variety of configurations depending on their oxidation states and coordination numbers. Vanadium exists in oxidation states ranging from -3 to +5 and generally converts between states via a one-electron redox process. The properties permit the development of a wide range of organic reactions by controlling the redox processes of vanadium compounds.

Radical species are useful intermediates in organic synthesis. A variety of methods have been developed for the selective generation of radical species. One-electron reduction or oxidation of organic compounds using the redox process of metals provides a useful and practical route to generate anion radicals or cation radicals, respectively. Particularly, the redox properties of early transition metals including vanadium, titanium, and manganese, have been employed from this point of view, as exemplified by Scheme 1 (*1,2*). The redox function is tuned by a ligand or a solvent, allowing a more efficient interaction through the orbitals of metals and substrates for facile electron transfer.

Scheme 1

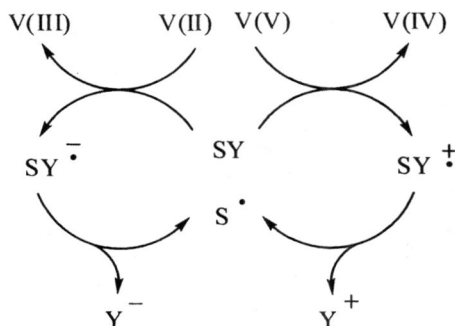

The redox process of V(II) to V(III) is known to induce one-electron reduction reactions. Vanadium(I) species also can serve as similar reductants. Pentavalent vanadium compounds, which exist in one of five possible configurations (e.g., $VOCl_3$, tetrahedral; VF_5, octahedral), are generally considered to be one-electron oxidants which utilize the V(V)-V(IV) couple (Scheme 1). The redox potential of this couple increases with acidity, so the reactions are usually carried out in acidic aqueous media. One-electron oxidation is also possible with the V(IV)-V(III) couple (E_0, 0.38 V), but the V(V)-V(III) couple (E_0, 0.68 V) is less useful for organic oxidation.

One of the synthetic limitations exists in the use of stoichiometric or excess amounts of metallic reductants or oxidants to complete the reaction. A catalytic system should be constructed to avoid the use of stoichiometric, expensive and/or toxic metallic reagents. Selection of stoichiometric co-reductants or co-oxidants is essential for the reversible cycle of a catalyst (Scheme 2). An alternative method for recycling a catalyst is achieved by electrolysis (*3*). A metallic co-reductant is ultimately converted to the corresponding metal salt in a higher oxidation state, which may work as a Lewis acid to facilitate the reduction reaction or, on the contrary, impede the reaction. Taking these interactions into account, the requisite catalytic system is envisioned to be formed by multi-component interaction. Steric control by means of coordination may permit the stereoselective and/or stereospecific transformations.

Scheme 2

Substrate ⟶ V(II)L ⟶ Oxized form of Coreductant

Product ⟶ V(III)L ⟶ Coreductant

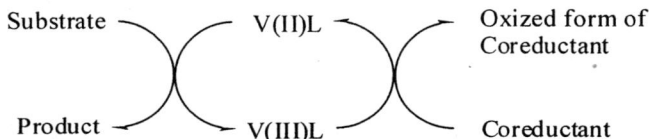

Additive Effect on Reduction Susceptibility

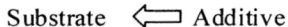

Substrate ⟸ Additive

Ligand Effect on Reduction Capability

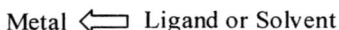

Metal ⟸ Ligand or Solvent

Redox interaction between a metal and a metal or a ligand (or an organic group) through inner- or outer-sphere electron transfer is considered to change the oxidation state of these components. Precise regulation of the redox process based on the electronic interaction between HOMO and LUMO orbitals is required for this purpose, which allows development of novel and selective methods for reductive or oxidative transformations under redox potential control. This chapter describes the synthetically useful redox reactions promoted or catalyzed by vanadium compounds.

Catalytic Transformations via One-Electron Reduction

Reduction with low valent vanadium compounds is utilized in various reductive transformations of organic compounds, especially organic halides and carbonyl compounds. Although a stoichiometric reaction has been investigated so far, the catalytic reaction is less investigated due to the difficulty in the construction of a catalytic redox system. This kind of catalytic reaction is very important biologically. Investigation on nitrogen fixation models is challenging from this point of view (*4*).

Dehalogenative Reduction

Homolytic cleavage of organic halides through one-electron reduction affords the corresponding radical intermediates. A catalytic dehalogenation is achieved in the vanadium-induced one-electron reduction. The highly stereoselective monodebromination of *gem*-dibromocyclopropane **1** to monobromide **2** proceeds with a catalytic amount of a low-valent vanadium species generated from VCl$_3$ or CpV(CO)$_4$ and Zn as a stoichiometric co-reductant. The co-existence of diethyl phosphonate or triethyl phosphite is essential for the reduction reaction (Scheme 3) (*5*). The phosphonate or phosphite is likely to play an important role in the debromination step. The coordination of the P-ligand to vanadium leads to a bulky and stronger reductant. The former effect is related to the stereoselectivity, since the bulky reductant is liable to approach the bromide from the less hindered side. Furthermore, a hydrogen source is assumed to be available from the phosphonate or phosphate in the coordination sphere (*6*). In this manner, the ternary reductant system contributes to the stereoselective catalytic debromination. The multi-component redox system is formed in an appropriate combination of a catalyst, a co-reductant, and an additive promoter.

Scheme 3

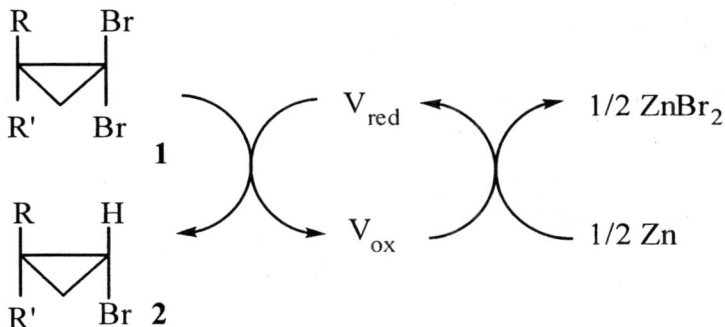

Pinacol Coupling

The reductive dimerization of carbonyl compounds is a useful synthetic method for forming vicinally functionalized carbon-carbon bonds. One-electron transfer from a metal to a carbonyl group generates the corresponding anion radical, which can dimerize to give 1,2-diols. For the stoichiometric reductive dimerization, low-valent metals such as aluminum amalgam, zinc, titanium, vanadium, and samarium have been employed conveniently. For example, the pinacol coupling reaction using $TiCl_3/Zn$-Cu and $[V_2Cl_3(THF)_6]_2[Zn_2Cl_6]$ was developed successfully for the synthesis of paclitaxel and C_2-symmetrical HIV protease inhibitors, respectively (7). To synthesize such compounds, the stereochemistry should be controlled (8). Furthermore, a catalytic system is to be constructed. The ternary system consisting of a vanadium catalyst, a chlorosilane, and a stoichiometric co-reductant provides a catalytic protocol for the pinacol coupling. A vanadium catalyst is essential although the combination of Zn and Me_3SiCl allows the reductive dimerization of aldehydes. The low-valent vanadium species mediating the electron transfer is generated *in situ*, and a reversible redox cycle is formed in the presence of Zn as a stoichiometric co-reductant (Scheme 4).

Scheme 4

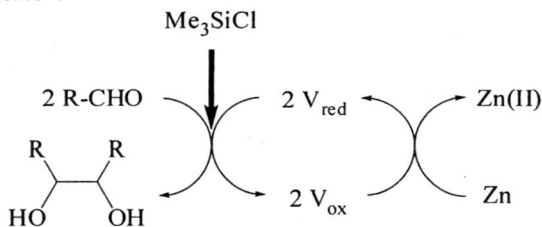

Cp$_2$VCl$_2$-catalyzed reaction of secondary aliphatic aldehydes in THF leads to the highly diastereoselective formation of the *dl*-1,2-diols **3** (eq 1) (9). The diastereoselectivity also depends on chlorosilanes: PhMe$_2$SiCl is superior to Me$_3$SiCl. Cp$_2$TiCl$_2$ can be similarly employed as a catalyst (10). The reaction in DME gives the 1,3-dioxolanes **4** via the pinacol coupling and acetalization (eq 2) (11). Fürstner has independently developed a similar catalytic method for the McMurry coupling of the oxoamide to the indoles in the presence of a catalytic amount of TiCl$_3$, Zn, and a chlorosilane (12).

$$2 \text{ R-CHO} \xrightarrow[\text{2) Workup}]{\substack{\text{1) cat. Cp}_2\text{VCl}_2 \text{ / R'}_3\text{SiCl / Zn} \\ \text{THF}}} \quad \underset{\mathbf{3}}{\text{HO, R / R, OH}} \quad (1)$$

$$3 \text{ R-CHO} \xrightarrow[\text{DME}]{\text{cat. Cp}_2\text{VCl}_2 \text{ / R'}_3\text{SiCl / Zn}} \quad \underset{\mathbf{4}}{\text{(dioxolane)}} \quad (2)$$

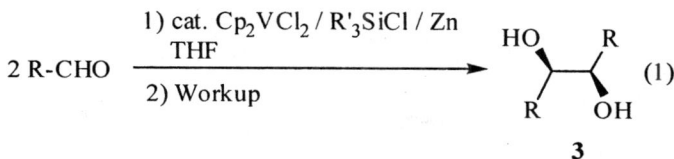

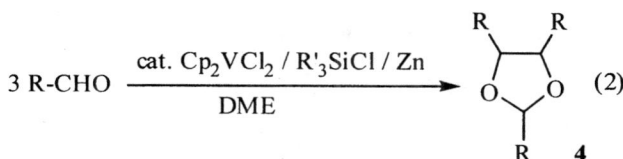

In the absence of a chlorosilane, a catalytic reaction is not observed. Silylation is considered to liberate the catalyst. The Lewis-acidic-like interaction of a chlorosilane with a carbonyl oxygen is suggested to facilitate the electron transfer to the carbonyl group, generating the stabilized silyloxyalkyl radical for dimerization. Another interaction with the vanadium catalyst is possible since the UV-vis spectrum of Cp_2VCl_2 changes on the addition of Me_3SiCl. The diastereoselectivity depends on the substituent of chlorosilanes, which implies its steric effect in the coupling step. Based on these observations (9-12), a variety of modified catalytic systems have been investigated for the diastereoselective carbon-carbon bond formation (13, 14, 15). A catalytic system for the enantioselective pinacol coupling is also achieved by a titanium or chromium catalyst with a chiral ligand (16).

This method is applied to the selective intramolecular coupling reaction of the 1,5-diketone **5** (eq 3) (9a).

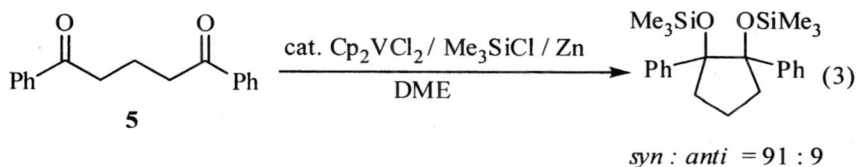

$$\underset{\mathbf{5}}{\text{Ph-C(O)-CH}_2\text{CH}_2\text{CH}_2\text{-C(O)-Ph}} \xrightarrow[\text{DME}]{\text{cat. Cp}_2\text{VCl}_2 \text{ / Me}_3\text{SiCl / Zn}} \quad \underset{}{\text{Me}_3\text{SiO \quad OSiMe}_3 / \text{Ph} \quad \text{Ph (cyclopentane)}} \quad (3)$$

$$syn : anti = 91 : 9$$

Ac_2O or $AcCl$ can be utilized instead of a chlorosilane in the $VOCl_3$-catalyzed pinacol coupling reaction of aromatic aldehydes to give the diacetate **6** (eq 4) (17). A low-valent vanadium species appears to be generated in situ.

$$\text{PhCHO} \xrightarrow[\text{DME}]{\text{cat. Cp}_2\text{VCl}_2 / \text{Ac}_2\text{O} / \text{Zn}} \quad \underset{\text{AcO} \quad \text{Ph}}{\overset{\text{Ph} \quad \text{OAc}}{\diagup\!\!\!\diagdown}} \qquad (4)$$

6

dl : *meso* = 64 : 36

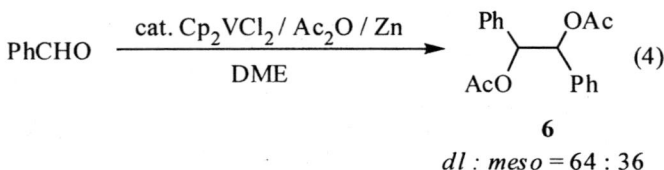

The reaction in water does not require the coexistence of a chlorosilane as an additive, probably since a proton source is available in this catalytic system (eq 5) (*18*).

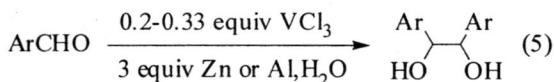

$$\text{ArCHO} \xrightarrow[\text{3 equiv Zn or Al,H}_2\text{O}]{\text{0.2-0.33 equiv VCl}_3} \quad \underset{\text{HO} \quad \text{OH}}{\overset{\text{Ar} \quad \text{Ar}}{\diagup\!\!\!\diagdown}} \qquad (5)$$

The $\text{Cp}_2\text{VCl}_2/\text{R}_3\text{SiCl}/\text{Zn}$ catalytic system can be also applicable to the reductive coupling of aldimines **7** (eq 6) (*19*). *meso*-Diamines **8** are obtained as a major product. The diastereoselectivity depends on the substituents on both the nitrogen and silane atoms. The allyl or benzyl group on the nitrogen atom is advantageous for *meso* selection.

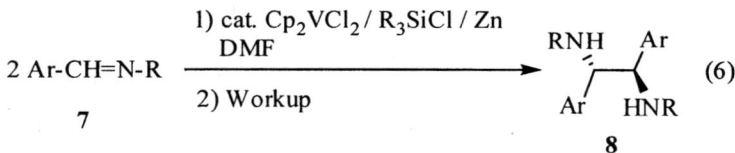

$$2\ \text{Ar-CH=N-R} \xrightarrow[\text{2) Workup}]{\substack{\text{1) cat. Cp}_2\text{VCl}_2 / \text{R}_3\text{SiCl} / \text{Zn} \\ \text{DMF}}} \quad \underset{\text{Ar} \quad \text{HNR}}{\overset{\text{RNH} \quad \text{Ar}}{\diagup\!\!\!\diagdown}} \qquad (6)$$

7

8

Radical Cyclization

The above-mentioned catalytic system is of synthetic potential in controlled radical reactions. The reaction of the δ,ε-unsaturated aldehyde **9** with cat. $\text{Cp}_2\text{VCl}_2/\text{Me}_3\text{SiCl}/\text{Zn}$ is conducted in THF to afford the cyclic alcohol **10** with excellent diastereoselectivity (eq 7) (*9*), probably through 5-*exo*-cyclization of the radical anion intermediate, followed by chlorination.

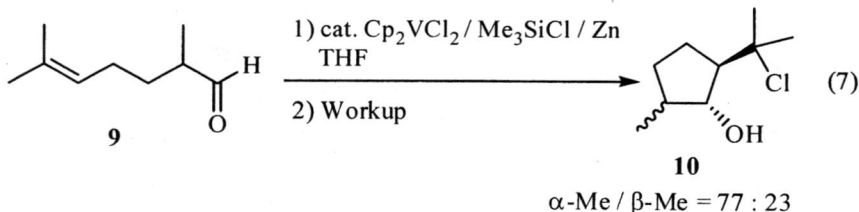

$$\xrightarrow[\text{2) Workup}]{\substack{\text{1) cat. Cp}_2\text{VCl}_2 / \text{Me}_3\text{SiCl} / \text{Zn} \\ \text{THF}}} \qquad (7)$$

9

10

α-Me / β-Me = 77 : 23

Arylidene malononitriles **11** undergoes diastereoselective catalytic cyclodimerization in the presence of chlorotrimethylsilane as shown in eq 8 (*20*).

$$ ArCH=C{\overset{\displaystyle CN}{\underset{\displaystyle CN}{}}} \quad \xrightarrow[\text{DMF}]{\text{cat. } Cp_2VCl_2,\ Me_3SiCl,\ Zn} \quad \text{(structure)} \quad (8) $$

11

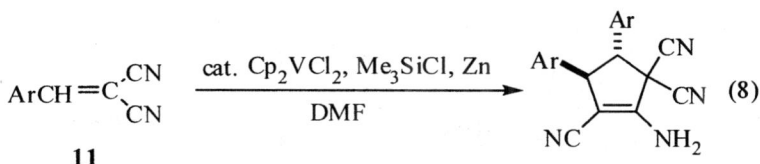

Oxovanadium(V)-induced oxidative transformations of main-group organometallics

Vanadium compounds in a high oxidation state are capable of inducing oxidative transformations. A variety of oxidative transformations with vanadium oxidants have been developed in an organic solvent (*2a*). Especially, oxidation of alcohols (*21*) and halide ions (*22*), epoxidation (*23*), and oxygenation of sulfides (*24*), hydrocarbons and arenes have been investigated to provide useful synthetic tools. The redox properties are utilized in the enantioselective oxidative coupling of naphthols (*25*).

The scope of synthetic reactions can be broadened even further by exploiting the versatility of vanadium compounds as an Lewis acidic oxidant (*26*). $VO(OR)X_2$ can act as such an oxidant towards carbonyl compounds. Scheme 5 summarizes its synthetic utility, which permits versatile dehydrogenative aromatization, allylic oxidation, ring-opening oxygenation, oxidative decarboxylation, and ring-opening oxidation and so on (*27, 28*).

Oxidation of main-group organometallic compounds with a metallic oxidant is considered to proceed via redox interaction between them, affording the corresponding radical species (Scheme 6). Another reaction path lies in the transmetallation. These reactions are considered to provide a new route to a reactive intermediate for novel oxidative transformations.

Oxidative transformation of organosilicon compounds

Silyl enol ethers are susceptible to one-electron oxidation (*29*). $VO(OR)Cl_2$ (*30*) is capable of inducing chemoselective homo- or cross-coupling of silyl enol ethers **12** to give the corresponding 1,4-diketones **14** via the regioselective formation of the radical species (*31, 32*). The reaction course is explained by one-electron mechanism shown in Scheme 7. The one-electron oxidation process depends on the redox potentials of the substrates and oxovanadium(V) compounds. The more readily oxidizable silyl enol ether **12** works as a radical precursor and the less oxidizable one **13** is a radicophiles. The silyl ketene acetals **15** undergo more facile oxidation for the cross-coupling with the silyl enol ethers **13** to give the γ-keto esters **16** (eq 9).

Scheme 5

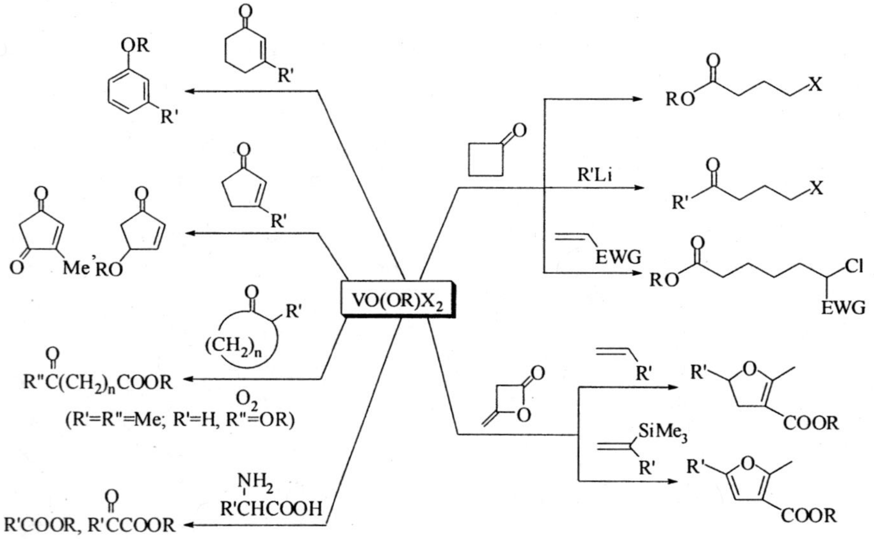

Scheme 6

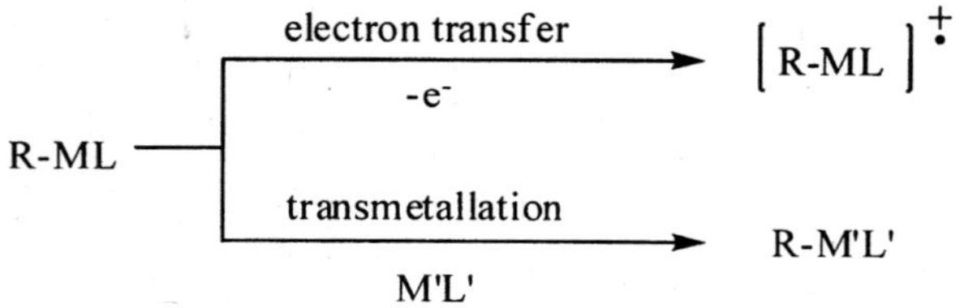

Scheme 7

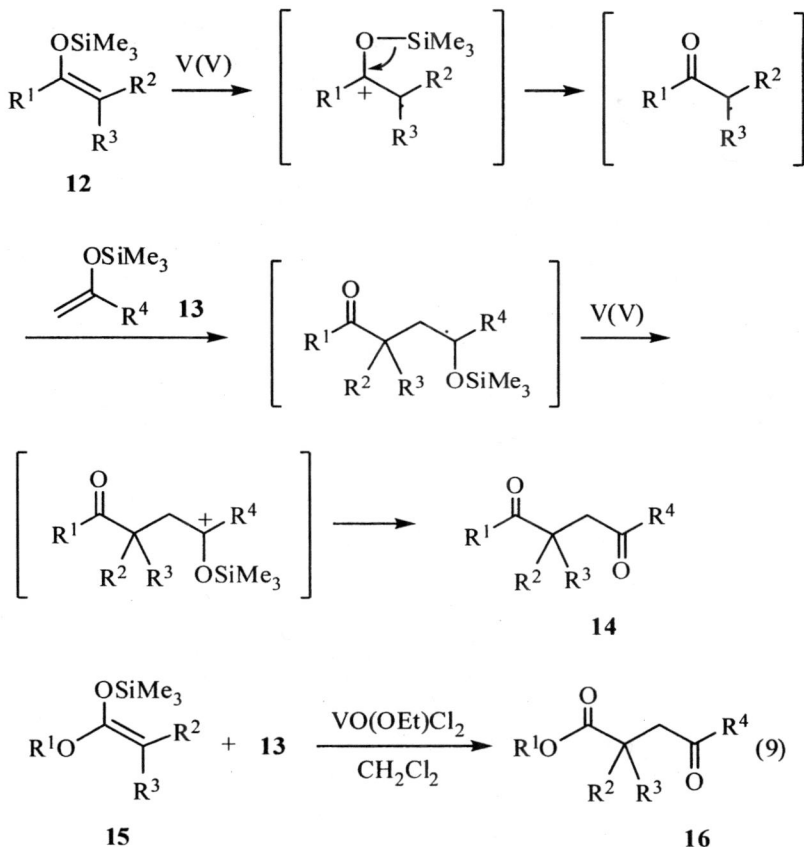

This method is applied to the chemoselective oxovanadium(V)-induced cross-coupling reaction of silyl enol ethers and allylic silanes depending on their oxidation susceptibility (Scheme 8) (*33*).

Similar chemoselective oxidative desilylation is also observed with benzylic silanes. The electron-donating group on the arene ring lowers the ionization potential of the benzylsilane **17**, thus activating it for desilylation. The intermolecular regioselective coupling occurs between such a substituted benzylic silane and an allylic silane or silyl enol ether (eq 10) (*34*).

The reaction of 4-methoxybenzyltrimethylsilane (**17a**) with VO(OR)Cl$_2$ is carried out in *t*-BuOH under oxygen atmosphere to give the benzaldehyde (eq 11) (*35*). VO(OCH$_2$CF$_3$)Cl$_2$ (*32*) exhibits higher reactivity than VO(OEt)Cl$_2$.

12

Scheme 8

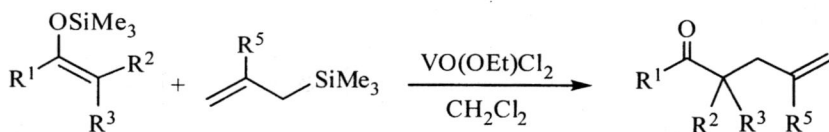

12

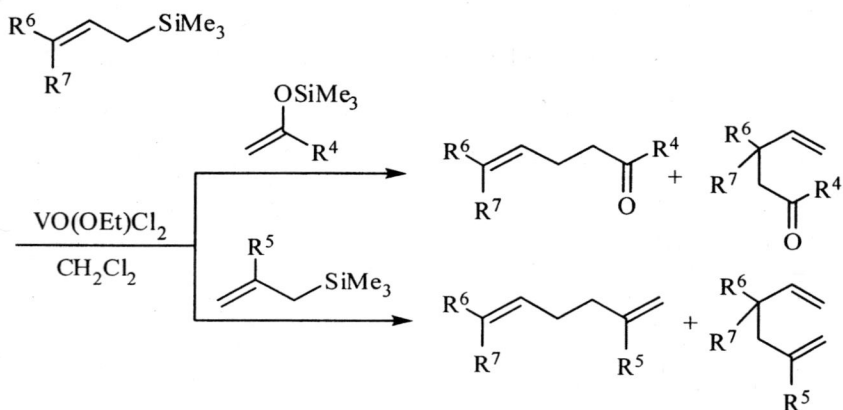

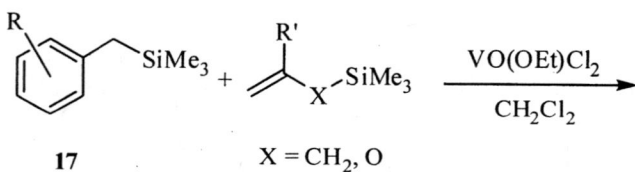

17   X = CH$_2$, O

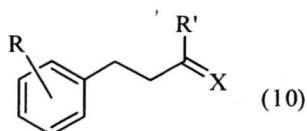

(10)

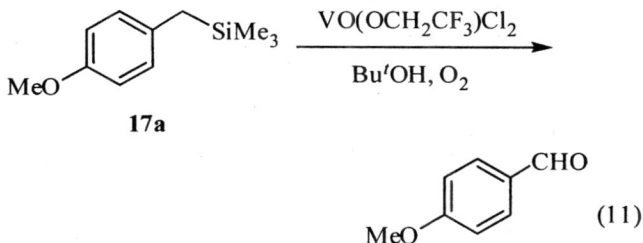

$$\text{(11)}$$

The redox potentials of organosilicon compounds, which are predicted from the ionization potentials calculated by MOPAC, determine whether they act as radical precursors or acceptors. The oxovanadium(V) reagent, which differentiates the oxidation process effectively, is a versatile oxidant to promote the chemoselective coupling via oxidative desilylation of organosilicon compounds under controlled conditions. The reactivity order depends on the acidity; $VO(OR)Cl_2$-AgOTf or Me_3SiOTf > $VO(OR)Cl_2$ > $VO(OR)_2Cl$ > $VO(OR)_3$ (27). Thus, the oxidation of organosilicon compounds provides a useful methods for the generation of cationic synthetic equivalents in the selective carbon-carbon bond formation.

Oxidative transformation of organotin compounds

Allylic and benzylic tins are known to undergo more facile oxidation than the corresponding silicon compounds (36), and are converted to the alcohols or their derivatives by a metallic oxidant such as Mn(IV), Ce(IV), Tl(III), or Fe(III) (37). The oxidation of benzyltributyltins 18 with $VO(OCH_2CF_3)Cl_2$ in t-BuOH under oxygen atmosphere leads to the aldehydes (ketones) 19 and the corresponding carboxylic acids 20 (eq. 12) (35, 38). The susceptibility to the oxidation depends on the substituent on the arene ring of 18. $VO(OPr\text{-}i)_2Cl$ and $VO(OCH_2CF_3)Cl_2$ are superior to a weaker oxidant such as $VO(acac)_2$, $VO(OPr\text{-}i)_3$, or $VO(OEt)_3$. Furthermore, the catalytic reaction proceeds with $VO(OCH_2CF_3)Cl_2$.

Cyclopropyl-1-phenylmethyltributyltin (18a) is oxidized to the cyclopropyl ketone 19a as a major product (eq 13). As for ring-opened compounds, neither the aldehyde nor the carboxylic acid is detected, although only a small amount of the chloride 21a is obtained. These results do not suggest the electron transfer mechanism. Although benzyltin compounds undergo more facile oxidation than benzylsilanes, the ketone 19a may be derived by transmetallation. A low- valent vanadium species generated *in situ* is oxidized under oxygen atmosphere to regenerate a vanadium species in its high oxidation state for catalytic reaction (38).

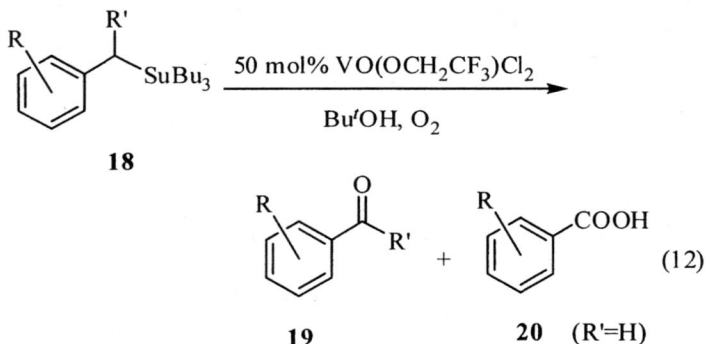

$$(12)$$

18 → **19** + **20** (R'=H)

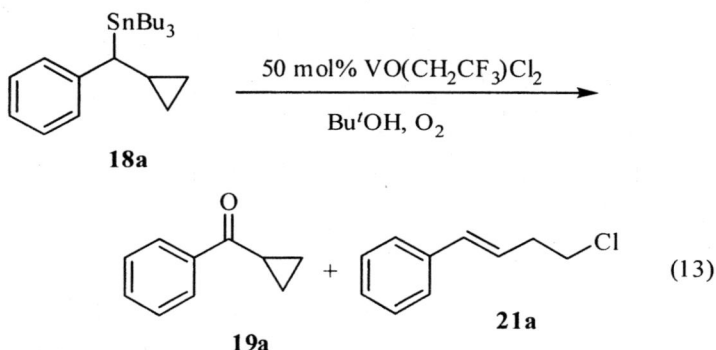

$$(13)$$

18a → **19a** + **21a**

Oxidative transformation of organoaluminum compounds

The redox processes of organometallics are important in synthetic transformations, especially in transition-metal-induced reactions. The transition-metal-catalyzed cross-coupling reactions of main-group organometallic compounds with organic halides provide versatile synthetic tools in organic synthesis. The reaction pattern lies in the coupling reaction between an electrophile and an nucleophile. As mentioned above, the transmetallation and/or electronic interaction can widen the reactivities of main-group organometallic compounds. One-electron oxidation of d^0-transition metal complexes such as titanocene, zirconocene, and platinum complex with a metallic oxidant permits synthetically useful transformations via organometallic cation radical species as key intermediates (*39*). In these reactions, selective coupling is achieved between two ligands on the transition metal, providing a useful method for carbon-carbon bond formation. From this point of view, the usage of main-group organometallics is limited due to their restricted redox processes. Oxidation of organoaluminum compounds usually affords alcohols (*40*), but to the best of our knowledge, selective carbon-carbon bond formation of organic substituents on aluminum has not been investigated.

Treatment of the aryldiethylaluminums **22**, obtained from aryllithiums and diethylaluminum chloride, with $VO(OEt)Cl_2$ results in ethylation to ethylarenes (Scheme 9) (*41*). This method can be applied to a wide variety of arylaluminums bearing an electron-donating group, which permits the selective ligand coupling of the organic groups on aluminum.

Scheme 9

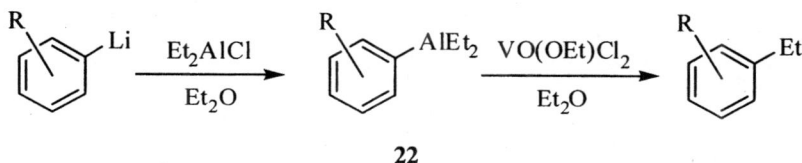

22

The ate complexes can be similarly used in the coupling of organic substituents on aluminum. The addition of 1-alkynyllithium to the 1-alkenylaluminum **23**, followed by treatment with $VO(OEt)Cl_2$, leads to the corresponding *trans*-enyne **24** via chemoselective and stereoselective bond formation between *sp* and sp^2 carbons (Scheme 10) (*42*).

Scheme 10

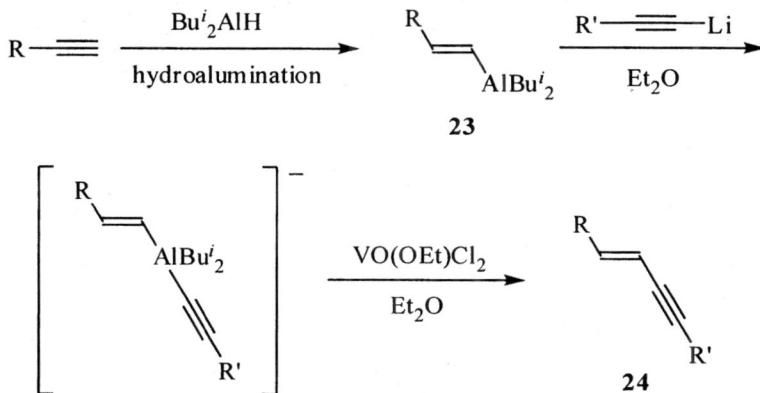

23

24

The ate complex **26** derived from the aryl-substituted aluminum **25** also undergoes the selective ligand coupling to the aryl-1-alkynyl coupling product (Scheme 11).

Scheme 11

25

26

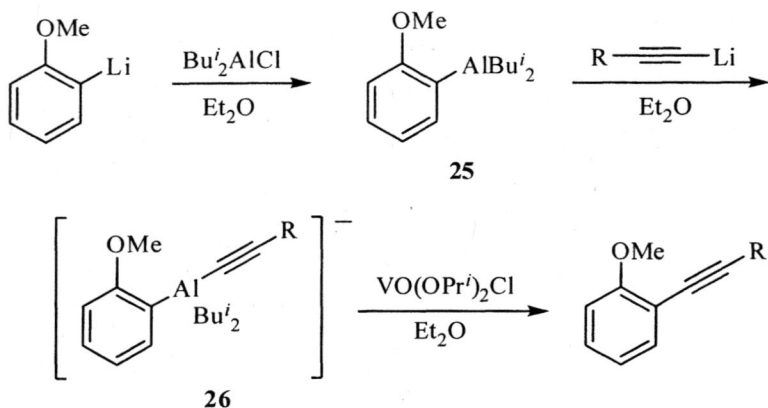

Although the precise reaction mechanism is ambiguous, coordination of the oxo atom of the oxovanadium(V) species to organoaluminums is considered to promote electron transfer or transmetallation for the oxidative coupling. The transformation is evaluated to be the first example for the formal reductive elimination on aluminum (*43*).

Oxidative transformation of organoboron compounds

The organoboron compound **27**, prepared from dicyclohexylborane and phenylacetylene or acetylenecarboxylic ester, is oxidized with $VO(OEt)Cl_2$ to give the (*E*)-ethenylcyclohexane **28** chemoselectively without the formation of dicyclohexyl and 1,4-butadiene derivative (eq 14) (*44*). A cross-over reaction suggests that the coupling reaction proceeds mostly in an intramolecular way.

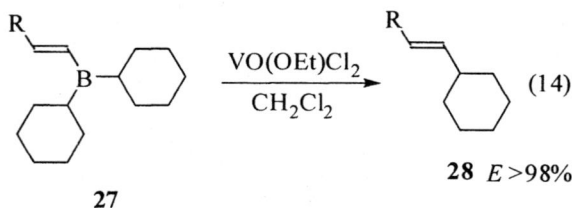

28 *E* >98%

27

The ate complexes **29** of the organoborons undergo more facile oxidation with $VO(OEt)Cl_2$, as observed in the oxidation of the aluminum ate complexes. The organic groups are effectively differentiated in the coupling reaction. Although small amount of the *Z* isomer is obtained in the case of the BuLi-

Scheme 12

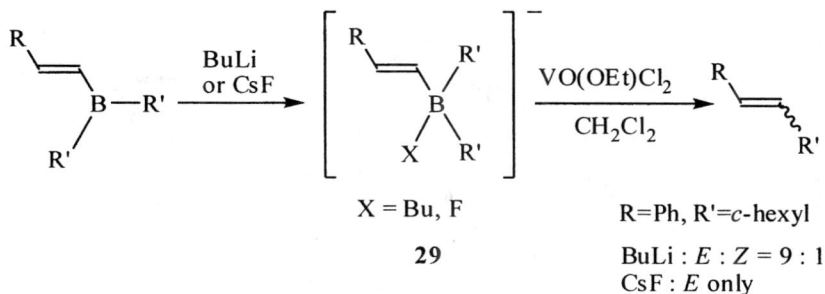

X = Bu, F

29

R=Ph, R'=*c*-hexyl

BuLi : *E* : *Z* = 9 : 1
CsF : *E* only

derived ate complex, use of the organoborate derived from CsF (*45*) improves the stereoselectivity, giving the *E* isomer exclusively (Scheme 12).

Alkenyltrialkylborates have been reported to be oxidized to alkylated alkenes with I_2 or BrCN (*46*). Biaryl formation also occurs by electrochemical, photochemical, and chemical oxidation, for example with Ir(IV), of tetraarylborates (*47*). The oxovanadium(V)-induced ligand coupling provides another promising method for the carbon-carbon bond formation on boron.

Catalytic oxidative ligand coupling of the organoborates **30** is promoted by VO(OEt)Cl$_2$ under oxygen atmosphere, leading to a versatile method for the selective synthesis of symmetrical or unsymmetrical biaryls (Scheme 13).

Scheme 13

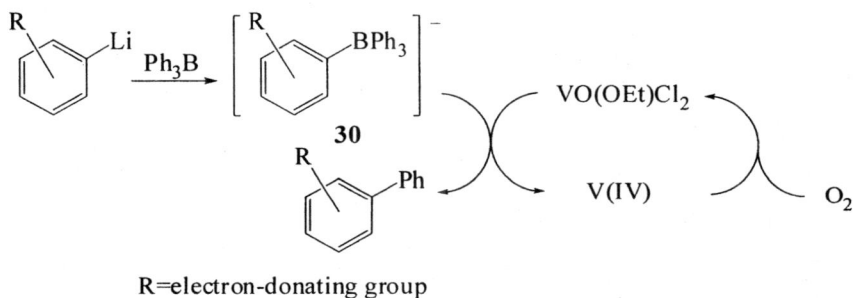

30

R=electron-donating group

Oxidative transformation of organozinc compounds

Organozinc compounds can tolerate a broad range of functional groups. Cross-coupling reactions between organozinc reagents and electrophiles such as organic halides are catalyzed by transition metal complexes (*48*). However, examples for the selective cross-coupling of two ligands of organozinc compounds are limited to a few cases, which include 1,2-migration of zincate carbenoids and intramolecular coupling of organozinc compounds by organocopper reagents (*49*).

The organozinc compound **31a**, prepared *in situ* by transmetallation of the methylzinc chloride with the aryllithium, is oxidized with Cp_2FePF_6, to give the homo-coupling compound **33a** selectively (Scheme 14). $AgBF_4$ serves as a useful oxidant to give the desired cross-coupling compound **32a**, probably via a one-electron oxidation process. Using $VO(OEt)Cl_2$ instead of $AgBF_4$, the cross-coupling reaction proceeds in preference to the homo-coupling reaction (*50*). Higher selectivity for the cross-coupling is observed with $VO(OEt)Cl_2$ than with $VO(OPr\text{-}i)Cl_2$ or $VO(OPr\text{-}i)_2Cl$.

Scheme 14

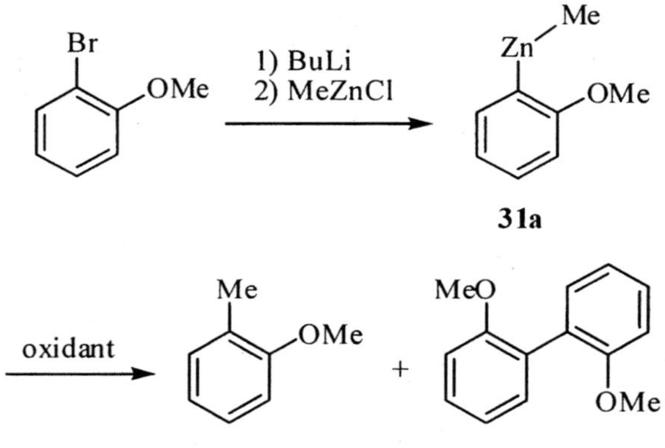

The coupling reaction of organozinc compounds **31** bearing an *o*-methoxy, *o*-phenyl, or *o*-methylthio group on the arene ring proceeds smoothly, but the *o*-cyano-substituted alkylarylzinc exhibits a lower reactivity, although organoaluminum compounds bearing an electron-withdrawing substituent do not undergo oxidative coupling under the similar conditions. Alkyl and 1-alkynyl groups can couple with the aryl group successfully (Scheme 15).

Triorganozincates **34** are readily oxidized with $VO(OEt)Cl_2$ smoothly to give the cross-coupling compounds **32** (Scheme 16) (*50*).

If the conversion to **34** is not complete, the homo-coupling product derived from the oxidation of aryllithium compound with oxovanadium(V) compound is accompanied (*42*). Such a step is able to be avoided by the preparation of the ate complex by iodine-zinc exchange. Arylzincate **34**, prepared from R_3ZnLi (*51*) and aryl iodide, is oxidized to give the cross-coupling product **32** exclusively (Scheme 17) (*50*).

Scheme 15

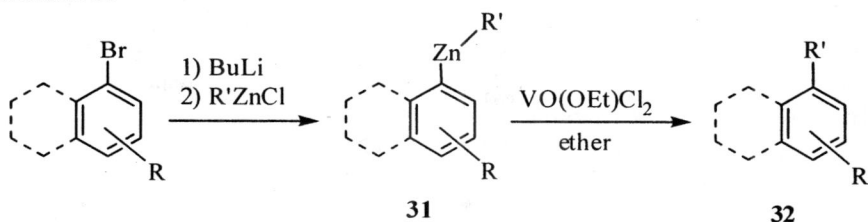

31   **32**

Scheme 16

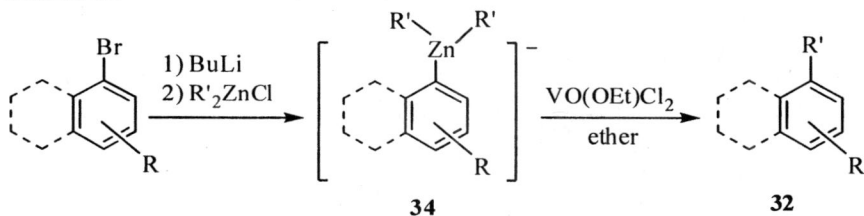

34   **32**

Scheme 17

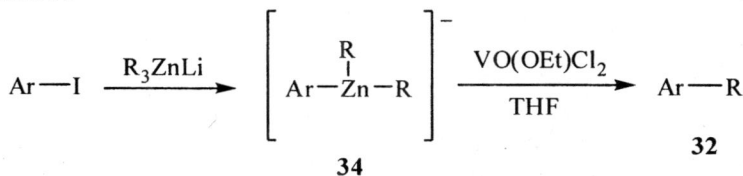

$$Ar-I \xrightarrow{R_3ZnLi} \left[Ar-\overset{R}{\underset{}{Zn}}-R \right]^- \xrightarrow[THF]{VO(OEt)Cl_2} Ar-R$$

34   **32**

The organozincates **35**, obtained from Me_4ZnLi_2 and various bromoarenes (*51*), are similarly oxidized with $VO(OEt)Cl_2$ to give the methylarene **32a** (Scheme 18) (*52*). Thus, the coupling between sp^2-carbon (aryl group) and sp^3-carbon (methyl group) of aryltrimethylzincates is achieved chemoselectively.

Scheme 18

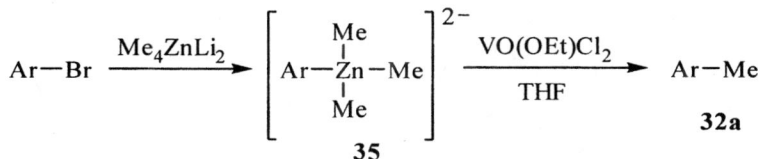

35

The above-mentioned method is applied to the selective carbon-carbon bond formation between sp^3-carbons (*52*). A bromine-zinc exchange reaction of **36** selectively occurs at the position *cis* to the phenyl group by treatment with Me_4ZnLi_2 (*53*). The oxidation of the thus-obtained zincate **37** with $VO(OEt)Cl_2$ leads to the stereoselective formation of 1-bromo-1-methyl-2-phenylcyclo-propane, **38**. On the other hand, when the reaction mixture is warmed up to 0 °C, followed by treatment of $VO(OEt)Cl_2$, dimethylation takes place to give the dimethylcyclopropane **40** via the organozinc **39** (Scheme 19).

Scheme 19

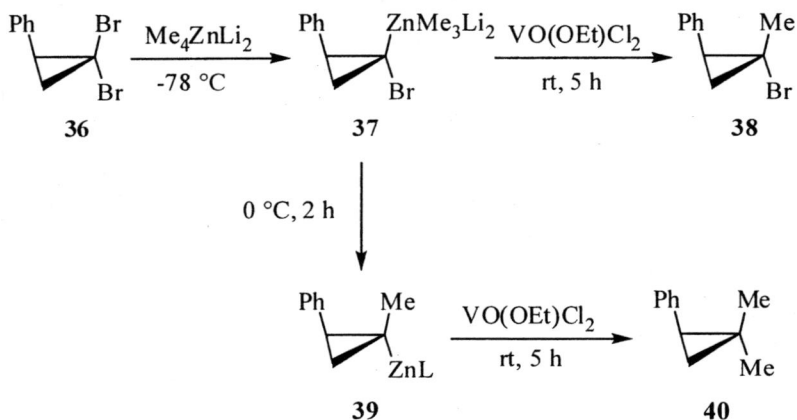

Dialkylzinc reagent (R_2Zn) is known as a mild nucleophile in the presence of an additional promoter such as a Lewis acid (*54*). Characteristics of oxovanadium(V) compounds as a Lewis acid and one-electron oxidant permit vicinal dialkylation at both the α and β positions of α,β-unsaturated carbonyl compounds with dialkylzinc reagents through conjugate addition and ligand coupling (Scheme 20) (*55*). For example, reaction of 2-cyclohexenone with dimethylzinc in the presence of $VO(OEt)Cl_2$ affords 2,3-dimethylcyclohexanone (eq 15). Use of a stronger Lewis acid, $VO(OEt)Cl_2$ > $VO(OPr-i)Cl_2$ > $VO(OPr-i)_2Cl$, gives a better result.

Scheme 20

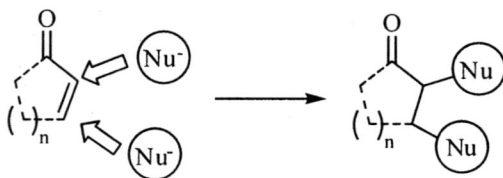

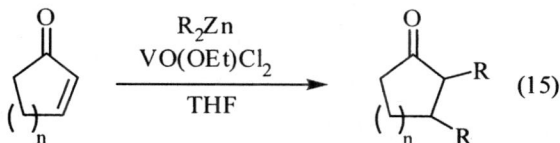

$$\text{(15)}$$

This method can be applied to vicinal alkylation with triethylaluminum or triethylborane compounds as shown in eq 16 (55). In contrast, a similar oxidative dialkylation is not observed with organocuprate reagents.

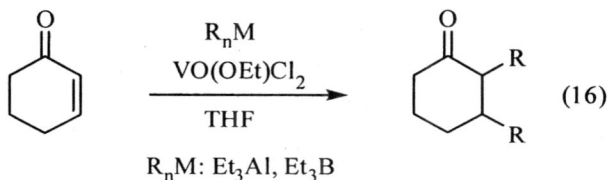

$$\text{(16)}$$

R_nM: Et_3Al, Et_3B

Contrary to dialkylzincs, lithium trialkylzincates (R_3ZnLi) have enough nucleophilicity for conjugate addition without the aid of a Lewis acid. The vicinal dialkylation with $R_2R'ZnLi$ occurs as shown in Scheme 21. Use of the organozinc compound prepared from a $ZnCl_2$-TMEDA complex and BuLi results in a better yield. It is noteworthy that introduction of alkyl groups is differentiated regioselectively. For example, use of $BuMe_2ZnLi$ results in the selective formation of 3-butyl-2-methylcycloalkanone, **41**.

Scheme 21

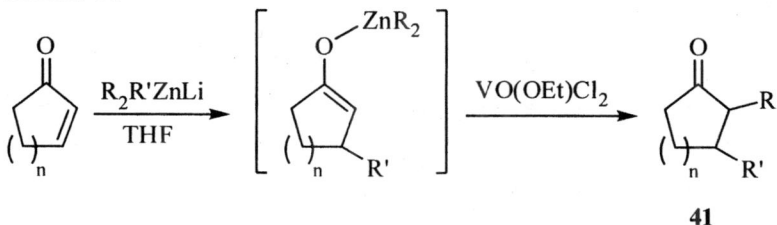

41

Oxidative transformation of organozirconium compounds

The oxidation reaction of (E)-1-alkenylchlorozirconocenes with VO(OPr-i)$_2$Cl leads to the stereoselective homo-coupling product, the (E,E)-diene **42** (Scheme 22) (56). (E)-1-Alkenyl-1-alkynylzirconocenes undergo the oxovanadium(V)-induced cross-coupling of organic substituents on zirconium, to give the (E)-enynes stereoselectively.

Scheme 22

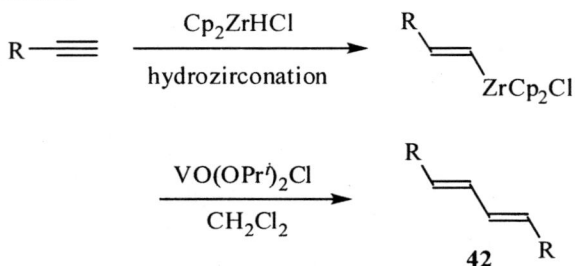

Conclusion

The multi-component system consisting of a catalyst, a co-reductant, and an additive promoter cooperates with each other to construct the catalytic systems for efficient one-electron reduction (*2, 57*). In this system, the active catalyst is smoothly regenerated by redox interaction with the co-reductant. The selection of the co-reductant is, of course, important from this point of view. Furthermore, the oxidized form of the co-reductant should not interfere with, but assist the reduction reaction or at least, be tolerant under the conditions. Additive promoters, which are considered to contribute to the redox cycle, possibly facilitate the electron transfer and liberate the catalyst from the reaction adduct.

Oxovanadium(V) compounds are potential Lewis acids with oxidation ability to induce one-electron oxidation reactions based on their characteristics. Oxidation capability and redox potential are effectively controlled by the substituent of oxovanadium(V) compounds. A catalytic system is allowed to be realized by the redox interaction with molecular oxygen. The oxidative ligand coupling proceeds via the intermetallic interaction between vanadium species and main-group organometallics.

Both synthetic methods are expected to be promising as synthetic tools. These methods are complementary and useful to generate radical intermediates (Scheme 1). The higher stereoselectivity is attained. This chapter focuses on the synthetic utility, but the redox properties of vanadium compounds are also important in the design of functional materials (*58*).

Acknowledgment

The work in this review was mostly done by the members in our laboratory, whose names are cited in the references.

References

1. a) Ho, T.-L. *Synthesis* **1979**, 1. b) Sheldon, R. A.; Kochi, J. K. *Metal-Catalyzed Oxidations of Organic Compounds*; Academic Press: New York, 1981. c) Freeman, F. In *Organic Syntheses by Oxidation with Metal Compounds*; Mijs, W. J.; de Jonge, C. R. H. I., Eds.; Plenum Press: New York, 1986; Chapter 1. d) Pons, J.-M.; Santelli, M. *Tetrahedron* **1988**, *44*, 4295. e) Rehder, D.; Gailus, H. *Trends in Organometallic Chemistry* **1994**, *1*, 397. f) Iqbal, J.; Bhatia, B.; Nayyar, N. K. *Chem. Rev.* **1994**, *94*, 519. g) Dalko, P. I. *Tetrahedron* **1995**, *51*, 7579.
2. a) Hirao, T. *Chem. Rev.* **1997**, *97*, 2707. b) Hirao, T. *Synlett* **1999**, 175.
3. a) Scheffold, R In *Modern Synthetic Methods*, Vol 3; Otto Salle Verlag GmbH: Frankfurt, 1983. b) Scheffold, R.; Busato, S. *Helv. Chim. Acta* **1994**, *77*, 92. c) Ozaki, S.; Horiguchi, I.; Matsushita, H.; Ohmori, H. *Tetrahedron Lett.* **1994**, *35*, 725. d) Ozaki, S.; Matsui, E.; Waku, J.; Ohmori, H. *Tetrahedron Lett.* **1997**, *38*, 2705. e) Rollin, Y.; Derien, S.; Duñach, E.; Gebehenne, C.; Perichon, J. *Tetrahedron* **1993**, *49*, 7723. f) Mubarak, M. S.; Pagel, M.; Marcus, L. M.; Peters, D. G. *J. Org. Chem.* **1998**, *63*, 1319. g) Léonard, E.; Duñach, E.; Périchon, J. *J. Chem. Soc., Chem. Commun.* **1989**, 276.
4. Recent work, for example: a) Davies, S. C.; Hughes, D. L.; Janas, Z.; Jerzykiewicz, L. B.; Richards, R. L.; Sanders, J. R.; Silverston, J. E.; Sobata, P. *Inorg. Chem.* **2000**, *39*, 3485. b) Chu. W.-C.; Wu, C.-C.; Hsu, H.-F. *Inorg. Chem.* **2006**, *45*, 3164.
5. Hirao, T.; Hirano, K.; Hasegawa, T.; Ikeda, I.; Ohshiro, Y. *J. Org. Chem.* **1993**, *58*, 6529.
6. Barton, D. H. R.; Jang, D. O.; Jaszberenyi, J. C. *Tetrahedron Lett.* **1992**, *33*, 2311.
7. a) Nicolaou, K. C.; Liu, J.-J.; Yang, Z.; Ueno, H.; Guy, R. K.; Sorensen, E. J.; Claiborne, C. F.; Hwang, C.-K.; Nakada, M.; Nantermet, P. G. *J. Am. Chem. Soc.* **1995**, *117*, 634. b) Shiina, I.; Nishimura, T.; Ohkawa, N.; Sakoh, H.; Nishimura, K.; Saitoh, K.; Mukaiyama, T. *Chem. Lett.* **1997**, 419. c) Kammermeier, B.; Beck, G.; Holla, W.; Jacobi, D.; Napierski, B.; Jendralla, H.; *Chem. Eur. J.* **1996**, *2*, 307. d) Kammermeier, B.; Beck, G.; Jacobi, D.; Jendralla, H. *Angew. Chem., Int. Ed. Engl.* **1994**, *33*, 685.
8. a) Freudenberger, J. H.; Konradi, A. W.; Pedersen, S. F. *J. Am. Chem. Soc.* **1989**, *111*, 8014. b) Park, J.; Pedersen, S. F. *J. Org. Chem.* **1990**, *55*, 5924. c) Konradi, A. W.; Pedersen, S. F. *J. Org. Chem.* **1992**, *57*, 28. d) Konradi, A. W.; Kemp, S. J.; Pedersen, S. F. *J. Am. Chem. Soc.* **1994**, *116*, 1316.
9. a) Hirao, T.; Asahara, M.; Muguruma, Y.; Ogawa, A. *J. Org. Chem.* **1998**, *62*, 4566. b) Hirao, T.; Ogawa, A.; Asahara, M.; Muguruma, Y.; Sakurai, H. *Organic Synthesis*, **2005**, *81*, 26.

24

10. Hirao, T.; Hatano, B.; Asahara, M.; Muguruma, Y.; Ogawa, A. *Tetrahedron Lett.* **1998**, *39*, 5247.
11. a) Hirao, T.; Hasegawa, T.; Muguruma, Y.; Ikeda, I. *J. Org. Chem.* **1996**, *61*, 366. b) Hirao, T.; Hasegawa, T.; Muguruma, Y.; Ikeda, I. *Abstracts for the 6th International Conference on New Aspects of Organic Chemistry*, 1994; p 175.
12. Fürstner, A.; Hupperts, A. *J. Am. Chem. Soc.* **1995**, *117*, 4468.
13. Gansäuer, A. *Chem. Commun.* **1997**, 457.
14. Gansäuer, A. *Synlett* **1997**, 363.
15. a) Halterman, R. L.; Zhu, C.; Chen, Z.; Dunlap, M. S.; Khan, M. A.; Nicholas, K. M. *Organometallics* **2000**, *19*, 3824. b) Bensari, A.; Renaud, J.-L.; Riant, O. *Org. Lett.* **2001**, *3*, 3863. c) Chatterjee, A.; Bennur, T. H.; Joshi, N. N. *J. Org. Chem.* **2003**, *68*, 5668.
16. a) Bensari, A.; Renaud, J.-L.; Riant, O. *Org. Lett.* **2001**, *3*, 3863. b) Chatterjee, A.; Bennur, T. H.; Joshi, N. N. *J. Org. Chem.* **2003**, *68*, 5668. c) Takenaka, N.; Xia, G.; Yamamoto, H. *J. Am. Chem. Soc.* **2004**, *126*, 13198. c) Li, Y.-G.; Tian, Q.-S.; Zhao, J.; Feng, Y.; Li, M.-J.; You, T.-P. *Tetrahedron: Asymmetry* **2004**, *15*, 1707.
17. Hirao, T.; Takeuchi, H.; Ogawa, A.; Sakurai, H. *Synlett* **2000**, 1658.
18. Xu, X.; Hirao, T. *J. Org. Chem.* **2005**, *70*, 8594.
19. Hatano, B.; Ogawa, A.; Hirao, T. *J. Org. Chem.* **1998**, *63*, 9421.
20. Zhou, L.; Hirao, T. *Tetrahedron Lett.* **2000**, *41*, 8517.
21. Recent papers, for example: a) Maeda, Y.; Kakiuchi, N.; Matsumura, S.; Nishimura, T.; Kawamura, T.; Uemura, S. *J. Org. Chem.* **2002**, *67*, 6718. b) Li, C.; Zheng, P. Li, J.; Zhang, H.; Cui, Y.; Shao, Q.; Ji, X.; Zhang, J.; Zhao, P.; Xu, Y. *Angew. Chem., Int. Ed.* **2003**, *42*, 5063. c) Velusamy, S.; Punniyamurthy, T. *Org. Lett.* **2004**, *6*, 217. d) Radosevich, A. T.; Musich, C.; Toste, F. D. *J. Am. Chem. Soc.* **2005**, *127*, 1090.
22. Recent papers, for example: a) Smith, T. S., II; Pecoraro, V. L. *Inorg. Chem.* **2002**, *41*, 6754. b) Martinez, J. S.; Carroll, G. L.; Tschirret-Guth, R. A.; Altenhoff, G.; Little, R. D.; Butler, A. *J. Am. Chem. Soc.* **2002**, *123*, 3289. c) Carter-Franklin, J. N.; Butler, A. *J. Am. Chem. Soc.* **2004**, *126*, 15060.
23. Recent papers, for example: a) Jacobsen, E. N.; Marko, I.; Mungall, W. S.; Schröder, G.; Sharpless, K. B. *J. Am. Chem. Soc.* **1988**, *110*, 1968. b) Bolm, C. Kuhn, T. *Synlett* **2000**, 899. c) Wu, H.-L.; Uang, B.-J. *Tertahedron: Assymetry* **2002**, *13*, 2625. d) Makita, N.; Hoshino, Y.; Yamamoto, H.; *Angew. Chem., Int. Ed.* **2003**, *42*, 941. e) Zhang, W.; Basak, A. Kosugi, Y.; Hoshino, Y.; Yamamoto, H. *Angew. Chem., Int. Ed.* **2005**, *44*, 4389. f) De la Pradilla, R. F.; Castellanos, A.; Femandez, J.; Lorenzo, M.; Manzano, P.; Mendez, P.; Priego, J.; Viso, A. *J. Org. Chem.* **2006**, *71*, 1569.
24. Recent papers, for example: a) Pelotier, B.; Anson, M. S.; Campbell, I. B.; MacDonald, S. J. F.; Priem, G.; Jackson, R. F. W. *Synlett* **2002**, 1055. b)

Ohta, C.; Shimizu, H.; Kondo, A.; Katsuki, T. *Synlett* **2002**, 161. c) Jeong, Y.-C.; Choi, S.; Hwang, Y. D.; Ahn, K.-H. *Tetrahedron Lett.* **2004**, *45*, 9249. d) Sun, J.; Zhu, C.; Dai, Z.; Yang, M.; Pan, Y.; Hu, H. *J. Org. Chem.* **2004**, *69*, 8500. e) Du, G.; Espenson, J. H. *Inorg. Chem.* **2005**, *44*, 2465. f) Drago, C.; Caggiano, L.; Jackson, R. F. W. *Angew. Chem., Itl. Ed.* **2005**, *44*, 7221.

25. Recent papers, for example: a) Hwang, D.-R.; Chen, C.-P.; Uang, B.-J. *Chem. Commun.* **1999**, 1207. b) Hon, S.-W.; Li, C.-H.; Kuo, J.-H.; Barhate, N. B.; Liu, Y.-H.; Wang, Y.; Chen, C.-T. *Org. Lett.* **2001**, *3*, 869. c) Chu, C. Y.; Hwang, D. R.; Wang, S. K.; Uang, B.-J. *Chem. Commun.* **2001**, 980. d) Barhate, N. B.; Chen, C.-T. *Org. Lett.* **2002**, *4*, 2529. e) Luo, Z.; Liu, Q.; Gong, L.; Cui, X.; Mi, A.; Jiang, Y. *Chem. Commun.* **2002**, 914. f) Chu, C.-Y.; Uang, B.-J. *Terahedron: Asymmetry* **2003**, *14*, 53. g) Somei, H; Asano, Y.; Yoshida, T.; Takizawa, S.; Yamataka, H.; Sasai, H. *Tetrahedron Lett.* **2004**, 45, 1841. h) Luo, Z.; Liu, Q.; Gong, L.; Cui, X.; Mi, A.; Jiang, Y. *Angew. Chem., Int. Ed.* **2004**, *41*, 4532.

26. Recent papers, for example: a) Kirihara, M.; Ichinose, M.; Takizawa, S.; Momose, T. *Chem. Commun.* **1998**, 1691. b) Piao, D.-g.; Inoue, K.; Shibasaki, H.; Taniguchi, Y.; Kitamura, T.; Fujiwara, Y. *J. Organometal. Chem.* **1999**, *574*, 116. c) Hwang, D. R.; Chu, C. Y.; Wang, S. K.; Uang, B.-J. *Synlett* **1999**, 77. d) Chen, C. T.; Hon, S. W.; Weng, S. S. *Synlett* **1999**, 816. e) Hartung, J.; Schmidt, P. *Synlett* **2000**, 367. f) Bora, U.; Bose, G.; Chaudhuri, M. K.; Dhar, S. S.; Gopinath, R.; Khan, A. T.; Patel, B. K. *Org. Lett.* **2000**, *2*, 247. g) Gopinath, R.; Patel, B. K. *Org. Lett.* **2000**, *2*, 577.

27. a) Hirao, T.; Mori, M.; Ohshiro, Y. *Bull. Chem. Soc. Jpn.* **1989**, *62*, 2399. b) Hirao, T.; Mori, M.; Ohshiro, Y. *J. Org. Chem.* **1990**, *55*, 358. c) Hirao, T.; Mori, M.; Ohshiro, Y. *Chem. Lett.* **1991**, 783.

28. a) Hirao T.; Ohshiro, Y. *Tetrahedron Lett.* **1990**, *31*, 3917. b) Hirao, T.; Mikami, S.; Mori, M.; Ohshiro, Y. *Tetrahedron Lett.* **1991**, *32*, 1741. c) Hirao, T.; Fujii, T.; Ohshiro, Y. *J. Organometal. Chem.* **1991**, *407*, C1. d) Hirao, T.; Fujii, T.; Miyata, S.-i.; Ohshiro, Y. *J. Org. Chem.* **1991**, *56*, 2264. e) Hirao, T.; Fujii, T.; Tanaka, T.; Ohshiro, Y. *J. Chem. Soc., Perkin Trans.1* **1994**, 3. f) Hirao, T.; Fujii, T.; Tanaka, T.; Ohshiro, Y. *Synlett* **1994**, 845. g) Hirao, T.; Sakaguchi, M.; Ishikawa, T.; Ikeda, I. *Synth. Commun.* **1995**, *25*, 2579.

29. a) Ito, Y.; Konoike, T.; Saegusa, T. *J. Am. Chem. Soc.* **1975**, *97*, 649. b) Kobayashi, Y.; Taguchi, T.; Morikawa, T.; Tokuno, E.; Sekiguchi, S. *Chem. Pharm. Bull.* **1980**, *28*, 262. c) Baciocchi, E.; Casu, A.; Ruzziconi, R. *Synlett* **1990**, 679.

30. Funk, H.; Weiss, W.; Zeising, M. *Z. Anorg. Allg. Chem.* **1958**, *36*, 296.

31. Fujii, T.; Hirao, T.; Ohshiro, Y. *Tetrahedron Lett.* **1992**, *33*, 5823.

32. Ryter, K.; Livinghouse, T. *J. Am. Chem. Soc.* **1998**, *120*, 2658.

26

33. a) Fujii, T.; Hirao, T.; Ohshiro, Y. *Tetrahedron Lett.* **1993**, *34*, 5601. b) Hirao, T.; Fujii, T.; Ohshiro, Y. *Tetrahedron* **1994**, *50*, 10207.
34. Hirao, T.; Fujii, T.; Ohshiro, Y. *Tetrahedron Lett.* **1994**, *35*, 8005.
35. Hirao, T.; Morimoto, C.; Takada, T.; Sakurai, H. *Tetrahedron Lett.* **2001**, *42*, 1961.
36. a) Pereyre, M.; Quintard, J-P.; Rahm, A. *Tin in Organic Synthesis*, Butterworths: London, 1987. b) Harrison, P. G. *Chemistry of Tin*, Blackie & Son: New York, 1989.
37. a) Baciocchi, E.; Del Giacco, T.; Rol, C.; Sebastiani, G. V. *Tetrahedron Lett.* **1989**, *30*, 3573. b) Still, W. C. *J. Am. Chem. Soc.* **1977**, *99*, 4186. c) Ochiai, M.; Fujita, E.; Arimoto, M.; Yamaguchi, H. *Chem. Pharm. Bull.* **1984**, *32*, 887. d) Ochiai, M.; Fujita, E.; Arimoto, M.; Yamaguchi, H. *Chem. Pharm. Bull.* **1984**, *32*, 5027. e) Corey, E. J.; Walker, J. C. *J. Am. Chem. Soc.* **1987**, *109*, 8108.
38. Hirao, T.; Morimoto, C.; Takada, T.; Sakurai, H. *Tetrahedron* **2001**, *57*, 5073.
39. a) Jordan, R. F.; JaPointe, R. E.; Bajgure, C. S.; Echols, S. F.; Willet, R. *J. Am. Chem Soc.* **1987**, *109*, 4111. b) Borkowsky, S. L.; Baenziger, N. C.; Jordan, R. F. *Organometallic* **1993**, *12*, 486. c) Burk, M. J.; Tumas, W.; Ward, M. D.; Wheeler, D. R. *J. Am. Chem. Soc.* **1990**, *112*, 6133. d) Sato, M.; Mogi, E.; Kumakura, S. *Organometallics* **1995**, *14*, 3157. e) Hayashi, Y.; Osawa, M.; Wakatsuki, Y. *J. Organomet. Chem.* **1997**, *542*, 241.
40. a) Eisch, J. J. In *Comprehensive Organometallic Chemistry II*; Abel, E. W., Stone, F. G. A., Wilkinson, S. G., Eds.; Pergamon Press: Oxford, 1994; Vol. 11, pp 277. b) Zietz, J. R., Jr.; Robinson, G. C.; Lindsay, K. L. In *Comprehensive Organometallic Chemistry II*; Wilkinson, S. G., Stone, F. G. A., Abel, E. W., Eds.; Pergamon Press: Oxford, 1982; Vol. 7, p 354.
41. Ishikawa, T.; Ogawa, A; Hirao, T. *J. Am. Chem. Soc.* **1998**, *120*, 5124.
42. Ishikawa, T.; Ogawa, A.; Hirao, T. *Organometallics* **1998**, *17*, 5713.
43. Feher, F. J.; Blanski, R. L. *J. Am. Chem. Soc.* **1992**, *114*, 5886.
44. Ishikawa, T.; Nonaka, S; Ogawa, A; Hirao, T. *Chem. Commun.* **1998**, 1209.
45. a) Negishi, E.; Idacavage, M. J.; Chiu, K.-W.; Yoshida, A.; Abramovitch, A.; Goettel, M. E.; Silveira, A.; Bretherick, H. D. *J. Chem. Soc., Perkin Trans. 2* **1978**, 1225. b) Garnier, L.; Plunian, B.; Morter, J.; Vaultier, M. *Tetrahedron Lett.* **1996**, *37*, 6699. c) Reetz, M. T.; Niemeryer, C. M.; Harms, K. *Angew. Chem., Int. Ed. Engl.* **1991**, *30*, 1472 and 1474.
46. Pelter, A.; Smith, K.; Brown, H. C. *Borane Reagents*, Academic Press: London, 1988.
47. a) Geske, D. H. *J. Phys. Chem.* **1959**, *63*, 1062; **1962**, *66*, 1743. b) Pelter, A.; Pardasani, R.; Pardasani, P. *Tetrahedron* **2000**, 56, 7339. c) Abley, P.; Halpern, J. *J. Chem. Soc., Chem. Commun.* **1971**, 1238.
48. a) Oshima, K. *Transition Metal Catalyzed Reactions of Organozinc Compounds*, In *Advance in Organometallic Chemistry*, Liebeskind, L. S.,

Eds.; JAI Press: London, 1991, Vol. 2, p. 101. b) Erdik, E. *Tetrahedron* **1992**, *48*, 9577.

49. a) Iyoda, M.; Kabir, S. M. H.; Vorasingha, A.; Kuwatani, Y.; Yosihda, M. *Tetrahedron Lett.* **1998**, *39*, 5393. b) Harada, T.; Iwazaki, K., Hara, D.; Hattori, K.; Oku, A. *J. Org. Chem.* **1997**, *62*, 8966. c) Harada, T.; Katsuhira, T.; Takeshi, K.; Atusshi, O.; Katsuhiro, I.; Keiji, M.; Oku, A. *J. Am. Chem. Soc.* **1996**, *118*, 11377.

50. Hirao, T.; Takada, T.; Ogawa, A. *J. Org. Chem.* **2000**, *65*, 1511.

51. Uchiyama, M.; Kameda, M.; Mishima, O.; Yokoyama, N.; Koike, M.; Kondo, Y.; Sakamoto, T. *J. Am. Chem. Soc.* **1998**, *120*, 4934.

52. Takada, T.; Sakurai, H.; Hirao, T. *J. Org. Chem.* **2001**, *66*, 300.

53. Harada, T.; Katsuhira, T.; Hattori, K.; Oku, A. *J. Org. Chem.* **1993**, *58*, 2958.

54. a) Knochel, P.; Perea, J. J. A.; Jones, P. *Tetrahedron* **1998**, *54*, 8275. b) Knochel, P.; Singer, R. D. *Chem. Rev.* **1993**, *93*, 2117 and references therein.

55. Hirao, T.; Takada, T.; Sakurai, H. *Org. Lett.* **2000**, *2*, 3659.

56. Ishikawa, T.; Ogawa, A.; Hirao, T. *J. Organometal. Chem.* **1999**, *575*, 76.

57. Fürstner, A. *Chem. Eur. J.* **1998**, *4*, 567.

58. a) Hirao, T. *J. Inorg. Biochem.* **2002**, *80*, 27. b) Hirao, T.; Fukuhara, S.; Otomaru, Y.; Moriuchi, T. *Synth. Met.*, **2001**, *123*, 373.

Chapter 2

Vanadium-Catalyzed Oxidation in Ionic Liquids

Valeria Conte[*], Barbara Floris, and Adriano Silvagni

Dipartimento di Scienze e Tecnologie Chimiche, Università di Roma "Tor Vergata", via della ricerca scientifica, 00133 Roma, Italy
[*]Corresponding author: valeria.conte@uniroma2.it

Among the various interesting features of vanadium(V) there is its ability to activate hydrogen peroxide in diverse oxidation reactions. In this respect, we proposed a simple and effective procedure for oxybromination of unsaturated substrates in a biphasic system H_2O/CH_2Cl_2. Then, with the aim of rendering such a procedure even more interesting from the sustainability point of view, we have substituted chlorinated solvents with ionic liquids, ILs. Interestingly enough, we observed a remarkable increase of both yield and selectivity with styrene and ethynylbenzene, obtaining synthetically interesting results. Here we report results with other substrates in order to enlarge the scope of the reaction. Very preliminary data of peroxovanadium catalyzed hydroxylation of arenes in ILs are also mentioned.

© 2007 American Chemical Society

One of the most important goals for 21st century chemistry is the achievement of sustainable synthetic procedures; this is particularly appropriate when upgrading of known processes is considered. Several reactions, in fact, have been recently tailored either by employing catalysts with better performances, in terms of both yields and selectivities, and eventually reusing them in subsequent cycles, or significantly decreasing the environment impact of the solvents.

In the perspective of a more sustainable reaction medium, and considering the improvement obtained by reducing the amount of volatile organic solvents (VOCs) used in an industrial process, numerous research groups have proposed new reaction media ranging from fluorinated solvents (1,2), to supercritical CO_2 (3,4,5) and ionic liquids (ILs); combination of the latter two media has also been proposed.

Our attention in this field is devoted to the use of ILs which, because of their peculiar properties, have been quite recently proposed as novel and environmentally benign reaction media for several organic syntheses. N-alkylpyridium and N,N'-dialkyl imidazolium cations coupled with a variety of inorganic anions have been, in fact, used as suitable solvents for several organic reactions as well as in catalytic processes: Friedel Craft reactions (6), Diels Alder cycloadditions (7), metal catalyzed hydrogenations (8), Mn-catalyzed asymmetric epoxidation (9) and bromination of double and triple bonds (10) are just a few examples in this quite new field.

In the course of our studies we directed our attention toward the use of some hydrophilic and hydrophobic ILs in V(V)-catalyzed oxidations with peroxides.

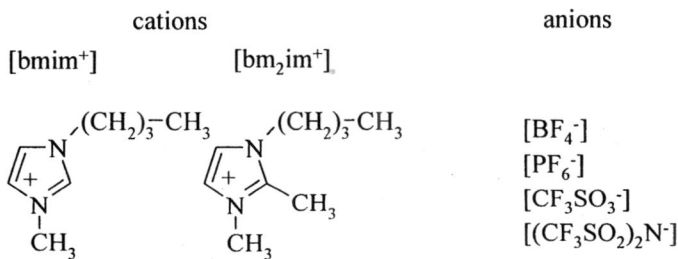

[bmim+][BF4-]; [bmim+][CF3SO3-] = hydrophilic

[bmim+][PF6-]; [bm2im+][PF6-]; [bmim+][(CF3SO2)2N-] = hydrophobic

Figure 1. Selected hydrophilic and hydrophobic ionic liquids.

The hydrophilic and hydrophobic nomenclature is normally used (*11*) in order to identify ILs which either form a single phase with water or dissolve very small quantity of water, thus forming two-phase systems. In particular we considered 1-methyl-3-butylimidazolium [bmim$^+$], and 1,2-dimethyl-3-butyl-imidazolium [bm$_2$im$^+$] cations with tetrafluoborate, [BF$_4^-$], hexafluorophosphate, [PF$_6$-], triflate, [CF$_3$SO$_3^-$], and bistrifluoromethane-sulfonimide anions, [(CF$_3$SO$_2$)$_2$N$^-$]. Accordingly, the solvents indicated in Figure 1 were tested in the title reactions, i.e. [bmim$^+$][BF$_4^-$], [bmim$^+$][PF$_6^-$], [bm$_2$im$^+$][PF$_6^-$], [bmim$^+$][CF$_3$SO$_3^-$] and [bmim$^+$][(CF$_3$SO$_2$)$_2$N$^-$].

Our interest is, since long time, mainly focused toward the oxidative functionalization of double bonds in the presence of metal catalysis (*12,13*). In this respect, we proposed a simple and effective procedure for oxybromination of unsaturated substrates in a biphasic system H$_2$O/CH$_2$Cl$_2$ where the V-catalyst, oxidant, and bromide are dissolved in the aqueous phase, while the substrate is dissolved in the organic phase. With that procedure, from one side we mimicked the activity of V-dependent haloperoxidase enzymes (*13*), and from another we obtained an interesting synthetic method for functionalization of organic substrates in mild conditions using a sustainable oxidant, H$_2$O$_2$, and a sustainable source of positive bromine such as KBr. Furthermore, by-products are bromide and water, either reusable or non-polluting.

More to the point, in the course of our studies, by combining reactivity analysis with spectroscopic techniques, mainly [51]V-NMR, we have been able to identify monoperoxo vanadium complexes as competent oxidants of Br$^-$, while diperoxo vanadium species likely act as reservoir of the active oxidant. Without entering too much in details, a good deal of mechanistic studies carried out in these last years, offered us a detailed description of our system. In particular we observed that: (*14*)

a. selectivity is strongly dependent on the rate of stirring
b. with other brominating systems only dibromide is obtained
c. kinetic evidence indicates substrate coordination to vanadium
d. more nucleophilic substrates favor formation of bromohydrin
e. Hammett correlations from competitive experiments indicate two parallel processes
f. V-bound hypobromite intermediate can be detected via ESI-MS (*15*)

On this basis our mechanistic proposal involves two intermediates: the first one, i.e a V-bound hypobromite ion, is responsible for the formation of bromohydrin, the second one is simply Br$_2$, whose reaction with double bonds produces selectively dibromo derivative.

More recently, a modification of the two-phase procedure for oxybromination of double and triple bonds with M/H$_2$O$_2$/Br$^-$ (M=V(V) or Mo(VI)) by substitution of the chlorinated solvent with hydrophobic ILs has been presented (*16,17*), and evidence was offered that such variation gives rise to even more sustainable processes characterized by higher rates and better selectivities.

Results and Discussion

The oxybromination reactions were carried out by using the different hydrophilic and hydrophobic ILs in order to identify, first of all, which system, homogeneous versus a two-phase one, is superior. The two-phase option is dictated, as in the case of the chlorinated solvents (18), by the use of aqueous solutions of hydrogen peroxide as primary oxidant. The ILs indicated in Figure 1 were chosen on the basis of their availability and stability in an aqueous-oxidative medium. Furthermore, in order to have ILs with appropriate purity we synthesized our solvents. In fact in several instances irreproducible results were obtained by using commercial ILs (16).

Reactions carried out with styrene as model substrate and hydrophilic ILs ([bmim$^+$][BF$_4^-$]; [bmim$^+$][CF$_3$SO$_3^-$]), at 25°C, were disappointing in terms of both yield and selectivity even though a shortening of the reaction time was observed. This approach was therefore abandoned (16).

Thus, we concentrated our efforts in exploring the two-phase procedure employing [bmim$^+$][PF$_6^-$], [bm$_2$im$^+$][PF$_6^-$], and [bmim$^+$][(CF$_3$SO$_2$)$_2$N$^-$].

Figure 2 reports the schematic representation of the reaction and of the substrates used, while Table I collects the most significant results obtained. Optimization of the reaction conditions has been already reported (16) and it was made by using styrene as model substrate and [bmim$^+$][PF$_6^-$] as typical IL.

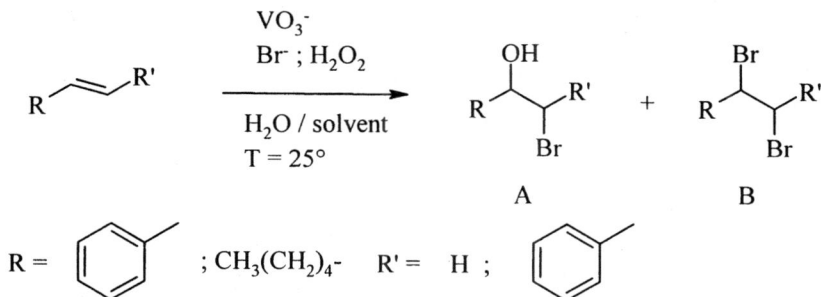

Figure 2. Oxybromination of alkenes in a two-phase system.

Data collected in Table I clearly indicate that substitution of the chlorinated solvent with hydrophobic ILs, regardless to the nature of the substituent present on the double bond, results in faster reactions and higher selectivity toward the formation of bromohydrins. Very impressive is the case of the oxybromination of 1-octene, the less nucleophilic substrate used. In this case the selectivity A:B = 9:91 obtained in CH_2Cl_2 is completely reversed both in [bmim$^+$][PF$_6^-$], and [bmim$^+$][(CF$_3$SO$_2$)$_2$N$^-$], reaching a value of A:B = 87:13 rarely achievable with other reagents and in such mild conditions.

Table I. V(V)-catalyzed Oxybromination of alkenes with Br$^-$/H$_2$O$_2$ in a two-phase System H$_2$O/solvent^a

Substrate	Solvent	V(V)	Time	Yield	A:B
		mol l^{-1}	hrs	%	
styrene	CH_2Cl_2	0.02	2	81	43 : 57
"	[bmim$^+$][PF$_6^-$]	0.02	2	>99	98 : 2
"	"	0.01	4	96	98 : 2
"	"	0.001	24	87	96 : 4
"	[bm$_2$im$^+$][PF$_6^-$]	0.01	4	>99	94 : 6
"	[bmim$^+$][(CF$_3$SO$_2$)$_2$N$^-$]	0.01	6	92	97 : 3
1-octene	CH_2Cl_2	0.01	6	76	9 : 91
"	[bmim$^+$][PF$_6^-$]	0.02	4	74	87 : 13
"	[bmim$^+$][(CF$_3$SO$_2$)$_2$N$^-$]	0.02	4	70	87 : 13
t-stilbene	CH_2Cl_2	0.02	4	64^b	b
	[bmim$^+$][PF$_6^-$]	0.02	1	>99	42 : 58
	[bmim$^+$][(CF$_3$SO$_2$)$_2$N$^-$]	0.01	1	81	61 : 39

^{a} H$_2$O/Solvent 1: 1 mL, T = 25°C, rpm 1000, Substrate 0.02 molL^{-1}; KBr 0.1 molL^{-1}, H$_2$O$_2$ 0.02 molL^{-1}.

^{b} Quantitave HPLC analysis was possible only for dibromide, likely both stereoisomers of bromohydrin were formed and overlapping of the peaks did not allow a precise determination of them, estimated amount of both bromohydrins is ca.10%.

Explanation to the results here presented, provided that our reaction mechanism still holds in the ionic environment, can be found taking into consideration that transfer of molecular bromine in the IL phase, where the oxidation takes place, is disfavoured by the medium.

In addition, on the basis of the results shown above for the oxybromination of double bonds, we have now further support to the proposal that the ionic environment produces both a higher concentration of the active species in the IL phase (favouring the first equilibrium indicated in Figure 3) (*16*), and facilitates, because of a slower escape from the organized solvent cage, the reaction between the bromiranium species (formed upon reaction of the vanadium-bound hypobromite intermediate and the substrate (*19*)), and the vanadium coordinated water molecule, to form the bromohydrin. The net result is thus a more selective and faster formation of bromohydrins.

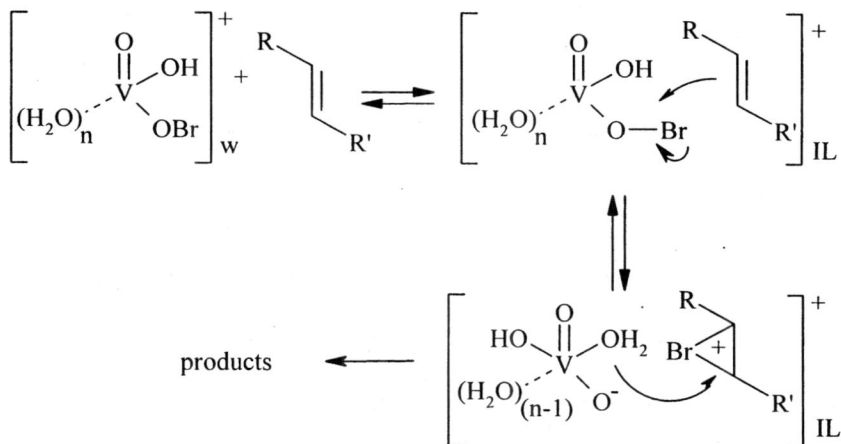

Figure 3. Essential steps for the V-catalysed oxybromination of double bonds in a two-phase system water/IL

In this respect the data obtained in the reaction with *trans*-stilbene in molecular solvent, where both stereoisomers of the halohydrin were likely formed, suggest that in chlorinated solvent the more stable carbocation is formed, while, the formation of bromiranium is more favored in ionic liquids, even though the selectivity achieved (A:B) is lower than that observed with the other substrates.

In view of these results, investigation of a similar reaction with phenylethyne, as model alkyne, appeared very interesting. However, since bromination of a triple bond is expected to be more difficult than that of a similarly substituted double bond, the more reactive Mo(VI) species was used as catalyst (*17*) even though also V(V) was tested. Here we report only the data obtained with vanadium. Being the process not previously reported, oxybromination reaction of phenylethyne was first carried out at room temperature in a two-phase system H_2O/CH_2Cl_2. Figure 4 sketches the outcome of the reaction, while Table II collects the experimental results. A number of interesting aspects emerge. First, the reaction performed in water/[bmim$^+$][PF$_6^-$]

34

is faster than that in the halogenated solvent. Second, the selectivity is definitely shifted toward the dibromoketone F.

This is a significant result, because α,α-dibromoacetophenone is a key molecule, with antibacterial, fungicidal and algicidal properties. Moreover, it is a valuable intermediate for further transformations, for example, to α-haloenolates or biologically active heterocyclic compounds.

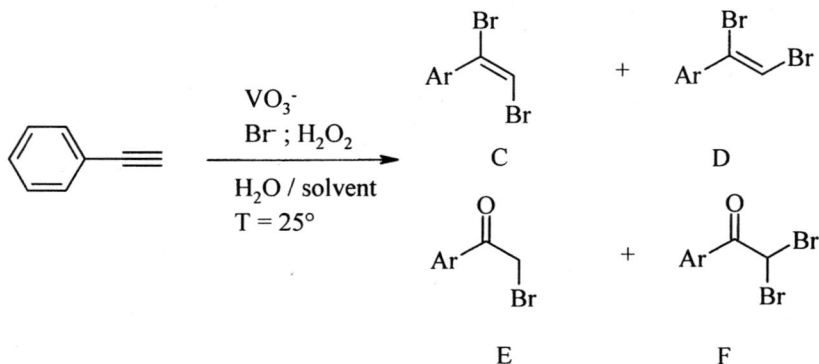

Figure 4. V-catalysed oxybromination of phenylacetylene in a two-phase system: Molecular solvent vs ILs

Table II. V(V)-catalyzed Oxybromination of phenylacetylene with Br⁻/H₂O₂ in a two-phase System H₂O/solvent[a]

Solvent	Sub.	V(V)	KBr	H₂O₂	Time	Yield	C:D:E:F
	$moll^{-1}$	$moll^{-1}$	$moll^{-1}$	$moll^{-1}$	hrs	%	
CH_2Cl_2	0.01	0.01	0.05	0.01	22	30	14:43: 4:39
[bmim⁺][PF₆⁻]	0.02	0.02	0.1	0.02	4	22	16:24:16:43
"	0.02	0.02	0.1	0.02[b]	24	70	13:15: 9:63

[a] H₂O/Solvent 1: 1 mL, T = 25°C, rpm 1000; KBr 0.05 $molL^{-1}$, H₂O₂ 0.01 $molL^{-1}$. [b] H₂O₂ added in two portions

It is however important to underline that, in order to reach an interesting total yield of products of 70%, hydrogen peroxide needs to be added portionwise, so that its vanadium catalyzed decomposition can be kept under control.

The pathways of phenylethyne oxybromination have been proposed (17) in analogy to that found in alkene oxybromination (15,16). That mechanistic scheme well explains the experimental results. In fact, the disfavoured formation of 1,2-dibromostyrene on going from dichloromethane to [bmim⁺][PF₆⁻] can be ascribed to a slowed down – or even inhibited – formation of molecular bromine in the latter medium, where the functionalization of the substrate is likely to occur. Moreover, the internal structure of IL may help in keeping the reactive species within the organized solvent cage, thus rendering the reaction faster.

Accordingly, also with phenylethyne the oxybromination reaction by H_2O_2/KBr was successful the main points being: catalysis by Mo(VI) is more effective than that by V(V); conversion of phenylethyne can be made synthetically interesting by adding H_2O_2 in portions; and selectivity can be diverted from 1,2-dibromoalkene to the synthetically useful dibromoketone, changing the solvent from dichloromethane to ionic liquids.

To finish this short survey regarding the vanadium catalyzed oxidation with hydrogen peroxide in ionic liquids is worthy of mention here some very preliminary, yet disappointing up to now, data obtained in the attempt to extend this approach also to the hydroxylation of benzene.

Such a reaction has been very well studied (20) since the initial work of Mimoun (21) who published the synthesis and the reactivity of the peroxo vanadium picolinato complex $VO(O_2)pic$, which reacts with benzene, in acetonitrile, producing good yields of phenol (up to 70% in stoichiometric conditions).

At the same time, by using an appropriate phase transfer agent, namely 4-(3-heptyl)-pyridine-2-carboxylic acid, we settled a two-phase procedure (22) where benzene and substituted benzenes are hydroxylated to monophenols with fair yields. Considering that inorganic salts as well as organic compounds are generally much more soluble in ionic liquids than in molecular solvents, we hoped that hydroxylation of aromatics could take place in the biphasic system H_2O/ILs where the aqueous phase contains V-catalyst and picolinic acid and benzene is dissolved in the IL phase, while slow addition of hydrogen peroxide would assure a better performance in terms of products vs peroxide decomposition.

Unfortunately, very small amounts of phenol were detected both in [bmim⁺][PF₆⁻], and [bmim⁺][(CF₃SO₂)₂N⁻]. Also attempts to use hydrophilic ILs, i.e. [bmim⁺][BF₄⁻] [bmim⁺][CF₃SO₃⁻], thus obtaining homogeneous reaction mixtures, failed. These results may be due to the fact that hydroxylation reaction proceedes through a radical mechanism, at variance with the polar mechanism of the oxybromination of double and triple bonds.

However, making the effort to understand why $VO(O_2)pic$ was unreactive in the presence of IL we have observed that vanadium peroxocomplexes in ILs change dramatically their spectral (UV-VIS and ⁵¹V-NMR) behavior. Therefore we are currently carrying out speciation studies in order to understand which kind of interactions are taking place between vanadium species, both in the presence and in the absence of hydrogen peroxide, and the ionic components of

36

ILs, with the ultimate aim to synthesize an ionic liquid with appropriate characteristics for the oxidative functionalization of benzene.

Acknowledgment

Financial support from MIUR, Prin 2003 Project "Development of new recyclable catalysts for oxidation processes with hydrogen peroxide" is gratefully acknowledged. Work carried out in the frame of ESF COST D29 action: Sustainable/Green Chemistry and Chemical Technology; WG 0016-04 "Novel Sustainable Metal Catalyzed Oxidations with Hydrogen Peroxide and Molecular Oxygen". Experimental contribution by the undergraduate students A. Coletti and M.L. Naitana is acknowledged.

References

1. Horvath, I.T., *Pure Appl. Chem.*, **1998**, *31*, 641.
2. de Wolf, E.; Van Koten, G. ; Deelman, B.-J. *Chem. Soc. Rev.*, **1999**, *28*, 37.
3. Jessop, P.G.; Ikariya, T.; Noyori, R. *Chem. Rev.*, **1999**, *99*, 475.
4. Leitner, W. *Acc. Chem. Res.*, **2002**, *35*, 746.
5. Oakes, R.S.; Clifford, A.A.; Rayner, C.M. *J. Chem. Soc. Perkin Trans. 1*, **2001**, 917.
6. Boon, J. A.; Levisky, J. A.; Pflug, J. L.; Wilkes J. S. *J. Org. Chem.* **1986**, *51*, 480.
7. Howarth, J.; Hanlon, K.; Fayne, D.; Mc Cormac P. *Tetrahedron Lett.* **1997**, *38*, 3097.
8. Chauvin, Y.; Mussmann, L.; Oliver H. *Angew. Chem. Int. Ed.* **1995**, *34*, 2698.
9. Song, C.E.; Roh E.J. *Chem. Commun.* **2000**, 837.
10. Chiappe, C.; Conte, V.; Pieraccini D. *Eur. J. Org. Chem.* **2002**, 2831.
11. *Ionic Liquids in Synthesis*, Wasserscheid, P.; Welton, T. Wiley-VCH, Weinheim, 2003.
12. Conte, V.; Di Furia, F.; Moro, S. *J. Phys. Org. Chem.* **1996**, *9*, 329.
13. Conte, V.; Di Furia, F.; Moro, S. in *Vanadium Compounds: Chemistry Biochemistry and Therapeuthic Applications*, (Eds. D.C. Crans, A. Tracey), ACS Symposium Series 711, **1998**, chap. 10 and refs. cited therein
14. Bortolini, O.; Conte, V. *J. Inorg. Biochem.* **2005**, *99*, 1549, and refs. cited therein.
15. Bortolini, O.; Conte, V.; Carraro, M.; Moro, S. *Eur. J. Inorg. Chem.* **2003**, 42.
16. Conte, V.; Floris, B.; Galloni, P.; Silvagni, A. *Pure & Appl. Chem.* **2005**, *77*, 1575.

17. Conte, V.; Floris, B.; Galloni, P.; Silvagni, A. *Adv. Syn.& Catal.* **2005**, *347*, 1341.
18. Conte, V.; Di Furia, F.; Moro, S. *Tet. Lett.* **1994**, *35*, 7429.
19. O. Bortolini, C. Chiappe, V. Conte, M. Carraro. *Eur. J. Org. Chem.* **1999**, 3237.
20. Bonchio, M.; Conte, V.; Di Furia, F.; Modena, G.; Moro, S. *J. Org. Chem.* **1994**, *59*, 6262.
21. Mimoun, H.; Saussine, L.; Daire, E.; Postel, M.; Fisher, J.; Weiss, R. *J. Am. Chem. Soc.* **1983**, *105*, 3101.
22. Bianchi, M.; Bonchio, M.; Conte, V.; Coppa, F.; Di Furia, Modena, G.; Moro, S.; Standen, S. *J. Molec. Catal.* **1993**, *83*, 107.

Chapter 3

The Synthesis of Functionalized Tetrahydrofurans via Vanadium(V)-Catalyzed Oxidation of Alkenols

Jens Hartung, Arne Ludwig, Mario Demary, and Georg Stapf

Department of Chemistry, Technische Universität Kaiserslautern, Erwin-Schrödinger Straße, D–67663 Kaiserslautern, Germany

Vanadium(V) complexes formed from tridentate Schiff-bases or 2,6-dihydroxymethyl-substituted piperidines are able to catalyze the synthesis of β-hydroxyl-substituted tetrahydrofurans from alkenols and *tert*-butyl hydroperoxide (TBHP). (Schiff-base)vanadium(V) complexes showed their highest catalytic activity in anhydrous chlorinated methanes. Selective alkenol, aniline, bromide, and thianthrene *S*-oxide oxidation in protic solvents (e.g. EtOH, H$_2$O), however, was feasible using piperidine-derived vanadium(V) complexes for TBHP activation. Data from ligand substitutions indicate that auxiliaries, in which the O,N,O-donor atom sites are part of a conjugated π-electron system, bind stronger to vanadium(V) than those which lack this structural motif.

© 2007 American Chemical Society

A considerable number of naturally occurring tetrahydrofurans has been selected in recent years for pursuing more detailed pharmacological studies (*1*). This assay-guided lead selection, however, faces at some stage of the investigation the challenge, how to cover the supply of the active compound in adequate amounts. Extraction from natural sources may be feasible, if crops or other renewable herbal sources serve as raw material. Partial or total synthesis have to be considered, if the active compound originates from a rare or even protected organism (*2*). In view of this background, the synthesis of tetrahydrofuran-derived building blocks has received considerable attention within the last decade. Major advances originated in particular from the field of transition metal-catalyzed oxidation (*3,4,5*). This concept is based on an in situ adduct formation between the central ion, an *O*-atom donor ligand, and the substrate. Differences in steric congestion associated with competing reaction channels frequently serve as means for controlling regio- and/or stereoselectivity of the oxygenation step. This strategy has been applied, for instance, in order to develop an efficient synthesis of β-hydroxyl-substituted mid-sized ethers from alkenols and *tert*-butyl hydroperoxide (TBHP). Activation of the primary oxidant in this case occurs via peroxide binding to a (Schiff-base)vanadium(V) complex. This step provides a novel peroxy reagent that may bind, e.g., substrate (*R*)-**1**, to afford "loaded" complex **I**. Intramolecular *O*-atom transfer in intermediate **I** thus affords tetrahydrofuran (*2S,5R*)-**2** (Figure 1) with an excellent regio- and stereoselectivity (*6*).

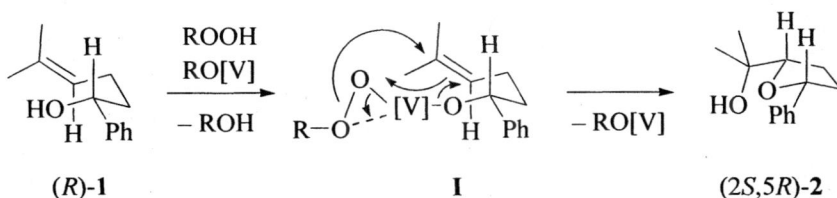

Figure 1. Stereoselective tetrahydrofuran formation via vanadium-catalyzed alkenol oxidation. RO[V] = (alkoxo)(Schiff-base)vanadium(V) complex (6, 7).

Other alkenols than (*R*)-**1** are oxidized in a similar predictable way to afford functionalized tetrahydrofurans in synthetically useful diastereoselectivities and yields (Figure 2) (*7,10*). In spite of the recognition this method has received in organic synthesis (*7,8*), its incompatibility with aqueous solvents is a severe restriction these days (*6,9*). Another drawback originates from the activity of

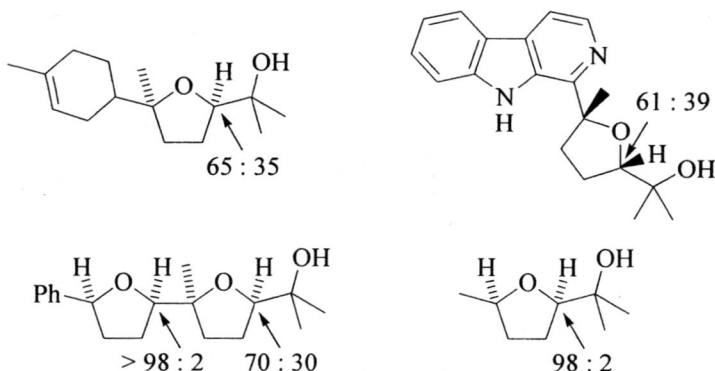

Figure 2. Natural products and synthetic analogues (7,10). Arrows denote cis/trans-ratios associated with oxidative, vanadium(V)-catalyzed alkenol cyclizations.

the catalyst, which fades after approximately 50 cycles (*6*). Both issues thus have to be addressed in order to broaden the scope of the method.

Following guidelines from a computational analysis (*11*), Schiff-bases **II** formed from substituted salicylic aldehydes (S^1 = e.g. NEt$_2$, NO$_2$) and/or substituted β-aminophenols (S^2 = e.g. OMe, NO$_2$; Figure 3) were considered to bind stronger to vanadium(V) than the unsubstituted derivative **3a** (Figure 4). Also, a formal hydrogenation of π-bonds between the O,N,O donor atom entity in auxiliary **II** was predicted (*11*) to furnish ligands that show a stronger affinity toward vanadium(V) than the dianion of **3a**. In the first part of this chapter, results from the synthesis of (Schiff-base)vanadium(V) complexes, their properties, and their performance in a benchmark oxidation for stereoselective tetrahydrofuran formation (*7*) are summarized. The second part of the chapter deals with the synthesis of vanadium(V) complexes starting from 2,6-dihydroxymethyl-substituted piperidines **III** and their use in oxidation catalysis. An emphasis has been laid on reactions in protic media.

(Schiff-base)vanadium(V) Complexes

Preparation and Ligand Substitutions

Imines **3a–f** and **3h–i** (Figure 4) were prepared upon treatment of an appropriate *ortho*-(hydroxy)arylcarbaldehyde with a β-aminoalcohol in hot EtOH (**3a**: quant; **3b**: 78 %; **3c**: 94 %; **3d**: 84 %; **3e**: 74 %; **3f**: 26 %; **3h**: 94 %;

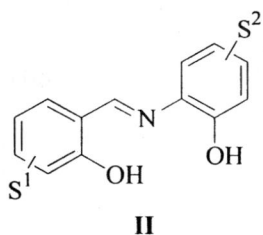

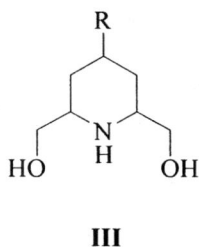

Figure 3. Constitution formulae of novel tridentate auxiliaries for vanadium(V)-based oxidation catalysts (R = H, tert-butyl; S^1, S^2 = +M- or –M-substituents).

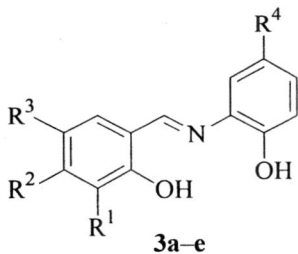

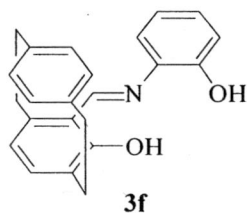

3	R^1	R^2	R^3	R^4
a	H	H	H	H
b	H	NEt$_2$	H	NO$_2$
c	H	H	H	OMe
d	NO$_2$	H	NO$_2$	H
e	H	NEt$_2$	H	H

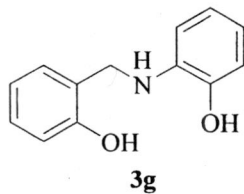

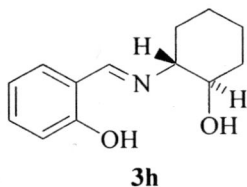

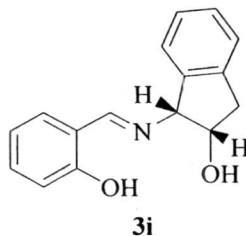

Figure 4. Structure formulae of tridentate Schiff-base-derived ligands 3a–f, 3h–i, and secondary amine 3g.

3i: quant) *(6,10,12)* NaBH$_4$-reduction of *N*-salicylidene-2-aminophenol (**3a**) furnished amine **3g** (Figure 4) in 51 % yield.

The reaction of triethyl vanadate (**4**) with NO$_2$-donor ligands **3a–i** in hot EtOH provided complexes **5a–i** (**5a**: quant; **5b**: 92 %; **5c**: 19 %; **5d**: 99 %; **5e**: 95 %; **5f**: 93 %; **5g**: 60 %; **5h**: 33 %; **5i**: quant) as dark brown to red brown cristalline solids (Figure 5) ,*13*). Compound identification was achieved via UV/Vis-spectroscopy, IR-analysis [$\nu_{V=O}$ /cm^{-1} (KBr) = **5a**: 990, **5b**: 994, **5c**: 983; **5d**: 962; **5e**: 959, **5f**: 988, **5g**: 988, **5h**: 976, **5i**: 997], combustion analysis, ^{51}V-NMR [in CDCl$_3$, referenced versus VOCl$_3$ in CDCl$_3$: **5a**: δ = –529, **5b**: –531, **5c**: –524; **5d**: –535; **5e**: –519, **5g**: –535, **5h**: –510/–531 = 1/1.7 (CD$_3$OD); **5i**: –534/–538 = 1/1.2 (CD$_3$OD)] and X-ray diffraction analysis (**5a, 5d, 5i**).

$$H_2L^n + VO(OEt)_3 \xrightarrow[78\,°C]{EtOH} VOL^n(OEt)(EtOH)_q + (2-q)\,EtOH$$

$$\mathbf{3a–i} \qquad \mathbf{4} \qquad\qquad\qquad \mathbf{5a–i}$$

*Figure 5. Synthesis of vanadium(V) complexes **5a–i** from auxiliaries **3a–i** (see also Figure 4; q = 0 or 1) (6).*

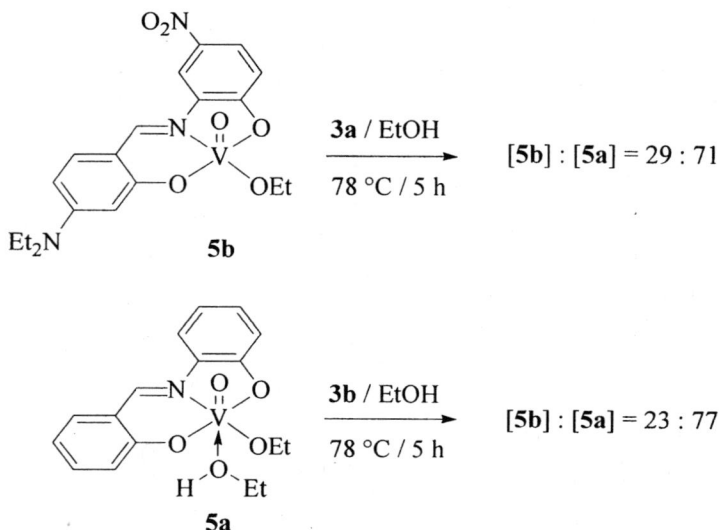

3a / EtOH
———————→
78 °C / 5 h

[5b] : [5a] = 29 : 71

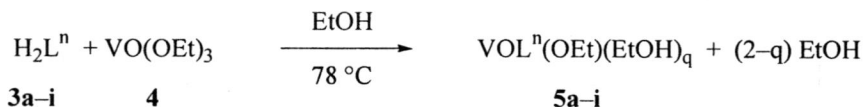

3b / EtOH
———————→
78 °C / 5 h

[5b] : [5a] = 23 : 77

Figure 6. Substituting tridentate ligands in vanadium(V) complexes.

The ground state stability of vanadium(V) complexes **5a–i** was explored via ligand substitutions (Figure 6). This approach reflects the concept of isodesmic reactions (*14*) that has been applied for theoretically predicting relative stabilities of structurally related compounds (*11*). *N*-Salicylidene-2-aminophenol (**3a**) and derived (Schiff-base)vanadium(V) complex **5a** (*6,7*) served as reference, based on a long term experience with the compounds.

Heating of an equimolar ratio of Schiff-base **3a** and diethylamino-nitro-substituted vanadium(V) complex **5b** for 5 h in EtOH furnished a 29/71-mixture of starting material **5b** and compound **5a** (^{51}V-NMR) (Figure 6). This ratio is considered to reflect relative stability of vanadium(V) compounds **5b** and **5a** because approximately the same numbers (**5b**:**5a** = 23:77) are obtained, if complex **5a** is treated under identical conditions with an equimolar amount of Schiff-base **3b** (Figure 6). The affinity for binding of auxiliaries **3c–i** to vanadium(V) was assessed in a similar set of reactions, to furnish a sequence of relative complex stability, which decreases along the progression **5a** > **5b**, **5c** > **5d–i**. According to ^{51}V-NMR analysis, complexes **5d–i** are entirely decomposed in the presence of auxiliary **3a**, to furnish compound **5a** as only detectable vanadium(V) product. The yield of the latter in ligand substitutions has so far not been determined.

Interpretation of Relative Vanadium(V) Complex Stability

The central ion in vanadium(V) complexes **5a**, **5d**, and **5i** is slightly offset from the basal plane of a distorted square pyramidal coordination polyhedron, which is formed by the three donor atoms of the chelate ligand and an ethoxide *O* (*6,15*). The structure of complexes **3b**, **3c**, **3e–h** is assumed to follow this guideline (*16,17*). The affinity of auxiliaries **3a–i** to bind to vanadium(V), however, differs significantly, as deduced from data of ligand substitutions. This observation is explicable on the basis of valence bond (VB) and molecular orbital (MO) theory (Figures 7 and 8).

According to VB treatment, effects originating from either +M- or –M-substituents located in positions, which are relevant for the present study, indicate that a shift of electron density from the periphery to the central ion or vice versa has the potential to lower bond strengths between vanadium(V) and phenolato *O*-atoms. The degree of these changes may be subtle but could suffice to favor binding of the unsubstituted auxiliary **3a** to substituted derivatives **3b–e**.

The notable driving force for binding of *N*-salicylidene-2-aminophenol (**3a**), if compared to derivatives **3g–i** that lack in π-electron conjugation starting from one phenolato *O* across the imine *N* to the second phenolato *O*, may be modelled using an orbital scheme similar but not identical to the Ballhausen-Gray orbital transformation scheme for C_{4v} species (*18*). Thus, ligand coordination in sp^{2}d^2-hybridized complexes of the type **5a–i** is proposed to occur in the following way:

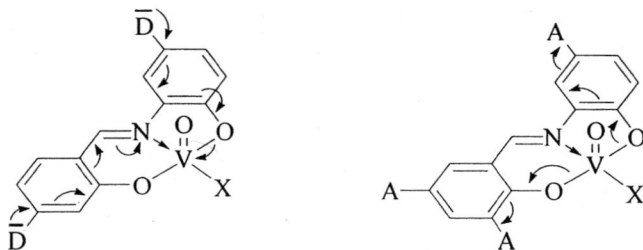

Figure 7. Valence bond model for correlating polar substituent effects with ground state stability of vanadium(V) complexes (A = NO$_2$, D = e.g. NEt$_2$).

- two σ-bonds form along the y-axis between one $2sp^3$-orbital of both phenolato O and two $4sp_y$-hybridized orbitals at vanadium;
- one σ-bond forms between the $2sp^2$-hybridized lone pair at the imine N and the vanadium $3d_{x^2-y^2}$ orbital along x; a second bond along x is formed via $2sp^3 \rightarrow 4p_x$ overlap, in case of X^- refering to a ligand with a second row donor atom such as EtO^-;
- the σ-contribution of the V=O bond originates from $2sp^2 \rightarrow 3d_{z^2}$-overlap, the π-bond forms via $2p_x \rightarrow 3d_{xz}$ interaction;
- a neutral ligand with a second row donor atom, e.g. an alkenol, forms a $2sp^3 \rightarrow 4p_z$-σ-bond, to coordinate along z in trans position of the oxo ligand.

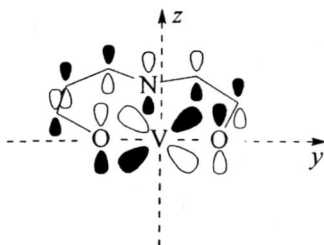

Figure 8. Proposed π-interaction between the HOMO$_{-1}$ of the 1,8-dioxa-4-azaocta-2,4,6-triene dianion and the $3d_{yz}$-orbital of vanadium(V) in, e.g., Schiff-base complex 5a (MO-model).

The proposed orbital scheme for ligand binding in, e.g., complex 5a leads to 14 valence electrons at vanadium. The hitherto unoccupied $3d_{xy}$ orbital may participate in binding of, for instance, *tert*-butyl hydroperoxide, in particular, once the monodentate ligand along x has been realased from the central ion. η^2-Peroxide binding therefore is considered to occur within the xy-plane (*19*). If the chelate ligand in **5a** is approximated as 1,8-dioxa-4-azaocta-2,4,6-triene dianion (Figure 8), the $HOMO_{-1}$ of this entity shows proper symmetry to undergo a π-type interaction with the vanadium $3d_{yz}$-orbital thus increasing the valence electron count at vanadium by a number of 2.

Oxidation Catalysis

Vanadium(V) complexes **5a–i** catalyze diastereoselective tetrahydrofuran formation starting from bishomoallylic alcohols and TBHP (*10*), as evidenced from the oxidation of 1-phenyl-4-penten-1-ol (**6**) as reporter reaction (Table I). In the absence of a vanadium(V) complex, no oxidation of alkenol **6** with TBHP in anhydrous chlorinated methanes, e.g. $CHCl_3$, is observed. Addition of 10 mol % of aminoindanol-derived complex **5i** to this solution affords 47% of 2-hydroxymethyl-5-phenyl tetrahydrofuran (**7**) (*cis:trans* = 39:61), after complete consumption of substrate **6** (48 h, 20 °C; Table 1, entry 5) (*6*). In all other instances, the yield of heterocycle **7** and its diastereoselectivity remained below this value (entries 1–4, Table 1). The issue of mass balance and 3-benzoyl-γ-butyrolactone formation in the majority of reactions is under current investigation.

Table I. (Schiff-base)vanadium(V)-Catalyzed Alkenol Oxidations

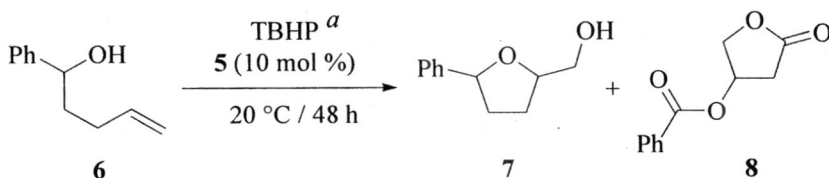

entry	5	7 [%] (cis:trans)	8 [%]	conversion of 6 [%]
1	a	38 (43 : 57) [b]	7	quant.
2	f	37 (44 : 56) [c]	– [d]	54
3	g	11 (54 : 46) [b]	18	quant.
4	h	14 (50 : 50) [b]	19	quant.
5	i	47 (39 : 61) [b]	– [e]	quant.

[a] 1.5 equiv. of a 5.5 M solution in nonane; [b] in $CHCl_3$; [c] in $CHCl_3$; [d] not detected (^{1}H-NMR); [e] not determined.

The data compiled in Table I in combination with those reported in the literature for the oxidation of structurally related alkenols, lead to guidelines for an application of (Schiff-base)vanadium(V) complexes, e.g. **5a** or **5i**, in stereoselective tetrahydrofuran formation (*6,7,10*). The compounds are

- valuable catalysts for the oxidation of ω-substituted bishomoallylic alcohols (e.g. **1**). The reactions proceed highly 2,5-*cis*-, 2,4-*trans*-, or 2,3-*trans*-stereoselectively to furnish the *exo*-cyclized compound, i.e, the functionalized tetrahydrofuran in synthetically useful yields;

- poor catalysts for an efficient and stereoselective oxidative cyclization of terminal unsubstituted alkenols (e.g. **6**);

- so far incompatible with tetrahydrofuran formation in protic media (e.g. EtOH, H_2O).

Piperidine-Derived Complexes

Concept and Synthetic Work

In view of the unsatisfactory performance of (Schiff-base)-derived catalysts in oxidative alkenol ring closures using protic reaction media, 2,6-substituted piperidines (Figure 9) were assessed as alternative auxiliaries for this purpose. Heterocycles **9a–c** were selected as structurally simple achiral reagents (*20*), in order to prove the concept.

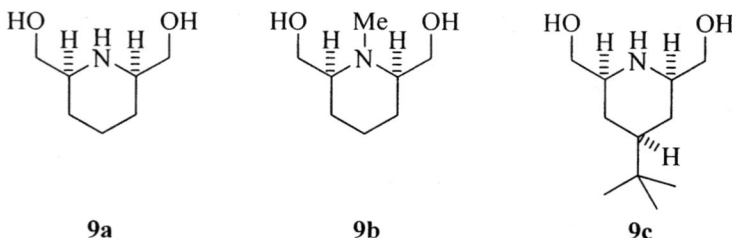

9a   **9b**   **9c**

Figure 9. Structural formulae of 2,6-substituted piperidines **9a–c**.

2,6-Disubstituted piperidines **9a–c** react with triethyl vanadate(V) to form hitherto unknown vanadium(V) complexes **10a–c**, as colorless (**10a**), light green (**10b**), light brown (**10c**) crystalline compounds (Table II).

Oxidation Catalysis

In view of the low solubility of complex **10a** in several organic media, the experiments outlined below were restricted to the use of vanadium(V)

Table II. Formation of Piperidine-Derived Vanadium(V) Complexes

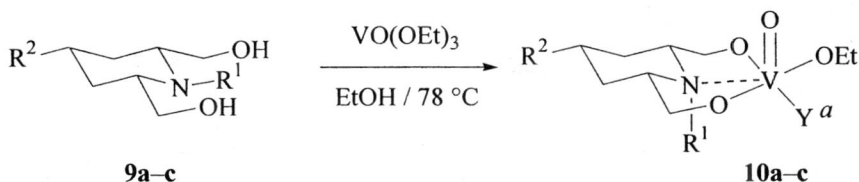

9a–c   10a–c

entry	10	R^1	R^2	yield [%]	V=O [cm^{-1}]
1	a	H	H	39	951
2	b	CH$_3$	H	64	922
3	c	H	C(CH$_3$)$_3$	71	952

[a] Y = H$_2$O according to C/H/N analytical data

compounds **10b** and **10c**. Both complexes are able to activate *tert*-butyl hydroperoxide for selective oxidation of 5-methyl-1-phenyl-4-hexenol (**1**). The reactions furnish *cis*-2,5-disubstituted tetrahydrofuran *cis*-**2** as major product. Its diastereomeric excess ranges between 86% and 90% (Table III). Tetrahydropyran **11** and lactone **12** are formed as minor components. The latter product probably is formed from tetrahydrofuran *cis*-**2** under these conditions (*6*).

Yields and ratios of compounds **2**, **11** and **12** varied, depending on the reaction time and the amount of vanadium(V) catalyst applied. In general, 5 mol % of catalyst were found to be adequate to obtain a total yield of e.g. 87% of oxidation products **2**, **11**, and **12**, if taken together (entry 3, Table III). According to a stereochemical analysis, heterocycles *cis*-**2** and *cis*-**11** originate from a pathway that has been termed as *syn*-selective *O*-atom transfer onto alkenol **1** (see Figure 1). A reversal of facial selectivity in π-bond oxygenation thus provides cyclic ethers *trans*-**2** and *trans*-**11**. If lactone **12**, in extension to previously reported results, is exclusively formed from *cis*-**2a**, the information provided in Table III points to *syn/anti*-selectivities of 87/13 (entry 1) to 78/23 (entries 3 and 5). It is noteworthy that all oxidations of substrate **1** were feasible in a 1/1-solvent mixture of CHCl$_3$/EtOH. The data furthermore indicate that the catalytic activity of **10b**, or an appropriate derivative that is formed under turnover conditions, did not fade considerably. Oxidation of **1** thus continued upon subsequent addition of two equivalents of the substrate and the required amount of TBHP (entry 4).

In extension to tetrahydrofuran formation, heteroatom oxygenations were assessed in protic reaction media (equations 1–3). The reactions were restricted to the use of 4-*tert*-butylpiperidine-derived catalyst **10c**, for reason of its solubility.

Table III. (Piperidine)vanadium(V)-Catalyzed Alkenol Oxidation

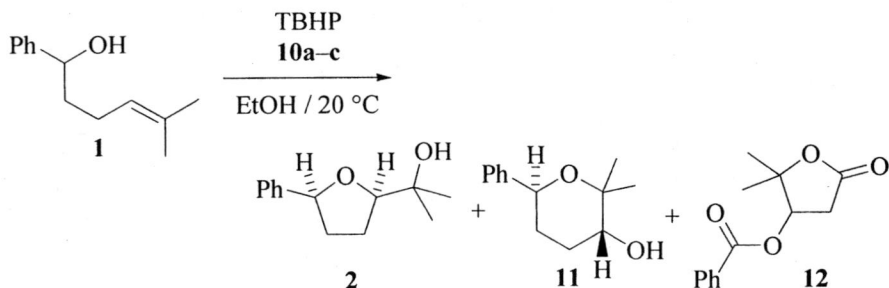

entry	1 [equiv.]	10 [mol %]	time [h]	2 [%] (cis:trans)	11 [%] (cis:trans)	12 [%]
1	1	10b [10]	24	48 (95 : 5)	9 (32 : 68)	5
2	1	10b [5]	24	35 (92 : 8)	13 (40 : 60)	1
3	1	10b [5]	48	51 (95 : 5)	17 (22 : 78)	19
4	3 × 1	10b [5]	3 × 12	47 (92 : 8)	13 (31 : 69)	8
5	1	10c [5]	24	48 (95 : 5)	18 (17 : 83)	12

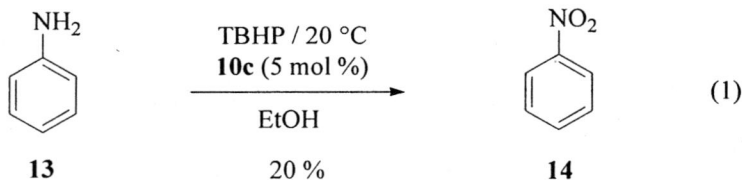

(1)

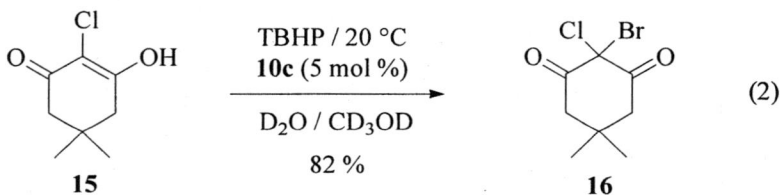

(2)

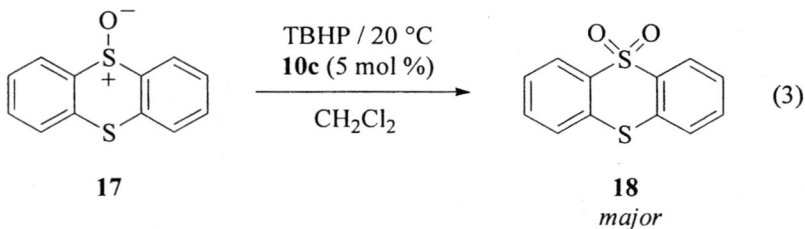

(3)

In a preliminary study, aniline (13) was converted by TBHP in the presence of 10c into nitrobenzene (14) (equation 1). Nitrosobenzene, which so far has not been completely purified from the reaction mixture and thus is omitted from equation 1, is formed as sole side product in approximately similar yields (^{1}H-NMR). Vanadium(V) compound 10c is further able to catalyze the oxidation of bromide with TBHP, as evident from a chlorodimedone assay (equation 2) (21). The synthesis of bromochlorodimedone (16) from substrate 15 in the absence of piperidine-derived catalyst 10c failed. Thianthrene-*S*-oxide (17) is oxidized by TBHP in the presence of 10c preferentially at the sulfoxide sulfur to afford sulfone 18. This selectivity was unexpected, since reagent combinations of TBHP and vanadium(V) complexes generally exhibit electrophilic rather than nucleophilic properties. This issue is under further investigation.

In conclusion, 2,6-dihydroxymethyl-substituted piperidines 9a–c bind to vanadium(V) to provide coordination compounds 10a–c that are able to activate TBHP in protic solvents for an application in selective π-bond and heteroatom oxygenations.

Acknowledgements

This work was generously supported by the Deutsche Forschungsgemeinschaft (grant Ha1705/8–2) and the Fonds der chemischen Industrie.

References

1.   Faul, M.M; Huff, B.E. *Chem. Rev.* **2000**, *100*, 2407–2473.
2.   Anatas, P.; Kirchhoff, M.M. *Acc. Chem. Res.* **2002**, *35*, 686–694
3.   Hartung, J.; Greb. M. *J. Organomet. Chem.* **2002**, *661*, 67–84.
4.   Klein, E.; Rojahn, W. *Tetrahedron* **1965**, *21*, 2353–2358.
5.   (a) Donohoe, T.J.; Winter, J.J.G.; Helliwell, M.; Stemp, G. *Tetrahedron Lett.* **2001**, *42*, 971–974. (b) Bilfulco, G.; Caserta, T.; Gomez-Paloma, L.; Piccialli, V. *Tetrahedron Lett.* **2003**, *44*, 5499–5503.
6.   Hartung, J.; Drees, S.; Greb, M.; Schmidt, P.; Svoboda, I.; Fuess, H.; Murso, A.; Stalke, D. *Eur. J. Org. Chem.* **2003**, 2388–2408.
7.   Hartung, J. *Pure Appl. Chem.* **2005**, *77*, 1559–1574.
8.   Blanc, A.; Toste, F.D. *Angew. Chem. Int.* Ed. **2006**, *45*, 2096–2099.
9.   van der Felde, F.; Arends, I.W.C.E.; Sheldon, R.A. *Top. Catal.* **2000**, *13*, 259–265.
10. Hartung, J.; Ludwig, A.; Demary, M. *manuscript in preparation.*

11. Sturm, H.C. *Diploma Thesis*, Universität Würzburg, Würzburg, 2002.
12. Westland, A.D.; Tarafder, M.T.H. *Inorg. Chem.* **1981**, *20*, 3992–3995.
13. Mimoun, H.; Mignard, M.; Brechot, P.; Saussine, L. *J. Am. Chem. Soc.* **1986**, *108*, 3711–3718.
14. George P.; Trachtman, M.; Brett, A.M.; Bock, C.W. *J. Chem. Soc., Perkin Trans. 2*, **1977**, 1036–1046.
15. Hartung, J.; Ludwig, A.; Svoboda, I.; Fuess, H. *manuscript in preparation for Acta Cryst. Sect. E.*
16. Bashirpoor, M.; Schmidt, H.; Schulzke, C.; Rehder, D. *Chem. Ber./Receuil* **1997**, *130*, 651–657.
17. Correia, I.; Pessoa, J.C.; Duarte, M.T. de Piedade, M.F.M.; Jackusch, T.; Kiss, T.; Castro, M.M.C.A.; Geraldes, C.F.G.C.; Avecilla, F. *Eur. J. Inorg. Chem.* **2005**, 732–744.
18. (a) Selbin, J. *Coord. Chem. Rev.* **1966**, 1, 293–314; (b) Ballhausen, C.J.; Gray, H.B. *Inorg. Chem.* **1962**, *1*, 111–122.
19. Mimoun, H.; Chaumette, P.; Mignard, M.; Saussine, L.; Fischer, J.; Weiss, R. *Nouv. J. Chim.* **1983**, *7*, 467–475.
20. Henderson, N.; Plumb, J.; Robins, D.; Workman, P. *Anti-Cancer Drug Design* **1996**, *11*, 421–438.
21. Hager, L.P.; Morris, D.R.; Brown, F.S.; Eberwein, H. *J. Biol. Chem.* **1966**, *241*, 1769–1777.

Chapter 4

Vanadium-Catalyzed Alkane Functionalization Reactions under Mild Conditions

Armando J. L. Pombeiro

Centro de Química Estrutural, Complexo I, Instituto Superior Técnico, Av. Rovisco Pais, 1049–001 Lisboa, Portugal (email: pombeiro@ist.utl.pt)

The conversion of alkanes into more valuable organic products, under mild conditions, is a current challenge to modern chemistry, and the partial oxidation of the former with transition metal catalysts is a rather promising approach. We now discuss the use of a number of vanadium compounds, commonly complexes with N,O- or O,O-ligands, including Amavadine and its model, as catalysts for the hydroxylation, oxygenation, halogenation or carboxylation of alkanes, gaseous (in particular methane) and liquid (both linear and cyclic) ones, to afford, under mild or moderate conditions, the corresponding alcohols, ketones, organohalides or carboxylic acids. Some of the systems operate in peroxidative liquid biphasic media, whereas others involve dioxygen as the oxidant and V-catalysts supported on modified silica gel. Evidence for the involvement of radical mechanisms is presented.

© 2007 American Chemical Society

The functionalization of alkanes, under mild conditions, toward the synthesis of organic products with commercial value (e.g., alcohols, ketones or carboxylic acids) is a challenge (1-7) to modern Chemistry. Currently, alkanes are mainly used as fuels, but more noble applications are being looked for, what has been hampered by their unreactivity which, however, can be promoted by the use of a suitable metal centre which can activate (i) the alkane itself or (ii) a reagent (like O_2 or H_2O_2) that can thus generate a highly reactive species (e.g. a metal-oxo radical, a metal-peroxo or a metal-hydroperoxo species or an hydroxo radical) capable of reacting with the alkane. The latter general case (ii) appears to account for the biological oxidation of alkanes to the corresponding alcohols by cytochrome P-450 or methane-monooxygenase.

Durng the last few years, we have been applying a number of vanadium complexes, in particular with biological or pharmacological significance, as catalysts or catalyst precursors for the oxidation and carboxylation of alkanes, and now some aspects of this research, which has been undertaken partially with the cooperation of Prof. J.J.R. Fraústo da Silva, are shortly reviewed.

Our work in this field was initiated by searching for peroxidase and haloperoxidase behavior (toward alkanes) of Amavadine, a natural base vanadium complex, $[V(HIDPA)_2]^{2-}$ [1, HIDPA = basic form of 2,2'-(hydroxyimino)dipropionic acid], present in some Amanita toadstools, and its model $[V(HIDA)_2]^{2-}$ [2, HIDA = basic form of 2,2'-(hydroxymino)diacetic acid] (Figure 1). This followed our recognition that Amavadine acts as (i) an electron-transfer mediator in the electrocatalytic oxidation of thiols according to a Michaelis-Menten mechanism (8), thus behaving like an enzyme in spite of its simplicity, and (ii) a peroxidase toward thiols by catalysing their oxidation by H_2O_2 (9). The work was also inspired on the well known (10,11) fact that V-haloperoxidases catalyse the peroxidative halogenation of alkenes and aromatic compounds, and we intended to test less reactive substrates, i.e. alkanes, for which such a behavior had not been reported.

In view of their synthetic relevance, carboxylations of alkanes were also investigated mainly toward the formation of carboxylic acids. Moreover, various vanadium(IV or V) complexes with N,O- or O,O-bonded ligands other than Amavadine [e.g. 3 – 5, Figure 1; $VO(CF_3COO)_2$, $VO(CF_3SO_3)_2$ and $VOSO_4$] were also tested and shown to exhibit, like the latter species, peroxidase and haloperoxidase behaviors toward alkanes, and to catalyse their carboxylations (Scheme 1). However, the most active catalysts are usually 1 - 3, thus comprising Amavadine and its model.

Hydroxylation and Halogenation

Liquid Biphasic Systems

Alkanes (typically cycloalkanes such as cyclohexane and cyclooctane) are oxidized catalytically, at room temperature, in a liquid biphasic system

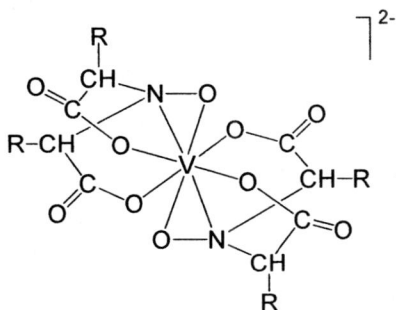

[VO(HIDPA)$_2$]$^{2-}$ (**1**, R = Me)

[VO(HIDA)$_2$]$^{2-}$ (**2**, R = H)

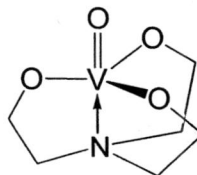

[VO(tea)] (**3**)

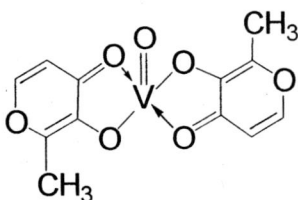

[VO(ma)$_2$] (**4**)

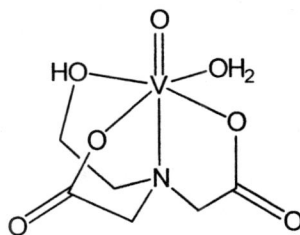

[VO(Hheida)(H$_2$O)] (**5**)

Figure 1. Amavadine (1), its model (2) and examples of related vanadium complexes with N,O- or O,O-ligands that catalyse alkane functionalization reactions. HIDPA, HIDA = basic forms of 2,2'-(hydroxyimino)dipropionic or – diacetic acid, respectively. H$_3$tea = triethanolamine. Hma = maltol. H$_3$heida = N-(2-hydroxyethyl)iminodiacetic acid.

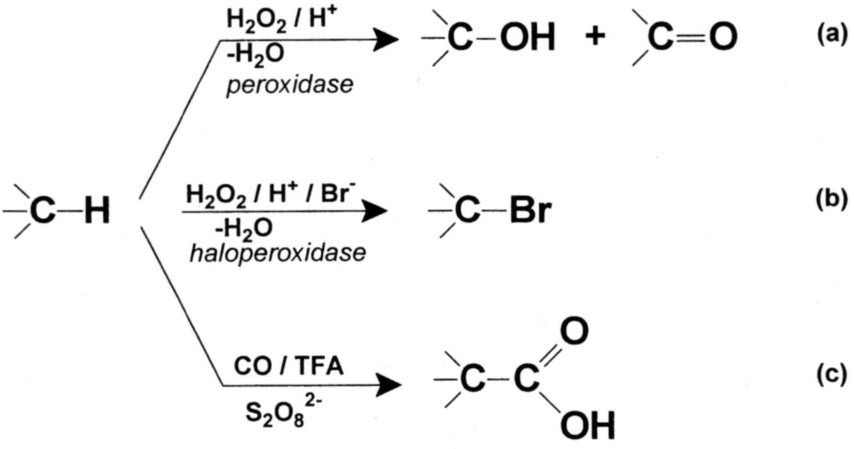

Scheme 1. Peroxidative oxidation (a), haloperoxidation (b) and carboxylation (c) of alkanes catalysed by vanadium complexes.

composed of any of the above vanadium complexes and aqueous H_2O_2 in acidic medium (HNO_3/NCMe) to give the corresponding alcohols and ketones (Scheme 1a), the former being the main products (*12*). Turnover numbers (TONs) and yields, after 6h reaction, can reach values up to ca. 50 mols products/mol catalyst and 10% respectively (12).

If the reaction is carried out in the presence of KBr or KCl, catalytic and selective monohalogenation of the cycloalkane occurs to yield the corresponding organohalide (Scheme 1b) (*12*).

The activity increase with the amounts of H_2O_2 and of the acid but only until certain limits beyond which it does not grow further or even decreases. Those V-complexes constitute, to our knowledge, the first vanadium catalysts to be reported for the peroxidative halogenation of alkanes.

Such complexes also catalyse (*13*) the selective peroxidative oxidation of benzene to phenol and of mesitylene (1,3,5-trimethylbenzene) to the aldehyde (3,5-dimethylbenzaldehyde) via the corresponding alcohol.

Although the mechanisms of the above reactions have not yet been established, radical processes are conceivably involved, as suggested by radical trap experiments (see below). As possible hydrogen-atom abstractors from the alkane (RH) to form the R˙ radical, one can postulate oxo- or peroxo-vanadium species or the hydroxyl (HO˙) radical. It is noteworthy that in Amavadine **1** and its model **2**, the hydroxyimino(1-) groups of the ligands, η^2-(O-N)⁻, are isolectronic with peroxo(2-), thus the complexes relate to bis(peroxo)vanadium species.

In the case of the peroxidative halogenation of alkanes and in view of the analogy of our V complexes, in particular [VO(tea)] **3**, with the metal centre of the V-haloperoxidases (*14*), one is attracted to postulate a mechanism based on that proposed (*15*) for the halide oxidation by such peroxidases to "X$^+$" (e.g. in the hypohalous HOX form) by an hydroperoxo-V^v species.

The role of the chelating N,O and O,O-ligands has not has not yet been identified but one can consider their involvement in proton-transfer steps, as suggested (*16-18*) for some vanadium/H$_2$O$_2$/O$_2$ systems.

We have also achieved peroxidative oxidation of cycloalkanes to the corresponding alcohols and ketones, under similar reaction conditions as those of the V-complexes, by using other metal catalysts, such as benzoylhydrazido- and benzoyldiazenido-complexes of rhenium with N,O-ligands (*19*), multinuclear copper triethanolamine complexes (*20, 21*), and iron(III)-chromium(III) hydroxo-complexes or heterogeneous hydroxides of these metal ions (*22, 23*).

The multinuclear copper complexes are of particular biological significance in terms of mimicking particulate methane monooxygenase (pMMO) (*24*). They exhibit a high activity (overall yields and TONs up to 32% and 380, respectively) and also catalyze, although less effectively, the   peroxidative oxidation of gaseous alkanes such as methane and ethane (*20*).

Supported Vanadium Catalysts

Heterogenization or immobilization of active metal complexes on a support constitutes a convenient method for combining the advantages of homogeneous and heterogenous catalysts and we have applied this approach to some bis(maltolato)oxovanadium complexes upon anchoring them to carbamate modified silica gel (*25,26*).

Hence, [VO(ma)$_2$] **4** (ma = maltolate) (Figure 1) and related complexes, such as [VO(py)(ma)$_2$] (py = pyridine) and *cis*-[VO(OR)(ma)$_2$] (R = Me, Et), supported on carbamated silica gel, catalyse the oxidation, with dioxygen, of either cyclic (*25*) or linear (*26*) alkanes (typically cyclohexane or *n*-pentane and *n*-hexane, respectively), under relatively mild reaction conditions (150-175 °C, 10-15 atm O$_2$), to the corresponding alcohols and ketones. Remarkably overall TONs (above 10^3) and quite significant overall yields (ca. 12-17%) can be achieved, with a good selectivity that can be optimised by using the most adequate temperature, reaction time, catalyst amount, O$_2$ pressure, etc. (*25, 26*). After being used, the catalyst can be reactivated by heating and recycled, yet displaying a considerable activity.

Although the mechanistic details are still unknown, a carbon- and oxygen-centered free-radical process is corroborated by the high yield drop when the reaction is carried out in the presence of the liquid carbon-radical trap CBrCl$_3$ or of the oxygen-radical trap Ph$_2$NH. Hence, in a way related to that proposed (*27*) for catalysts in which the metal has two available oxidation states of comparable stabilities, the alkylperoxy radical ROO$^.$ (which initially can be formed upon H-atom abstraction from the alkane RH by O$_2$ followed by oxidation of the derived R· radical by O_2) can play a key role (26). The derived hydroperoxide ROOH (formed by H-abstraction from RH) can undergo vanadium-catalyzed decomposition to RO· and ROO· from which the alcohol and the ketone are obtained.

Carboxylation

Metal-catalyzed carboxylation of alkanes to give the corresponding carboxylic acids (Scheme 1c) constitutes a field of high current interest (3, 4, 6, 7, 28-31) and particular attention has been paid to the single-pot conversion of methane into acetic acid in view of the high commercial value of this acid and of the much more complicated common industrial route involving three separate stages: the metal catalysed high temperature steam reforming of methane (to give CO and H_2), the conversion of the synthesis gas into methanol, and the carboxylation of this alcohol with CO and an expensive Rh (Monsanto process) or Ir (BP-Amoco "Cativa" process) catalyst. Hence, the search for more direct and less energy demanding routes for acetic acid is understandable and we have found (31, 32) that Amavadine and other vanadium(IV or V) complexes (see above) can catalyse that reaction (conversion of CH_4 into CH_3COOH), in trifluoroacetic acid (TFA) and using peroxodisulfate ($K_2S_2O_8$) as the oxidizing agent. An yield of 54% and a TON close to 30 were initially obtained at a moderate temperature (80 °C) and low gas pressure (ca. 5 atm), but current developments in our group (patents pending) have allowed to reach yields above 90% and TONs above 10^4.

The carboxylation, by these systems, does not require the use of the noxious CO gas since the TFA solvent can behave as a carbonylating agent. However, the formation of acetic acid can be enhanced, in some cases, by the presence of CO (at sufficient low pressures) which thus also acts as a carboxylating agent. However, higher CO pressures usually result in an inhibiting effect possibly due to coordination of CO with a resulting lowering of the catalyst activity (31).

Methane is the carbon source for the methyl group of acetic acid as shown by using ^{13}C-labeled CH_4 which yields $^{13}CH_3COOH$ as indicated by the $^{13}C\{^1H\}$ and ^{13}C NMR spectra of the reaction solution (reaction 1) (31). The reactions do not appear to proceed via free CO_2 (this can be formed by radical reactions of TFA with $K_2S_2O_8$ derivatives) or free methanol.

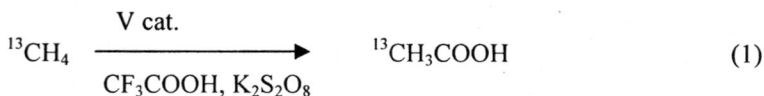

$$^{13}CH_4 \xrightarrow[\text{CF}_3\text{COOH, K}_2\text{S}_2\text{O}_8]{\text{V cat.}} {}^{13}CH_3COOH \qquad (1)$$

Our system is a development of that pioneered by Fujiwara (3, 4, 33, 34) which consisted of a metal compound/$K_2S_2O_8$/CO/TFA. It always required CO, and palladium(II) or copper(II) acetate was typically used as the metal species

(7) although [VO(acac)$_2$] (acac = acetylacetonate) was applied latter on (34); however, low TON values were then obtained.

Another rare system in which CH$_4$ can be converted into acetic acid without deliberate use of CO is provided by NaVO$_3$/pyrazine-2-carboxylic acid/H$_2$O$_2$ in aqueous solution (18), which however exhibits a low activity.

Our V-catalysts, in the K$_2$S$_2$O$_8$/TFA system, also operate for the carboxylation of higher alkanes, either gaseous or liquid ones, to afford usually mixtures of the corresponding carboxylic acids (35). CO in some cases is required, but the use of either too low or too high CO pressures should be avoided since the former promote the formation of trifluoroacetate esters and the latter, above a certain level, do not lead to higher yields or TONs of the acids (35). The effects of other factors (e.g., oxidizing agent, temperature, time, solvent) were also investigated (35).

The carboxylation of the linear alkanes gives a mixture of carboxylic acids but the main ones (2-methylpentanoic and 2-ethylbutanoic acids from pentane, and 2-ethylpentanoic acid from hexane) are derived from carboxylation of a secondary carbon atom, while the acids formed by carboxylation of a primary carbon are obtained in much lower yields (35). This points out to the involvement of a radical mechanism, what is also corroborated by the suppression of the catalytic activity when the reaction is undertaken in the presence of either a carbon- or an oxygen-radical trap (35).

The peroxide S$_2$O$_8{}^{2-}$ is a source of the sulfate radical SO$_4{}^{\cdot-}$, formed either by thermal decomposition or as a product of the reduction of the former when behaving as a single-electron oxidant. H-abstraction from alkane could form the alkyl radical R$^{\cdot}$ which on carbonylation would lead to the acyl radical RCO$^{\cdot}$. Oxidation of the latter, e.g. by a V^V complex, would form RCO$^+$ which, on further reaction with TFA, could lead to the carboxylic acid RCOOH. Theoretical studies are under way to check this proposal and alternative ones, toward the establishment of the mechanism.

We have extended the peroxodisulfate/TFA system to other metal catalysts, namely various (i) oxo-rhenium complexes with N,O-ligands which catalyse the ethane conversion into propionic and acetic acids (overall yields up to 30%) (36), and (ii) pyrazole and trispyrazolymethane Re complexes and their precursors, which catalyze the ethane oxidation to acetic acid (up to 40% yield) (unpublished work). Those reactions provide an unprecedented use of Re complexes as catalysts in alkane functionalization.

Final Remarks

Vanadium complexes with N,O- or O,O-ligands act as remarkably active catalysts or catalyst precursors for a variety of alkane functionalization reactions under mild conditions, including their (i) hydroxylation or oxygenation with H$_2$O$_2$ or O$_2$ ("green" oxidants without environmental pollution drawbacks) to

58

give alcohols and ketones, (ii) peroxidative halogenation with H_2O_2 to afford organohalides, and (iii) carboxylation, not only with CO but also in the absence of this noxious gas when the trifluoroacetic acid solvent behaves as the carbonyl source.

These catalytic processes can involve liquid biphasic systems or supported catalysts and proceed, at least in part, via radical mechanisms in which the V^{IV}/V^{V} redox interconversion conceivably plays an important role. Radical processes are also followed by the biological oxidation of alkanes by iron-containing oxidases, i.e. cytochrome P-450 and methane-monoooxygenase, but in our V-systems mechanistic details are still lacking.

The reactions are not only of commercial significance in view of the high added value of the functionalized products relatively to the starting alkanes, but also of biological meaning. In fact, the hydroxylation of alkanes is catalysed by some oxygenases. They are not based on vanadium but, in view of the high activity of some of our vanadium systems and of the growing recognition of involvement of this metal in Biology, it will not be surprising if a biological alkane functionalization role will also be discovered for vanadium.

Amavadine appears as one of the most active catalyst precursors and, although its biological role still remains unknown, the studies suggest it can act as a peroxidase and/or a haloperoxidase eventually involved in the protection system of the Amanita fungi.

Hence, the vanadium catalyzed functionalization of alkanes appears to constitute an emerging and promising field of research with a multi- and interdisciplinary character that deserves to be further explored.

Acknowledgements

The work concerning Amavadine has been carried out with the cooperation of Prof. João J.R. Fraústo da Silva, to whom the author expresses his gratitude. Acknowledgements are also due the other co-authors cited in the joint publications, in particular (i), at the Instituto Superior Técnico, Dr. José A.L. da Silva, Mrs. Marina Kirillova, Mr. Alexander Kirillov, Dr. Gopal Mishra, Dr. Maximilian N. Kopylovich, Prof. António Palavra, Dr. M. Fátima C. Guedes da Silva, Dr. Luísa Martins, Dr. Elisabete Alegria and Mr. Jenya Karabach, and (ii), at the Kyushu University, Prof. Yuzo Fujiwara. The work has been partially supported by the Fundação para a Ciência e a Tecnologia (FCT) and its POCI 2010, POCTI and PRAXIS programmes (FEDER funded), and by a Human Resources and Mobility Marie-Curie Research Training Network (AQUACHEM project, CMTN-CT-2003-503864).

References

1. *Catalytic Activation and Functionalization of Light Alkanes*; Derouane, E.G.; Haber, J.; Lemos, F.; Ramôa Ribeiro, F.; Guinet, F., Eds.; NATO ASI Series, vol. 44, Kluwer Academic Publ., Dordrecht, 1998.
2. Shul'pin, G.B. in *Transition Metals for Organic Synthesis*; Beller, M.; Bolm, C., Eds.; Wiley-VCH, Weinheim/New York, Vol. 2, 2nd ed., 2004, Ch. 2.2, pp. 215-242.
3. Jia, C.; Kitamura, T. Fujiwara, Y. *Acc. Chem. Res.* **2001**, *34*, 633.
4. Fujiwara, Y.; Takaki, K.; Taniguchi, Y. *Synlett* **1995**, 591.
5. Crabtree, R.H. *J. Chem. Soc., Dalton Trans.* **2001**, 2437.
6. Sen, A. *Acc. Chem. Res.* **1998**, *31*, 550.
7. Periana, R.A.; Mironov, O.; Taube, D.; Bhalla, G.; Jones, C.J. *Science* **2003**, *301*, 814.
8. Guedes da Silva, M.F.C.; Silva, J.A.L.; Fraústo da Silva, J.J.R.; Pombeiro, A.J.L.; Amatore, C.; Verpeaux, J.-N. *J. Am. Chem. Soc.*, **1996**, *118*, 7568.
9. Matoso, C.M.M.; Pombeiro, A.J.L.; Fraústo da Silva, J.J.R.; Guedes da Silva, M.F.C.; Silva, J.A.L.; Baptista-Ferreira, J.L.; Pinho-Almeida, F. in ref. 13, Ch. 18, pp. 241-247.
10. *Vanadium Compounds*; Tracey, A.S.; Crans, D.C., Eds.; ACS Symposium Series no. 711, ACS, Washington, 1998.
11. Crans, D.C.; Smee, J.J. ; Gaidamanskas, E. ; Yang, L.Q. *Chem. Rev.* **2004**, *104*, 849.
12. Reis, P.M.; Silva, J.A.L.; Fraústo da Silva, J.J.R.; Pombeiro, A.J.L. *Chem. Commun.* **2000**, 1845.
13. Reis, P.M.; Silva, J.A.L.; Fraústo da Silva, J.J.R.; Pombeiro, A.J.L. *J. Mol. Cat. A: Chem.* **2004**, *224, 189.*
14. Messerschmidt, A.; Prade, L.; Wever, R., in ref.13, Ch. 14, pp. 186-201.
15. Pecoraro, V.L.; Slebodnick, C.; Hamstra, B., in ref.13, Ch. 12, pp. 157-167.
16. Shul'pin, G.B. *J. Mol. Catal. A: Chem.* **2002**, *189*, 39
17. Shul'pin, G.B.; Kozlov, Y.N.; Nizova, G.V.; Süss-Fink, G.; Stanislas, S.; Kitaygorodskiy, A.; Kulikova, V.S. *J. Chem. Soc. Perkin Trans. 2* **2001**, 1351.
18. Nizova, G.V.; Süss-Fink, G.; Stanislas, S.; Shul'pin, G.B. *Chem. Commun.* **1998**, 1885.
19. Kirillov, A.M.; Haukka, M.; Guedes da Silva, M.F.C.; Pombeiro, A.J.L. *Eur. J. Inorg. Chem.* **2005**, 2071.
20. Kirillov, A.M.; Kopylovich, M.N.; Kirillova, M.V.; Haukka, M.; Guedes da Silva, M.F.C.; Pombeiro, A.J.L. *Angew. Chem. Int. Ed.* **2005**, *44, 4345.*
21. Pombeiro, A.J.L.; Kirillov, A.M.; Kopylovich, M.N.; Kirillova, M.V.; Haukka, M.; Guedes da. Silva, M.F.C. *PT 103225 (2005).*

22. Kopylovich, M.N.; Kirillov, A.M.; Baev, A.K.; Pombeiro, A.J.L. *J. Mol. Cat. A: Chem.* **2003**, *206*, 163.
23. Pombeiro, A.J.L.; Kopylovich, M.N.; Kirillov, A.M. *PT 103033 (2003).*
24. Lieberman, R.L.; Rosenzweig, A.C. *Nature* **2005**, *434*, 177.
25. Mishra, G.S.; Pombeiro, A.J.L. *J. Mol. Cat. A: Chem.* **2005**, *239*, 96.
26. Mishra, G.S.; Pombeiro, A.J.L. *Appl. Cat. A: Gen.* **2006**, *304*, 185.
27. Hartman, M.; Ernst, S. *Angew. Chem. Int. Ed.* **2000**, *39*, 888.
28. Nizova, G.V.; Süss-Fink, G.; Stanislas, S.; Shul'pin, G.B. *Chem. Commun.* **1998**, 1885.
29. Bagno, A.; Bukala, J.; Olah, G.A. *J. Org. Chem.* **1990**, *55*, 4284.
30. Zerella, M.; Mukhopadhyay, S.; Bell, A.T. *Org. Lett.* **2003**, *5*, 3193.
31. Reis, P.M.; Silva, J.A.L.; Palavra, A.F.; Fraústo da Silva, J.J.R.; Kitamura, T.; Fujiwara, Y.; Pombeiro, A.J.L. *Angew Chem. Int. Ed.* **2003**, *42*, 821.
32. Pombeiro, A.J.L.; Fraústo da Silva, J.J.R.; Fujiwara, Y.; Silva, J.A.L.; Reis, P.M.; Palavra, A.F. *PT 102859 (2002), WO 2004/037416 A3, 03748820.2-2104-PT0300015* (Europe), *10/532.387* (USA), *2004-546574* (Japan), *850332MP* (China).
33. Nakata, K.; Yamaoka, Y.; Miyata, T.; Taniguchi, Y.; Takaki, K.; Fujiwara, Y. *J. Organometal. Chem.* **1994**, *473*, 329.
34. Taniguchi, Y.; Hayashida, T.; Shibasaki, H.; Piao, D.-G.; Kitamura, T.; Yamaji, T.; Fujiwara, Y. *Org. Lett.* **1999**, *1*, 557.
35. Reis, P.M.; Silva, J.A.L.; Palavra, A.F.; Fraústo da Silva, J.J.R.; Pombeiro, A.J.L. *J. Cat.* **2005**, *235*, 333.
36. Kirillov, A.M.; Haukka, M.; Kirillova, M.V.; Pombeiro, A.J.L. *Adv. Synth. Catal.* **2005**, *347*, 1435.

Chapter 5

Biomimetic Vanadium Complexes and Oxo Transfer Catalysis

Pingsong Wu, Gabriella Santoni, Cornelia Wikete, Falk Olbrich, and Dieter Rehder*

Institut für Anorganische und Angewandte Chemie, Universität Hamburg, 20146 Hamburg, Germany

Chiral allylamino-diethanols H₂L, prepared from allylamine and styreneoxide, react with VO(O*i*Pr)₃ to form trigonal-bipyramidal complexes of composition [VO(OMe)L]. The complexes model the structure of the active center of vanadate-dependent haloperoxidases, and they also model the sulfideperoxidase activity of these enzymes. Reaction of H₂L with NaH/CH₃I yields (CH₃)₂L, which forms chloro-{*bis*(oxyphenylethyl)propylamine}silicate with HSiCl₃.

Introduction

Vanadate-dependent haloperoxidases from marine macro-algae (such as *Ascophyllum nodosum* (*1*) and *Corallina officinalis* (*2*)), and the fungus *Curvularia inaequalis* (*3*) contain vanadate(V), linked covalently to a histidine, and by hydrogen bonds to a variety of amino acid residues in the active site pocket, Figure 1 (a). Vanadium is center of a trigonal bipyramid, with an oxo group in the trigonal plane and an OH in one of the apical positions. The second apical position is occupied by the Nε of the His. The enzyme catalyzes the oxidation, by peroxide, of halide to a Hal⁺ species, presumably hypohalous acid (eqn 1), which can halogenate non-enzymatically a large variety of organic substrates.

© 2007 American Chemical Society

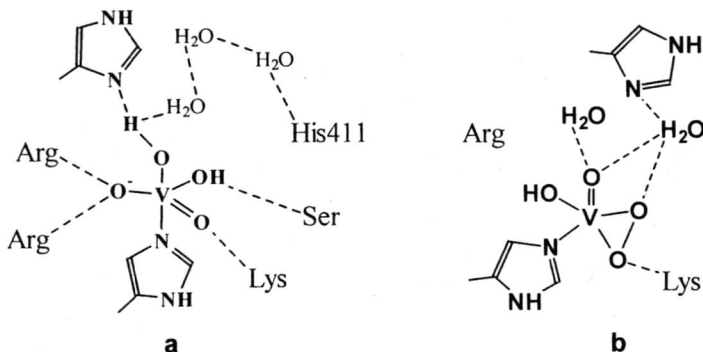

*Figure 1. Native (**a**) and peroxo (**b**) forms of the bromoperoxidase from the A. nodosum enzyme. His 411 is replaced by Phe in the chloroperoxidases from the fungus C. inaequalis.*

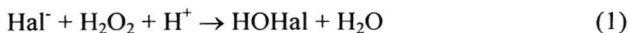

$$Hal^- + H_2O_2 + H^+ \rightarrow HOHal + H_2O \tag{1}$$

$$\tag{2}$$

In the course of the catalytic reaction, a peroxo intermediate is formed, which attains the structure of a distorted square pyramid (*4, 5*), Figure 1 (b). The peroxo group is subject to protonation (*5*) and thus provides a center for nucleophilic attack of the substrate, e. g. bromide, leading to the generation of a hypobromito intermediate (*6*) which is then released in the form of hypobromous acid.

The haloperoxidases also exhibit a sulfideperoxidase activity (*7*) in as far as (prochiral) sulfides are oxygenated to (chiral) sulfoxides plus some sulfone; eqn. 2. This latter reaction is of particular interest, since chiral sulfides are important synthons in organic synthesis (*8*). Consequently, several groups have been working on vanadium-based model systems of the haloperoxidases during the past decade (*9-12*), aiming at the synthesis of chiral enantio-pure sulfoxides. Our approach to this challenge has been the development of chiral, di- to tetradentate aminoalcohols as suitable ligands for the oxovanadium(V) moiety (*13*). In the present communication, we introduce results on aminoalcohols containing an allyl substituent attached to the nitrogen, allowing for further variations of the ligand and thus in the complex periphery. Selected results on the catalytic potential of the vanadium complexes are also reported.

Results and Discussion

Synthesis and Characteristics of Ligands

The reaction of R-styreneoxide and allylamine (molar ratio 2:1) in *iso*propanol yields, after refluxing the mixture for a couple of days and work-up on silicagel (elutant: hexane/ethylacetate, with gradual increase of the polarity) the ligands H_2L_A and H_2L_B (cf. Scheme 1), and thus the products of the 1,3- and 1,2/1,3-cleavage of styreneoxide in 56 and 23% yields, respectively. The crystal structure of H_2L_A (Figure 2, left) indicates that the stereo centers remain in the R configuration.

Reaction of H_2L_A with sodiumhydride plus methyliodide affords the dimethyl ether Me_2L_A, which reacts with trichlorosilane under N_2 in CH_2Cl_2 and in the presence of catalytic amounts of $H_2[PtCl_6]$/*iso*propanol to form the hydrosilylation product with simultaneous formation of two silylether bonds, generating the monochlorosilicon compound $ClSiL_A$, Scheme 1, again with retention of the stereo centers; Figure 2, right. $ClSiL_A$ has also been obtained together with other silicon compounds by hydrosilylation of H_2L_A. Characteristic ^1H NMR data of the four amines are collated in Table 1. The ^{29}Si NMR of $ClSiL_A$ shows a resonance at -112 ppm. The bond lengths $d(Si-Cl) = 2.1699(7)$, $d(Si-N) = 2.0661(17)$, $d(Si-C) = 1.869$ and $d(Si-O) = 1.6546(13)/1.6637(14)$ Å are in the expected range. The silicon center is in a trigonal-bipyramidal environment, with Cl and N in the axis ($\angle Cl-Si-N = 175.98(5)°$); Si deviates from the plane spanned by the two oxygens (O1 and O2) and the carbon C11 by only 0.111 Å towards Cl.

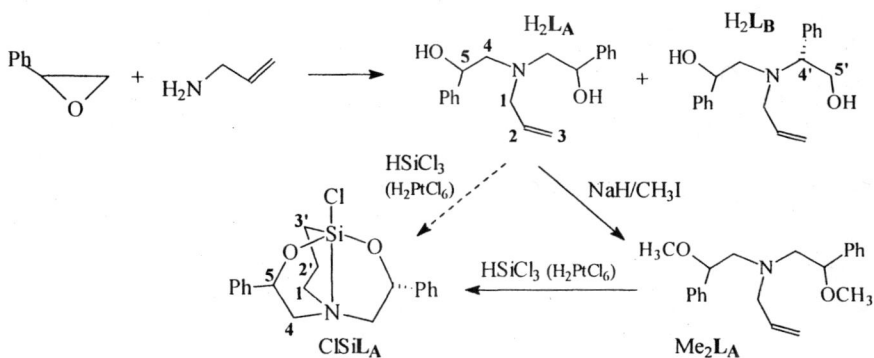

Scheme 1. Synthesis of ligands. Numbers refer to the assignment of ^1H NMR signals (Table 1).

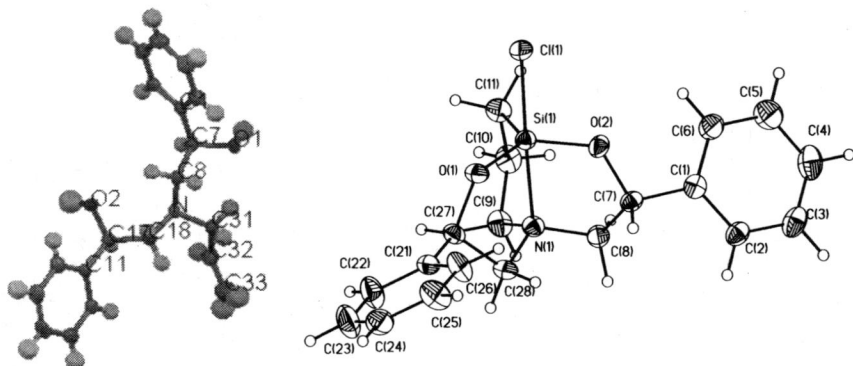

Figure 2. Molecular structures of R,R-H₂Lₐ (left) and R,R-ClSiLₐ (right).

Table 1. ¹H NMR chemical shifts in CDCl₃ or (H₂Lₐ) CD₃CN, and multiplet structures

	1/1'	*2/2'*	*3*	*4/4'*	*5/5'*
H₂Lₐ	3.296 ddd	5.921 m	5.168 dd	2.628 m	4.712 dd
H₂L_B	3.152 ddd	5.775 m	5.116 dd	2.712 ddd / 3.871 dd	4.658 dd / 3.771 ddd
Me₂Lₐ	3.277 ddd	5.746 m	5.065 t	2.840 ddd	4.262 dd
ClSiLₐ	2.896 ddd	1.454 dm	1.620 dm	3.017 ddd	5.409 ddd

NOTE: For signal assignments, confer the atom numbering in Scheme 1. m = multiplet, dd = doublet of doublets, ddd = doublet of doublets of doublets, dm = doublet of multiplets, t = pseudo-triplet.

Synthesis and Characterization of Complexes

The reaction between H₂Lₐ and VO(O*i*Pr)₃ (molar ratio 1:1) in THF under N₂ yields a green solution, from which a geen sticky residue is obtained after removal of the solvent *in vacuo*. The residue is redissolved in dry methanol, undissolved product is filtered off, and the filtrate kept at -20 °C to yield yellow needle-shaped crystals of *orthorhombic* R,R-[VO(OMe)Lₐ] (**1a**) or yellow cubes of *monoclinic* R,R-[VO(OMe)Lₐ] (**1b**). The complex R,S-[VO(OMe)L_B] (**2**) has been prepared accordingly. Both complexes exhibit characteristic ν(V=O) bands at 978 cm⁻¹ (KBr pellets).

Complex **1a** crystallizes in the space group $P2_12_12_1$ ($a = 6.4827(4)$, $b = 7.7631(5)$, $c = 38.15(2)$ Å), **1b** in the space group $P2_1$ ($a = 10.7426(5)$, $b =$

8.5187(4), c = 20.8825(9) Å, β = 97.8070(10)°), and **2** in the monoclinic space group $P2_1$ (a = 11.1327(13), b = 5.9643(7), c = 14.6191 Å, β = 93.347(2)°). The asymmetric unit of **1b** contains two independent molecules. The molecular structures (Figure 3) and structure parameters (Table 2) indicate that, in principle, the complexes attain a trigonal-bipyramidal geometry, with the methoxo group and the nitrogen function in the axis. The angular parameter τ (0 for an ideal square pyramid, 1 for an ideal trigonal bipyramid) amounts to 0.66 for complex **1a**, 0.75 and 0.77 for the two independent molecules of **1b**, and 0.53 for complex **2**; the complexes are thus somewhat distorted towards the square pyramid, comparable to compounds derived from aminocarboxylates (*13b*) or benzylamines (*13a*). The distances d(V-N) are comparatively long as a consequence of the *trans* influence imparted by the methoxy group. The particularly long d(V-N) in the case of complex **2** (the complex with the asymmetrical ligand **L$_B$**), along with the particularly large deviation from the trigonal-bipyramidal geometry, may reflect steric crowding.

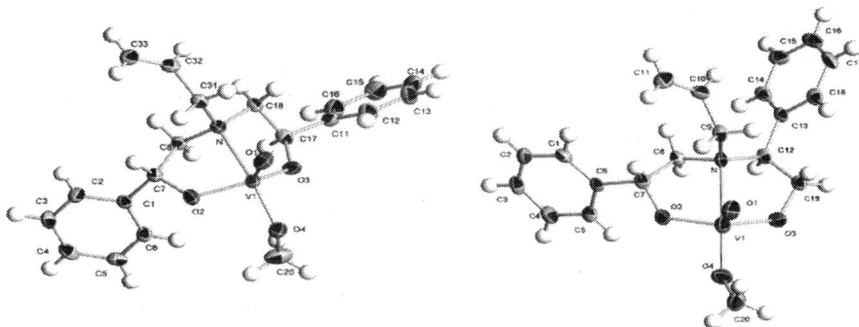

Figure 3. Molecular structures of the complexes
[VO(OMe)L$_A$] (1, left) and [VO(OMe)L$_B$] (2).
(See page 1 of color inserts.)

In CDCl$_3$ solution, the complexes exhibit three ^{51}V NMR signals in the case of complex **1** (at -418, -447 and -455 ppm relative to VOCl$_3$; intensity ratio 1:0.24:0.13), and two signals (at -443 and -458, intensity ratio 1:0.5) in the case of complex **2**. As we have shown previously (*13*), observation of more than one signal is a common feature for this family of oxovanadium complexes, denoting the presence of isomers and thus structural flexibility in solution. DFT calculations carried out on comparable complexes carrying the CH$_2$CO$_2$Me substituent on the nitrogen function suggest the presence of the isomers [VO$_{eq}$(OMe)$_{ax}$L] and [VO$_{ax}$(OMe)$_{eq}$L], Scheme 2, which differ in energy by only a few kJ/mol (*13b*).

**Table 2. Selected bond lengths (Å) and angles (°) for the complexes
[VO(OMe)L$_A$] (1) and [VO(OMe)L$_B$] (2)**

	1a	1b	2
V-O1	1.5914(14)	1.6037(15) / 1.5966(14)	1.600(3)
V-O2	1.8251(13)	1.8086(13) / 1.8133(13)	1.830(3)
V-O3	1.8127(14)	1.8002(13) / 1.8066(13)	1.820(3)
V-O4	1.7930(14)	1.7849(15) / 1.7775(14)	1.759(3)
V-N	2.2251(16)	2.3526(17) / 2.3421(17)	2.413(3)
O4-V-N	164.75(7)	168.60(6) / 169.54(6)	163.88(14)
O2-V-O1	113.33(7)	114.16(7) / 114.96(7)	110.92(15)
O2-V-O3	125.15(7)	123.72(7) / 123.52(7)	132.19(14)
O2-V-O4	96.83(6)	97.91(7) / 98.10(6)	98.61(14)
O2-V-N	77.94(6)	77.27(6) / 77.20(6)	75.48(12)

NOTE: The two entries for **1b** (the monoclinic variant of **1**) correspond to the two independent molecules in the asymmetric unit

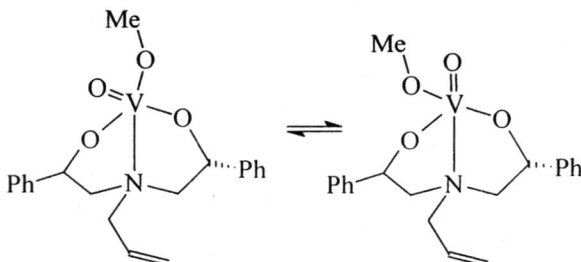

Scheme 2. The complex [VO(OMe)L$_A$] with the methoxo substituent in the (energetically slightly more stable) axial or equatorial position.

Sulfoxygenation reactions

The overall reaction scheme for the sulfoxygenation of benzylphenylsulfide (BzPhS) by cumylhydroperoxide (CHP) in the presence of vanadium catalyst is depicted in Scheme 3

The catalytically conducted reaction in dichloroethane leads to the sulfoxide as the main products and varying amounts of sulfone as a by-product. The catalyst {V}, either the genuine complex [VO(OMe)L$_A$] or the *in situ* system H$_2$L + VO(OiPr)$_3$ (intermediately forming [VO(OiPr)L]) primarily reacts with the oxidant CHP to form the cumylperoxo complex, which transfers an oxo anion to the sulfide sulfur. The catalyst ends up as [VO(Ocumyl)L]. There is [51]V NMR evidence for all of the intermediates formulated in Scheme 3. In particular, the resonances for the peroxo intermediates are shifted to higher field with respect to the *iso*propoxo complexes by ca. 70 ppm. The end-on coordination of the cumylperoxo group in our complexes is suggested by DFT calculations (*13b*). In contrast, the organic peroxo ligand in the seven-coordinate vanadium complex [VO(OO*i*Pr)(dipic)(H$_2$O)] (dipic = 2,6-dipicolinate(2-)), has been shown to coordinate in the side-on fashion (*14*).

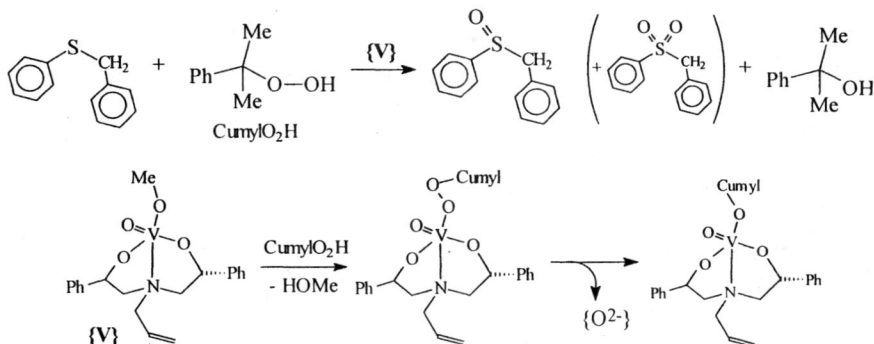

Scheme 3. Conversion of sulfide to sulfoxide (and sulfone) in the presence of the catalyst {V} (upper trace), and the conversion of the catalyst (lower trace).

Table 3 provides selected results on the sulfoxygenation of BzPhS, at 0 °C. With the *in situ* systems, selectivity is better than with the genuine complex. On the other hand, the enentiomeric excess (*e.e.*) is better for [VO(OMe)L$_A$] than for VO(O*i*Pr)$_3$ + H$_2$L$_A$, but compares to the *e.e.* for the system VO(O*i*Pr)$_3$ + H$_2$L$_B$. Reaction times are considerably faster at room temperature, at the expense, however, of the *e.e.*s (not shown). The overall moderate *e.e.*s reflect the fact that – as in the case of the alkoxo complexes (Scheme 2) – the active peroxo complexes are present in solution in the form of (at least) two isomers, viz. [VO$_{eq}$(O$_2$R)$_{ax}$L] (the thermodynamically more stable isomer) and [VO$_{ax}$(O$_2$R)$_{eq}$L].

Conclusion

The complexes [VO(OMe)L], where L is a tridentate ligand providing two alkoxo and one tertiary amine function, are *structural* models of the active center of vanadate-dependent haloperoxidases (VHPO), because they

- Form (distorted) trigonal bipyramids;
- Contain an O$_4$N donor set, with the axial ligands, MeO$^-$ and NR$_3$, mimicking the axial HO$^-$ and N(His) in VHPO;
- Provide chiral centers otherwise present in the active site protein pocket of the VHPO.

Table 3. Results of the catalytically conducted sulfoxygenation of benzylphenylsulfide.

Catalyst	BzPhSO:BzPhSO$_2$	e.e. (%)	Consumption of sulfide (%)	Time (min)
H$_2$L$_A$ + VO(OiPr)$_3$	88:12	13	100	35
[VO(OMe)L$_A$]	74:26	57	100	25
H$_2$L$_B$ + VO(OiPr)$_3$	90:10	67	98	75

NOTE: Molar ratio catalyst:sulfide:oxidant = 0.1:1:1, initial concentration of sulfide 0.1 M; solvent: 1,2-dichloroethane, temperature 0 °C. The enantiomeric excess of the sulfoxide (e.e. of the R-enantiomer) was determined by NMR after addition of pirkle alcohol.

In addition, the complexes are *functional* models, since they mimic the oxygenation, by peroxide, of prochiral sulfides to chiral sulfoxides. This is achieved through a peroxo intermediate, providing the site for nucleophilic attack by the substrate, the approach of which is possibly facilitated by hydrophobic contacts between the phenyl substituents in the ligand periphery of the complex and on the substrate. In contrast to the enzyme, our complexes and the peroxo intermediates formed thereof are structurally flexible in solution, deteriorating the e.e.s found for the oxygenation of RSR' by VHPO (*15*), drastically ameliorating, however, the turn-over rates observed for the sulfoxygenation by the enzyme (*15*).

Acknowledgement

This work was supported by the German Research Community (DFG) in the frame of the Graduate School 611.

References

1. Weyand, M.; Hecht, H.-J.; Kieß, M.; Liaud, M.-F.; Vilter, H.; Schomburg, D. *J. Mol. Biol.* **1999**, *293*, 595-611.
2. Isupov M. I.; Dalby, A. R.; Brindley, A. A.; Izumi, Y.; Tanabe, T.; Murshudov, G. N.; Littlechild, J. A. *J. Mol. Biol.* **2000**, *299*, 1035-1049.
3. Messerschmidt, A.; Wever, R.; Proc. Natl. Acad. Sci. USA 1996, 93, 392-396.
4. Messerschmidt, A.; Prade, L; Wever, R. *Biol. Chem.* **1997**, *378*, 309-315.

5. Zampella, G.; Fantucci, P.; Pecoraro, V. L.; De Gioia, L. *J. Am. Chem. Soc.* **2005**, *127*, 953-960.

6. Conte, V.; Di Furia, F.; Moro, S. *J. Phys. Org. Chem.* **1996**, *9*, 329-336.

7. ten Brink, H. B.; Holland, H. L.; Schoemaker, H. E.; van Lingen, H.; Wever, R. *Tetrahedron: Asymmetry.* **1999**, *10*, 4563-4572.

8. (a) Andersen, K. K. in: Patai, S.; Rappoport, Z.; Stirling, C. J. M. (Eds.) *The Chemistry of Sulfones and Sulfoxides*, John Wiley Sons, Chichester, **1998**, ch. 3 and 16. (b) Carreño, M. C. *Chem. Rev.* **1995**, *95*, 1717.

9. Nakajima, K.; Kojima, M.; Kojima, K.; Fujita, J. *Bull. Chem. Soc. Jpn.* **1990**, *63*, 2620.

10. Bolm, C.; Bienewald, F. *Angew. Chem. Int. Ed.* **1995**, *34*, 2640.

11. Kagan, H. B. in: Ojima, I. (Ed.) *Asymmetric Oxidation of Sulfides*, Wiley-VCH, New York, **2000**, ch. 6c.

12. Smith, II, T. S.; Pecoraro, V. L. *Inorg. Chem.* **2002**, *41*, 6754-6780.

13. (a) Santoni, G.; Licini, G.; Rehder, D. *Chem. Eur. J.* **2003**, *9*, 4700-4708. (b) Wikete, C.; Wu, P.; De Gioia, L.; Licini, G.; Rehder, D. *Inorg. Chem.*, submitted.

14. Mimoun, H.; Chaumette, P. ; Mignard, M.; Saussine, L.; Fischer, J.; Weiss, R. *Nouv. J. Chim.* **1983**, *7*, 467-475.

15. (a) Martinez, J. S.; Croll, G. L.; Tschirret-Guth, R. A.; Altenhoff, G.; Little, R. D.; Butler, A. *J. Am. Chem. Soc.* **2001**, *123*, 3289-3294. (b) ten Brink, H. B.; Holland, H. L.; Schoemaker, H. E.; van Lingen, H.; Wever, R. *Tetrahedron: Asymmetry* **1999**, *10*, 4563. (c) ten Brink, H. B.; Schoemaker, H. E.; Wever, R. *Eur. J. Biochem.* **2001**, *268*, 132. (d) Andersson, M. A.; Allenmark, S. G. *Tetrahedron* **1998**, *54*, 15293.

Chapter 6

Coordination Chemistry and Applications of Vanadium Alkoxides in Catalysis

Esther C. E. Rosenthal, Huiling Cui, and Juliane Koch

Department of Chemistry, Technische Universität Berlin, 10623 Berlin, Germany

The synthesis of oxovanadium alkoxides with 2-aryloxy- and 2-alkoxyethanol ligands and their use for the polymerization and oxidation of unfunctionalized olefins is described. The catalytic epoxidation of the complexes with *tert*-butyl hydroperoxide is discussed. Catalytic polymerizations in combination with different cocatalysts and methyl trichloro acetate as promoter generate different reaction products than without promoter.

Introduction

The present paper contains new results on the synthesis and application of oxovanadium alkoxides with ether alcohol ligands. For a better understanding of the recent results a summary of earlier work by the authors (*1-3*) is included, completed by an extended discussion. The ligands studied (Figure 1) were chosen because of their proposed ability to stabilize the obtained vanadium compounds in their high oxidation states by forming chelating structures with the additional oxygen donor when complexed to vanadium. With this approach we were able to synthesize compound classes, which were previously inaccessible and to prove their existence by single crystal x-ray diffraction (*2-4*). The presence of remaining chloride ligands in the mixed alkoxo chloro complexes obtained makes the metal atom functionalizable e.g. to introduce alkyl groups.

70   © 2007 American Chemical Society

Chloro ligands either in the catalysts or in the cocatalysts are also essential for olefin polymerization (5). Heterometallic alkyl alkoxides derived from the vanadium alkoxide precatalyst and the alkyl aluminum cocatalysts have been postulated to act as the active catalysts in Ziegler-Natta polymerization with such systems (6). With regard to an approximation to the so far unidentified active species in vanadium-based homogeneous Ziegler-Natta polymerization, alkylation reactions as well as reactions of the vanadium complexes with Lewis bases have been studied. Our catalytic investigations on this class of compounds revealed activity for the oligomerization of styrene, 1-decene and 1-hexene and for the polymerization of styrene and ethene as well as for the epoxidation of unfunctionalized olefins.

In the following part, the syntheses of the oxovanadium complexes as well as the catalytic reactivities of the vanadium complexes will be discussed in detail. For a more complete description of the systems, including discussion of spectral and structural data, and for experimental details, the reader is referred to references 1–3.

Chlorooxovanadium Alkoxides

Although mixed oxovanadium(V) alkoxide chlorides are known since 1913 (7), knowledge of the structural features of the systems in the solid state was limited prior to our work as the chloro alkoxides usually are moisture and light sensitive liquids, which have only a limited lifetime even when stored under nitrogen (8–14). Therefore no such compound derived from a mono alcohol has been structurally characterized without additional donor coordination and the corresponding rise of the coordination number to five or six (15–19). This prompted us to choose Lewis base free reaction media for the following reactions in order to avoid the coordination of external solvent molecules to the vanadium center.

Indeed tetrahedral coordination at the V(V) center could be generated (Figure 1) by equimolar reaction of VOCl$_3$ with 2-phenoxyethanol in hexane giving the moderately air-sensitive, orange [VOCl$_2$(OCH$_2$CH$_2$OPh)] 1, which is the first simply four-coordinated dichlorooxovanadium(V) alkoxide to be structurally characterized (1). Electronic saturation of the Lewis acidic V(V) center in this neither sterically nor coordinatively saturated complex is reached by rather short bond lengths and through a second, non-bonding coordination sphere formed by 3 neighboring molecules and the phenyl ether, whose electronegative atoms lie above the faces of the coordination tetrahedron.

The analogous reaction with 2-alkoxyethanols yields the octahedrally coordinated chloride bridged dimers 2–7 in Figure 1 (2, 3). Use of the corresponding lithium alcoholates is a synthetic alternative in all cases. With exception of the benzyl derivative 5 the solid state structures of all alkoxides have been proven by single crystal x-ray diffraction and are the first of chloride bridged oxovanadium(V) alkoxides. Whereas in the monomeric **1** electronic saturation of the Lewis acidic V(V) center results from the second coordination sphere complexes **2–7** dimerize and draw additional electron density from intramolecular donor coordination: the two bridging chloride ligands build a shared edge of 2 VO$_3$Cl$_3$ octahedrons, which incorporate a five-membered chelate ring. The unequal coordination behaviour results from the different σ donor bond strengths of the ether functions. No bonding interaction is observed with the weakest σ donor, the phenyl ether. Consistently, the shortest vanadium ether oxygen bond within the series of dimeric complexes **2–7** is found for the strongest σ donor, the cyclic ether.

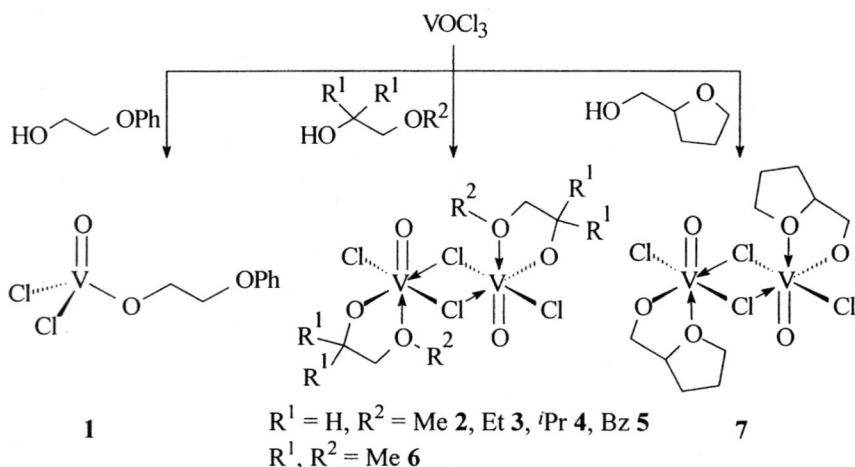

Figure 1. Different coordination geometries for equimolar stoichiometry.

The chloride bridges in the dimeric complexes **2–7** can easily be split with donor solvents (*3*). The compounds react with THF (Figure 2) forming monomeric octahedrally coordinated V(V) complexes **8–11** by breaking of the chloride bridge and simultaneous coordination of a THF molecule. This reaction is completely reversible. As the crystals of all THF complexes contain only the *cis* isomers as racemic mixtures the mechanism of the reaction is believed to be associative. With this information in mind the high solubility of the monomeric phenyl ether derivative **1** in polar organic solvents like ethers or pyridine should also be ascribable to the coordination of these Lewis bases to the Lewis acid V(V). On the other hand, solvolysis of the V–Cl bonds in **1** is observed in alcohols, which becomes clearly visible by the high field shift of the original ^{51}V NMR signal (*1*, *14*, *20–22*). Although the phenoxy complex reacts with water and alcohols to form substituted compounds no crystallizable material could be isolated from the reaction with THF. On storing the reaction mixture at –30 °C the previously orange solution slowly turns blue, clearly indicating reduction of the V(V) to lower valent vanadium species.

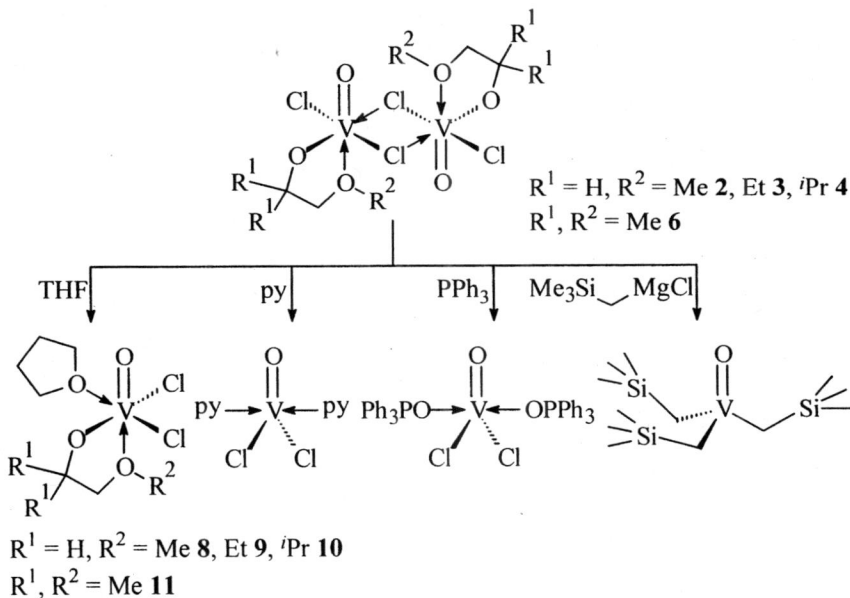

R^1 = H, R^2 = Me **2**, Et **3**, *i*Pr **4**
R^1, R^2 = Me **6**

R^1 = H, R^2 = Me **8**, Et **9**, *i*Pr **10**
R^1, R^2 = Me **11**

Figure 2. Reactions of the chloride bridged dimers.

In opposition to ethers as Lewis basic molecules nitrogen and phosphorus donors reduce the V(V) to low valent species and completely separate it from the alkoxide ligand (*3*). In case of pyridine as nitrogen donor the alkoxide ligand is oxidized to the corresponding aldehyde, whereas in case of triphenylphosphine as phosphorus donor the phosphorus is oxidized to give triphenylphosphine oxide as ligand coordinated to the V(IV) center (Figure 2). Besides Lewis bases the dimers **2–7** have been reacted with different alkylation reagents: against phenyl ethyne the dimers show inert behaviour, alkyl lithium and potassium compounds reduce the V(V) to complex product mixtures. Reaction with Me_3SiCH_2MgCl yields the tris(alkyl) complex [$VO(CH_2SiMe_3)_3$] (Figure 2) again by complete separation of the alkoxide ligand. For this compound the first structural characterization of an oxovanadium tris(alkyl) was achieved.

Catalytic Investigations

Oxidation

Vanadium(V) centers are usually strong Lewis acids as a consequence of the low radius/charge ratio, which makes them suitable for the activation of peroxidic reagents (*23*). Accordingly, V(V) complexes have been found to act as catalyst precursors in various oxidation reactions (*24*). The monomeric tetrahedral **1** and the dimeric octahedral complexes **2–4**, **6** and **7** catalyze the epoxidation of unfunctionalized olefins with *tert.*-butyl hydroperoxide with moderate to good yield (Table I) without any significant dependency on the coordination number at vanadium, the nuclearity of the complexes and the ether function involved. One explanation of the similar activities of the distinct catalysts might be that the structural discrepancy of the single compounds is located too far from the reactive center, especially if the coordination of the ether function is not maintained under the reaction conditions applied.

Table I. Epoxidation of *cis*-Cyclooctene in Dichloromethane

Catalyst	Conversion (%)
1 [VOCl$_2$(OCH$_2$CH$_2$OPh)]	62
2 [VOCl$_2$(OCH$_2$CH$_2$OMe)]$_2$	60
3 [VOCl$_2$(OCH$_2$CH$_2$OEt)]$_2$	68
4 [VOCl$_2$(OCH$_2$CH$_2$O^{i}Pr)]$_2$	70
6 [VOCl$_2$(OCMe$_2$CH$_2$OMe)]$_2$	72
7 [VOCl$_2$(OCH$_2$*cyclo*-{CHO(CH$_2$)$_3$})]$_2$	53

Conditions: catalyst/olefin/^{t}BuOOH = 0.01:1:1.5, t = 24 h, T = 40 °C.

Polymerization and Oligomerization

Introductory Remarks

Since Carrick employed vanadium for ethene polymerization (*25*), an increased interest focused on vanadium compounds as catalysts in Ziegler-Natta polymerization of olefins (*5*, *26*). Vanadium compounds in combination with alkyl aluminum halides were shown to produce soluble active species possessing desirable properties as single site catalysts (*27*). Oxygen donors are able to modify the active vanadium centers and produce polyethenes with narrower molecular mass distributions (*28*). Due to their overall low activity arising from ligand abstraction and reduction to V(II) utilization of vanadium catalysts is highly restricted. Therefore polymerization has to be performed either at low temperature or in the presence of a promoter able to continuously reoxidize V(II) to vanadium species of higher oxidation states during the polymerization (5).

1-Hexene

Conversion of 1-hexene has been conducted over 24 h at room temperature without additional solvent employing different precatalysts (Figure 3), diethyl aluminum chloride (DEAC) as cocatalyst and methyl trichloro acetate (MTCA) as promoter. Oligomerization rather than polymerization took place since oils were the only isolated products. Reaction times have not been optimized, but shorter reaction times lead to significantly lower yields of oligomers. Whereas the nuclearity of the precatalysts seems to play no role concerning the activity, the different behaviour of the *iso*-propyl derivative 4 is remarkable.

Styrene

The polymerization of styrene was first investigated without the use of any promoter. Indeed monomeric 1 initiates styrene polymerization without promoter at room temperature on adding a variety of cocatalysts (Table II) to the toluene solution of precatalyst and monomer. Data from gel permeation chromatography analysis show high polydispersities for the polystyrene obtained. With $^n Bu_2 Mg$ as cocatalyst the polydispersity is especially high. One reason for the high values is the bimodal distribution of the polymers. Two maxima are found for the cocatalysts based on magnesium and aluminum, the one at higher molecular mass being the one with lower percentage. In case of the Grignard reagent as cocatalyst the yield can be significantly increased by extended reaction times (1) indicating very slow polymerization. Similar results were obtained with the dimeric ethyl derivative 3 as precatalyst (c(V) = 3 mmol l^{-1}, 3/AlEt$_3$/olefin = 1:2:220), which gave only 4 % polystyrene (M_w = 17400 g mol^{-1}, M_n = 8700 g mol^{-1}, M_w / M_n = 2.0) even after 12 h reaction time.

The overall very low yields imply a high optimization potential of the styrene polymerizations under investigation. Therefore a promoter has been employed for further studies. Indeed yields for dimeric methyl derivative 2 as precatalyst with methyl alumoxan (MAO) as cocatalyst and MTCA as promoter lie at significantly higher values—quantitative conversion is already reached at room temperature, but low molecular masses show that only styrene oligomers have been obtained (2). In opposition to the promoter-free reactions the catalytic activity decreases with rising temperatures. The monomeric precatalyst 1 behaves similar. When both were tested for their activity for 1-decene

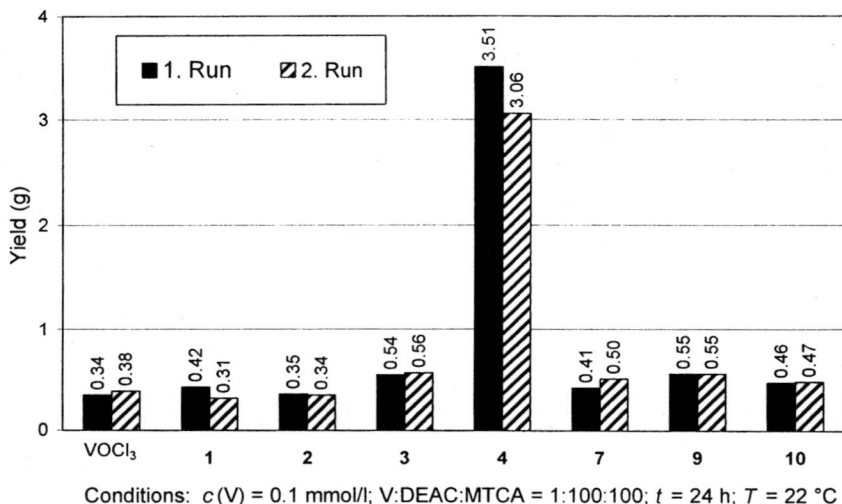

Figure 3. Oligomerization of 1-hexene. (Data are in part from reference 3.)

Table II. Polymerization of Styrene with [VOCl₂(OCH₂CH₂OPh)] 1

Cocatalyst	Yield (%)	M_w (g mol^{-1})	M_n (g mol^{-1})	M_w / M_n
Me₃SiCH₂MgCl	5	258000	65000	4.0
ⁿBu₂Mg	20	475000	36200	13.1
MAO	6	181000	39800	4.5
Me₂Zn	4	250000	78000	3.2

Conditions: $c(V)$ = 10 mmol l^{-1}, 1/cocatalyst/olefin = 1:2:150, t = 3 h, solvent: toluene.

oligomerization, it turned out that it is only ten percent of that for styrene oligomerization under the same conditions even with the use of a promoter.

Again analogous behaviour of the corresponding dimeric ethyl derivative **3** is observed in styrene oligomerization with MAO as cocatalyst. The results in Table III show that with MTCA as promoter high yields could be obtained even after short reaction times. The best yield is reached at the slightly elevated temperature of 40 °C. Values drop again significantly, when the temperature is raised further. The polydispersities for the oligomers lie around 1.5 and therewith at lower values than for the corresponding methyl derivative **2** (*2*), which might be due to the increased steric demand of the ethyl group compared with the methyl one.

Table III. Oligomerization of Styrene with [VOCl$_2$(OCH$_2$CH$_2$OEt)]$_2$ 3

Temperature (°C)	Yield (%)	M_w (g mol^{-1})	M_n (g mol^{-1})	M_w / M_n
23	65	3400	2400	1.5
40	71	3000	1900	1.5
60	34	3600	2400	1.5
80	30	2500	1800	1.4

Conditions: c(V) = 2 mmol l^{-1}, **3**/MAO/styrene = 1:160:170, t = 1 h, solvent: toluene.

In conclusion the product of the catalytic conversion of styrene depends on the utilization of a promoter: use of MTCA as promoter gave oligomers with moderate yields, while abandonment of the promoter produced polymers with very low yields. In no case has polystyrene or oligostyrene been obtained using the different cocatalysts in the absence of any vanadium species, whilst the use of **1–3** in the absence of a cocatalyst afforded only negligible amounts of conversion products. Altogether the influence of the promoter on the product characteristics, which points in the opposite direction than the yield increasing properties, prevents the polymerization of 1-alkenes. On the other hand, polymerization activities without promoter are too inefficient. Catalysts functioning without a promoter would be a solution.

Ethene

For ethene polymerization with the oxovanadium(V) complexes **1–10** with the cocatalyst MAO higher molecular masses of polyethene above 10^6 g mol^{-1} but lower polymerization activities were found as for DEAC as cocatalyst in direct comparison. This phenomenon has already been described in the literature for other vanadium precatalysts (*5*). Without the promoter MTCA activities are negligible in all cases. Due to the frequently observed at least bimodal molecular mass distributions polydispersities of the polyethene are very broad. The best values concerning activity and polydispersity were obtained for the system 2/DEAC/MTCA = 1:100:500 with a starting concentration of vanadium of 0.02 mmol l^{-1} and a reaction time of 30 min at 2 bar ethene pressure at a temperature of 30 °C, which polymerized ethene with a rate of 560 kg polyethene per mol vanadium, hour and bar to give polyethene with M_w = 1962000 g mol^{-1}, M_n = 443000 g mol^{-1} and a polydispersity of 4.4. Over extended polymerization times the polydispersity declines. Obviously the systems are no true single site catalysts and more than one type of active species is present, which might even change during the course of the polymerization. There is almost no dependency of the polymerization activities on the ether function in the ligand or the nuclearity of the complexes (Figure 4). A very open catalytic site, which is probably far removed from the ether alkoxide ligands, might be responsible for this result. On the other hand, the ether alkoxide ligands are suitable ligands for vanadium-based catalysis of ethene polymerization and have positive effects on the polymerization reaction. As expected from the cylovoltammetric proof of higher stability of the synthesized vanadium alkoxides against reduction in comparison with $VOCl_3$ (3) the production of polyethene with higher molecular mass, lower polydispersity and better polymerization activity has been made possible.

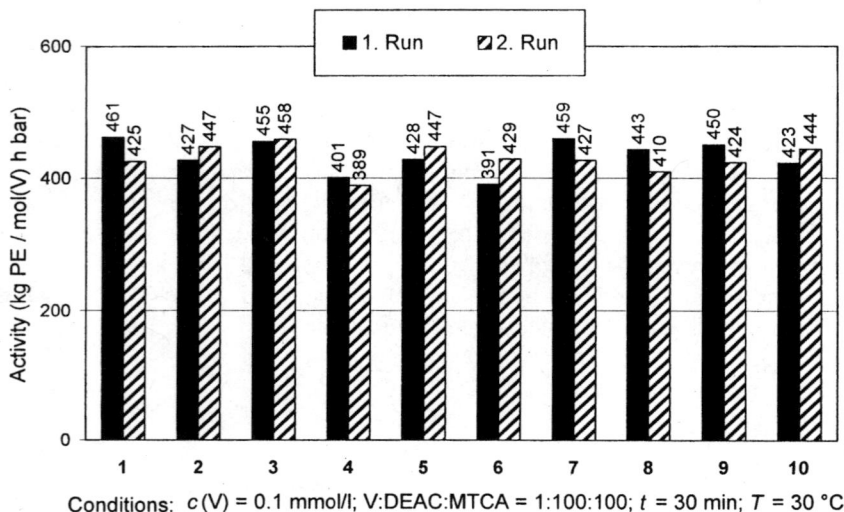

Conditions: c(V) = 0.1 mmol/l; V:DEAC:MTCA = 1:100:100; t = 30 min; T = 30 °C

Figure 4. Polymerization of ethene. (Data are in part from reference 3.)

References

1. Rosenthal, E. C. E.; Girgsdies, F. *Z. Anorg. Allg. Chem.* **2002,** *628,* 1917.
2. Rosenthal, E. C. E.; Cui, H.; Lange, K. C. H.; Dechert, S. *Eur. J. Inorg. Chem.* **2004,** 4681.
3. Rosenthal, E. C. E.; Cui, H.; Koch, J.; Escarpa Gaede, P.; Hummert, M.; Dechert, S. *Dalton Trans.* **2005,** 3108.
4. Foulon, G.; Foulon, J.-D.; Hovnanian, N. *Polyhedron* **1993,** *12,* 2507.
5. Hagen, H.; Boersma, J.; van Koten, G. *Chem. Soc. Rev.* **2002,** *31,* 357.
6. Shuke, J.; Dingshing, Y. *Polymer J.* **1985,** *17,* 899.
7. Prandtl, H.; Hess, L. *Z. Anorg. Chem.* **1913,** *82,* 103.
8. Funk, H.; Weiss, W.; Zeising, M. *Z. Anorg. Allg. Chem.* **1958,** *296,* 36.
9. Mittal, R. K.; Mehrotra, R. C. *Z. Anorg. Allg. Chem.* **1964,** *332,* 189.
10. Choukroun, R.; Dia, A.; Gervais, D. *Inorg. Chim. Acta* **1979,** *34,* 211.
11. Hillerns, F.; Rehder, D. *Chem. Ber.* **1991,** *124,* 2249.
12. Bürger, H.; Smrekar, O.; Wannagat, U. *Monatsh. Chem.* **1964,** *95,* 292.
13. Miles, S. J.; Wilkins, J. D. *J. Inorg. Nucl. Chem.* **1975,** *37,* 2271.
14. Priebsch, W.; Rehder, D. *Inorg. Chem.* **1985,** *24,* 3058.
15. Glas, H.; Köhler, K.; Herdtweck, E.; Maas, P.; Spiegler, M.; Thiel, W. R. *Eur. J. Inorg. Chem.* **2001,** 2075.
16. Hagen, H.; Bezemer, C.; Boersma, J.; Kooijman, H.; Lutz, M.; Spek, A. L.; van Koten, G. *Inorg. Chem.* **2000,** *39,* 3970.
17. Crans, D. C.; Felty, R. A.; Anderson, O. P.; Miller, M. M. *Inorg. Chem.* **1993,** *32,* 247.
18. Chang, Y.; Chen, Q.; Khan, M. I.; Salta, J.; Zubieta, J. *J. Chem. Soc. Chem. Commun.* **1993,** 1872.
19. Crans, D. C.; Felty, R. A.; Miller, M. M. *J. Am. Chem. Soc.* **1991,** *113,* 265.
20. Rehder, D. *Bull. Magn. Reson.* **1982,** *4,* 33.
21. Rehder, D. *Z. Naturforsch.* **1977,** *32b,* 771.
22. Paulsen, K.; Rehder, D. *Z. Naturforsch.* **1982,** *37a,* 139.
23. Conte, V.; Di Furia, F.; Moro, S. *J. Phys. Org. Chem.* **1996,** *9,* 329.
24. Butler, A.; Clague, M. J.; Meister, G. E. *Chem. Rev.* **1994,** *94,* 625.
25. Carrick, W. L. *J. Am. Chem. Soc.* **1958,** *80,* 6455.
26. Gambarotta, S. *Coord. Chem. Rev.* **2003,** *237,* 229.
27. Murphy, V. J.; Turner, H. *Organometallics* **1997,** *16,* 2495.
28. Karol F. J.; Kao S.-C. *New J. Chem.* **1993,** *18,* 97.

Insulin-Enhancing Agents: Compound Design and Mechanism of Action

Chapter 7

Enhancement of Insulin Action by bis(Acetylacetonato)oxovanadium(IV) Occurs through Uncompetitive Inhibition of Protein Tyrosine Phosphatase-1B

Marvin W. Makinen[1,*], Stephanie E. Rivera[1], Katherine I. Zhou[1], and Matthew J. Brady[2]

[1]Department of Biochemistry and Molecular Biology, Gordon Center for Integrative Science, The University of Chicago, 929 East 57th Street, Chicago, IL 60637
[2]Department of Medicine and Committee on Molecular Metabolism and Nutrition, The University of Chicago, 5841 South Maryland Avenue, Chicago, IL 60637

We have examined the influence of *bis*(acetylacetonato)oxo-vanadium(IV) [VO(acac)$_2$] on the catalytic activity of protein tyrosine phosphatase-1B (PTP1B). In the presence of p-nitro-phenylphosphate as the substrate, VO(acac)$_2$ exhibited mixed inhibition. However, VO(acac)$_2$ exhibited uncompetitive inhibition of the enzyme with the undecapeptide substrate DAD-EpYLIPQQG, in which the sequence corresponds to residues 988-998 of the epidermal growth factor receptor and pY indicates the phosphotyrosine residue. These results are consistent with our earlier observations, on the basis of phos-photyrosine immunoblots, showing that VO(acac)$_2$ potentiates tyrosine phosphorylation of the insulin receptor synergistically with insulin. Because uncompetitive inhibitors of PTP1B have not been described heretofore, we discuss the importance of uncompetitive inhibition with respect to design of inhibitors of the enzyme for therapeutic purposes.

© 2007 American Chemical Society

Introduction

Because impaired insulin action is the underlying cause of type 2 diabetes, the search for suitable targets of pharmacologic agents to alleviate the pathophysiology of this condition has necessarily focused on the insulin signaling pathway (*1-3*). Activation of the insulin receptor (IR) on the cell surface by binding of insulin initiates phosphorylation of tyrosine residues through the autophosphorylative and tyrosine kinase activities of the receptor, culminating in the translocation of glucose into the cell from the blood stream by glucose transporter systems (*4-6*). While a series of further downstream phosphorylative events continues the signaling events to result in glycogen formation or fatty acid synthesis, the activated IR must be returned to a basal state to respond anew to the binding of insulin. Protein tyrosine phosphatase-1B (PTP1B) catalyzes the hydrolysis of phosphotyrosine containing peptides (*7-9*), and, as a member of the protein tyrosine phosphatase superfamily of enzymes (*10*), is thought to have a major role in deactivation of the IR (*11,12*). Because there are over 100 known members of this enzyme superfamily, sharing common structural motifs and similar active site residues involved in signal transduction and cellular processes such as growth, differentiation, and proliferation, inhibition of PTP1B must be not only potent but also highly specific.

In earlier studies of the insulin enhancing effects of organic chelates of the vanadyl (VO^{2+}) ion to facilitate glucose uptake by cultured 3T3-L1 adipocytes, we observed that the behavior of *bis*(acetylacetonato)oxovanadium(IV) [VO-acac)$_2$] was synergistic with added insulin, in contrast to that of *bis*(maltolato)oxovanadium(IV) [VO(malto)$_2$] (*13,14*). This observation was made on the basis of phosphotyrosine immunoblots of cell lysates showing a dose-dependent increase in phosphotyrosine levels of the IR and the insulin receptor substrate-1 (IRS-1) after the cells were challenged for glucose uptake in the presence of VO^{2+}-chelates. Because VO(acac)$_2$ elicits an increase in the phosphotyrosine content of the IR and IRS-1 in the presence of wortmannin, a specific inhibitor of phosphatidyl inositol 3'-kinase, we concluded that VO(acac)$_2$ acts directly on one or more enzymes that regulate the phosphotyrosine content of the IR (*14*). To this end, we have investigated the influence of VO(acac)$_2$ on the catalytic properties of human recombinant PTP1B. Our results demonstrate that VO-(acac)$_2$ acts as a mixed inhibitor of PTP1B when *p*-nitrophenylphosphate (pNPP) is employed as the substrate. However, in the presence of the phosphotyrosine containing undecapeptide analog of the epidermal growth factor receptor, residues 988 – 998 having the amino acid sequence DADEpYLIPQQG, VO(acac)$_2$ acts as an uncompetitive inhibitor. Not only is the observation of uncompetitive inhibition unusual, but also the results explain the synergism of VO(acac)$_2$ with insulin and the lack of insulin synergism with VO(malto)$_2$. Because no uncompetitive inhibitor of PTP1B has been reported in the literature, to our knowledge, we discuss the importance of these observations with respect to inhibitor design for therapeutic purposes.

Experimental Procedures

General

Crystalline VO(acac)$_2$ was purchased from Sigma-Aldrich (Milwaukee, WI 53209) and used without further purification. The di-sodium salt of pNPP and D,L-dithiothreitol were obtained from Fluka Chemical Company (Milwaukee, WI 53233) and used directly. All other chemicals were of analytical reagent grade, and doubly distilled, deionized water was used throughout.

The DNA corresponding to the soluble portion of wild type human PTP1B (residues 1-321), kindly provided by Professor Z.-Y. Zhang, was cloned into *E. coli* BL21(DE3) cells following the protocol described from the Zhang laboratory (*8,15*). Engineered cells were grown in liquid Luria-Bertani medium at 37 °C for 12 hours following induction with isopropyl-1-thio-β-galactoside; lysis of cells was achieved by several freeze-thaw cycles using liquid nitrogen followed by 12 – 15 cycles of sonication, each cycle lasting 20 – 30 s. For sonication the cell suspension was kept submerged in ice-water to prevent heating. The cell suspension was then centrifuged at 5000 x *g* for 20 min, the supernatant carefully decanted and mixed with pre-swollen CM-Sephadex C50 according to Zhang and co-workers (*15*). Elution of the enzyme from CM-Sephadex C50 was achieved with a 0.04 M – 0.5 M gradient of NaCl. The pooled fractions containing enzyme determined on the basis of the optical density at 280 nm and the presence of catalytic activity were pooled and concentrated with an Ultrafree centrifugal filter device with a 10,000 molecular weight cut-off (Millipore Corp., Bedford, MA 01730).

Kinetic Studies

Initial velocity data were collected with a Cary 15 spectrophotometer modified by On-Line Instrument Systems, Inc. (Bogart, GA 30622) for microprocessor-controlled data acquisition and equipped with an efficient mixing device for kinetic studies (*16*). Hydrolysis of the substrate pNPP was monitored at 349 nm, experimentally determined to yield the largest change in absorption for monitoring the reaction at pH 5, the pH optimum for enzyme activity (*8*). The ($\varepsilon_{substrate} - \varepsilon_{product}$) difference extinction coefficient was 1911 M^{-1}cm^{-1}. Only freshly prepared stock solutions of pNPP, dissolved in 0.1 M NaCl buffered to pH 5.0 with 0.01 M sodium acetate, were used. The pNPP concentration was varied from $(0.25 - 5)$ x K_M. Crystalline VO(acac)$_2$ was directly dissolved in 0.1 M sodium chloride buffered to pH 5 with 0.01 M acetate and extensively purged with N_2 prior to use. The highest VO(acac)$_2$ concentration in reaction mixtures was 90.0 x 10^{-6} M. Kinetic data were analyzed with use of Origin 5.0, released by Microcal Software, Inc. (Northampton, MA 01060).

The influence of VO(acac)$_2$ on the enzyme-catalyzed reaction was also evaluated with use of the Protein Tyrosine Phosphatase 1B Assay Kit (Calbiochem, La Jolla, CA 92039) based on colorimetric Malachite Green detection of inorganic phosphate released as the product. In this reaction, however, the substrate is the phosphotyrosine containing undecapeptide DADEpYLIPQQG where pY represents the phosphotyrosine residue. The amino acid sequence of this peptide corresponds to residues 988-998 of the epidermal growth factor receptor. Reaction mixtures were set up at 30 °C in 96-well plates and the amount of inorganic phosphate released as product was measured at 620 nm with a Tecan Safire[2] microplate scanner (Tecan AG, CH-8708 Mannedorf, Switzerland) running on XFluor software. Assay measurements were made in triplicate, and reaction conditions at ambient temperature were established to ensure phosphate release within the linear portion of the standard curve for measurement of phosphate concentration. An incubation time of 30 minutes was employed for the reaction prior to quenching of the reaction and addition of Malachite green for color formation.

Results

Steady-State Kinetic Assays

For the steady-state kinetic parameters k_{cat} and K_M, initial velocity data for the hydrolysis of pNPP catalyzed by PTP1B yielded values of 36.5 ± 2.0 s^{-1} and 4.2 ± 0.1 x 10^{-4} M, respectively, at ambient room temperature (~22 °C) . While the value of K_M at pH 5 agrees well with that of ~ 4 x 10^{-4} M reported by Zhang and coworkers (8), our value of k_{cat} is higher. Zhang et al. report a value of ~37 s^{-1} at 37 °C, requiring that the value at 22 °C would be lower. Our experiments were carried out with use of an efficient mixing device (16) with mixing time of $\leq$ 6 s at room temperature. Micro cuvettes were employed in experiments reported by Zhang and co-workers with no mention of the stirring assembly for initiating the reaction (8). Inefficient mixing causes an apparent decrease in initial velocity, and this decrease may be in part the origin of the difference in k_{cat} values.

Figure 1 illustrates the inhibitory behavior of VO(acac)$_2$ in hydrolysis of pNPP catalyzed by PTP1B. As in experiments carried out in the absence of VO-(acac)$_2$, the Lineweaver-Burk plot shows evidence for substrate inhibition. To best evaluate the influence of VO(acac)$_2$, therefore, the data at high substrate concentrations, represented with open symbols, were not used to estimate slope and intercept values. The results indicate that VO(acac)$_2$ acts as a mixed inhibitor. From these results, the K_I of this organic chelate of VO^{2+} is estimated to lie in the $15 - 20 \times 10^{-6}$ M range. In contrast, VO(malto)$_2$ is reported to act as a competitive inhibitor under steady-state conditions with use of the fluorescent substrate 6,8-difluoro-4-methylumbeliferyl phosphate (17). Of special interest in the study by Peters and coworkers is the observation that the hydrated VO^{2+} cation stripped of its organic ligands acts as the competitive inhibitor (17).

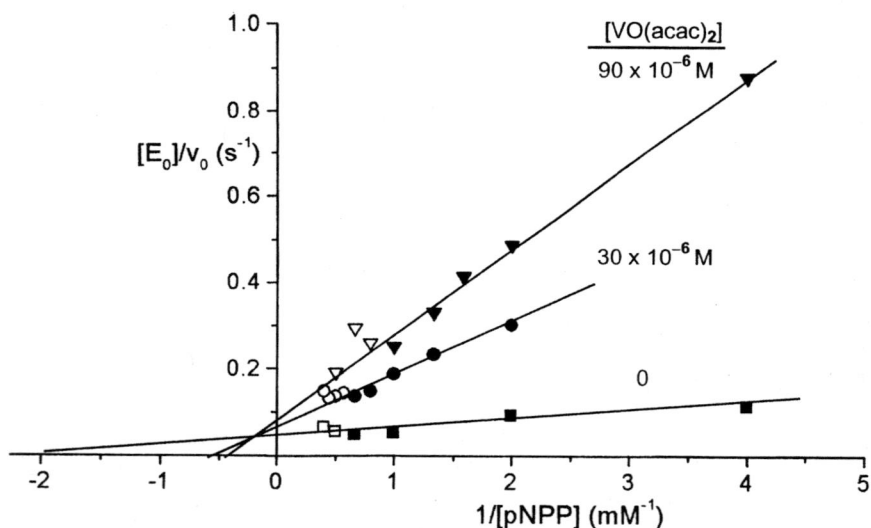

Figure 1. Double reciprocal plots of initial velocity data collected for the hydrolysis of pNPP catalyzed by PTP1B at pH 5. VO(acac)$_2$ concentrations are indicated adjacent to the line of the corresponding least-squares linear fit to each data set.

Malachite Green Colorimetric Assays

Figure 2 illustrates a double reciprocal plot of phosphate release from the phosphotyrosine containing undecapeptide DADEpYLIPQQG, as measured with the Malachite Green colorimetric method. The enzyme commercially supplied with the assay kit similarly showed a tendency towards substrate inhibition. As with use of pNPP as the substrate, data at high substrate concentrations were not used to calculate the slope and intercept of the straight-line graphs. In sharp contrast to the use of pNPP as the substrate, VO(acac)$_2$ exhibits uncompetitive inhibition against the undecapeptide substrate. From these results, the K_I of the VO^{2+}-chelate was estimated to lie in the $5 - 8 \times 10^{-6}$ M range. It is of considerable interest to note that no uncompetitive inhibitor of PTP1B has been previously described.

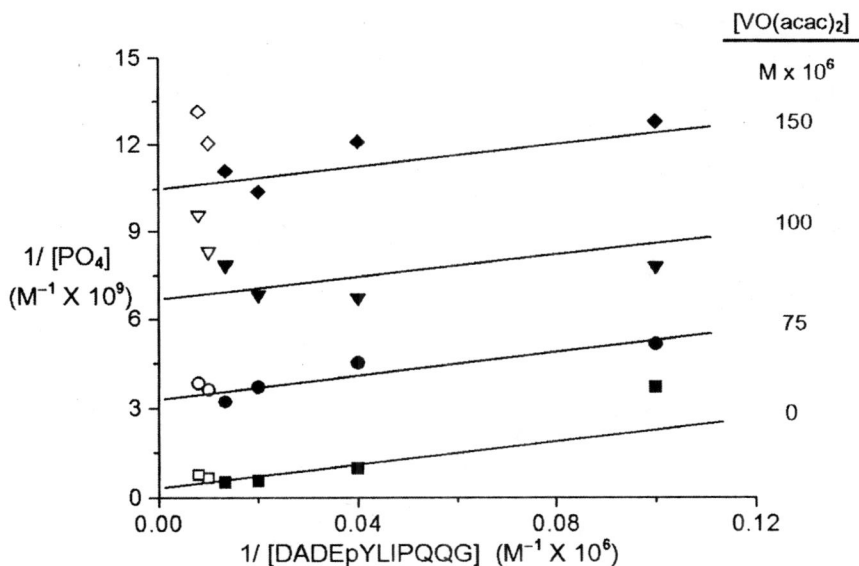

Figure 2. Comparison of the influence of VO(acac)$_2$ on PTP1B catalyzed hydrolysis of the phosphotyrosine containing undecapeptide substrate, as evaluated with use of a Malachite Green colorimetric method. VO(acac)$_2$ concentrations are indicated on the right for each double-reciprocal plot.

Discussion

In contrast to our results in Figure 1, illustrating mixed inhibition of PTP1B by VO(acac)$_2$, Peters and co-workers observed competitive inhibition of PTP1B by VO(malto)$_2$ (*17*). While a small molecule, organic phosphate ester was used as the substrate in the study by Peters and co-workers, as for the results in Figure 1, the significant difference lies in the observation that the hydrated VO^{2+} ion is the competitive inhibitor when VO(malto)$_2$ is used. This result, as noted by Peters *et al.* (*17*), indicates that the VO^{2+} ion is stripped of its organic ligands upon binding to the enzyme. Since the pattern of inhibition in Figure 1 requires a mixture of classical competitive and noncompetitive inhibition, we cannot exclude the possibility that a small fraction of VO(acac)$_2$ is dissociated, releasing some of the VO^{2+} ion to act as a competitive inhibitor in the active site, as in the case of VO(malto)$_2$. However, because the patterns in Figures 1 and 2 are so different from each other and because VO(acac)$_2$ is significantly more stable in aqueous solution than VO(malto)$_2$ (*14,18*), we conclude that the VO(acac)$_2$ complex added to the enzyme reaction mixture remains intact and accounts for the noncompetitive and uncompetitive inhibitory behavior illustrated in Figures 1 and 2, respectively, dependent on the substrate. This observation requires that VO(acac)$_2$ binds as an organic chelate, either as an intact complex or possibly as a hemi-liganded [VO(acac)]$^+$ complex.

Hydrolysis of a phosphotyrosine containing polypeptide substrate catalyzed by PTP1B requires that the phosphate moiety esterified to the hydroxyl group of the tyrosine side chain is positioned into the active site for nucleophilic attack by Cys-215 (*8,15,19*). The steric properties of the phosphate group of pNPP, esterified to the hydroxyl group of *p*-nitrophenol, are essentially identical, ensuring a similar stereochemical approach to the nucleophilic sulfhydryl group. The observation through Figures 1 and 2 that the pattern of inhibition is dependent on substrate structure requires, therefore, that the site occupied by VO(acac)$_2$ as a noncompetitive inhibitor in Figure 1 differs from that occupied by VO(acac)$_2$ as an uncompetitive inhibitor in Figure 2. This conclusion is supported by estimates of the inhibitor constant of VO(acac)$_2$ as a noncompetitive inhibitor (15-20 x 10^{-6} M) and as an uncompetitive inhibitor (5-8 x 10^{-6} M) extracted from the graphical plots. While VO(acac)$_2$ as a noncompetitive inhibitor may bind near the pNPP substrate, the more extended steric volume of the undecapeptide substrate probably blocks this binding site, forcing the VO^{2+}-chelate to find a different binding location on the enzyme, resulting in uncompetitive inhibition. While definition of the binding site of VO(acac)$_2$ on the enzyme as an uncompetitive inhibitor awaits X-ray crystallographic analysis, its identification through Figure 2 suggests that this site may be important as a focus of inhibitor design for therapeutic purposes.

Because the structural relationships of backbone and side chain atoms located in the immediate vicinity of catalytically active residues of PTP1B are well defined through X-ray crystallography, the focus of inhibitor design studies has been entirely directed towards antimetabolites that compete with the substrate in the active site (*1,2,19-22*). This approach constitutes a large fraction of the literature on structure based drug design in general. It has been demonstrated, however, that in *open* systems, in which there is constant input of substrate and removal of products, as in cells, this strategy is unlikely to be effective (*23,24*). An approach based on uncompetitive inhibition is far superior when the efficacy of inhibition of metabolic reactions is considered. With respect to kinetic relationships, the differences between these two approaches and the greater effectiveness of an uncompetitive inhibitor over competitive inhibitors are illustrated in Figure 3.

Figure 3 compares the inhibition patterns of competitive and uncompetitive inhibitors and their characteristic hyperbolic plots of relative reaction velocity *vs.* substrate concentration. The decrease in reaction velocity at any given substrate concentration is also illustrated according to the ratio of the concentration of the inhibitor to the value of the inhibitor binding constant. The hallmark of competitive inhibition is that the inhibitor bound in the active site can be displaced by substrate, particularly when there is constant input of substrate into the system, as occurs in the cell, with only a fixed amount of inhibitor. Under these conditions, competitive inhibitors cannot be expected to provide effective, long-term inhibition. On the other hand, the kinetic scheme for uncompetitive inhibition shows that the inhibitor has affinity for *only* the enzyme-substrate complex. Because open systems are distinguished by constant input of substrate, correspondingly, with build-up of enzyme-substrate complex, the inhibition gains in effectiveness in the long-term because there is an increasing amount of the enzyme-substrate complex as the only target of the inhibitor. For competitive inhibition with a fixed inhibitor concentration, the relative velocity increases as the substrate concentration increases because the inhibitor is displaced. On the other hand, for a fixed amount of an uncompetitive inhibitor, the relative velocity quickly reaches a plateau value as the substrate concentration increases, resulting in sustained inhibition as a long-term effect.

The different patterns illustrated by the hyperbolic plots in Figure 3 for competitive and uncompetitive inhibition help to explain our observations that the insulin-enhancing capacity of $VO(acac)_2$ is synergistic with insulin while that of $VO(malto)_2$ is not. In measurements of glucose uptake by cultured 3T3-L1 adipocytes, we observed that the level of the phosphorylated form of the IR, detected on the basis of phosphotyrosine-specific antibodies, was greater when elicited by $VO(acac)_2$ and insulin together in the incubation medium than when elicited by either $VO(acac)_2$ or insulin alone (*14*). According to Figure 3, a ternary ESI complex is formed through the action of $VO(acac)_2$ as an uncompetitive inhibitor, in which the substrate is the (tyrosine) phosphorylated form of the IR.

Designating phosphorylated IR as pY(IR) and the ESI ternary complex as $[VO(acac)_2 : PTP1B : pY(IR)]$, we can readily see that addition of insulin generates pY(IR) which becomes "trapped" in the form of the ternary complex

90

Competitive Inhibition

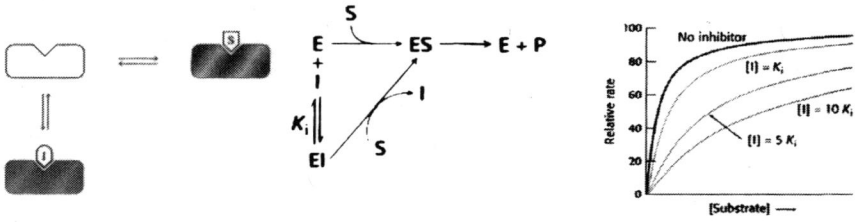

Uncompetitive Inhibition

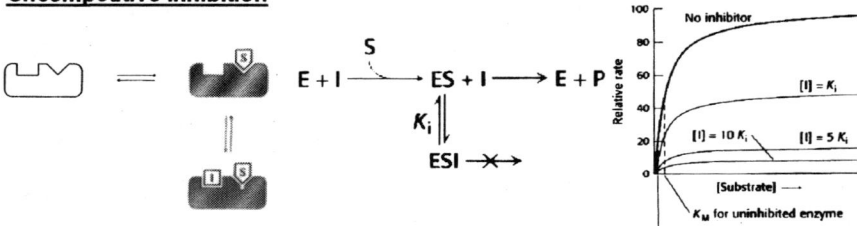

Figure 3. Comparison of reaction schemes and hyperbolic plots of relative reaction velocity vs. substrate concentration for competitive and uncompetitive inhibition of an enzyme catalyzed reaction.

[VO(acac)$_2$: PTP1B : pY(IR)] on top of that formed in the presence of VO-(acac)$_2$ alone. On the other hand, because VO(malto)$_2$ is effectively a competitive inhibitor (*17*), addition of insulin to the incubation medium containing this VO^{2+}-chelate generates more pY(IR) that simply displaces VO^{2+} from the enzyme, allowing the increased quantity of pY(IR) to be hydrolyzed by PTP1B. Thus, as confirmed through experiment in the cell, the level of pY(IR), detected on the basis of phosphotyrosine-specific immunoblotting, is unchanged with increased addition of insulin in the presence of VO(malto)$_2$ and remains essentially identical to that detected in the presence of insulin alone (*14*).

This very different pattern of uncompetitive inhibition arises because the inhibitor has affinity *only* for the enzyme-substrate complex. Because such complexes are not usually described through X-ray structural analysis, design of an uncompetitive inhibitor of an enzyme is invariably more difficult than design of a competitive inhibitor. Nonetheless, the effort is likely to be more rewarding. As pointed out by Cornish-Bowden (*23*), an uncompetitive inhibitor of an enzyme that makes only a very small contribution to setting the flux of metabolites through a pathway is much more likely to have a major pharmacological effect than a competitive inhibitor.

Acknowledgments

We thank Professor Z.-Y. Zhang for providing the DNA for overexpression of PTP1B. This work was supported by a grant of the National Institutes of Health (DK064772) awarded to M. J. B.

References

1. Sarmiento, M.; Wu, L.; Keng, Y.-F.; Song, L. Luo, Z.; Huang, Z.; Wu, G.-Z.; Yuan, A.K.; Zhang, Z.-Y. *J. Med. Chem.* **2000**, *43*, 146-155.
2. Doman, T. N.; McGovern, S. L.; Witherbee, B. J.; Kasten, T. P.; Kurumbail, R.; Stallings, W. C.; Connolly, D. T.; Shoichet, B. K. *J. Med. Chem.* **2002**, *45*, 2213-2221.
3. Ramachandran, C.; Kennedy, B. P. *Curr. Topics Med. Chem.* **2003**, *3*, 749-757.
4. Sun, X. J.; Rothenberg, P. C.; Kahn, R.; Backer, J. M.; Araki, E.; Wilden, P. A.; Cahill, D. A.; Goldstein, B. J.; White, M. F. *Nature* **1991**, *352*, 73-77.
5. Miralpeix, M.; Sun, X. J.; Backer, J. M.; Myers, M. G., Jr.; Araki, E.; White, M. F. *Biochemistry* **1992**, *31*, 9031-9039.
6. Yenush, L.; White, M. F. *BioEssays* **1997**, *19*, 491-500.
7. Guan, K.; Dixon, J. E. *J. Biol. Chem.* **1991**, *266*, 17026-17030.
8. Zhang, Z.-Y.; Thieme-Sefler, A. M.; Maclean, D.; McNamara, D. J.; Dobrusin, E. M.; Sawyer, T. K.; Dixon, J. E. *Proc. Natl. Acad. Sci. USA* **1993**, *90*, 4446-4450.
9. Sarmiento, M.; Zhao, Y.; Gordon, S. J.; Zhang, Z.-Y. *J. Biol. Chem.* **1998**, *273*, 26368-26374.
10. Wang, W.-Z.; Sun, J.-P.; Zhang, Z.-Y. *Curr. Topics Med. Chem.* **2003**, *3*, 739-748.
11. Steely, B. L.; Staubs, P. A.; Reichart, D. R.; Berhanu, P.; Milarski, K. L.; Saltiel, A. R.; Kusari, J.; Olefsky, J.M. *Diabetes* **1996**, *45*, 1379-1385.
12. Wang, X. Y.; Bergdahl, K.; Heijbel, A.; Liljebris, C.; Bleasdale, J. E. *Mol. Cell. Endocrinol.* **2001**, *173,* 109-119.
13. Makinen, M. W.; Brady, M. J. *J. Biol. Chem.* **2002**, *277*, 12215-12220.
14. Ou, H.; Yan, L.; Mustafi, D.; Makinen, M. W.; Brady, M. J. *J. Biol. Inorg. Chem.* **2005**, *10*, 874-886.
15. Puius, Y. A.; Zhao, Y.; Sullivan, M.; Lawrence, D. S.; Almo, S. C.; Zhang, Z.-Y. *Proc. Natl. Acad. Sci. USA* **1997**, *94*, 13420-13425.
16. Churg, A.K.; Gibson, G.; Makinen, M. W. *Rev. Sci. Instrum.* **1978**, *49*, 212-214.
17. Peters, K. G.; Davis, M. G.; Howard, B.W.; Pokross, M.; Rastogi, V.; Diven, C.; Greis, K. D.; Ey-Wilkens, E.; Maier, M.; Evdokimov, A.; Soper, S.; Genbauffe, F. *J. Inorg. Biochem.* **2003**, *96*, 321-330.

18. Mustafi, D.; Makinen, M. W. *Inorg. Chem.* **2005**, *44*, 5580-5590.
19. Sarmiento, M.; Puius, Y. A.; Vetter, S. W.; Keng, Y.F.; Wu, L.; Zhao, Y.; Lawrence, D. S.; Almo, S. C.; Zhang, Z. Y. *Biochemistry* **2000**, *39*, 8171-8179.
20. Park, J.; Pei, D. *Biochemistry* **2004**, *43*, 15014-15021.
21. Guo, X. L.; Shen, K.; Wang, F.; Lawrence, D. S.; Zhang, Z. Y. Probing the *J. Biol. Chem.* **2002**, *277*, 41014-41022.
22. Asante-Appiah, E.; Patel, S.; Desponts, C.; Taylor, J. M.; Lau, C.; Dufresne, C.; Therien, M.; Friesen, R.; Becker, J. W.; Leblanc, Y.; Kennedy, B. P.; Scapin, G. *J. Biol. Chem.* **2006**, *281*, 8010-8015.
23. Cornish-Bowden, A. *FEBS Letters* **1986**, *203*, 3-6.
24. Westley, A.; Westley, J. *J. Biol. Chem.* **1996**, *271*, 5347-5352.

Chapter 8

Comparing Administration Route in Rats with Streptozocin-Induced Diabetes and Inhibition of Myoblast Growth of Vanadium [V(III), V(IV), and V(V)] Dipicolinic Acid Complexes

Gail R. Willsky[1], Michael E. Godzalla, III[1,2], Paul J. Kostyniak[2], Lai-Har Chi[1], Rohit Gupta[1], Violet G. Yuen[3], John H. McNeill[3], Mohammad Mahroof-Tahir[4], Jason J. Smee[4], Luqin Yang[4], Aaron Lobernick[4], Shari Watson[4], and Debbie C. Crans[4]

Departments of [1]Biochemistry and [2]Pharmacology, University at Buffalo (SUNY), Buffalo, NY 14214
[3]Faculty of Pharmaceutical Sciences, University of British Columbia, Vancouver, Canada
[4]Department of Chemistry, Colorado State University, Fort Collins, CO 80523

The V(V) dipicolinic acid complex was the most potent complex in two administration routes for reducing diabetic hyperglycemia or hyperlipidemia. After V(V) dipic treatment in both routes an inverse correlation of the blood/serum vanadium pool versus blood/plasma glucose pool was found. This is the first vanadium complex to show such a correlation in any diabetic population. The distribution in blood of V(V) and V (IV) dipic complexes was best modeled with a 2-compartment model after acute administration. Only the V(III) and V(V) dipic complexes stimulated myoblast growth at 1 μM complex, while all complexes inhibited growth at higher concentrations. These experiments show that the oxidation state of the vanadium in dipic complexes affects biological efficacy and that processing of the V(V) dipic complex is different fom that of the lower oxidation states. These results

© 2007 American Chemical Society

also show that effects of vanadium complexes on rat myoblasts are similar to those observed in diabetic animals administered compounds intraperitoneally.

Vanadium (abbreviated V) has been shown to exist in three oxidation states V(III), V(IV), and V(V) in the biosphere, with V(IV) and V(V) being most prevalent in mammalian systems. V(IV) is the most frequently used oxidation state in studies of anti-diabetic properties of V (1,2,3,4). Common metabolites found in all cells such as ascorbate and glutathione are capable of reducing V(V) to V(IV) inside the living cell (5,6). Oxidative stress is now known to be involved in the etiology of diabetes (7) and the development of diabetic complications (8,9). Paradoxically, reactive oxygen and nitrogen species, previously thought of as only being toxic, are now known to be involved in normal cellular regulation processes (10). The ability of V to interfere with signal transduction pathways via inhibition of protein tyrosine phosphatases is central to the anti-diabetic mechanism of action of V (11). Recent studies have

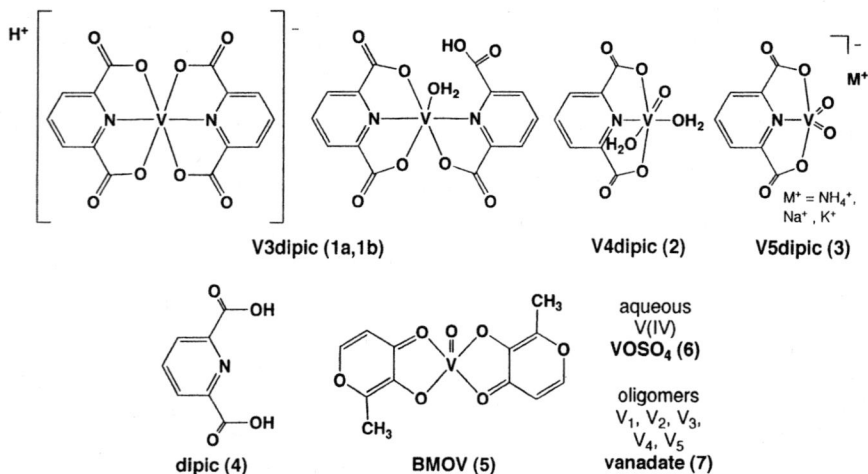

Figure 1. Structures of vanadium compounds used. dipic is the abbreviation for 2,6-pyridinedicarboxylate 1a. H[V(dipic)$_2$(H$_2$O)], b. H[V(dipic)$_2$] – both abbreviated as V3dipic, 2. VO(dipic)(H$_2$O)$_2$ - V4dipic, 3. VO$_2$(dipic) – V5dipic, 4.dipic ligand, 5. bis(maltolato)oxovanadium – BMOV, 6,7 vanadium salts used: 6 vanadyl sulfate at acidic pH exists as oxovanadyl cation, 7. vanadate oligomers –several anionic species.

implicated the interaction of V compounds with cellular redox systems as also being part of the mechanism of action of V in alleviating the symptoms of diabetes *(12)*.

The effects on biological systems of a series of dipicolinic acid (2,6-pyridinedicarboxylic acid, abbreviated dipic) complexes was examined (see Figure 1) in which the oxidation state varies while the ligand remains constant. The following abbreviations will be used: anionic vanadium(III) complex $[V(dipic)_2]^-$ is V3dipic, the neutral vanadium(IV) complex $[VO(dipic)(H_2O)_2]$ is V4dipic and the anionic vanadium(V) complex $[VO_2(dipic)]^-$ is V(V)dipic. The chemistry and synthesis of these compounds has been previously described *(13,14,15)*. The effects of chronic administration of V3dipic, V4dipic and V5dipic on elevated blood glucose and the ability of these complexes to be taken up into the serum in rats with STZ-induced diabetes has been previously reported *(15)*.

The purpose of this current report is to compare the effects of V oxidation state by adding V(III),V(IV) and V(V) dipic complexes to different biological systems. The results obtained in a tissue culture growth experiment are compared to different routes of administration of complexes to diabetic rats to investigate if cell culture work resembles either of the whole animal models. The previous work on hyperglycemia with chronic oral administration of the complexes has been expanded to include effects on hyperlipidemia, while the role of administration route on the efficacy of the complexes in the same animal model was studied using acute administration via intraperitoneal (ip) injection. In the experiments presented here, the studies in cell culture are similar to animal studies using intraperitoneal administration of vanadium complexes.

Methods and Materials

Acute Administration Animal Protocol

Male Wistar rats (190-210 grams) were obtained from the Animal Care Center at the University of British Columbia. Animals were cared for in accordance with the principles and guidelines of the Canadian Council on Animal Care as described previously *(16)*. Diabetes was induced by a single tail vein injection of streptozotocin (Sigma) at a dose of 60 mg/kg dissolved in 0.9% NaCl under light halothane anesthesia. Animals with blood glucose levels greater than 23 mM (234 mg/dl) as determined three days after STZ injection with a blood glucometer were considered diabetic and used in this study. Seven days post-STZ injection, animals were randomly assigned to treatment or control groups (n=10 for all groups). V complexes were given as a single *ip* injection at a dose of 0.1mmol/kg. For glucose determination plasma samples were analyzed for glucose content on a Beckman Glucose II Analyzer. Blood samples for V analysis were collected and stored at $-70°C$. Animals were killed at 72 hours with an overdose of pentobarbital.

Chronic Administration Animal Protocol

Previously published procedures for the induction of diabetes and subsequent animal care were used in a protocol approved by the SUNY at Buffalo IACUC *(15)*. Male Wistar outbred rats (186-252g) were obtained from Harlan (Indianapolis, IN). Diabetes was induced by administration of 60 mg/mL freshly prepared STZ injected intravenously (iv) in 0.9% saline at a dose of 60 mg/kg body weight. Four days after STZ injection animals with blood glucose levels above 300 mg/dl (measured with an Accu-Chek monitor) were randomly divided into groups (n=8). These experiments are part of a multiple series. Controls (normal, diabetic and vanadyl sulfate treated groups) are pooled from all experiments leading to larger n values for these groups. The amount of fluid consumed and body weight were monitored daily. The amount of V in the drinking water varied from 0.5 to 2 mg/ml. V dose ingested, calculated as mmol metal or compound/kg/day, was monitored. After 4 weeks the animals were killed by decapitation and serum collected for lipid and V analysis. End point serum triglyceride levels were determined in randomly selected animals using Sigma (St. Louis, MO) Procedure No. 337, serum cholesterol was determined using Sigma (St. Louis, MO) Procedure No. 332, and serum NEFA were determined using the Wako (Richmond, VA) NEFA-C test kit.

Tissue Culture Growth Protocol

L6 rat myoblasts, obtained from the ATCC were maintained and treated as previously described *(17)*. Cells were grown in normal growth medium (DMEM) supplemented with 1.5g/L sodium bicarbonate, 10% FBS, 50,000 I.U./L penicillin 50,000 µg/L streptomycin mixture, and 15 mM Hepes at 37°C and an atmosphere of 5% carbon dioxide. About 2×10^4 L6 cells per well were inoculated into 6-well culture plates. Normal growth medium was supplemented with V compounds and cells were grown at 37 °C. When cells had achieved 30 to 40 % confluence (3 to 4 days) or 70-80% confluence (7 or 8 days) adhered cells were washed twice with phosphate buffered saline, treated with trypsin and an approximate 2 fold volume of fresh media added. Viable cell counts were done using a hemocytometer and trypan-blue exclusion.

Vanadium Determination in Blood

Blood V concentration was determined by Graphite Furnace Atomic Absorption Spectroscopy (Perkin Elmer 4110ZL unit) after drying and ashing as previously described for serum V determination *(15)*. V standards were prepared in sheep blood.

Pharmacokinetic Analysis

Whole blood V was initially modeled to a first order, one-compartment model of type $C_t = C_o e^{-kt}$, where C_t is the blood concentration at time t, C_o is the initial concentration at time t_o, and k is the first order elimination constant. A linear least squares algorithm was used to obtain the fit and to estimate C_o, k, and R^2. Half-lives were calculated from relationship $t_{1/2} = 0.693/k$. If $R^2 < 0.970$ for the one compartment model, a biphasic exponential function described by the sum of two first order functions with the apparent rate constants α and β; *i.e.*, $C_t = Ae^{-\alpha t} + Be^{-\beta t}$ was applied, where A and B are proportionality constants and α and β are rate constants with units of reciprocal time. A first order fit representing the β-phase was generated as described above to obtain values for B, β and R^2. Fits with an $R^2 > 0.970$ were accepted. Where the two-compartment model was applied, fits for the 24, 48, and 72 hr points gave the best R^2 values and were used in the β-phase modeling. To obtain the α phase values extrapolated values from the β curve were subtracted from the earlier time points at 2, 12, and 18 hr. This subtraction generated the α-phase data points which were fit to a single exponential decline function to obtain values for A, α and R^2. The two-compartment model was only deemed valid if $R^2 > 0.750$ for the α-phase fit.

Statistics

Data are presented as the mean $\pm$ the standard error of the mean (SEM). Statistical analysis was done by a one-way ANOVA followed by a Bonferonni multiple comparisons test for the chronic study and a repeated measures GLM-ANOVA followed by the Neuman-Keuls *ad hoc* test for the acute model. Correlation analyses were done using bivariate Spearman tests. $P \leq 0.05$ was considered to be statistically significant.

Compounds

The V dipic complexes were prepared according to procedures described previously from dipicolinic acid and appropriate vanadium precursor [V(V) -

NaVO$_3$ *(13)*; V(IV) - VOSO$_4$ *(14,18)*;V(III) - VCl$_3$ *(15,19)*]. The charges of the systems vary with the V(V) dipic being anionic (using the monoammonium salt NH$_4$[VO$_2$dipic]), the V(IV) dipic being neutral ([VO(H$_2$O)$_2$dipic]) and the V(III) dipic being anionic (the acid form was isolated and the form administered was H[V(dipic)$_2$]. Solutions of the compounds were prepared just before use and for the chronic administration studies the solutions were changed every other day.

Results

Acute intraperitoneal administration of vanadium dipicolinic acid complexes in rats with STZ-induced diabetes.

The effects of three V-dipicolinate compounds and bis(maltolato)-oxovanadium(IV) (BMOV) on reducing plasma glucose levels in STZ-diabetic rats were evaluated after a single intraperitoneal injection (0.1 mmol/kg) using saline-treated diabetic animals as controls as shown in Figure 2A. The V(IV)-containing BMOV was used as a reference compound in this experiment. BMOV is a benchmark compound for alleviating STZ-induced diabetic hyperglycemia in rats *(20)*. There was a significant reduction in plasma glucose levels over time in the BMOV group as compared to saline. Statistically significant reductions were seen at 18 ($p < 0.05$), 24 ($p < 0.01$), and 48 ($p < 0.01$) hours. The V3dipic group also showed significant reduction at three time points as compared to saline. Reductions were seen at 8 ($p < 0.05$), 24 ($p < 0.05$), and 48 ($p < 0.01$) hours. No significant reductions were observed when the V4dipic group was compared to the saline control, although at early time points there was a drop in plasma glucose levels. Among the dipic complexes, the largest decline in plasma glucose was found in the V5dipic treated group, where reductions were observed at five time points. Significant reductions in plasma glucose were seen at 8 ($p < 0.05$), 12 ($p < 0.05$), 18 ($p < 0.05$), 24 ($p < 0.01$) and 48 ($p < 0.05$) hours when compared to saline controls.

Mean blood V concentrations for the treatment groups are plotted in Fig 2B. Statistics shown are comparing the V-dipicolinate groups to the BMOV positive control. Both V5dipic and V3dipic treated animals had significantly higher total blood V levels compared to BMOV treated animals at four out of the six time points with V5dipic exhibiting this difference early at 2 hours and lasting until 24 hours while V3dipic starts later at 12 hours but lasts until 48 hours. The V(IV) dipic group exhibits significantly higher blood V levels at only two time points, 12 and 18 hours, as compared to BMOV control. Since all treatment groups received 0.1mmol/kg V, the data show that the V-dipicolinate compounds accumulate V into blood more effectively than BMOV. However, as seen with V4dipic's relative ineffectiveness in lowering plasma glucose levels as compared to BMOV, higher absolute blood V levels do not seem to correlate with the glucose-lowering effect of these compounds. To examine the relationship between blood V and plasma glucose levels within a treatment group, individual animal blood V and plasma glucose levels (n= 8 or 9) were plotted against one another at 18 and 24 hours after drug administration and were fit to a linear function. The 18 and 24-hour time point was chosen because plasma glucose-lowering effects were established (Figure 2A) and substantial amounts of V were still in the blood (Figure 2B). A non-linear relationship between blood V and plasma glucose is suggested by the correlation coefficients for the BMOV (R^2 = 0.173 and 0.059 respectively for 18 and 24 hours), V3dipic (R^2 = 0.229 and 0.173) and V4dipic (R^2 = 0.254 and 0.311) groups. A linear relationship between these variables is only suggested for the data from the V5dipic group (R^2 = 0.721 and 0.774 for 18 and 24 hours respectively). Moreover, inspection of the data for V5dipic (Figure 3 for the 24 hour data, p=0.000787) shows that the V5dipic group's blood V and plasma glucose levels exhibit an inverse relationship, i.e., high blood V correlates with low plasma glucose. These data, in which a linear correlation was only found for one compound, agree with other reports in the literature in which serum or blood V and blood or plasma glucose levels usually do not correlate after administration of V to rats *(20)* or humans *(21)*.

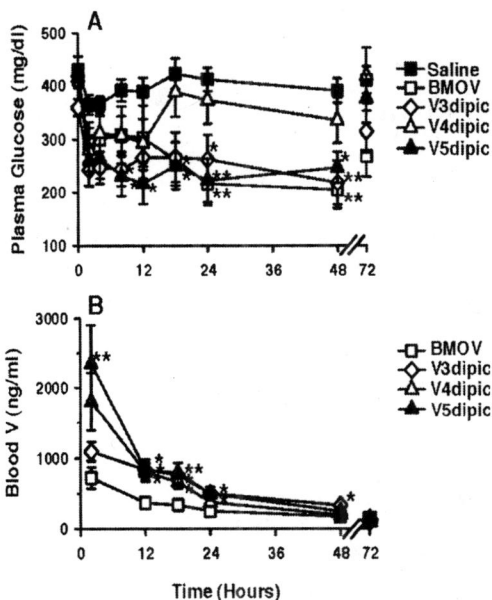

Figure 2. Effect of acute intraperitoneal administration of vanadium dipicolinic acid complexes in rats with STZ-induced diabetes. A. Plasma glucose levels. Diabetic rats treated with (■)saline, (□)BMOV, (◇)V3dipic, (△)V4dipic, (▲)V5dipic. B. Blood vanadium levels. Diabetic rats treated with (□)BMOV, (◇)V3dipic, (△)V4dipic, (▲)V5dipic.

Mean blood V levels were fit to either a one or two-compartment model as outlined above in Methods. Pharmacokinetic parameters and the correlation coefficient obtained for the first order fit of V3dipic's blood V levels are given in Table I. Using the general rule that steady state will be achieved after five half-lives, the $t_{1/2}$ of 22.6 hours for V3dipic suggests it's blood V levels will reach steady state in about five days. The pharmacokinetic parameters and the correlation coefficients obtained for the two-compartment fits of BMOV, V4dipic, and V5dipic are also given in Table 1. Because the calculation for

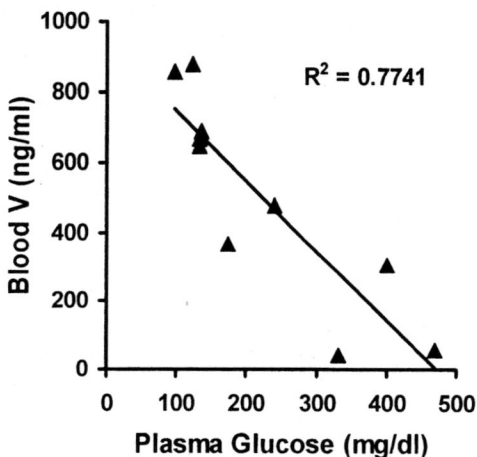

Figure 3. Correlation of blood vanadium and plasma glucose 24 hours after acute administration of V5dipicolinic acid..

when steady state will be achieved in a two-compartment model is based on the slower phase, the β half times of 25.0 and 25.5 hours for V4dipic V5dipic respectively, suggest their blood V levels will reach steady state in about five days. The $t_{1/2\beta}$ of 56.8 hours for BMOV suggests its blood V levels will not reach steady state until about twelve days. The kinetic constants for the two compartment model for BMOV are similar to reported values (22,23).

Oral chronic administration of vanadium dipicolinic acid complexes in rats with STZ-induced diabetes.

The effect of these complexes on lowering the elevated blood glucose levels in STZ-induced diabetic rats when chronically administered for four weeks in the

Table I. Kinetic constants for modeling study[a]

Compound	Compartments	Kinetic Constants		R_2
V3dipic	one	$t_{1/2} = 22.6$ hr^{-1}	k =0.0306 hr^{-1}	0.976,
V4dipic	two	$t_{1/2\alpha} = 6.5$ hr^{-1}	$\alpha = 0.106$hr^{-1}	0.941,
		$t_{1/2\beta} = 25$ hr^{-1}	$\beta = 0.028$hr^{-1}	0.999,
V5dipic	two	$t_{1/2\alpha} = 5.3$ hr^{-1}	$\alpha = 0.130$hr^{-1}	0.759
		$t_{1/2\beta} = 25.5$ hr^{-1}	$\beta = 0.0272$ hr^{-1}	0.999
BMOV	two	$t_{1/2\alpha} = 6.3$ hr^{-1}	$\alpha = 0.111$hr^{-1}	0.900,
		$t_{1/2\beta} = 56.8$ hr^{-1}	$\beta = 0.0122$ hr^{-1}	0.971

[a] Diabetic animals were treated with the indicated compounds and kinetic and modeling studies done as described in Methods.

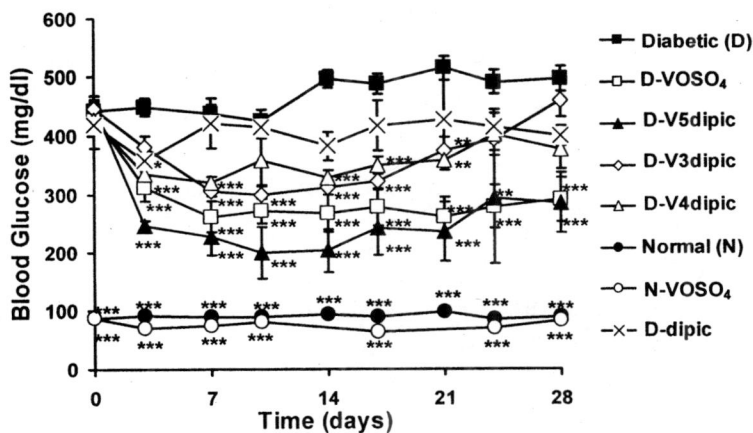

Figure 4. Effect of chronic oral administration of vanadium dipicolinic acid complexes on blood glucose in diabetic rats. Diabetic rats treated with (■)water only, (□)VOSO₄, (◇)V3 dipic, (△)V4 dipic,(▲)V5dipic, (×)dipic. Normal rats treated with(●)water only, (○)VOSO₄.

drinking water is shown in Figure 4, extending the two week data previously published for these animals *(15)*. The positive control and V(IV) containing reference compound for these experiments was $VOSO_4$. Vanadyl sulfate is commonly used as a reference compound, although the speciation at neutral pH is complex. Aqueous V(IV) is typically regarded to contain cationic forms of vanadium, however, recently it was recognized that at subnanomolar concentration, the most prevalent form in aqueous solution of V(IV) was an anionic form of vanadium (IV) *(3)*. All V compounds lowered blood glucose levels as compared to diabetic controls. V5dipic showed the most promise clinically since it lowered blood glucose levels to a level that could not be distinguished from the normal controls on days 10, 14, and 21 (p=0.183, 0.210, and 0.269 when compared to normal controls, respectively). V3dipic and V4dipic showed similar blood glucose-lowering effects and both showed a loss of activity in the final two weeks of the experiment. Although a lowering trend was observed with the dipicolinate ligand with respect to the diabetic controls, this lowering was only statistically significant on day three of treatment (p = 0.013). At no time were the blood glucose values of normal animals treated with $VOSO_4$ statistically different from that of the untreated normal controls.

Endpoint serum lipid levels are shown in Table II for this experiment. All V dipic complexes lowered serum triglyceride and cholesterol while only the V(IV) containing V4dipic and $VOSO_4$ lowered serum free fatty acids when compared to diabetic controls. Treatment with the dipicolinic acid ligand did show a slight but statistically insignificant lowering of serum lipid levels.

Table II. Serum lipid levels after chronic oral administration of vanadium compounds in normal and diabetic rats. [a]

Group	Cholesterol (mg/dl)	Triglyceride (mg/dl)	Free Fatty Acid (mg/dl)
Normal (N)	90 ± 4 (21)***	128 + 14 (21)***	0.36 ± 0.04(21)***
N/VS	79 ± 6***	113 ± 18***	0.32 ± 0.07***
Diabetic (D)	216+14(34)	1381+ 121 (34)	1.67 ± 0.19 (31)
D/VS	94 ± 4(27)***	165+14 (27)***	0.31 ± 0.03(22)***
D/V3dipic	104 ± 8(8)***	245+33 (8) ***	1.37 ± 0.17 (8)
D/ V4dipic	114 ± 5(5)***	154 ± 22 (5)***	0.45 ± 0.09(7)***
D/ V5dipic	100 ± 6(6)***	121 ± 13 (5)***	0.80 ± 0.08(3)
D/dipic	195 + 21(6)	1755 + 483 (6)	1.15 + 0.13

[a]Data is presented as the mean ± SEM   *** p<0.001 compared to diabetic

The number of animals tested is given in parenthesis after the values.

Spearman correlations of blood glucose, serum triglyceride, serum cholesterol, and serum free fatty acids with ingested metal dose or serum metal concentration were performed within each treatment group. The only dipic treatment group to show significant correlations among these variables was the V5dipic group where cholesterol level was inversely correlated with dose (spearman's rho correlation coefficient (srcc) = -1.000, p< 0.001 n=4) and blood glucose was inversely correlated with serum V (srcc = -1.000, p< 0.001 n=4). Cholesterol was correlated with increasing dose in animals treated with ligand alone (srcc= 943m p< 0.01 n=6), while in the normal animals treated with VOSO$_4$, cholesterol inversely correlated with dose (srcc = -929, p< 0.01, n=7).

Effects of vanadium dipicolinic acid complexes on the growth of rat myoblasts in tissue culture.

The effects of these V dipic complexes on myoblast cell growth for different confluence levels are shown in Figure 5. At the 30 to 40% confluence levels all three complexes showed similar effects on growth with little growth effects at 1 µM and approximately 30 to 40% growth compared to control at 10 µM. The V dipicolinic acid complexes showed similar growth effects as the positive control vanadate. When the cells had reached 70 to 80% confluence exposure to V3dipic and V5dipic caused significant growth stimulation at 1 µM of complex. In the presence of V3dipic there was only 50% cell growth

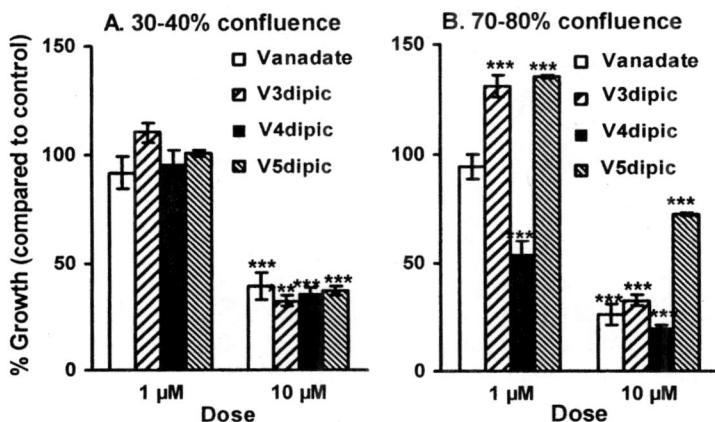

Figure 5. Effect vanadium dipicolinic acid complexes on growth of rat myoblasts. A separate control was done for each vanadium compound.
A. 30% to 40% confluence, untreated 100% cell number range (3 - 7 x 10^5).
B. 70% to 80% confluence, untreated 100% cell number range (1.6 - 2.5 x 10^6).

compared to control. Vanadate, the reference compound used in these studies, is known to form several labile oxovanadates in solutions including the monomeric, dimeric, tetrameric and pentameric forms *(3)*. The presence of the most prevalent form near neutral pH the cyclic tetrameric form $V_4O_{12}^{4-}$ was observed by ^{51}V-NMR spectroscopy in the growth media. However, as the concentration of V(V) is lowered the concentration of this and other oligmers change. Furthermore, they interconvert rapidly in solution making it difficult to attribute the observed effects to one specific isomer. We will thus refrain from attributing the observed effects to any specific isomer and refer to vanadate in these studies. The vanadate treated control showed only a slight lowering of cell number compared to the control after exposure to 1μM compound. At 10 μM vanadate, in the presence of V3dipic and V4dipic only 20% to 30% of cell growth was observed compared to the control, while V5dipic addition resulted in 70% of the growth of control cells. At 100 μM and 1000 μM all V compounds inhibited growth by more than 99% (data not shown).

Discussion

Data is presented on the effects of three V complexes having the same dipic ligand but differing in oxidation state in three different biological systems. In the diabetic rats both acute and chronic administration were investigated and those studies are compared to the effects on myoblast cells in tissue culture. Under the physiological reducing conditions found in cells one V compound would be oxidized (the V3dipic), another V compound would be reduced (the V5dipic) and the third V compound would be redox inert in the normal reducing cellular environment (the V4dipic). Redox cycling among vanadium complexes in different oxidation states is generally anticipated in part because the reduction of V(V) to V(IV) in living cells was demonstrated by paramagnetic electron resonance spectroscopy (EPR) *(5)*. Although cellular metabolites such as ascorbate and glutathione reduce V(V) to V(IV) in the laboratory *(3,24)* the details of the metabolism of these compounds in cells are complex and several metabolites can reduce V in cells. It was hypothesized, based on the redox potentials of the complexes, that redox cycling would take place. However, we recently showed that the V-dipic compounds had different effects in the chronic diabetic rats *(15)*. The biological effects of the V-dipic complexes would be the same if the three complexes underwent redox cycling before exerting their biological effects. The experiments described here tested the possibility that interactions with cellular oxidation and reduction processes in three biological systems would result in different responses, and thus demonstrate that biological processing varied in these different systems.

A differential effect on all the biological parameters studied would demonstrate that biological action depends on the oxidation state of the V-dipic complex. In all three systems there was a differential effect on the biological parameters studied related to the oxidation state of the V-dipic complex. The V5dipic complex was either the most effective or one of the two most effective of the three V dipic complexes studied for most of the biological parameters measured. There was no parameter that significantly correlated with increasing or decreasing oxidation state. These results demonstrate that the biological activity of V-dipic complexes is modified by the oxidation state of the V in the complex. These results suggest that the changes in the cellular redox system that result from the intracellular reduction of V5dipic to V4dipic could be in part responsible for the therapeutic action of this complex. Perhaps, the reduction process may be key to maximizing the anti-diabetic activity of the V5dipic.

Although the V in the complexes is anticipated to dissociate from laboratory studies, this apparently does not occur until a differential biological effect has manifested itself. Rapid dissociation of the V from the dipic in these complexes in the same cellular environment should result in the formation of the corresponding V salts. Although several forms of vanadate and vanadyl sulfate have been studied, the organic ligands are generally found to enhance efficacy and reduce toxicity *(20)*, which would not occur if the V complexes immediately dissociated. The studies shown here embrace these potentially conflicting observations, because the complex that is the least likely to stay intact under biological conditions is the most active, the V5dipic. Interestingly, this complex is most stable under acidic conditions. The higher acid stability of the V5dipic might secure the high efficacy of this complex in the chronic oral administration studies by allowing this complex to remain intact in the stomach. However, the fact this compound is also active after intraperitoneal injection demonstrates that this complex is active also at neutral pH where the ligand should rapidly dissociate from the vanadium.

A very important result from these studies is that after administration of V5dipic an inverse correlation was found for both blood V and plasma glucose in the acute model, and for serum V and blood glucose in the chronic model. V5dipic is the first V complex for which such a correlation has been found in any diabetic model. This data implies that the blood/serum pool after V5dipic administration is in equilibrium with the biological target for this complex. Such a correlation was not found for V3dipic or V4dipic complexes after either acute or chronic administration. The lack of correlation of blood V with the ability to lower blood glucose in acutely treated animals has been previously reported for the vanadyl complexes BMOV, bis(ethylmaltolato) oxovanadium and bis(isopropylmaltolato) oxovanadium *(23)* and chronically treated animals treated with dipicolinic acid complexes *(15)*. Lack of correlation after vanadyl sulfate treatment in humans has also been found *(21)*. Documentation of such correlation is important for future studies and furthermore demonstrate that the V5dipic complex is unique and varies from others studied, regardless of its stability profile in pure solutions.

The effect of oxidation state appears to be ligand specific. Using maltol as a ligand the V(IV) maltol complex was more successful as an insulin-enhancing agent than the V(V) maltol complex *(20)*. A V(III) maltol complex has also been shown to be effective in lowering diabetic hyperglycemia when administered chronically by oral gavage or acutely via ip injection *(25)* The maltol ligand in BMOV has been show to dissociate from the V after acute administration by following the distribution of the radioactive [14]carbon in the maltol group and the radioactive [48]V *(22)*. Given the speciation reactions of the V-dipic complexes *(13,14,15)* the V would also be expected to dissociate from the ligand in the cell or animal. Surprisingly, the maltol and dipic ligands confer different properties upon the ligated V. This result is important and is not anticipated based on the expectation that both series of complexes are believed to fall apart to form the same free forms of V. Previously a complex of V(IV) and picolinate ([VOpic$_2$]) has been shown to lower blood glucose after acute administration and after chronic administration for a 6 week period *(26)*. The V(IV) picolinate complex behaved differently in those experiments than the V4dipic complex studied here with respect to lowering diabetic hyperglycemia.

Interesting a similar response pattern is observed in the acute model for lowering diabetic hyperglycemia and that observed in cell culture. Specifically, only the V3dipic and V5dipic complexes stimulate growth in the myoblast cell culture system, the V4dipic complex only inhibits cell growth. The V3dipic and V5dipic complexes both cause lowered plasma glucose after intraperitoneal administration, in contrast to the relatively ineffective V4dipic. In contrast, after processing through the gastrointestinal tract, only the V5dipic complex retains maximum effectiveness in lowering diabetic hyperglycemia. The myoblast growth system is similar to the acute diabetic model in that the cellular environment in the acute model or the cells themselves in the tissue culture system directly interact with the administered V-dipic complex without any processing via the gastrointestinal system. The studies presented here thus provide a rationale for why many studies fail to show a correlation between cell culture studies and chronic animal studies and supports the idea that cell culture may be a good model in which to study the effects of V compounds inside the body.

In a study of diabetic and normal rats treated with VOSO$_4$ in which the diabetic rats responded with lowered hyperglycemia and hyperlipidemia, analysis of global gene expression showed differential vanadate-induced gene expression for normal and diabetic animals. In addition when the V altered gene expression in the diabetic animals was compared to the gene expression pattern seen in the normal animals in that study there was little difference; demonstrating that V treatment of diabetics normalized gene expression *(12)*. The DNA microarray results extend to global gene expression the previously reported differences between diabetic and normal animals treated with V salts or complexes *(11,27)*. Expanding those published gene experiments to include gene expression data from diabetic animals treated with $VOSO_4$ in which only diabetic hyperlipidemia was altered, showed that the vanadate-induced changes in gene expression did not return gene expression towards the normal pattern. When the hyperlipidemia alone was lowered, the changes in gene expression were not the reversal of the diabetic changes in gene expression originally causing the abnormal lipid metabolism (Willsky and Crans unpublished). These gene expression studies show that in addition to ligand or vanadium oxidation state caused differences in biological effects of V, there are massive changes in the effect of the metal attributable to the metabolic state of the organism. These results were mirrored in the present studies where only the cells closer to the stage of myotube formation had the metabolic profile that allowed V to stimulate growth after treatment with V3dipic and V5dipic at low concentrations.

In summary, the effects of a series of V-dipic complexes with three different oxidation states were shown to have different effects in biological systems. These data show that the V-dipic complexes do not undergo redox cycling to form one V species prior to exerting a biological effect, which would have been the implication if all three of the V-dipic complexes had the same biological effects. V5dipic is the first V complex to show inverse correlation of the blood/serum pool of V with blood/plasma glucose levels after administration in any diabetic population, in both the acute and chronic treatment model. The results reported here, along with other studies using multiple ligands *(28)*, imply that the oxidation state of V ligand complexes is an important factor determining compound efficacy and how the V interacts in biological systems.

References

1. Shechter, Y.; Goldwaser, I.; Mironchik, M.; Fridkin, M.; Gefel, D. *Coordination Chem. Rev.* **2003**, *237*, 3-11.
2. Sakurai, H.; Yasui, H. *J. Tr. El. Exp. Med.* **2003**, *16*, 269-280.
3. Crans, D. C.; Smee, J. J.; Gaidamauskas, E.; Yang, L. Chem. *Rev.* **2004**, *104*, 849-902.
4. Tracey, A.; Willsky, G. R.; Takeuchi, E. *Vanadium: Chemistry, Biochemistry, Pharmacology and Practical Applications;* Taylor and Francis: Boca Raton, FL, **2007**; 328 pp.
5. Degani, H.; Gochin, M.; Karlish, S. J. D.; Shechter, Y. *Biochemistry* **1981**, *20*, 5795-5799.
6. Willsky, G. R.; White, D. A.; McCabe, B. C. *J. Biolog. Chem.* **1984**, *259*, 13273-13281.
7. Brownlee, M. *Diabetes* **2005**, *54*, 1615-1625.

108

8. Tabatabaie, T.; Vasquez-Weldon, A.; Moore, D. R.; Kotake, Y. *Diabetes.* **2003**, *52*, 1994-1999.
9. Robertson, R. P.; Harmon, J.; Tran, P. O.; Tanaka, Y.; Takahashi, H. *Diabetes.* **2003**, *52*, 581-587.
10. Goldstein, B. J.; Mahadev, K.; Wu, X. *Diabetes* **2005**, *54*, 1249.
11. Marzban, L.; McNeill, J. H. *J.Tr. Elem. Exp. Med.* **2003**, *16*, 253-267.
12. Willsky, G. R.; Chi, L.-H.; Gaile, D. P.; Hu, Z.; Crans, D. C. *Physiol. Genomics.* **2006**, *26*, 192-201.
13. Crans, D. C.; Yang, L.; Jakusch, T.; Kiss, T. *Inorg. Chem.* **2000**, *39*, 4409-4416.
14. Crans, D. C.; Mahroof-Tahir, M.; Johnson, M. D.; Wilkins, P. C.; Yang, L.; Robbins, K.; Johnson, A.; Alfano, J. A.; Godzala, M. E.; Austin, L. T.; Willsky, G. R. *Inorg. Chim. Acta* **2003**, *356*, 365-378.
15. Buglyo, P.; Crans, D. C.; Nagy, E. M.; Lindo, R. L.; Yang, L.; Smee, J. J.; Jin, W.; Chi, L.-H.; Godzala, M. E., III; Willsky, G. R. *Inorg. Chem.* **2005**, *44*, 5416-5427.
16. Yuen, V. G.; Orvig, C.; McNeill, J. H. *Can. J. Physiol.Pharm.* **1995**, *73*, 55-64.
17. Crans, D. C.; Yang, L.; Alfano, J. A.; Chi, L.-H.; Jin, W.; Mahroof-Tahir, M.; Robbins, K.; Toloue, M. M.; Chan, L. K.; Plante, A. J.; Grayson, R. Z.; Willsky, G. R. *Coor. Chem. Rev.* **2003**, *237*, 13-22.
18. Bersted, B. H.; Belford, R. L.; Paul, I. C. *Inorg. Chem.* **1968**, *7*, 1557-1562.
19. Chatterjee, M.; Maji, M.; Ghosh, S.; Mak, T. C. W. *J. Chem. Soc., Dalton Inorg. Chem.* **1998**, 3641-3646.
20. Cam, M. C.; Brownsey, R. W.; McNeill, J. H. *Can. J. Physiol.Pharmacol.* **2000**, *78*, 829-847.
21. Goldfine, A. B.; Patti, M. E.; Zuberi, L.; Goldstein, B. J.; LeBlanc, R.; Landaker, E. J.; Jiang, Z. Y.; Willsky, G. R.; Kahn, C. R. *Metabolism.* **2000**, *49*, 400-410.
22. Setyawati, I. A.; Thompson, K. H.; Yuen, V. G.; Sun, Y.; Battell, M.; Lyster, D. M.; Vo, C.; Ruth, T. J.; Zeisler, S.; McNeill, J. H.; Orvig, C. *J. Appl. Physiol..* **1998**, *84*, 569-575.
23. Thompson, K. H.; Liboiron, B. D.; Sun, Y.; Bellman, K. D. D.; Setyawati, I. A.; Patrick, B. O.; Karunaratne, V.; Rawji, G.; Wheeler, J.; Sutton, K.; Bhanot, S.; Cassidy, C.; McNeill, J. H.; Yuen, V. G.; Orvig, C, *J. Biolog. Inorg. Chem.* **2003**, *8*, 66-74.
24. Wilkins, P. C.; Johnson, M. D.; Holder, A. A.; Crans, D. C. *Inorg. Chem.* **2006**, *45*, 1471-1479.
25. Melchior, M.; Rettig, S. J.; Liboiron, B. D.; Thompson, K. H.; Yuen, V. G.; McNeill, J. H.; Orvig, C. *Inorg. Chem.* **2001**, *40*, 4686-4690.

26. Melchior, M.; Thompson, K. H.; Jong, J. M.; Rettig, S. J.; Shuter, E.; Yuen, V. G.; Zhou, Y.; McNeill, J. H.; Orvig, C. *Inorg. Chem.* **1999**, *38*, 2288-2293.

27. Willsky, G. R.; Goldfine, A. B.; Kostyniak, P. J. *ACS Symposium Series* **1998**, *711*, 278-296.

28. Thompson, K. H.; Tsukada, Y.; Xu, Z.; Battell, M.; McNeill, J. H.; Orvig, C. *Biolog. Tr. Elem. Res.* **2002**, *86*, 31-44.

Chapter 9

Structure–Activity Relationship and Molecular Action Mechanism in the Family of Antidiabetic *bis*(Picolinato)oxovanadium(IV) Complexes

Hiromu Sakurai

Department of Analytical and Bioinorganic Chemistry, Kyoto Pharmaceutical University, 5 Nakauchi-cho, Misasagi, Yamashina-ku, Kyoto 607–8414, Japan

Trials for the use of vanadium in treating diabetes mellitus (DM) was reported in 1899, which was 23 years before Banting and Best discovered insulin and used it to treat diabetic patients. However, pharmaceutical interest in vanadium as an insulin substitute started in 1980. Ten years after, in 1990, orally active insulin substitutes were found as chelated oxovanadium(IV) complexes such as bis–(methylcysteinato)oxovanadium(IV) and bis(malonato)–oxovanadium(IV). Since these findings, a wide variety of antidiabetic oxovanadium(IV) complexes have been proposed. In 1995, bis(picolinato)oxovanadium(IV) complexes [$VO(pa)_2$] was found to have excellent *in vitro* insulinomimetic and *in vivo* antidiabetic activities in type 1–like diabetic rats. This $VO(pa)_2$ complex has an advantage to study structure–activity relationships in both *in vitro* and *in vivo* levels, and produce many interesting results. This article reviews the recent new progress in our research groups on the structure–activity relationship and the molecular action mechanism in the family of $VO(pa)_2$ complex.

© 2007 American Chemical Society

The total number of patients with diabetes mellitus (DM) worldwide is estimated to rise from 171 million in 2000 to 366 million in 2030 (*1*).

Insulin−dependent type 1 DM is a result of the autoimmune destruction of the pancreatic β cells (*2*), thus the patients require daily insulin injections for survival. On the other hand, non−insulin−dependent type 2 DM is associated with metabolic syndromes with obesity, impaired glucose metabolism, insulin resistance, and other complications, thus the patients require several types of synthetic pharmaceuticals (*2*). In place of insulin injections with physical and mental pain and synthetic pharmaceuticals with severe side effects, development of therapeutics with new approaches is anticipated.

For many years, oxovanadium(IV) sulfate, $VOSO_4$, has been tested in the treatment of both type 1 and 2 DM since 1899 (*3−6*). In fact, treatment by $VOSO_4$ improved DM as well as hepatic, peripheral, and muscle insulin sensitivity (*7, 8*). However, because the bioavailability of $VOSO_4$ is low (*9, 10*), we proposed methods to enhance vanadium uptake by an encapsulation via enteric coating of $VOSO_4$ (*11*) and by a drug delivery system consisting of $VOSO_4$ and poly(γ−glutamic acid) (*12*). Alternatively, in 1990, we proposed first several types of orally active oxovanadium(IV) complexes such as bis(methylcysteinato)oxovanadium(IV) [$VO(cysm)_2$] and bis(malonato)−oxovanadium(IV) [$VO(mal)_2$] with different coordination environments involving $VO(S_2N_2)$ and $VO(O_4)$, that enhanced the bioavailability and efficacy of $VOSO_4$ (*13, 14*).

During these investigations (*4−6*), in 1995, an oxovanadium(IV)-picolinate complex with $VO(N_2O_2)$ coordination environment was found to exhibit excellent *in vitro* insulinomimetic activity and *in vivo* antidiabetic effect on daily oral administrations (*15*). The finding indicated that the complex has an advantage to examine not only the structure−activity relationship with regard to insulinomimetic and antidiabetic activities but also their molecular mechanisms for insulinomimetic and antidiabetic activities.

This article reviews our recent progress in the development of insulinomimetic antidiabetic oxovanadium(IV) complexes in terms of structure-activity relationship and action mechanism in molecular level focusing on the family of oxovanadium(IV)−picolinate complex.

Establishment of *In Vitro* Evaluation System of Insulinomimetic Compounds

Before subjecting metal ions and metal complexes *in vivo* hypoglycemic test, we had to establish a reliable *in vitro* appraisal system to evaluate them for their potential antidiabetic activity. The system was developed with respect to the interaction of metal ions and metal complexes with isolated rat adipocytes treated with adrenaline (epinephrine) (*16, 17*).

Because the molecular basis of the action mode of insulin and insulinomimetic compounds has gradually been revealed, there is strong evidence that insulin receptor is activated by inhibiting protein tyrosine phosphatase (PTP−1B), which in turn relates to the activation by phosphrylation of insulin receptor substrate 1 (IRS−1), activation of phosphatidylinosital 3 kinase (PIB−K), and then Akt phosphorylation, leading to glucose transporter−4 (GLUT4) translocation, as well as the activation of phosphodiesterase (PDE) (*4−6, 18*). In fact, insulin and $VOSO_4$ exhibited both incorporation of glucose (*13, 19*) and inhibition of free fatty acids (FFA) release in the rat adipocytes (*13, 16, 19*). Although insulin has sole action site, insulin receptor, $VOSO_4$ has been found to have multiple action sites in the cells, as named as "*Ensemble mechanism*" (*18*). Consequently, both glucose incorporation and suppression of FFA release in the adipocytes were measured to evaluate *in vivo* insulinomimetic activity of compounds by using commercially available determination kits.

Bis(picolinato)oxovanadium(IV) [VO(pa)$_2$] Complex

By using the *in vitro* evaluation system, in 1995, we found that bis(picolinato)oxovanadium(IV) [VO(pa)$_2$] complex (Figure 1) with $VO(N_2O_2)$ coordination mode exhibits excellent insulinomimetic activity and *in vivo* hypoglycemic ability in streptozocin−induced type 1−like diabetic rats (STZ−rats) on daily intraperitoneal (*ip*) injections and oral administrations (*15*). It is noteworthy that when this complex was given daily to STZ-rat by oral administrations, at the dose of 196 μmolV/kg body/weight/day for 14 days, the normoglycemic effect was observed within 7 days, and the effect continued as long as the complex was given to the animals. After the end of treatment, normoglycemic effect was maintained for approximately 80 days.

VO(pa)$_2$ was also revealed to promote the incorporation of not only 2−deoxy−D−[1−^{3}H]glucose in Ehrlich ascites tumor cells (*20*) but also glucose in isolated rat adipocytes (*16*) dose−dependently.

Structure−Activity Relationships of VO(pa)$_2$ Related Complexes

VO(pa)$_2$ with a partition coefficient (log*P*) of − 0.48 in n−octanol/buffer (pH 7.4) system has an advantage of preparing several analogs to examine the structure−activity relationship of insulinomimetic and antidiabetic activities, by introducing electron−donating or electron−withdrawing groups at different positions of the picolinate ligand.

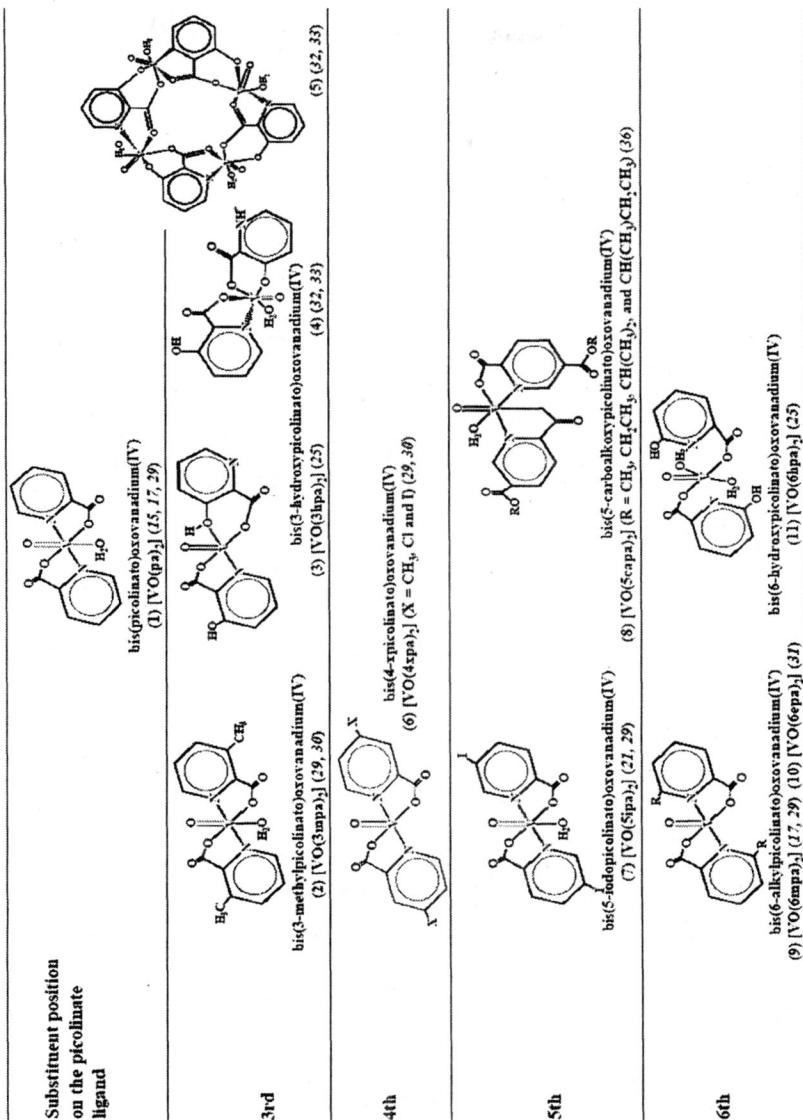
Substituent position on the picolinate ligand

bis(picolinato)oxovanadium(IV)
(1) [VO(pa)₂] (15, 17, 29)

3rd

bis(3-methylpicolinato)oxovanadium(IV)
(2) [VO(3mpa)₂] (29, 30)

bis(3-hydroxypicolinato)oxovanadium(IV)
(3) [VO(3hpa)₂] (25)

bis(3-hydroxypicolinato)oxovanadium(IV)
(4) [VO(3hpa)₂] (32, 33)

(5) (32, 33)

4th

bis(4-xpicolinato)oxovanadium(IV)
(6) [VO(4xpa)₂] (X = CH₃, Cl and I) (29, 30)

5th

bis(5-iodopicolinato)oxovanadium(IV)
(7) [VO(5ipa)₂] (27, 29)

bis(5-carboalkoxypicolinato)oxovanadium(IV)
(8) [VO(5capa)₂] (R = CH₃, CH₂CH₃, CH(CH₃)₂, and CH(CH₃)CH₂CH₃) (36)

6th

bis(6-alkylpicolinato)oxovanadium(IV)
(9) [VO(6mpa)₂] (17, 29) (10) [VO(6epa)₂] (31)

bis(6-hydroxypicolinato)oxovanadium(IV)
(11) [VO(6hpa)₂] (25)

Figure 1. Insulin substitutes as oxovanadium(IV) complexes inspired from bis(picolinato)ocovanadium(IV) [VO9pa] complex

Bis(6–methylpicolinato)oxovanadium(IV) [VO(6mpa)$_2$] prepared in 1997 (*17*) (Figure 1 (9)) and bis(5–iodopicolinato)oxovanadium(IV) [VO(5ipa)$_2$] prepared in 2001 (*21*) (Figure 1 (7)), both exhibited better *in vitro* insulinomimetic and *in vivo* hypoglycemic effect in STZ–rats than the leading VO(pa)$_2$ complex. The former was found to be long–term acting on oral administrations better than the effect of VO(pa)$_2$, when it was orally administrated in STZ–rats (*22*). The long–term effect of the complex was proposed by both accumulation of vanadium in the bone of animals as determined by neutron activation analysis (NAA) (*17*) and ternary complex formations composed of picolinate–oxovanadium(IV)–proteins or amino acids in the liver and kidney as detected by electron spin–echo envelope modulation (ESSEM) spectroscopy (*23*). Interestingly, VO(6mpa)$_2$ showed the hypoglycemic effect in hereditary type 2 diabetic KK–A^y mice with insulin resistance by both daily ip injections and oral administrations (*24*).

Based on these results, we performed structure–dependent metallokinetic analysis using 6 kinds of VO(pa)$_2$ complexes, such as VO(pa)$_2$, VO(3mpa)$_2$, VO(6mpa)$_2$, VO(3hpa)$_2$, (bis(3–hydroxypicolinato)oxovanadium(IV)), and VO(5ipa)$_2$ (Figure 1), in rats with *in vivo* blood circulation monitoring–electron spin resonance (BCM–ESR) (*25, 26*), which has been proposed as a new analytical method to evaluate the real–time pharmacokinetics of stable spin probes in a live animal, and was demonstrated to be very useful to determine the paramagnetic metal ions such as oxovanadium(IV) complexes in the circulating blood (*27, 28*).

The relationship between the *in vitro* insulinomimetic activity and the metallokinetic parameters in the family of VO(pa)$_2$ was observed; IC$_{50}$ value, which is 50% inhibitory concentration of the complexes on the FFA release from isolated rat adipocytes, was found to sufficiently correlate with area under the concentration curve (AUC), mean residence time (MRT), total clearance (CL$_{tot}$) and distribution volume at steady state (V$_{ss}$). Furthermore, the *in vivo* hypoglycemic activity of the complexes was enhanced with increasing exposure and residence of oxovanadium(IV) species in the blood of animals (*25, 26*).

Unfortunately, from this study we could not deduce which complexes are the best to exhibit antidiabetic effect. Thus we extended our study to examine the relationship among the physio–chemical properties, *in vitro* insulinomimetic activity and *in vivo* hypoglycemic activity of 13 complexes of VO(pa)$_2$ family. Obtained results are as follows; both bis(4–methylpicolinato)oxovanadium(IV) [VO(4mpa)$_2$] and bis(4–iodopicolinato)oxovanadium(IV) [VO(4ipa)$_2$] (Figure 1 (6)) are better complexes than the previously proposed VO(6mpa)$_2$ and VO(5ipa)$_2$ (*29*). The fact suggests the importance of 4th substituent position of picolinate ligand rather than the characteristics of electronic effect of substituent either electron donating or withdrawing, which makes the complex possible to interact with several enzymes relevant to insulin signaling pathway in cells.

Structures in Solid State of VO(pa)$_2$ Related Complexes

Because crystals in the family of VO(pa)$_2$ complexes suitable for X−ray structure analysis have been generally difficult to obtain, the structures of the complexes were characterized by electronic absorption, IR, ESR, EXAFS and MS spectra; VO(pa)$_2$, VO(3mpa)$_2$ and VO(5ipa)$_2$ complexes have a six−coordinate structure with an additional V−OH$_2$ bond, in contrast, VO(6mpa)$_2$ and VOSO$_4$ have no coordinate H$_2$O molecular and a five−coordinate structure as estimated by EXAFS method (30).

Fortunately, the structure of bis(6−ethylpicolinato)oxovanadium(IV) [VO(6epa)$_2$] (Figure 1 (10)) was analyzed by X−ray, where two distinct molecules are in an asymmetric unit. Each V in the complex is coordinated by two carboxylate oxygens, two pyridine nitrogens, one oxovanadium(IV) oxygen and one H$_2$O oxygen, forming a distorted octahedral geometry. The two carboxylate oxygens and pyridine nitrogens occupy an equatorial plane; the two ligand coordinate to the V center in a trans arrangement (31).

Other interesting oxovanadium(IV) complexes with 3−hydroxypicolinic acid, (H$_2$hpic), mononuclear and tetranuclear complexes, [VO(Hhpic−O,O)−(Hhpic−O,N)(H$_2$O)] · H$_2$O and [(VO)$_4$(μ−(hpic−O,O',N)$_4$(H$_2$O)$_4$] · 8H$_2$O (Figure 1 (4) and (5), respectively) were prepared and their X−ray structures were analyzed (32, 33). The structures of both complexes are quite different from that of VO(pa)$_2$; in the mononuclear complex, one of the ligands binds with oxovanadium(IV) through the phenolate and carboxylate O atoms, and another ligand binds to the oxovanadium(IV) through pyridinium N and carboxylate O, and the tetranuclear complex comprises four distorted octahedral six coordinate V centers. Each ligand is tetradentate, bidentate of each of two oxovanadium(IV) moieties which it bridges, and two remaining O atoms are due to oxovanadium(IV) and H$_2$O. Formation of such cyclic tetranuclear complex has been proposed in 2000 (34), and confirmed in 2003 (35). The *in vitro* insulinomimetic activity of the mononuclear complex was higher than that of VOSO$_4$, however, that of the tetranuclear complex was lower than that of VOSO$_4$, probably due to the high molecular weight to be incorporated in the adipocytes (32, 33).

The structure of oxovanadium(IV) complexes with 5−carboalkoxypicolinate ligand (5ROpaH, R = methyl, ethyl, isopropyl and isobutyl) were also analyzed by X−ray (36), where H$_2$O and one of the picolinate ligands are in the equatorial positions and the second picolinate occupies the equatorial (N) and axial (O) positions (Figure 1 (8)). The *in vitro* insulinomimetic activity of the complexes, VO(5capa)$_2$ (ca = CH$_3$O) is substantially higher than that of VOSO$_4$, being comparable to that of VO(pa)$_2$ complex (36).

The speciation in solution for VO(pa)$_2$, VO(6mpa)$_2$ and related complexes was studied in detail in relation to the structures of the complexes (36−38).

Molecular Action Mechanism of VO(pa)$_2$ Related Complexes

It is well known that insulin binds with its specific receptor. The binding mediates signaling transduction to the downstream targets such as IRS (insulin receptor substrate) and PI3K (phosphatidylinositol 3−kinase). When the PI3K is activated, Akt is phosphorylated, which in turn GSK3β (glycogen synthase kinase 3β) and GLUT4 (glucose transporter 4) are then phosphorylated. The translocation of the activated GLUT4 from the intracellular compartment vesicle to the plasma membrane leads to glucose uptake in the cells.

V compounds have been known to bind to PTP1B (protein tyrosine phosphatase 1B) and inhibit its phosphatase activity, therefore, the tyrosine phosphorylations of IRβ (insulin receptor β−subunit) and IRS are enhanced to activate the PI3K and Akt (39). It was also reported that VOSO$_4$ inhibits c−AMP−dependent protein kinase (PKA), which affects a hormone−sensitive receptor (40). However, the molecular action mechanism of oxovanadium(IV) and its complexes for glucose uptake and insulin signaling cascade are not yet studied. Then we examined the molecular mechanism of VO(pa)$_2$ related compounds, VOSO$_4$, VO(pa)$_2$, VO(3mpa)$_2$ and VO(6mpa)$_2$, in terms of insulin receptor signaling cascade and GLUT4 translocation in 3T3−L1 adipocytes in the following points; (1) phosphorylations of Akt and GSK3β, (2) tyrosine phosphorylation of both IRβ and IRS, and (3) translocation of GLUT4 to the plasma membrane (41).

Among four compounds examined, VO(3mpa)$_2$ was the highest potent activator of Akt phosphorylation, followed by the effect of insulin. Because Akt is regulated by upstream kinases such as PI3K and PDK1, the effect of Akt phosphorylation was then examined, in which the cells were pretreated with a PI3K inhibitor, wortmannin. The enhanced phosphorylation of Akt by VO(pa)$_2$ complexes was suppressed by wortmannin, suggesting that the complexes affect the upstream kinases of Akt such as PI3K and phosphorylations of IRβ and IRS. Furthermore, VO(pa)$_2$ complexes were found to phosphorylate GSK3β, IRβ and IRS.

Finally, we examined the translocation of GLUT4 protein and found that 50 μM VO(3mpa)$_2$ was able to highly translocate GLUT4 to the plasma membrane, similar to the effect of 100 μM insulin. Such GLUT4 translocations were inhibited by wortmannin, indicating that the induction of GLUT4 translocation by VO(3mpa)$_2$ is due to the PI3K−Akt pathway in the signaling cascade.

The highest antidiabetic activity of VO(3mpa)$_2$ was then proved by the hypoglycemic effect of this complex when the four compounds were given in STZ−mice by a single ip injection at the same dose (196 μmol V kg^{-1}body weight) (42).

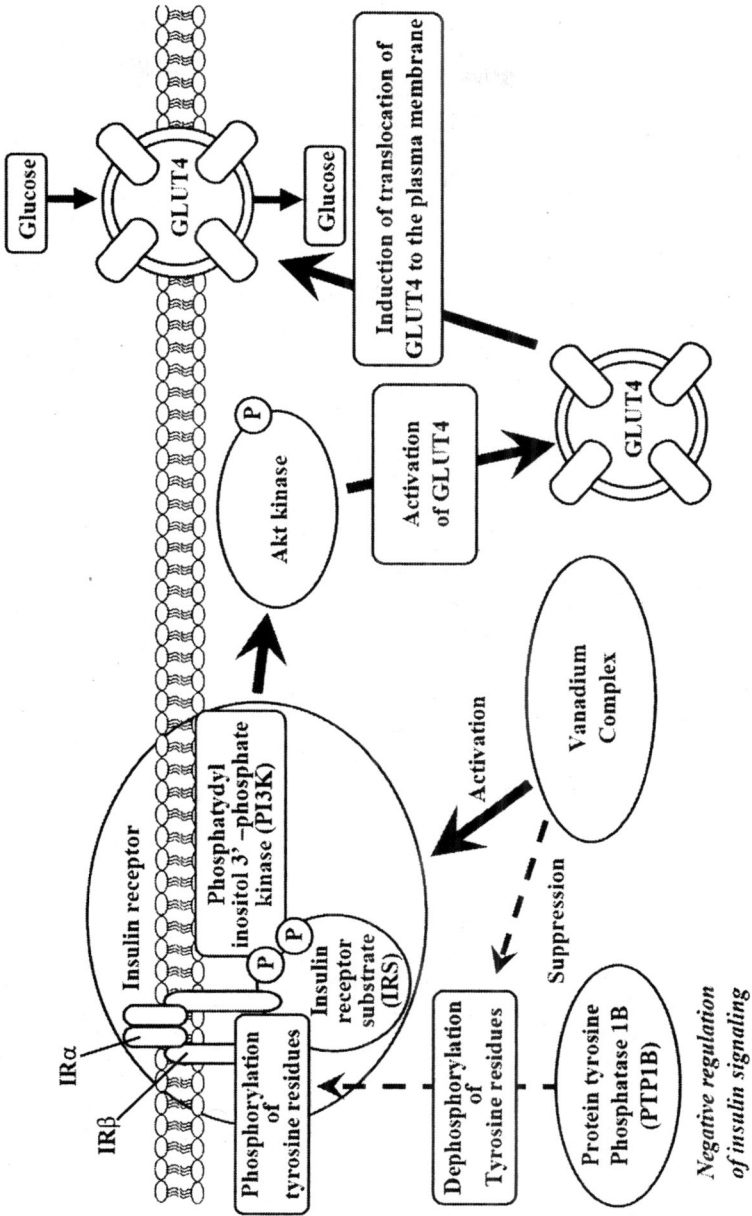

Figure 2. A proposed mechanism of vanadium complexes that induce the insulin signaling cascade

118

Based on the results, VO(3mpa)$_2$ was concluded to be a potent DM treating compound for activating the insulin signaling cascade. This complex acts on the IRβ and enhances IRS phosphorylation, which is followed by stimulation of the downstream PI3K/Akt pathway leading to GLUT4 translocation to the plasma membrane (Figure 2).

When vanadium complexes inhibit PTP1B or phosphorylate IRβ and IRS, then PI3K and Akt kinase are subsequently phosphorylated, which in turn activates GLUT4 and induces the translocation of GLUT4 to the plasma membrane. Therefore, glucose is incorporated in the cell.

Acknowledgment

I am grateful to Mrs. Fujii, Y., Mrs. Watanabe, H., Mr. Tamura, H., Mr. Fujimoto, S., Mrs. Uchida, K., Mrs. Tamiyama, J., Mrs. Fujisawa, Y., Mr. Takechi, K., Mrs. Tamura, A., Mr., Takino, T., Mr. Adachi, Y., Mr. Tayama, K., Ms. Basuki, W., Dr. Fugono, J., Dr. Yasui, H., Dr. Yoshikawa, Y., Dr. Kawabe, K., Dr. Yoshitake, A., Dr. Hamajima, Y. (Kyoto Pharmaceutical University), Dr. Takada, J. (Kyoto University), Prof. Tawa, R. (Hiroshima International University), Prof. Kanamori, K. (Toyama University), Dr. Fukui, K. (Yamagata Technopolis Institute), Prof. Ozutumi, K. (Ritsumeikan University), Dr. Kojima, Y. (Osaka City University), Prof. Rehder, D. (University of Hamburg, Germany), Prof. Kiss, T. (Szeged University, Hungary), Dr. Hattori, M. (Kyoto Medical Center) for their great contribution to the research for many years. The study was supported in part by grants from the Ministry of Education, Culture, Sports, Science and Technology (MEXT) of the Japanese Government (Grants–in Aid for Scientific Research (B), Scientific Research of Priority Area, and Specially Promoted Research to H. S.).

References

1. Wild, S.; Rogic, G.; Green, A.; Sicree, R.; King, H. *Diabetes Care* **2004**, *27*, 1047.
2. WHO, Definition, Diagnosis and Classification of Diabetes Mellitus and its complications, Geneva 1999, 1.
3. Lyonnet, B.; Martz, X.; Martin, E.; *Presse Med.* **1899**, *1*, 191.
4. Sakurai, H.; Kojima, Y.; Yoshikawa, Y.; Kawabe, K.; Yasui, H. *Coord. Chem. Rev.* **2002**, *226*, 187.
5. Sakurai, H. *Chem. Rec.* **2002**, *2*, 237.
6. Sakurai, H.; Yasui, H.; Adachi, Y. *Expert Opin. Investig. Durugs* **2003**, *12*, 1189.
7. Cohen, N.; Halberstam, M.; Shilimovich, P.; Chang, C. J.; Shamoon, H.; Rossetti, L. *J. Clin. Invest.* **1995**, *95*, 2501.

8. Cusi, K.; Cukier, S.; DeFronzo, R. A.; Torres, M.; Puchulu, F. M.; Redondo, J. C. P. *J. Clin. Endocrinol. Metab.* **2001**, *86*, 1410.
9. Fugono, J.; Yasui, H.; Sakurai, H. *J. Pharm. Pharmacol.* **2001**, *53*, 1247.
10. Sakurai, H.; Fugono, J.; Yasui, H. *Mini-Reviews in Med. Chem.* **2004**, *4*, 41.
11. Fugono, J.; Yasui, H.; Sakurai, H. *J. Pharm. Pharmacol.* **2002**, *54*, 611.
12. Karmaker, S.; Saha, T. K.; Yoshikawa, Y.; Yasui, H.; Sakurai, H. *J. Inorg. Biochem.* **2006**, *100*, 1535.
13. Sakurai, H.; Tsuchiya, K.; Nukatsuka, M.; Sofue, M.; Kawada, J. *J. Endocrinol.* **1990**, *126*, 451.
14. Sakurai, H.; Tsuchiya, K.; Nukatsuka, M.; Kawada, J.; Ishikawa, S.; Yoshida, H.; Komatsu, M. *J. Clin. Biochem. Nutr.* **1990**, *8*, 193.
15. Sakurai, H.; Fujii, K.; Watanabe, H.; Tamura, H. *Biochem. Biophys. Res. Commun.* **1995**, *214*, 1095.
16. Nakai, M.; Watanabe, H.; Fujiwara, C.; Kakegawa, H.; Satoh, T.; Takada, J.; Matsushita R,; Sakurai, H. *Biol. Pharm. Bull.* **1995**, *18*, 719.
17. Fujimoto, S.; Fujii, K.; Yasui, H.; Matsushita, R.; Takada, J.; Sakurai, H. *J. Clin. Biochem. Nutr.* **1997**, *23*, 113.
18. Kawabe, K.; Yoshikawa, Y.; Adachi, Y.; Sakurai, H. *Life Sic.* **2006**, *78*, 2860.
19. Adachi, Y.; Sakurai, H. *Chem. Pharm. Bull.* **2004**, *52*, 428.
20. Tawa, R.; Uchida, K.; Taniyama, J.; Fujisawa, Y.; Fujimoto, S.; Nagaoka, T.; Kanamori, K.; Sakurai, H. *J. Pharm. Pharmacol.* **1999**, *51*, 119.
21. Takino, T.; Yasui, H.; Yoshitake, A.; Hamajima, Y.; Matsushita, R.; Takada J.; Sakurai, H. *J. Biol. Inorg. Chem.* **2001**, *6*, 133.
22. Fugono, J.; Yasui, H.; Sakurai, H. *J. Pharm. Pharmacol.* **2001**, *53*, 1247.
23. Fukui, K.; Fujisawa, Y.; Ohya-Nishiguchi, H.; Kamada, H.; Sakurai, H. *J. Inorg. Biochem.* **1999**, *77*, 215.
24. Fujisawa, Y.; Sakurai, H. *Chem. Pharm. Bull.* **1999**, *47*, 1668.
25. Yasui, H.; Takechi K.; Sakurai, H. *J. Inorg. Biochem.* **2000**, *78*, 185.
26. Yasui, H.; Tamura, A.; Takino, T.; Sakurai, H. *J. Inorg. Biochem.* **2002**, *91*, 327.
27. Takechi, K.; Tamura, H.; Yamaoka, K.; Sakurai. H. *Free. Rad. Res.* **1997**, *26*, 483.
28. Sakurai, H,; Takechi, K,; Tsuboi, H,; Yasui, H. *J. Inorg. Biochem.* **1999**, *76*, 71.
29. Tayama, K.; Adachi, Y.; Yasui, H.; Sakurai, H. PACIFICHEM2005, Honolulu, USA., Dec. 2005.
30. Sakurai, H.; Tamura, A.; Takino, T.; Ozutsumi, K.; Kawabe, K.; Kojima, Y. *Inorg. Reac. Mechan.* **2000**, *2*, 69.
31. Sasagawa, T.; Yoshikawa, Y.; Kawabe, K.; Sakurai, H.; Kojima, Y. *J. Inorg. Biochem.* **2002**, *88*, 108.
32. Yano, S.; Nakai, M.; Sekiguchi, F.; Obata, M.; Kato, M.; Shiro, M.; Kinoshita, I.; Mikuriya, M.; Sakurai, H,; Orvig, C. *Chem. Lett.* **2002**, *9*, 916.

33. Nakai, M.; Obata, M.; Sekiguchi, F.; Kato, M.; Shiro, M.; Ichimura, A.; Kinoshita, I.; Mikuriya, M.; Inohara, T.; Kawabe, K.; Sakurai, H.; Orvig, C.; Yano, S. *J. Inorg. Biochem.* **2004**, *98*, 105.
34. Kiss, E.; Petrohan, K.; Sanna, D.; Garribba, E.; Micera, G.; Kiss, T. *Polyhedron* **2000**, *19*, 55.
35. Kiss, E.; Benyei, A.; Kiss, T. *Polyhedron* **2003**, *22*, 27.
36. Gätjens, J.; Meier, B.; Kiss, T.; Nagy, E. M.; Buglyó, P.; Sakurai, H.; Kawabe, K.; Rehder, D. *Chem. Eur. J.* **2003**, *9*, 4924.
37. Kiss, T.; Kiss, E.; Garribba, E.; Sakurai, H. *J. Inorg. Biochem.* **2000**, *80*, 65.
38. Kiss, E.; Garribba, E.; Micera, G.; Kiss, T.; Sakurai, H. *J. Inorg. Biochem.* **2000**, *78*, 97.
39. Srivastava, A.K.; Mehdi, M.Z. *Diabetic Med.* **2004**, *22*, 2.
40. Jelveh, K.A.; Zhande, R.; Brownsey, R.W. *J. Biol. Inorg. Chem.* **2006**, *11*, 379.
41. Basuki, W.; Hiromura, M.; Adachi, Y.; Tayama, K.; Hattori, M.; Sakurai, H. *Biochem. Biophy. Res. Commun.*, **2006**, *349*, 1163.

Chapter 10

Do Vanadium Compounds Drive Reorganization of the Plasma Membrane and Activation of Insulin Receptors with Lipid Rafts?

Deborah A. Roess[*], Steven M. L. Smith, Alvin A. Holder,
Bharat Baruah, Alejandro M. Trujillo, Daniel Gilsdorf,
Michelle L. Stahla, and Debbie C. Crans[*]

Departments of Biomedical Sciences and Chemistry, Colorado State
University, Fort Collins, CO 80523–1872

The enhancement by vanadium compounds of insulin-mediated signaling in RBL-2H3 cells was investigated. The studies were based on preliminary studies in which a hydrophobic vanadium compound facilitated the translocation of the insulin receptor into plasma membrane rafts. Such results led to the studies of a vanadium compound known to be insulin-enhancing in animal model studies. The proof of concept, namely that vanadium compounds could interact directly with lipid bilayers made up of amphipathic lipids, was supported by results in a very simple model system based on cholesterol-doped AOT reverse micelles. Studies adding vanadium compound to this simple model system demonstrated that the presence of the vanadium compound resulted in lipid reorganization.

© 2007 American Chemical Society

Introduction

The possibility for vanadium compounds facilitating any insulin-enhancing effects through reorganization of plasma membrane lipids was investigated in a cell line derived from rat baspophilic leukemia (RBL) cells and a simple model system. The rationale for these studies was based on the well-known insulin-enhancing properties of several lipophilic vanadium compounds (*1, 2*) and the recent report of a vanadium compound penetrating the lipid layer in a model system (*3*). Effects of vanadium-containing compounds are of interest because they can normalize both elevated blood glucose and lipid levels and may have long-term benefits to cardiovascular health, which is a frequent complication of diabetes (*4, 5*). The mechanism of insulin action on its target cells involves binding of insulin to its plasma membrane receptor (*6-9*). Evidence for translocation of the receptor to specialized lipid microdomains (rafts) in the plasma membrane has led to the suggestion, that these rafts may serve as platforms for insulin receptor-mediated signal transduction (*10, 11*). Although vanadium compounds are generally believed to act downstream of the insulin receptor (*12-16*), some effects of transition metal compounds such as vanadium compounds may be mediated through their actions on the plasma membrane and the organization of proteins and lipids within the lipid bilayer. Here we investigate whether insulin-enhancing vanadium compounds may evoke their effects through direct interactions with the plasma membrane of cells expressing insulin receptors. Such direct interactions could result in the perturbation of membrane lipid organization and facilitate the movement of insulin receptors or other signaling molecules into membrane rafts in the absence of an insulin signal. Thus, insulin-enhancing vanadium compounds may concentrate these molecules in rafts where they have ready access to other molecules involved in downstream signaling events and, in this fashion, enhance insulin-mediated cellular responses.

Insulin-enhancing lipophilic vanadium compounds were found to be more readily absorbed *in vivo*, and less toxic than the vanadium salts (*1, 2*). Since these compounds might interact with cell membrane lipids, we tested the effects of one highly lipophilic vanadium compound prepared from salicylaldehyde and tris(hydroxymethyl)aminoethane to form an adduct with $VO(acac)_2$ refered to as [VO(saltris)]$_2$, **1** here (*17, 18*). While these studies were underway vanadium-containing probes in reverse micelles (*3, 19-21*) e.g. $NH_4[VO_2(dipic)]$, **2** and $V_{10}O_{28}^{6-}$ (V_{10}), **3** were reported to be excellent probes of the water pool (*20, 21*) and the interfacial region (*3*) in the reverse micelle, respectively. The fact that one vanadium compound was found to intercalate in the interfacial layer of the AOT-based reverse micelle (*19*) further supported the concept that vanadium compounds could exert some of their actions at the plasma membrane of cells *in vivo*. Given the complexities of studies involving plasma membranes or micelles formed from plasma membrane components, reverse micelles provide an

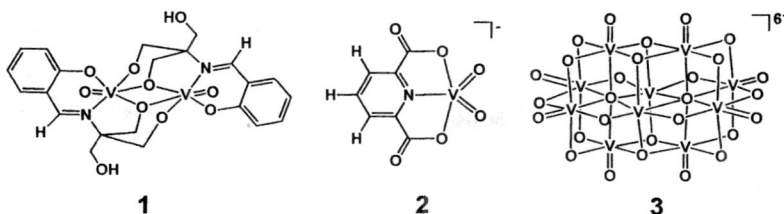

Figure 1. Structures of [VO(saltris)]$_2$, 1; [VO(dipic)]$^-$, 2 and V$_{10}$O$_{28}$$^{6-}$, 3.

attractive simple model system in which to evaluate vanadium compound interactions with lipids.

A cell line derived from RBL cells were used in the studies presented here for several reasons. We have had considerable experience in evaluating the localization of RBL-2H3 cell plasma membrane receptors in membrane rafts during cell signaling (*22-24*). Furthermore, we have used biophysical methods for evaluating molecular dynamics of membrane lipids and proteins and interactions between membrane molecules in this cell system (*22*). Importantly, RBL-2H3 cells have insulin receptors in addition to the Type I Fcε receptor (FcεRI), and are capable of downstream signaling in response to binding of these receptors' respective ligands. Moreover, vanadate has been reported to activate signaling in RBL-2H3 cells including release of Ca^{2+} from intracellular stores and plasma membrane flux although the mechanism is unclear (*25*). Mast cells, which like basophils are derived from a CD_{34}^+ precursor in the bone marrow and have basophil-like activity in tissues, may also be involved in the development of heart disease (*26*) although their role remains controversial. Thus, signaling mechanisms utilized by RBL-2H3 cells, a mast cell model, are, in and of themselves, of interest.

Materials and Methods

Chemicals. Sodium metavanadate, $NaVO_3$ (99.9%), salicylaldehyde (98%), tris(hydroxymethyl)aminoethane (99%), 2,6-pyridinedicarboxylic acid (99%), bis(acetylacetonato)oxovanadium(IV) (98%), AOT (sodium bis(2-ethylhexyl) sulfosuccinate (99%), cholesterol(98%), isooctane (99%) and carbon tetrachloride (CCl_4) (99.9%) were purchased from Aldrich and used without purification unless noted otherwise. The [VO(saltris)]$_2$ was prepared by condensation of salicylaldehyde and tris in methanol and reacting the pro-duct, 2-salicylideniminato-2-(hydroxymethyl)-1,3-dihydroxy-propane (H$_4$saltris) with VO(acac)$_2$ as reported previously (*17, 18*). Ammonium 2,6-dipico-linatodioxovanadium(V), NH$_4$[VO$_2$(dipic)] was synthesized as described previously (*27-29*). Decavanadate (V_{10}) solutions were prepared from solid metavanadate, which upon dissolution were acidified to pH 4 and thus converted completely to V_{10}. The compounds were characterized by routine methods including NMR and IR spectroscopy and elementary analysis. AOT was purified by dissolution in methanol and stirring overnight in the presence of activated charcoal. Subsequent filtration and removal of methanol by distillation under vacuum yielded AOT suitable for use. Cyclohexane-d_{12} (C_6D_{12}) and D_2O (Cambridge Isotope Laboratories Inc.) were used without further purification.

Stock Solution Preparation. The low solubility of [VO(saltris)]$_2$ in aqueous solution led to preparation of stock solutions in dmso. Aqueous stock solutions of NH_4[VO$_2$(dipic)] were prepared by dissolution in D_2O in a volumetric flask. All the freshly prepared stock solutions were of 50, 25 and 15 mM at pH 4.5. Decavanadate (V_{10}) solutions were prepared by dissolving $NaVO_3$ (50, 100 and 200 mM) in D_2O and reducing the pH to 4.0 to ensure complete conversion to V_{10}. The pH values of the aqueous V_{10} stock solutions were measured at 25°C using an Orion 420A pH meter calibrated with three buffers of pH 4, 7 and 10. The pH of the vanadate aqueous solutions was adjusted using NaOD/DCl.

Cell culture. RBL-2H3 cells were maintained in cell culture medium including Earle's Minimum Essential Medium (MEM), fetal bovine serum, and 5mM L-glutamine, penicillin, ampicillin and amphotericin B. Cells were harvested for experiments using 5 mM EDTA and washed in Hank's balanced salt solution (BSS).

Cell viability assay. Initial experiments characterized the effects of various concentrations of [VO(saltris)]$_2$ or [VO$_2$(dipic)]$^-$) on cell viability. RBL-2H3 cells were plated in petri dishes and grown to 80% confluence and then washed with BSS. Cells were then incubated with various concentrations of either [VO(saltris)]$_2$ or [VO$_2$(dipic)]$^-$ overnight. Cell aliquots were withdrawn from each culture, mixed with an equal volume of trypan blue, and examined microscopically. Both the dead cells stained by trypan blue and unstained living cells were scored.

Raft protocol. To examine the distribution of insulin receptors within membrane fractions exhibiting either low or high buoyancy, sucrose gradient ultracentrifugation methods were used. Briefly, 5 x 10^6 RBL-2H3 cells were obtained from cell culture and suspended in BSS. Cells were then treated with [VO(saltris)]$_2$ for one hour prior to cell lysis. The cell lysate was mixed with an equal volume of 80% sucrose to obtain a sample consisting of cell membranes and their components in 40% sucrose. A discontinuous sucrose gradient from 10-80% sucrose was constructed with the cell fraction comprising the 40% sucrose fraction. Samples were then subjected to isopycnic ultracentrifugation using an overnight spin at approximately 180,000 g. After centrifugation, 640 μL fractions are carefully colleted from the top down of the gradient downward. Aliquots from each fraction were probed for proteins of interest including the insulin receptor. The insulin receptor was identified using on Western blots using an anti-insulin receptor antibody (Sigma-Aldrich, St. Louis, MO).

Effects of vanadium compounds of RBL-2H3 degranulation. RBL-2H3 cells were grown in petri dishes and harvested as for other experiments. These cells, upon receiving signals via selected membrane receptors, can release the contents of their granules, which contain histamine and tryptase. The magnitude of a degranulation response was measured in RBL-2H3 cells by quantifying the amount of tryptase that is released using a mast cell degranulation assay kit from Chemicon (Temecula, CA) using Manufacturers' instructions.

Reverse Micelle Preparation. A 120 mM AOT stock solution was prepared by dissolving NaAOT in isooctane with a cholesterol concentration of 30 mM. Aliquots of aqueous stock solutions of concentration 5.0, 10 and 20 mM V_{10} were added to the AOT/cholesterol/isooctane stock solution to yield reverse micelle solution of $w_0 = 8$. A 50 mM AOT stock solution was prepared by dissolving AOT in cyclohexane-d_{12} (C_6D_{12}) under ambient conditions with a cholesterol concentration of 12.5 mM. Aliquots of aqueous stock solutions of concentration 50, 25 and 15 mM $NH_4[VO_2(dipic)]$ were added to the AOT/cholesterol/C_6D_{12} stock solution to yield reverse micelle solution of $w_0 = 8$. All samples were mixed by vortexing to yield an optically clear solution prior to ^1H NMR spectroscopic measurements. Formation of reverse micelles were confirmed by dynamic light scattering (DLS) experiments.

NMR spectroscopy. The samples were subjected to multinuclear NMR spectroscopy. The ^1H NMR results described in this manuscript were obtained using a Varian Inova-500 spectrometer at 500 MHz. Routine parameters were used for the 1D ^1H NMR experiments. ^1H NMR chemical shifts were referenced against a 3-(trimethylsilyl)propane sulfonic acid sodium salt (DSS) as an external reference.

Results

The rationale for compound selection of three different vanadium compounds [VO(saltris)]$_2$, [VO$_2$(dipic)]$^-$ and V_{10} was based on lipophilicity, insulin-enhancing effects (*29*) and previous studies carried out in reverse micelles (*3*). Together the results demonstrate that vanadium compounds have the potential to act through changes in lipid packing in cells and in simple model systems. The initial cellular experiments were carried out using the lipophilic vanadium compound ([VO(saltris)]$_2$) and provided results showing translocation of insulin receptors into high buoyancy membrane fractions in the absence of insulin. The recent report documenting intercalation of a vanadium compound ([VO$_2$(dipic)]$^-$) in the lipid interface (*3*) motivated additional studies with this compound that had already been demonstrated to have insulin-enhancing effects in an animal model system (*29*). The studies in the RBL-2H3 cells were supported by investigations in reverse micelles which showed that the vanadium-containing probe, V_{10} was able to impact the chemical shifts of some of the cholesterol H-atoms in the lipid interface. These effects can be interpreted as changes in lipid packing.

To examine effects of a hydrophobic vanadium compound on membrane localization of the insulin receptor in the absence of the insulin ligand, we treated RBL-2H3 cells for 1 hr with 30 μM [VO(saltris)]$_2$(17, 18). This compound is non-toxic at the concentrations used in these experiments (data not shown). As shown in Figure 2, the addition of [VO(saltris)]$_2$ to RBL-2H3 cells, resulted in movement of insulin receptor to higher buoyancy membrane fractions in the absence of insulin. Prior to [VO(saltris)]$_2$ treatment, 97% of IR was localized in sucrose fractions containing 39-52% sucrose. Following [VO(saltris)]$_2$ treatment, 97% of the receptor was localized in 34-39% sucrose. We believe that these structures are rafts based on previous reports showing that FcεRI on RBL-2H3 cells appearing in sucrose fractions containing less than 40% sucrose (30) following crosslinking of the receptor with antigen and similar results with the luteinizing hormone receptor in our laboratory (31). Furthermore, [VO(saltris)]$_2$ treatment caused degranulation of RBL-2H3 cells in a dose-dependant fashion, as demonstrated in Figure 3.

Given the problems associated with working with lipophilic compounds and to explore if insulin-enhancing vanadium compounds also exert similar effects, studies were continued using [VO$_2$(dipic)]⁻. Interestingly, treatment of cells with a vanadium compound that is much less lipophilic also enhanced the insulin-mediated effects on RBL-2H3 degranulation. Degranulation were measured

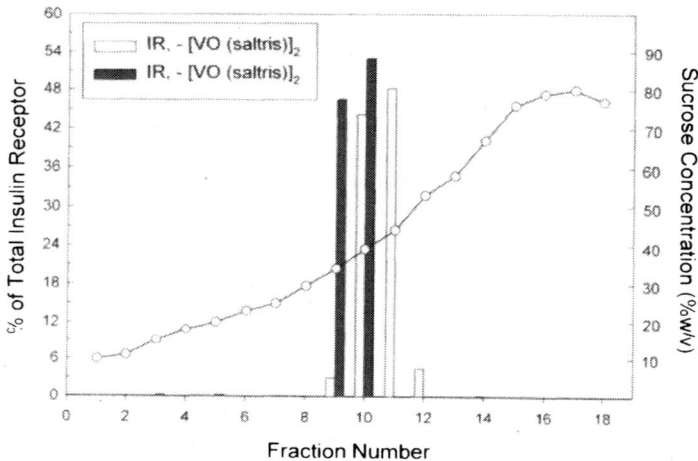

Figure 2. Translocation of IR from higher density sucrose fractions to lower density sucrose fractions upon treatment with approximately 30 μM [VO(saltris)]$_2$. The relative amount of IR in each fraction was measured from western blots using a Biorad calibrated densitometer. Sucrose concentrations (") for each fraction were measured using a Bausch and Lomb refractometer.

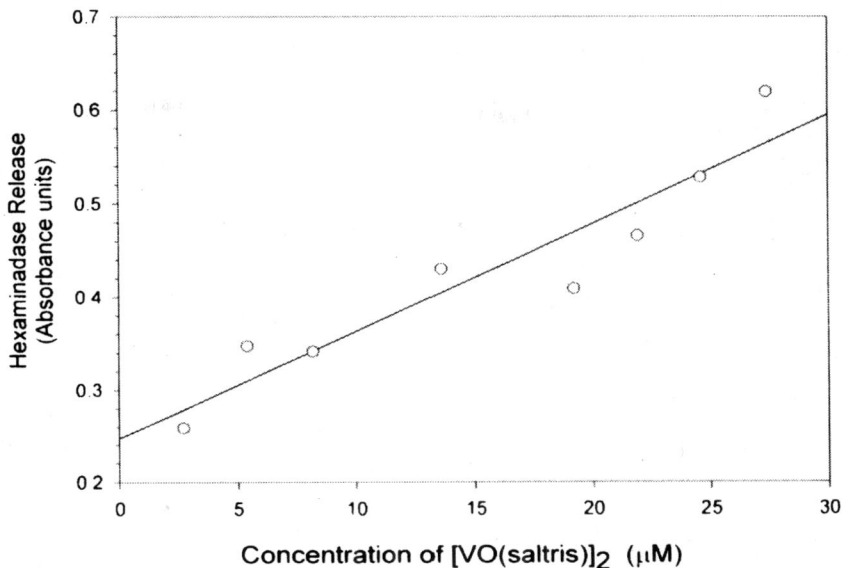

Figure 3. Degranulation of RBL-2H3 cells as measured by hexaminadase release from histamine-containing granules following incubation with increasing concentrations of [VO(saltris)]₂.

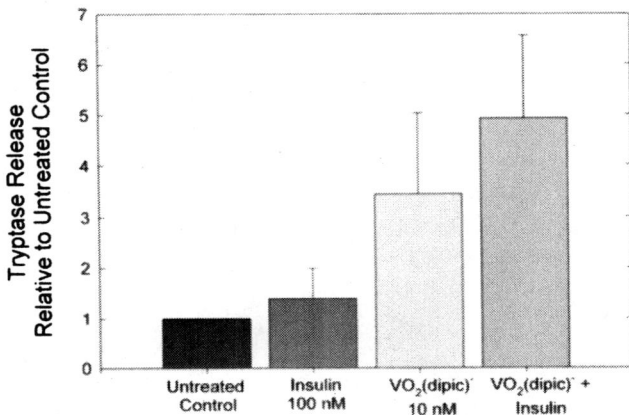

Figure 4. Degranulation of RBL-2H3 treated with [VO₂(dipic)]⁻ alone or with insulin. [VO₂(dipic)]⁻ caused increased degranulation of RBL-2H3 and enhanced insulin-mediated degranulation. Results shown are the mean and S.E.M. from three experiments performed in triplicate.

using hexamidase and tryptase release, respectively. There was a 3-fold increase in degranulation by RBL-2H3 cells compared to untreated cells following treatment with 10 nM [VO$_2$(dipic)]$^-$ alone (Figure 4). This concentration of [VO$_2$(dipic)]$^-$ was not toxic to cells on this timescale of these experiments (Figure 5). The extent of degranulation was 5-fold over untreated levels in response to 100 nM insulin and 10 nM [VO$_2$(dipic)]$^-$. Surprisingly, insulin alone had comparatively little effect on degranulation. These results suggest that [VO$_2$(dipic)]$^-$ enhances the activity of insulin which, by itself, has comparatively little effect on these cells.

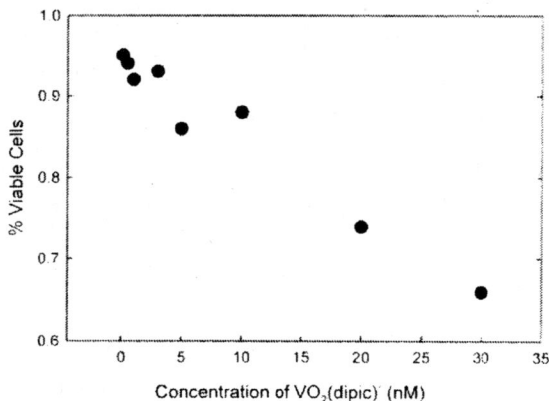

Figure 5. Cell viability as measured by trypan blue dye exclusion following treatment with [VO$_2$(dipic)]$^-$ at increasing concentrations.

Demonstration that two classes of vanadium compounds in the RBL-2H3 cells were able to facilitate insulin-mediated signal transduction led us to seek proof of concept in the simple cholesterol-containing AOT reverse micelles model system. Considering the stability of V$_{10}$ in aqueous solution, and the recent studies with this probe (*21*) led us to first carry out studies with this compound.

Cholesterol Environment Change in the Presence of a Vanadium-Containing Probe (V$_{10}$)

The ^{1}H NMR chemical shifts of cholesterol were followed when cholesterol was interchelated into reverse micelles and dispersed in solution. The structure and numbering system used is shown in Figure 6. Spectra recorded of cholesterol and AOT dispersed in CCl$_4$ were used as reference spectra for comparison.

Reverse micelles at a $w_o = 8$ were made containing a ratio of 1: 4 (30 mM : 120 mM) cholesterol:AOT using isooctane as the organic solvent. The nature of the interaction between cholesterol and AOT was investigated using 2D NOESY spectroscopy and will be detailed elsewhere; suffice to summarize here, that cholesterol appears to be deeply entrenced into the lipid interface (*32, 33*) as reported for membrane systems (*34, 35*).

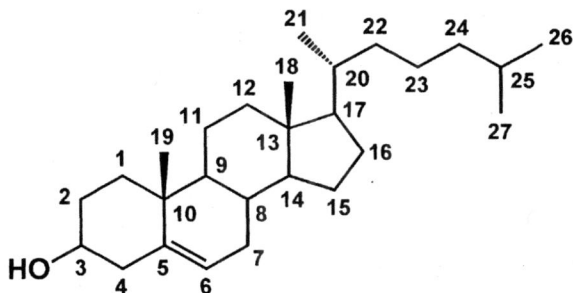

Cholesterol

Figure 6. Structure of cholesterol.

Reverse micelles were formed by the addition of aqueous solutions of preformed V_{10} (50 mM, 100 mM and 200 mM) for comparison with the reverse micelles in the absence of V_{10}. The chemical shifts for some of the ^1H in the cholesterol molecule changed as the V_{10} compound was added to the water pool and selected signals are shown in Fig. 6. The chemical shifts for the H-atoms on C_4, C_7 and C_{12} changed little, whereas, a large shift was observed for the H-atoms on C_3, C_6 and C_{18}.

The changes in chemical shifts reflect changes in environment and is consistent with changes in lipid packing near these H-atoms upon addition of V_{10} probe to the reverse micelle. However, other H-atoms change little indicating that their environments change little. Combined, these data support the possibility that the environment on one side (the one with the CH_3- group on C_{12}) of the cholesterol molecule is changed more than the side with the CH_3 group (C_{19}) on C_{10}. Future studies will explore in greater detail these effects, as well as studies in which different vanadium compounds have been investigated.

Discussion

The addition of $[VO(saltris)]_2$ to RBL-2H3 cells, resulted in movement of insulin receptor to slightly higher buoyancy membrane fractions. Although the shift in in location of the insulin receptor is modest, both the Fc receptor (*23, 24,*

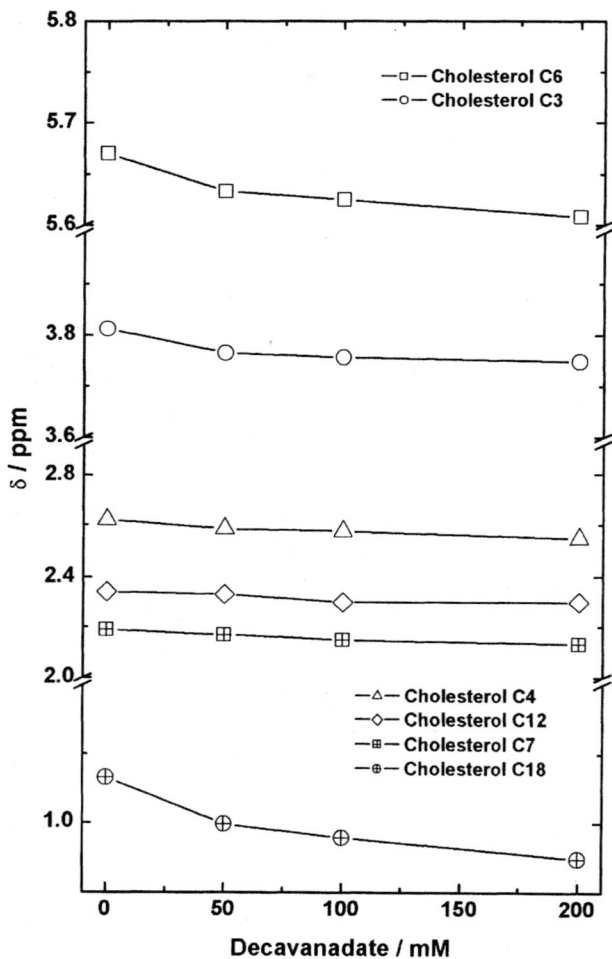

Figure 7. The ^1H NMR chemical shift of selected cholesterol protons prepared from cholesterol (30 mM) and AOT (120 mM) based w_o = 8 reverse micelles is shown as increasing concentrations of V_{10} is added to the water-pool of the reverse micelle. The chemical shifts for protons on carbons shown here are located as indicated in Figure 6 on the structure.

30) and the human luteinizing hormone receptor (*24*) have previously been shown to shift into higher buoyancy membrane fractions. Larger shifts were observed with the rat LH receptor (*31*). Additional work will document whether this modest shift is a characteristic of vanadium compounds effects on membranes in general and whether other compounds show greater shifts. We are continuing this work using a vanadium compound for which the insulin-enhancing effects have been documented in an animal model to determine if there is more profonounced movement to higher buoyancy membrane fractions.

The demonstration that $[VO_2(dipic)]^-$ facilitated increase in degranulation of RBL-2H3 cells is important and does imply that also vanadium compounds that are less lipophilic can exert these effects. The increase in insulin-mediated degranulation in the presence of $[VO_2(dipic)]^-$ could be the most significant result observed in these studies. However, verification using other treatment doses to confirm this interpretation of these observations with this complex and others are underway. Furthermore, experimentation to investigate whether the enhanced insulin response by $[VO_2(dipic)]^-$ treated cells is due to the translocation of the insulin receptor into rafts *prior to* binding of insulin is underway.

Vanadium compounds are generally not believed to bind to the insulin receptor (*36-38*) and are believed to exert their insulin-enhancing effects downstream of the insulin receptor (*12-16*). However, the likely effects on multiple pathways have recently been documented in, for example, the DNA microarray analysis of global gene expression levels documenting numerous changes in gene expression (*12*). The possibility that vanadium compounds interact directly with membranes or proteins closely associated with membranes seems high, particularly in light of the recent finding that the insulin-enhancing compound $[VO_2(dipic)]^-$ is located in the lipid layer of reverse micelles (*19*). This finding was particularly surprising considering the charge and polarity of this compound, but provide the precedent for other vanadium compounds partitioning in a imilar manner similarly in the cell membranes. However, the evidence presented needs substantiation in terms of a more exhaustive study. The observation that several classes of vanadium-compounds exert these responses may imply that the specific ligand is less important than anticipated (*39, 40*). This observation is in agreement with existing literature that show although effects of vanadium compounds vary with different oxidation state of the metal (*41, 42*), ligand really exert a more fine-tuning type effect.

Conclusions

Together these results suggest that insulin-mediated signaling in RBL-2H3 cells is enhanced by $[VO(saltris)]_2$ and $[VO_2(dipic)]^-$. Such a result may occur as the result of insulin receptor translocation into plasma membrane rafts where local concentrations of insulin receptors are high and downstream signaling molecules are readily available. Evidence for the possibility that a vanadium-containing probe would facilitate reorganization of the cholesterol arrangement in a model system is presented. Three different vanadium-containing probes were examined in this initial series of cells and a very simple model system, cholesterol-doped AOT reverse micelles. The results with the anionic probe V_{10} are presented here, and suggest that the presence of the vanadium-containing probe impacts the packing of the AOT-based reverse micelles containing cholesterol. Although the biological investigations have been carried out with vanadium-compounds capable of associating in lipid-environments, the studies in the model system show that even a strongly charged vanadium-containing probe affect the arrangements of the surfactant with respect to the cholesterol.

Acknowledgment

Financial support for this work by the American Heart Association (0650081Z to D.A.R.) and NSF 0314719 (DCC) is gratefully acknowledged. We thank Gabriel R. Harewood for providing some samples of [VO(saltris)]$_2$. We thnak Dr. Christopher D. Rithner for technical assistance and Dr. Nancy E. Levinger for stimulating discussions regarding the AOT-reverse system used in these studies.

References

1. Shechter, Y.; Karlish, S. J. D. *Nature* **1980**, *284*, 556-558.
2. Shechter, Y.; Meyerovitch, J.; Farfel, Z.; Sack, J.; Bruck, R.; Bar-Meir, S.; Amir, S.; Degani, H.; Karlish, S. J. D. In *Vanadium in Biological Systems: Physiology and Biochemistry*; Chasteen, N. D., Ed.; Kluwer Academic Publishers: Boston, 1990, pp 129-142.
3. Stover, J.; Rithner, C. D.; Inafuku, R. A.; Crans, D. C.; Levinger, N. E. *Langmuir* **2005**, *21*, 6250-6258.
4. Haffner, S. M. *Diabetes Care* **2003**, *26*, S83-S86.
5. Thom, T.; Haase, N.; Rosamond, W.; Howard, V. J.; Rumsfeld, J.; Manolio, T.; Zheng, Z.-J.; Flegal, K.; O'Donnell, C.; Kittner, S.; Lloyd-Jones, D.; Goff, D. C. J.; Hong, Y.; Adams, R.; Friday, G.; Furie, K.; Gorelick, P.; Kissela, B.; Marler, J.; Meigs, J.; Roger, V.; Sidney, S.; Sorlie, P.; Steinberger, J.; Wasserthiel-Smoller, S.; Wilson, M.; Wolf, P. *Circulation* **2006**, *113*, e85-e151.
6. Saltiel, A. R.; Kahn, C. R. *Nature* **2001**, *414*, 799-806.
7. Koricanac, G.; Isenovic, E.; Stojanovic-Susulic, V.; Miskovic, D.; Zakula, Z.; Ribarac-Stepic, N. *Gen. Physiol. Biophys.* **2006**, *25*, 11-24.

8. Dey, D.; Mukherjee, M.; Basu, D.; Datta, M.; Roy, S. S.; Bandyopadhyay, A.; Bhattacharya, S. *Cell. Physiol. Biochem.* **2005**, *16*, 217-228
9. Hegarty, B. D.; Bobard, A.; Hainault, I.; Ferre, P.; Bossard, P.; Foufelle, F. *P. Natl. Acad. Sci.* **2005**, *102*, 791-796.
10. Vainio, S.; Heino, S.; Mansson, J.-E. *EMBO reports* **2002**, *31*, 95-100.
11. Bickel, P. E. *Am. J. Physiol.* **2002**, *282*, E1-E10.
12. Willsky, G. R.; Chi, L.-H.; Liang, Y.; Gaile, D. P.; Hu, Z.; Crans, D. C. *Physiol. Genomics* **2006**, *26*, 192-201.
13. Basuki, W.; Hiromura, M.; Adachi, Y.; Tayama, K.; Hattori, M.; Sakurai, H. *Biochem. Bioph. Res. Co.* **2006**, *349*, 1163-1170.
14. Shechter, Y.; Goldwaser, I.; Mironchik, M.; Fridkin, M.; Gefel, D. *Coord. Chem. Rev.* **2003**, *237*, 3-11.
15. Thompson, K. H.; Orvig, C. *Dalton Trans.* **2006**, 761-764.
16. Liboiron, B. D.; Thompson, K. H.; Hanson, G. R.; Lam, E.; Aebischer, N.; Orvig, C. *J. Am. Chem. Soc.* **2005**, *127*, 5104-5115.
17. Asgedom, G.; Sreedhara, A.; Rao, C. P. *Polyhedron* **1995**, *13-14*, 1873-1879.
18. Asgedom, G.; Sreedhara, A.; Kivikoski, J.; Valkonen, J.; Kolehmainen, E.; Rao, C. P. *Inorg. Chem.* **1996**, *35*, 5674-5683.
19. Crans, D. C.; Rithner, C. D.; Baruah, B.; Gourley, B. L.; Levinger, N. E. *J. Am. Chem. Soc.* **2006**, *128*, 4437-4445.
20. Crans, D. C.; Baruah, B.; Levinger, N. E. *Biomed. Pharmacother.* **2006**, *60*, 174-181
21. Baruah, B.; Roden, J. M.; Sedgwick, M.; Correa, N. M.; Crans, D. C.; Levinger, N. E. *J. Am. Chem. Soc.* **2006**, *128*, 12758-12765
22. Roess, D. A.; Smith, S. M. L. *Biol. Reprod.* **2003**, *69*, 1765-1770.
23. Lei, Y.; Hagen, G. M.; Smith, S. M. L.; Barisas, B. G.; Roess, D. A. *Biochem. Bioph. Res. Co.* **2005**, *337*, 430-434.
24. Lei, Y.; Hagen, G. M.; Smith, S. M. L.; Barisas, B. G.; Roess, D. A. *Mol. Cell. Endocrinol.* **2006**, *in press*.
25. Ehring, G.; Kerschbaum, H.; Fanger, C.; Eder, C.; Rauer, H.; Cahalan, M. *J. Immunol.* **2000**, *164*, 679-687.
26. Pallandini, G.; Tozzi, R.; Perlini, S. *J. Hypertens.* **2003**, *21*, 1823-1825.
27. Wieghardt, K. *Inorg. Chem.* **1978**, *17*, 57-64.
28. Nuber, B.; Weiss, J.; Wieghardt, K. *Z. Naturforsch.* **1978**, *33B*, 265-267.
29. Crans, D. C.; Yang, L.; Jakusch, T.; Kiss, T. *Inorg. Chem.* **2000**, *39*, 4409-4416.
30. Sheets, E. D.; Holowka, D.; Baird, B. *Curr. Opin. Chem. Biol.* **1999**, *3*, 95-99.
31. Smith, S. M. L.; Lei, Y.; Liu, J.; Cahill, M. E.; Hagen, G. M.; Barisas, B. G.; Roess, D. A. *Endocrinology* **2006**, *147*, 1789-1795.
32. Maitra, A.; Dinesh; Patanjali, P.; Varshney, M. *Colloids Surf.* **1986**, *20*, 211-219.

33. Maitra, A.; Patanjali, P. K. *Colloids Surf.* **1987**, *27*, 271-276.
34. Tenchov, B. G.; MacDonald, R. C.; Siegel, D. P. *Biophys. J.* **2006**, *91*, 2508-2516.
35. Crockett, E. L.; Hazel, J. R. *J. Exp. Zool.* **1995**, *271*, 190-195.
36. Fantus, G.; Tsiani, E. *Mol. Cell. Biochem.* **1998**, *182*, 109-119.
37. Drake, P. G.; Posner, B. I. *Mol. Cell. Biochem.* **1998**, *182*, 79-89.
38. Srivastava, A. K.; Mehdi, M. Z. *Diabetic Med.* **2005**, *22*, 2.
39. Sakurai, H.; Kawabe, K.; Yasui, H.; Kojima, Y.; Yoshikawa, Y. *J. Inorg. Biochem.* **2001**, *86*, 94-94.
40. Sakurai, H.; Tamura, A.; Fugono, J.; Yasui, H.; Kiss, T. *Coord. Chem. Rev.* **2003**, *245*, 31-37.
41. Buglyo, P.; Crans, D. C.; Nagy, E. M.; Lindo, R. L.; Yang, L.; Smee, J. J.; Jin, W.; Chi, L.-H.; Godzala, M. E.; Willsky, G. R. *Inorg. Chem.* **2005**, *44*, 5416-5427.
42. Monga, V.; Thompson, K. H.; Yuen, V. G.; Sharma, V.; Patrick, B. O.; McNeill, J. H.; Orvig, C. *Inorg. Chem.* **2005**, *44*, 2678-2688.

Haloperoxidases: Mechanism and Model Studies

Chapter 11

Structural Studies of Vanadium Haloperoxidases: Insight into Halide Specificity, Stability, and Enzyme Mechanism

Jennifer Littlechild, Esther Garcia-Rodriguez, Elizabeth Coupe, Aaron Watts, and Mikail Isupov

Henry Wellcome Building for Biocatalysis, School of Biosciences, University of Exeter, Exeter, United Kingdom

Crystallographic studies of the vanadium haloperoxidase found in *Corallina* red algae has revealed details of the structure of these enzymes which has increased our understanding of halide specificity, stability, substrate binding and enzymatic mechanism. An efficient process to produce the enzyme in recombinant form after refolding of inclusion bodies has been developed. A novel truncated mutant dimeric form of the enzyme has been constructed for use in commercial biotransformation experiments.

Corallina officinalis is a red seaweed from the rhodophyta family, which also includes *Chondrus* and *Porphyra*. The red coloration is due to the presence of the pigment phycoerythrin that reflects red light and absorbs blue. This pigment allows the algae to photosynthesise at greater depths because blue light penetrates water to a greater depth than light of longer wavelengths. *C. officinalis* grows to about 5 inches in length and can be distinguished from other rhodophyta by the symmetrical branching of the thalli. *C. officinalis* can be seen throughout the North Western and North Eastern Atlantic including in rock pools in Ladram Bay, Devon, UK where it is seen growing approximately 1-5 inches below the water surface. A related *C. pilulifera* species is found off the coast of Japan.

The *Corallina* species have been shown to contain a bromoperoxidase that both binds and is dependent on vanadate for its activity. This enzyme is of particular industrial interest due to its known stability in organic solvents and its ability to withstand a wide pH range and temperatures up to 80°C (*1-3*).

© 2007 American Chemical Society

Structure of the *Corallina* bromoperoxidases

The structure of *C. officinalis* vanadium haloperoxidase was solved to 2.3Å in 2000 by Isupov *et al.*, (*4*) and showed a 595 amino acid chain folded into a single α + β type domain. The structure does not contain any disulfide bridges as found in the related *Ascophyllum* bromoperoxidase (*5*) although there are two cysteines present, and no post-translational modifications were observed. There are twelve monomers, each consisting of 19 α-helices which are 6 to 26 residues in length, eight 3_{10} helices and 14 β-strands which are mainly involved in β-hairpins. One surface of the monomer is flat and upon dimerisation this surface forms the central region resulting in two four-helical bundles at the centre of each dimer. The active site cleft uses residues from both monomers, with the residues of one predominantly being responsible for the bottom of the active site binding the vanadate, while the other constitutes the top region of the cleft . The dimers then interact to form a dodecamer with a 23 cubic point symmetry, which is approximately 150Å in diameter (Figure 1). The N-terminal helices point away from the main structure and form a central cavity in the dodecamer (Figure 2) which has no known function Two vanadate atoms are associated with each dimeric subunit (*4*) and are co-ordinated in a trigonal bipyramidal geometry with hydroxide and His496 in axial positions and three non-protein (solvent) oxygen atoms in equatorial positions. The original structure was solved with phosphate in place of the vanadium due to the high concentration of phosphate in the crystallisation conditions. Phosphate will prevent vanadate binding and it was observed that if phosphate binds to vanadium haloperoxidases phosphatase activity can be observed (*6*). Interestingly the structure of a novel acid phosphatase from *Escherichia blattae* has been published (*7*) and despite having relatively low amino acid sequence identity (18%) does have the same overall structural fold and conservation of most residues at the active site around the phosphate binding site. This reinforces the theory that both enzymes having a common origin (*8*). The only difference between the active site of this new acid phosphatase and that of the vanadium bromoperoxidase enzymes is His480, which in this case is an asparagine that lies completely out of the active site, facing the solvent. This change in position of an apparently key aminoacid in the active site reactivity of vanadium haloperoxidase is due to the smaller size of the phosphatase enzyme, causing the asparagine to be on the surface of the protein, interacting with the solvent and leaving a more open active site cleft. The same situation is found for the exposed position of Arg112 in the phosphatase (Arg397 in *Corallina* bromoperoxidase).

Later the structure of the vanadate bound form of the *Corallina* vanadium bromoperoxidase was solved using a different crystal form grown from polyethylene glycol (*9*). The structure of the enzyme's active site does not show any relevant difference to that of the phosphate-complexed structure, other than the coordination of vanadate with His553.

The *Corallina* vanadium bromoperoxidase is found to be stabilised by calcium and vanadium bound to the enzyme. One divalent cation (Ca^{2+} or Mg^{2+})

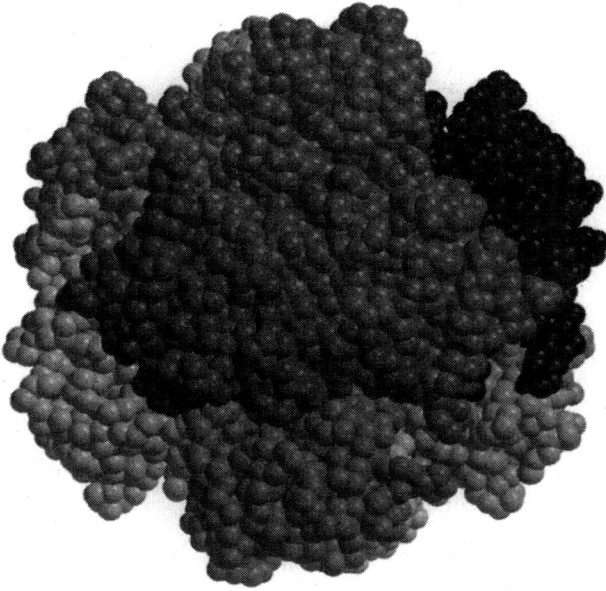

Figure 1. A CPK representation of the C. officinalis vanadium bromoperoxidase dodecamer (1QHB) colored by subunit. Figure constructed using Bobscript.

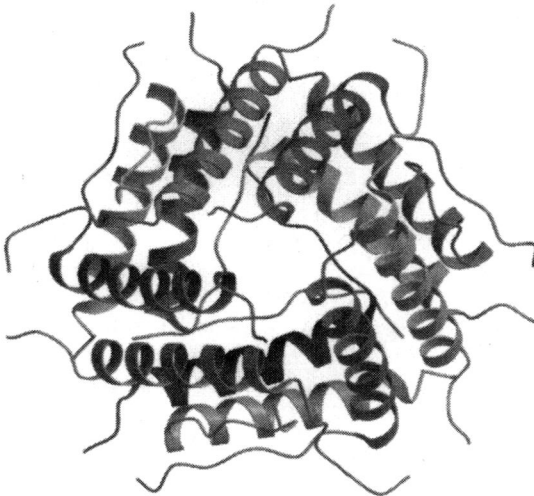

Figure 2. N-terminal helices from the central region of the C. officinalis dodecamer. Picture produced using Bobscript and rendered in Raster3D

is observed per monomer from the crystallographic analysis. A systematic study regarding the role of metal ions on the stability of the enzyme has been carried out (*10*). The structural analysis of the *C. pilulifera* recombinant enzyme over-expressed in yeast revealed that it contains one calcium atom per subunit which is found in the loop between the amino acid residues at 359 and 366, and that it is coordinated by the main chain oxygen atoms of Phe359, Gln361 and Gln368 and the carboxyl groups of Asp363 and Asp366 and a molecule of water (*9*). The coordination of a divalent metal has only been observed in bromoperoxidase from *Corallina* species. A sequence and structure comparison between the enzyme from *Ascophyllum nodosum* and that from *C. pilulifera* was carried out in order to study the differences that could lead to the acquisition of Ca^{2+} by the latter enzyme. Both alignments demonstrated that the calcium ion was bound in a loop that is shorter in the *A. nodosum* enzyme than in *Corallina* enzyme. The aminoacid sequence in the region of interest differs in length by three aminoacids. The thermal vibrations in this area are high (shown by a high B-factor in the crystallographic analysis) and a longer loop region would be expected to become more unstable. The interactions between the calcium atom and the protein molecule in the enzyme from *C. pilulifera* would be expected to stabilize this loop and lead to the stabilization of the whole structure. In fact, it can be observed that the corresponding loop in *C. pilulifera* shows relatively lower B-factors than that of *A. nodosum*. Other regions with high B-factors are solvent-exposed regions. It should be noted than differences between both enzymes other that the calcium-binding region are in those areas involved in the formation of the dodecameric arrangement in the *C. pilulifera* enzyme, in contrast with the more solvent-exposed dimeric arrangement for the *A. nodosum* bromoperoxidase which is stabilised by disulfide bonds.

The calcium content has been confirmed by inductively coupled plasma emission spectrometry experiments (*10*). The study of the effect of metal ions on the apo-enzyme stability has shown that the calcium ion significantly increased the enzyme stability. In addition, vanadate also increased the thermostability, and strontium and magnesium ions had similar effects to the calcium. Finally, the enzyme preincubated with both calcium and vanadate showed higher stability in methanol than that with vanadate alone. Calcium is commonly used in proteins for stabilisation and these studies confirm that this is the case for the *Corallina* bromoperoxidases.

Enzymatic Mechanism

Recent crystallographic studies have provided some information to help understand the mechanism of the *Corallina* vanadium dependent haloperoxidases The nature of the brominating agent for this enzyme has been described as a 'Br^{+}-like intermediate' being either a free halogenating agent such as HOBr, Br_2, or Br_3^{-} or an Enz-Br or V_{enz}-OBr species (*11a,b*). XAS studies (*12*) have clearly shown that the bromide binds within the active site in proximity to the vanadate centre. This would permit the bromide to directly attack the peroxyvanadate centre in the first step of the halide oxidation in the enzyme mechanism.

Spectroscopic studies recently carried out in Exeter (*13*) have utilised the comparative analysis of small molecule and enzyme catalysed reactions and have demonstrated that the enzyme produces a vanadium hypobromite intermediate as the reactive moiety. The analysis of the reaction of phenol red demonstrates a single multiple turnover pathway as the mechanism of bromination, with clear evidence for the formation of the intermediate bromophenol red. The comparison of the free hypobromite reaction with bromide and the reaction of the metallate/peroxide/bromide system shows inherent differences in the manner in which these react and the apparent protection of the hypobromite moiety conferred by the vanadate centre. Reactions with sulfanilic acid have demonstrated conclusively that the enzyme and the metallate/peroxide/bromide systems generate the same bromine like intermediate. Examination of the timecourse for this reaction results in the conclusion that the intermediate formed in the reaction is actually the vanadium-hypobromite intermediate and not a free bromine moiety. A reaction scheme as we understand this to date shows the key reaction intermediates (the resting state, the peroxovanadate intermediate and the vanadium hypohalite intermediate) in Figure 3.

Crystallographic studies on the native *C. pilulifera* bromoperoxidase and a R397W mutant form of the enzyme (see below) carried out by us have used bromide soaking experiments with the crystals. This has shown that there is a specific halogen-binding site involving Arg397 in the enzyme and other surrounding amino acids are also involved (Figure 4) (*14,15*). The crystal data sets of these halide complexes have been collected to a resolution of 1.78-2.5Å.

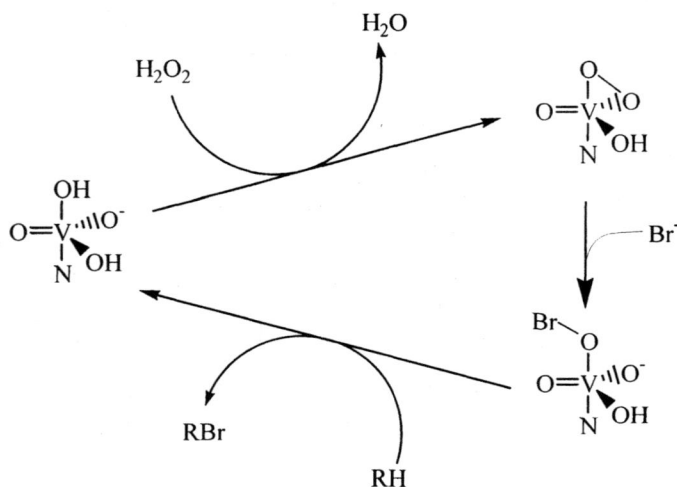

Figure 3. A reaction scheme of the vanadate centre for vanadium dependent bromoperoxidases. Charges have been omitted for clarity.

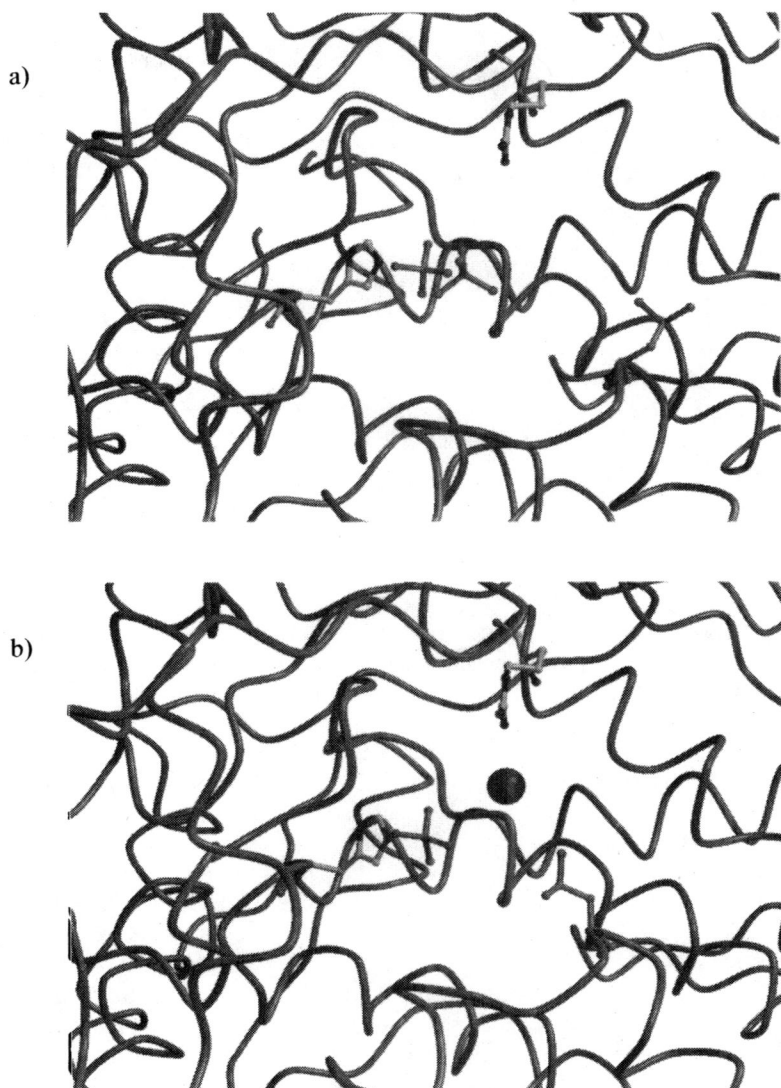

Figure 4. The active site of the Corallina bromoperoxidase a) Native enzyme b) Mutant enzyme showing residues His553, Arg/Trp397 and Leu337 in stick representation. The bromide ion is shown in CPK style. The figure was prepared using Bobscript.

142

Small   Molecule Mimics

The area of producing small molecule mimics of the vanadium haloperoxidase has been very active. Recently in collaboration with the group of Tucker at Exeter we have been able to construct two tripodal receptors that have been shown to bind phosphate and vanadate anions in organic solvents through H-bonding interactions. The compounds were synthesised in one step from their corresponding tetraamines, tris(2-aminoethyl)amine and tris(3-aminopropyl)-amine respectively.   Binding studies were carried out using the tetrabutyl-ammonium salts of dihydrogenphosphate and dihydrogenvanadate as guest species (16). This study has demonstrated the first successful complexation of vanadate by simple tripodal receptors in organic solvents and the first direct evidence of its binding by H-bonding interactions, as similarly observed at the active sites of the vanadium haloperoxidase enzymes. The characterisation of the halide binding site in the crystal structure of the *Corallina* vanadium haloperoxidase in close proximity to the vanadate site (see above) offers further scope for receptor design.

Substrate Binding

For many years it had been questioned as to whether the haloperoxidases bind their substrates, or whether they catalysed simply the formation of the brominating agent and allowed the reaction to occur non-specifically. It has been demonstrated for the bromoperoxidase from *A. nodosum* that there is a competition between indole and phenol red bromination (17). In addition this group also demonstrated that the indole was binding to the active site as evidenced by fluorescence quenching. Furthermore, the same enzyme will catalyse a regiospecific indole bromination (18). These results again are direct evidence that the reaction is occurring within the enzyme catalytic site.

We have used *Corallina* vanadium haloperoxidase enzymes and selected mutants that have been crystallised and subjected these to substrate soaking experiments with phenol red and   phoroglucinol. The substrate methyl-p-tolysulfide has been used to study sulfoxidation. These substrate binding sites have been determined in the crystal structures and are found to be located in a hydrophobic area in the active site channel of the enzyme (Figure 5) (14, 15). The crystal data sets of these substrate complexes have been collected to a resolution of 1.78-2.5Å.

Halide Specificity

Superposition of the active sites of the different vanadium halperoxidase enzymes has led to site-directed mutagenesis studies on the *C. pilulifera*

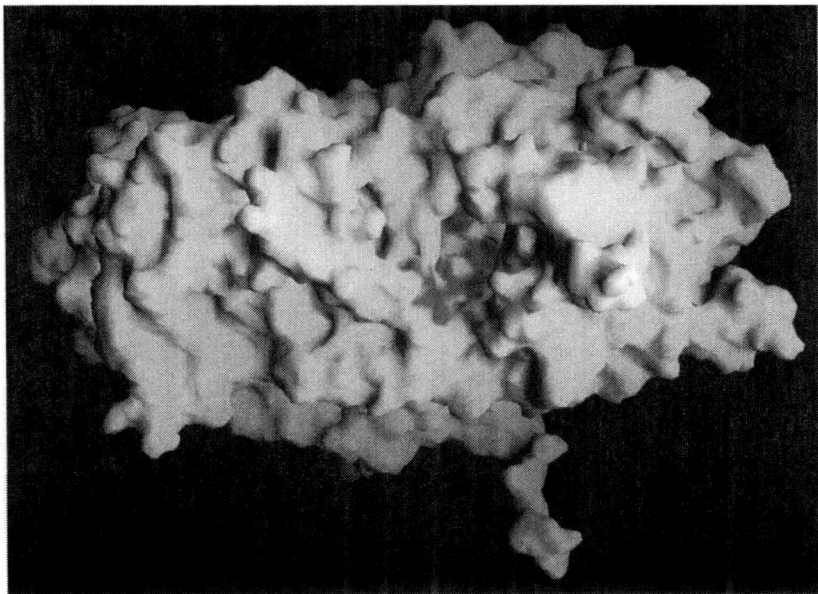

Figure 5. A representation of the area of hydrophobicity where the substrate binding is observed in the crystal structure. The figure is prepared using the program GRASP.

vanadium haloperoxidase with the aim to change its halogen specificity (*19*). It was suspected from the previous comparisons between vanadium bromoperoxidase and vanadium chloroperoxidase enzymes that the halide specificity should involve residue His480 and that its mutation to a phenylalanine would convert the halogenation reactivity of the algal enzyme to use a chloride ion. The fact that the enzyme from *A. nodosum* shows slight vanadium chloroperoxidase activity as well as its high vanadium bromoperoxidase activity points towards the involvement of other catalytic aminoacids. As mentioned previously, the position of Arg397 in *C. pilulifera* vanadium haloperoxidase is occupied by a tryptophan in the non-*Corallina* enzymes. Site-directed mutagenesis of Arg397 to every other aminoacid has been carried out and chloroperoxidase and bromoperoxidase activity assays have been performed. The results show slight chloroperoxidase activity in some of the mutants, particularly significant is the level of chloroperoxidase activity achieved when Arg397 is replaced with Trp and Phe. These values confirm the hypothesis of the involvement of this residue in the halogen selectivity. However, the specific chloroperoxidase activity of the R397W *C. pilulifera* mutant enzyme is much higher than that of the *A. nodosum* enzyme, which implies that other parts of the protein also play a role in the halide specificity of the reaction.

Refolding of the Recombinant *C. officinalis* Bromoperoxidase

The *Corallina* bromoperoxidases have already been cloned and over expressed including those from *C. officinalis* (*20*), and *C. pilulifera* (*21*). The expression of the *C. pilulifera bromoperoxidase* is low in both *Saccharomyces cerevisiae* and *E. coli* at 5.3mg and 0.3mg/litre respectively (*22, 21*). This is also the situation encountered with the related *C. officinalis* enzyme (*20*). In all cases the majority of the enzyme is expressed in inclusion bodies and the alteration of induction of expression does not improve solubility. The high propensity of inclusion body formation is thought to be due to the size and quaternary structure of the dodecameric enzyme.

A method to refold the recombinant *E.coli* over-expressed *C. officinalis* enzyme has been developed (*23*) using a systematic screen of re-folding conditions. This has yielded active enzyme from protein without a His-tag. It has been impossible to refold the tagged protein which can be explained by analysis of the crystal structure described above. It is likely that the histidines forming the tag are unable to pack into the central core of the decameric protein preventing refolding. Large scale refolding of the enzyme resulted in a yield of 40mg of enzyme representing a significant increase from the protein achieved by purifying the soluble recombinant protein. This represents a 59% recovery rate resulting in enzyme that does not require any further purification as the majority of contaminants have been removed by the inclusion body washing process. The activity of this enzyme was comparable to enzyme isolated directly from the algae using the monochlorodimedone (MCD) assay. The refolded protein has been characterised and compared to the native enzyme and was shown to be stable at temperatures of 80°C, over a pH range 5.5-10 and in organic solvents such as ethanol, acetonitrile, methanol and acetone. Circular Dichroism studies indicated that both the native and refolded enzyme have a similar structural conformation with the α-helical content showing a negative ellipticity at 230nm being the most predominant feature of the spectra.

Construction of a Truncated Mutant Protein

Several mutant enzymes were constructed which removed an increasing number of aminoacids from the N-terminal end of the *Corallina* bromoperoxidase protein. These were modelled on the crystal structure and constructs that resulted in no significant exposure of hydrophobic patches were investigated. The most successful mutant was that which had over 200 residues removed from the N-terminus. This mutant lacked residues up to the mid-region of α-helix 8. This length was chosen as α-helix 9 has been implicated in the active site cleft so the truncation did not want to extend into this region. Despite this mutation being severe with a large region in between the active site channel and the distal region, comprising of α-helices 14-16 being removed which could

Figure 6. Ribbon representation of the predicted structure for the N-terminal truncated C. officinalis mutant bromoperoxidase . Monomer one is pictured in brown, monomer two in blue. The vanadate at the active site is represented in CPK style. Picture produced using Bobscript and rendered in Raster3D.

disrupt the active site channel of the enzyme this proved to be the most successful mutant protein. The modelled structure of this mutant is shown in Figure 6.This enzyme retained activity in the standard MCD assay and was able to be over-expressed in *E. coli* in a soluble form (*24,25*). This dimeric form of the bromoperoxidase is being studied as to its substrate specificity since it remains a favored form of the enzyme for commercial applications in biocatalysis.

References

1. Sheffield, D.J.; Harry, T.; Smith, A.J.; Roger, L.J. *Phytochemistry* **1993**, *32*, 21-26.
2. (*2*) Rush, C.; Willetts, A.; Davies, G.; Dauter, Z.; Watson, H.; Littlechild, J. *FEBS Lett.* **1995**, *359*, 244-246.
3. Littlechild, J. *Current Opinions in Chemical Biology* **1999**, *3*, 28-34.
4. Isupov, M.; Dalby, A.; Brindley, A.; Izumi, Y.; Tanabe, T.; Murshudov, G.; Littlechild, J., *J. Molecular Biology* **2000**, *299*, 1035-1049.
5. .(*5*) Weyand, M.; Hecht, H.J.; Keiss, M.; Liaud, M.F.; Vilter, H.; Schomburg, D. *J. Molecular Biology* **1999**, *293*, 595-611.
6. (*6*) Hemrika, W.; Renirie, R.; Dekker, H.; Barnett, P.; Wever, R. *Proceedings of the National Academy of Science* **1997**, *94*, 2145-2149.
7. Ishikawa, K.; Mihara, Y.; Gondoh, K.; Suzuki, E-I.; Asano, Y. *EMBO Journal* **2000**, *19*, 2412-2423.
8. Littlechild, J.; Garcia-Rodriguez, E.; Dalby, A.; Isupov, M. *J. Molecular Recognition* **2002**, *15*, 291-296.
9. Littlechild, J.; Garcia-Rodriguez, E. *Coordination Chemistry Reviews* **2003**, *237*, 65-76.
10. Garcia-Rodriguez, E.; Ohshiro_T.; Iida, Y.; Kobayashi, T.; Izumi, Y.; Littlechild, J.; *J. Biological Inorganic Chemistry* **2005**, *10*, 275-282.
11. a. Butler, A. *Vanadium-dependent redox enzymes: Vanadium halo-peroxidase*, in *Comprehensive Biological Catalysis*, pp427-437, Academic Press.
    b. Raugei, S.; Carloni, P.; *J. Phys.Chem. B,* **2006**, *110*, 3747-3758.
12. Christmann, U.; Dau, H.; Haumann, M.; Kiss, E.; Liebisch, P.; Rehder, D.; Santoni, G.; Schulzke, C. *Dalton Trans.*, **2004**, 2534-2540.
13. Watts, A.B.; Coupe, E.E.; Littlechild, J.A. *J. Biological Inorganic Chemistry* **2006**, submitted.
14. Garcia-Rodriguez, E. PhD University of Exeter, **2005**.
15. Garcia-Rodriguez, E.; Isupov, M.; Littlechild, J. *Acta Cryst D.* **2006**, manuscript in preparation.
16. Tapper, S.; Littlechild, J.; Molard, J.; Prokes, I.; Tucker, J *J. Supramolecular Chemistry* **2006**, *18*, 55-58.

17. Tschirret-Guth, R.A.; Butler, A. *J. Am. Chem. Soc.*, **1994**, *116*, 411-412.
18. Martinez, J.S.; Carroll, G.L.; Tschirret-Guth, R.A.; Altenhoff, G.; Little, R.D. ; Butler, A. *J. Am. Chem. Soc.*, **2001**, *123*, 3289-3294
19. Ohshiro, T.; Littlechild, J.; Garcia-Rodriguez, E.; Isupov, M.N.; Iida, Y.; Kobayashi, T.; Izumi, Y. *Protein Science* **2004**, *13*, 1566-1571.
20. Carter, J.N.; Beatty, K.E.; Simpson, M.T.; Butler, A. *J. Inorganic Biochemistry* **2002**, *91*, 59-69.
21. Shimonishi, M.; Kuwamoto, S.; Inoue, H.; Wever, R.; Onishiro, T.; Izumi, Y.; Tanabe, T. *FEBS Lett.*, **1998**, *428*, 105-110.
22. Ohshiro, T.; Hemrika, W.; Aibara, T.; Wever, R.; Izumi, Y. *Phytochemistry*, **2002**, *60*, 595-601.
23. Coupe E.E, ; Smyth M. G.; Fosberry, A. ; Hall, R. M.; Littlechild, J. A. *Protein Expression and Purification* **2006**, in press
24. Coupe, E.E. PhD thesis, University of Exeter, **2005**.
25. Coupe, E.E.; Hall, R.M.; Littlechild, J.A. *Protein Engineering*, **2006**, submitted.

Chapter 12

Understanding the Mechanism of Vanadium-Dependent Haloperoxidases and Related Biomimetic Catalysis

Curtis J. Schneider[1], Giuseppe Zampella[2], Luca DeGioa[2], and Vincent L. Pecoraro[2,3]

[1]Department of Chemistry, University of Michigan,
Ann Arbor, MI 48109–1055
[2]Department of Biotechnology and Biosciences, University of Milano-Bicocca, Piazza della Scienza 2, 20126, Milano, Italy
[3]Biophysics Research Division, University of Michigan,
Ann Arbor, MI 48109–1055

Vanadium dependent haloperoxidases are a class of enzymes capable of the oxidation of halides and thioethers in the presence of hydrogen peroxide. We summarize herein DFT and QM/MM studies on models of the enzyme and small molecule mimics. These studies provide key insight into the enzyme mechanism of peroxide binding and substrate oxidation. A detailed mechanistic investigation has also been performed on the synthetic model systems confirming a nearly identical transition state to the enzyme for substrate oxidation. This oxo-transfer reaction occurs via an S_N2-like process with attack at the unprotonated peroxo oxygen of a hydroperoxo moiety. These results provide insight into the importance of acid/base catalysts involved in peroxide binding and substrate oxidation. Additionally, these studies allow for the development of asymmetric oxidation catalysts based on functional models for VHPO's.

© 2007 American Chemical Society

Introduction

Vanadium dependent haloperoxidases (VHPO) are metalloenzymes containing a mononuclear vanadium(V) co-factor that catalyze the oxidation of halides in the presence of hydrogen peroxide. The oxidized species (HOX, X_3^-, or X_2) subsequently adds to organic molecules, yielding halogenated marine products. The first example of this vanadoenzyme class was a bromoperoxidase isolated from the marine algae *Ascophyllum nodosum (1)*. The enzyme nomenclature is based upon the most electronegative halide that the enzyme is capable of oxidizing. VHPO's can also catalyze the stereospecific two electron oxidation of organic sulfides to chiral sulfoxides. A number of X-ray crystal structures are now available for VHPO's *(2-4)*, including peroxide bound *(3)* and unbound *(3,4)* forms of a vanadium chloroperoxidase from the fungus *Curvularia inaequalis*. The vanadate co-factor is covalently coordinated to an enzyme bound histidine. The coordination geometry is trigonal bipyramidal, with an oxygen atom in the axial position trans to N atom of histidine, three other oxygen atoms complete the coordination sphere. The peroxide bound form is best described as having a distorted tetragonal geometry when treating the peroxo moiety as a single η^2 donor (Figure 1).

Enzymological *(5)*, spectroscopic *(6)*, and, both synthetic *(7-10)* and computational model *(11-14)* studies have elucidated key steps in the major catalytic steps in this system *(15)*. Unlike heme peroxidases, which require the formation of highly oxidized compound I and compound II interemediates, there is no evidence for redox cycling of the vanadate co-factor during catalysis *(5,15)*. It is commonly believed that vanadium is acting as a Lewis acid to activate the terminal oxidant hydrogen peroxide by polarizing the peroxo bond. The kinetic mechanism is sequential ordered with peroxide binding to the vanadate co-factor followed by halide binding and oxidation. The importance of protonation of the active site for reactivity is suggested by the catalytically relevant pKa of 5.9 for a vanadium bromoperoxidase *(16)*. Synthetic models have shown that protonation of the peroxo-vanadium complex is critical for oxidation of halides and organic sulfides *(10,17)*. Hydrogen peroxide was shown to be the source of oxygen transferred to the thioether during sulfide oxidation via isotopic labeling studies using the enzyme *(18)*.

Coordination complexes of oxovanadium(V) and their peroxo analogues have been shown by a number of groups to be functional mimics *(8-10,19-23)* for VHPO's. In their own right, peroxo-oxovanadium complexes are excellent oxidation catalysts and have been used in a number of oxidative transformations *(24)* relevant to organic synthesis, including but not limited to alcohol oxidation, halide oxidation, and sulfoxidation. Of these reactions, the oxidation of achiral sulfides to chiral sulfoxides has sparked the most attention in recent years because of the potential applications of chiral sulfoxides *(25)*. A number of catalysts have been shown to be competent with respect to sulfide oxidation and

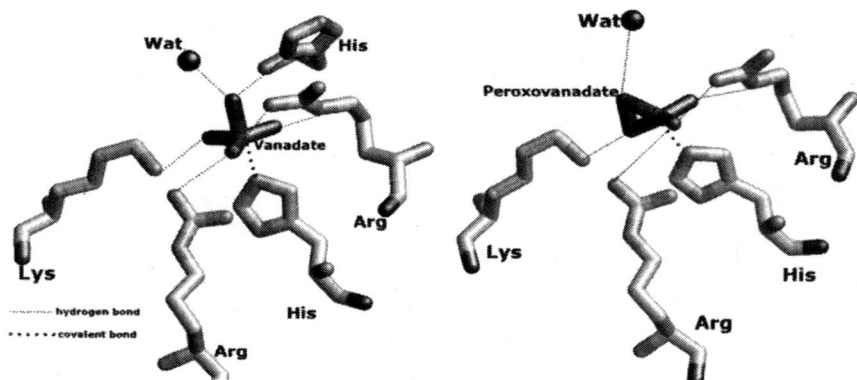

Figure 1. X-ray crystal structure of the active of VCPO depicting both the resting state (left) and the peroxo form (right). Reproduced from ref (42)

in some cases to be stereoselective *(9,17,24,26-28)*. Our group has synthesized a variety of tripodal amine complexes of oxovanadium(V) that are capable of mimicking the oxidation abilities of VHPO's *(10)*. These complexes have been used to probe the potential mechanism for VHPO's, in particular determining the role of protonation with respect to activity.

Despite these thorough studies, questions can still be raised about both the enzyme and model systems including: 1) what is the protonation state of vanadium species (both resting and active forms), 2) which oxygen of the peroxo ligand is transferred to the substrate and 3) what is the geometry of the transition state. Quantum chemical methods are useful tools to investigate structural, electronic and reactivity properties of transition metal complexes *(29,30)* and models of the active site of metallo-enzymes *(31-33)*. In fact, computational investigations of vanadium(V) complexes have been reported *(7,8,34)*. and recently theoretical models have been studied by our group to understand the catalytic properties of the enzyme and our functional models better. This paper will focus on the studies reported by our group that have provided critical information necessary for understanding the catalytic processes of the enzyme and functional models in their own right as oxidation catalysts.

Synthetic Models

Before the solving of the protein crystal structure, a number of groups including ours, synthesized and characterized functional models of the enzyme.

Our small synthetic models have been reported *(10,17,35)* and previously reviewed *(13,36)*. These complexes played key roles in determining the effects of complex protonation in both peroxide binding and substrate oxidation. Additionally, we have shown K[VO(O$_2$)Hheida] *(37)* (**Hheida**) to be an efficient catalyst for the oxidation of bromine and thioethers.

The proton dependence of peroxide binding was found to be very complex. There is a first order dependance on protons at stoichiometirc proton concentrations, but the presence of excess protons had no effect on the rate of peroxide coordination Additionally, the coordination of peroxide occurs in the absence of additional protons at a slower rate. This observation is consistent with the fact that hydrogen peroxide can act as a proton donor to the complex followed by rapid coordination of peroxide, alluding to the importance of acid/base catalysis within the active site of the enzyme.

Of the complexes studied by our group, **Hheida** was shown to be most efficient functional model. In the presence of one acid equivalent a protoned peroxo-vanadium complex is generated as demonstrated by shifts in the UV-Vis and ^{51}V NMR spectra. Upon protonation, this complex is capable of oxidizing bromide, iodide *(10)* and thioethers *(17)*. It was proposed that protonation of the vanadium complex led to a hydroperoxo intermediate that was responsible for catalysis.

These small molecule studies have provide important insights into the mechanism of VHPO's for both halide and thioether oxidation, allowing a detailed investigation of each catalytic step not available to enzymatic studies. From these studies we have established the importance of protons in both peroxide binding and the substrate oxidation, demonstrating the importance of acid base catalysis in the active site of the enzyme. As a catalyst, **1** has the potential to be an effective catalysis for oxidations of nucleophilic substrates and makes a suitable backbone ligandfor design of a chiral complex capable of the asymmetric oxidation of thioethers. However, before we design asmmyetric ligands, it was necessary to understand the specific details regarding the proposed catalytic mechanism better. These points include: 1) are there nearly isoenergetic species present in solution, 2) what is the actual site of protonation of the oxoperoxovanadium(V) complex, and 3) what is the transition state geometry. To address these questions, we shifted our focus to computational studies of the synthetic model systems.

QM Models of VO(O$_2$)Hheida^{-1}

Synthetic models have provided an excellent background to perform detailed computational analysis of **Hheida**. We began the computational study addressing what possible solution state species may be present using the

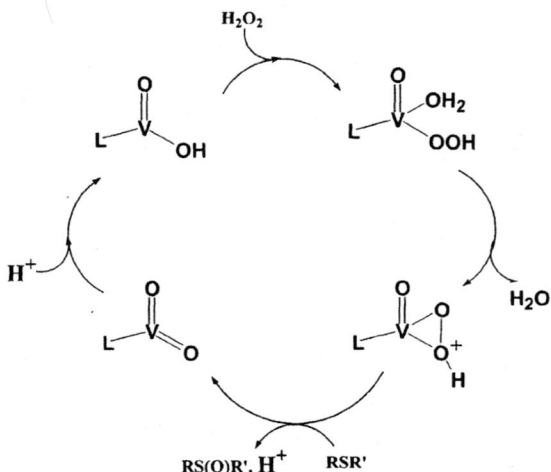

Figure 2. Proposed mechanism for thioether oxidation by Hheida.
Reproduced from ref (14)

previously reported crystal structure as a starting point. The geometry optimized structure of **Hheida** showed nearly identical bond lengths to those of the crystal structure. As there is only a single ^{51}V NMR resonance for **Hheida** dissolved in acetonitrile, we wanted to assess whether there was the possibility of two rapidly interconverting isomers that exhibited a coalesced resonance. We began by exploring the potential of exogenous donors coordinating to vanadium in place of the weakest donor of Hheida, the alcohol arm. Various exogenous donors were examined, including water, acetonitrile, and bromide. In all cases it was found that coordination of the alcohol arm of Hheida was energetically favored over an exogenous donor. The same conclusion was reached using the most stable protonated species to minimize the electronic repulsion *(38)*.

With the geometry optimized structure of **Hheida** we began exploring which site is most likely to be protonated. Given the presence of a pseudo mirror plane bisecting the peroxo bond, the left and right hand side of the Hheida ligand were treated as identical. The five sites which showed the highest electron density were: the uncoordinated carboxylate oxygen, coordinated carboxylate oxygen, oxo, and both oxygen's of the peroxide moiety. Each site was protonated and the geometry was optimized. The lowest energy structure revealed protonation of the peroxo moiety, on either side, as the most stable complex, while protonation of the oxo ligand generated a complex 2.6 kcal/mol higher in energy. Based on these results we can say that in the proposed mechanism formation of a hydroperoxo intermediate is energetically favorable over an alternative site of protonation. This result is consistent with the subtle changes in the peroxo to vanadium charge transfer band in UV/Vis spectrum upon protonation of **Hheida** in acetonitrile.

The mechanism based on the kinetic studies of **Hheida** shows that protonation of the the peroxo complex increase the rate of the reaction for sulfide and halide oxidation. Transition state searches were performed using dimethyl sulfide and halides (Cl⁻, Br⁻, and I⁻) to determine the effects of protonation on the barrier to activation. Transition states were located for all four substrates with the unprotonated form of **Hheida**. The located transitions show an oxo-transfer type of process with a nearly linear X-O-O bond angle for all substrates (Figure 3).

No transition states could be located for attack of the substrate on the oxo ligand of **Hheida**. Protonation of the peroxo ligand yielded a similar transition in which the substrate undergoes a nucleophilic attack on the uprotonated hydroperoxo oxygen. Protonation of the peroxo moiety yields an earlier transiton state in which the substrate peroxo bond distance is significantly longer than the unprotonated case. The barrier to activation drops significantly in the case of halide oxidation when compared to anionic transition states (aprox. 24 kcal/mol in all cases). The difference in the barrier to activation is overestimated due to the instability of the dianionic transition states located in the case of the unprotonated complex, this is evident by the extremely high barrier to activation in the unprotonated case. It is important to note that the difference in the energy of activation for all three halides is only 4.1 kcal/mol, this energy difference can not account for the inactivity of **Hheida** with respect to chloride oxidation observed in the experimental studies. These results are consistent with the hypothesis presented earlier that the basicity of the chloride in acetonitrile is responsible for the deactivation of the protonated peroxo complex.

Transiton states were also located for the hydroperoxo complex and dimethyl sulfide. The trend observed with the halide transition states with respect to protonation of the peroxo complex are identical to those observed when dimethyl sulfide is the substrate. The S-O-O angle is nearly linear with a slight twist of the peroxo ligand casuing a deviation from the expect 90° dihedral angle between the hydroperoxo ligand and the oxovanadium bond. The protonation of the peroxo moiety yields an earlier transition state with a longer sulfur peroxo oxygen bond distance. The barrier to activation decreased by 9.0 kcal/mol. Transition states could not be located for nuclephilic attack of the thioether on the protonated peroxo oxygen or the oxo moieties.

These results are consistent with the proposed kinetic mechanism and show that protonation of the peroxo ligand generates a more suitable leaving group for this S_N2-like process. Providing a confirmation that the functional models developed by our group proceed through a reaction coordinate nearly indentical to the enzyme system (vida infra). In addition the transition state geometries discussed provide the necessary information to design rationally ligand sets capable of imparting stereoselectivity on thioether oxidation.

154

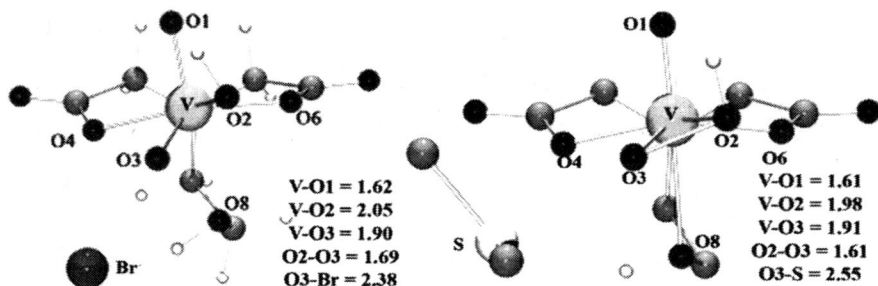

Figure 3. Ball-and-stick representation of the transition states located for bromide (left) and dimethyl sulfide (right) oxidation with relevant distances (Å).

Enzyme Models

Functional models developed by our group have played a key role in elucidating the mechanism by which VHPO's may function. We next turned to computational methods to explore in detail the electronic structure of the vandate co-factor. The X-ray structure of the resting state *(4,39)* form of the enzyme proposed a trigonal bypyramidal vanadate with three equatorial oxo's and an axial hydroxo. Given the importance of protons in the catalytic cycle for the functional models, the protonation state of the vanadate co-factor is crucial for understanding the mechanism by which hydrogen peroxide binds.

Kravitz and co-workers employed quantum chemical methods to explore the potential protonation states in the resting form of VHPO's. The first set of quantum chemical models were developed based on a simple vanadate model. The two reasonable cationic protonation states were found to be protonation of the axial hydroxo or one equatorial oxo.

These models provided a reasonable picture of what may occur in the active site but did not address the complex hydrogen bonding networks provided by the enzyme. To account for these differences, QM/MM calculations were performed on a truncated protein model of a vanadium chloroperoxidase (all bond angles and distances were initially taken from the crystal structure).

The final result showed two nearly isoenergetic structures shown in Figure 4. Interestingly the QM/MM models showed one species in which an axial aqua species with three equatorial oxo's was an energetically favorable species. It is likely that these two structures are in an equilibrium in the active site, this equilibrium may facilitate the proton transfer from nearby acid/base catalysis to assist in the coordination of hydrogen peroxide. Recently magic angle spinning [51]V NMR has confirmed the protonation of the resting form of the enzyme *(40)*, showing one axial and one equatorial hydroxo ions with two equatorial oxos. Using this well characterized resting state, we began studying the reactive

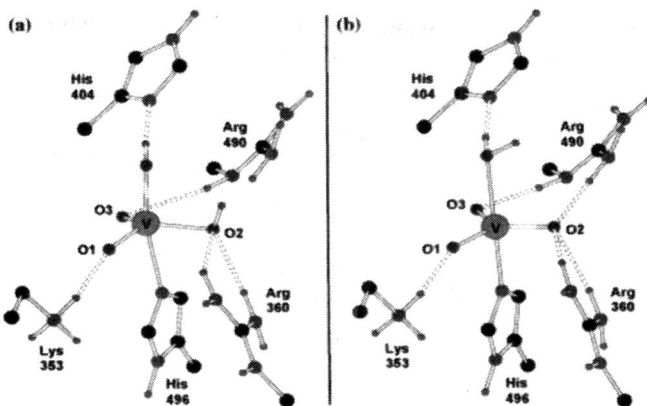

Figure 4. The two lowest energy minima located for the resting state configuration of VCPO. Reproduced from ref (14).

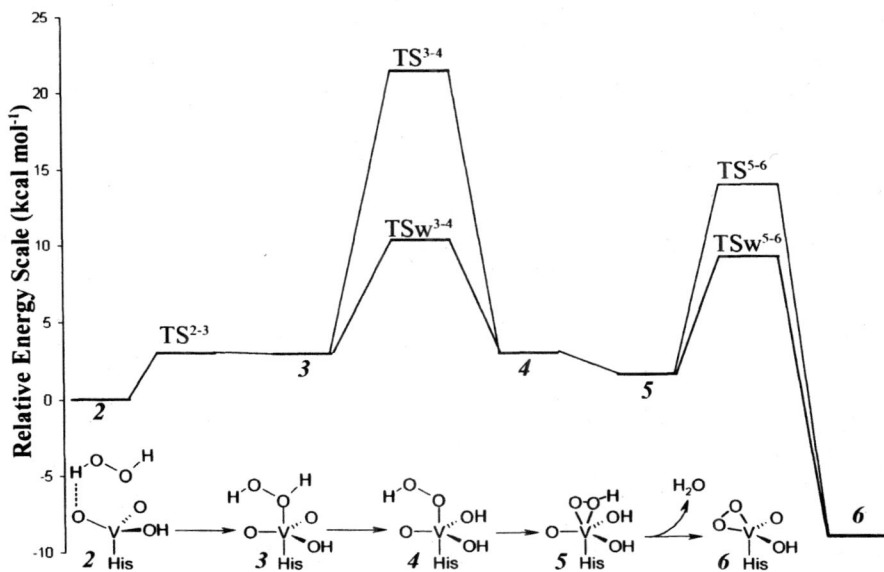

Figure 5. Relative energies for intermediates and transition states located for the binding of hydrogen peroxide to a model of the vanadate co-factor. A scheme of peroxide binding to the vanadate co-factor is located at the bottom.

forms of the enzyme, specifically the binding of hydrogen peroxide and the oxidation of a substrate molecule.

In light of the X-ray structures of the resting *(4,39)* and peroxo *(3)* forms of VHPOs, the reaction between H_2O_2 and the enzyme cofactor implies the stoichiometry:

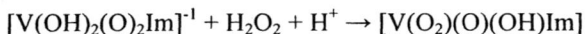

$$[V(OH)_2(O)_2Im]^{-1} + H_2O_2 + H^+ \rightarrow [V(O_2)(O)(OH)Im]$$

The DFT investigation of different reaction paths for the formation of the peroxo form led to the conclusion that the reaction between $[V(OH)_2(O)_2Im]^-$ and H_2O_2 is energetically unfeasible. Indeed, kinetic studies on synthetic functional models pointed out the role of protonation in labilizing oxo/hydroxo

Table 1. Reaction Energies and Barriers Relevant to Peroxide Binding and Substrate Oxidation.

$\Delta G, \Delta G^{\ddagger}$	$\varepsilon = 40$
$2 \rightarrow TS^{2\text{-}3}$	3.0
$2 \rightarrow 3$	2.9
$3 \rightarrow TS^{3\text{-}4}$	18.5
$3 \rightarrow TSw^{3\text{-}4}$	7.4
$3 \rightarrow 4$	0
$4 \rightarrow 5$	-1.4
$5 \rightarrow TS^{5\text{-}6}$	12.4
$5 \rightarrow TSw^{5\text{-}6}$	7.7
$5 \rightarrow 6$	-10.6
$*8 \rightarrow TS^{Br\text{-}}$	10.2
$*8 \rightarrow Product$	7.7

*calculated using BP86, all others calculated using B3-LYP

ligands of vanadium complexes, which allows hydrogen peroxide binding *(35)*. In fact, the protonated species $[V(H_2O)(OH)(O)_2Im]$ can react with H_2O_2 according to a dissociative reaction pathway, resulting in the formation of a tetrahedral intermediate species $([V(OH)(O)_2Im]$; **2** - Figure 5).

This tetrahedral form interacts with H_2O_2 leading to a transient species which has trigonal bipyramidal geometry (where H_2O_2 binds in an end-on fashion **3**; Figure 5). The latter complex spontaneously converts to a species in which a proton has moved from the peroxo OH group coordinated to vanadium to one of the equatorial oxo groups (**4**; Figure 5).

It is worth noting that proton transfer from H_2O_2 to the equatorial oxo group in **3** cannot take place directly but must be mediated by a suitable protic group. The presence of water in the reaction coordinate dramatically lowers the barrier to activation (Figure 5: TSw^{3-4} vs. TS^{3-4}) for the conversion of **3** to **4**. Barriers to activation can be found in **Table 1**. Remarkably, the analysis of the X-ray structure of vanadium haloperoxidase from *Ascophyllum nodosum* reveals that His418 and a water molecule (W772) are close to the metal cofactor and have the right spatial disposition to act as an acid/base catalyst *(2)*.

The end-on isomer **4** easily converts to the side-on form **5** (Figure 5). Then, release of a water molecule yields a species (**6**) whose structural features closely resembles the peroxo-form of the cofactor in the enzyme (Figure 5) *(3)*. This process is assisted by the presence of water acting as an acid base catalysis. The relevant transition states can be found in Figure 6.

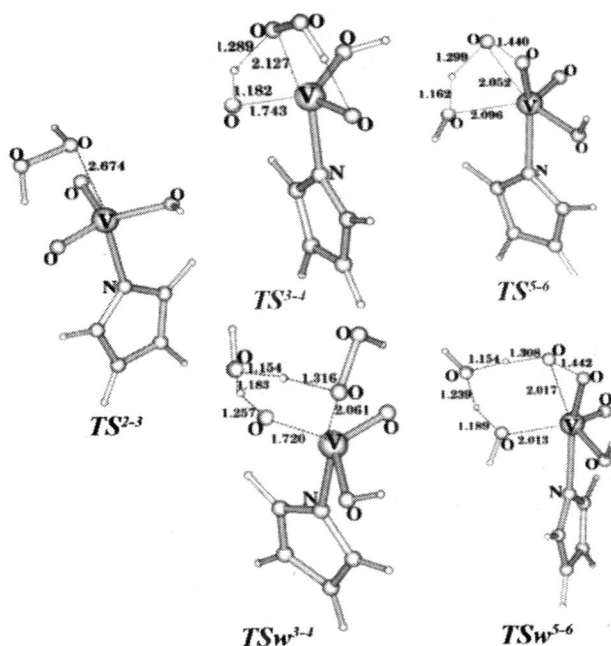

Figure 6. Transition states relevant to peroxide binding to the vanadate co-factor models.(42)

Table 1 contains the releative energies for all relevant steps related to peroxide binding. It is important to note the dramatic descrease in the activation barrier for transition states located containing a catalytic water. These results provide a further molecular basis supporting the proposal by Hamstra et al. concerning the role of acid-base catalysis not only for activation of the peroxo complex towards halide oxidation, but also to favor hydrogen peroxide binding to the cofactor.

With the aim of clarifying the catalytic mechanism of VHPO, we investigated by DFT the oxo-transfer reaction pathways involving either the equatorial or the axial peroxo oxygen atom using two typical VHPO substrates: Br⁻ and dimethylsulfide (DMS).

Even though structural data allow the assertion that **6** is the species most closely resembling the X-ray crystal structure of the peroxo form of the enzyme, the possibility of a reaction path occurring by means of an anionic form of **6** has also been evaluated. However, the reaction paths for the attack of Br⁻ on both the peroxo oxygen atoms of the anionic form of **6** turned out to be characterized by very large potential energy barriers, as expected considering a reaction between two anionic species. When considering the neutral substrate DMS, the computed potential energy barriers for the attack on the pseudo-axial and equatorial peroxo oxygen atoms were lower (21.2 and 29.8 kcal mol^{-1}, respectively), but still too large when compared to experimental values.

Differently, energy barriers computed for the reaction between Br⁻ or DMS and the neutral species **6** showed that the oxo-transfer step involving the pseudo-axial peroxo atom is more favored than the corresponding equatorial pathway for both substrates. The transition state resembles a classic SN_2 mechanism, in which the attack of the nucleophile and the cleavage of the O-O bond are simultaneous.

An analysis of the electronic properties of **8**, **9** and **10** (**Figure 7**) showed that the peroxo bond in **8** is very weakly polarized, whereas in **9** protonation of the equatorial peroxo atom leads to a significant polarization of the O-O bond, with the pseudo-axial oxygen more electrophilic than the protonated equatorial peroxo atom. A similar behavior was observed for **10**, in which the unprotonated peroxo atom is more electrophilic, even if the charge separation is smaller. Protonation of the peroxo causes a lengthening of the peroxo oxygen-vanadium bond and the peroxo oxygen-oxygen bond (**Figure 8a**). It should be noted that the LUMO of the peroxo from the vanadate co-factor shows a significant amount of atomic p orbital character on only the axial unprotonated peroxo oxygen atom (**Figure 8b**). This *p* character is maintained upon protonation of the peroxo moiety. Protonation causes a rotation of peroxo dihedral angle with respect to the V-N bond of nearly 25° and a small change in the oritentaiton of the *p* orbital character associated with axial oxygen (**Figure 8**). The effect of these factors contributes to aligning the orbital lobes of the LUMO associated with peroxo in the correct position for nucleophilic attack by a substrate molecule.

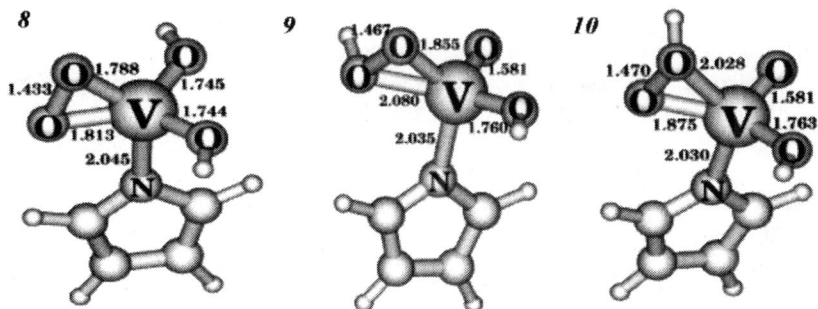

Figure 7. Ball-and-stick representation of the three nearly isoenergetic protonation states of 6.

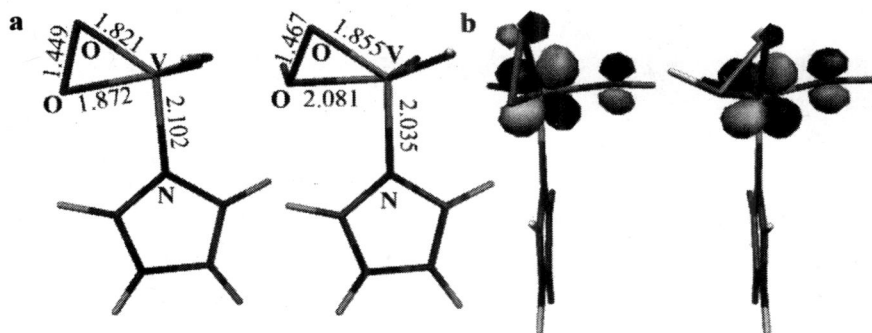

Figure 8. Relevant bond distances (a) and the LUMO(b) for the peroxo (left) and the hydroperoxo (b) forms of the vandate cofactor. The dihedral angle O-O-V-N is 6.5° for the peroxo form and 31.9° for the hydroperoxo form.

On the basis of these results, we conclude that the oxo-transfer reaction path involves the unprotonated axial peroxo oxygen atom. Moreover, protonation of the peroxo moiety plays a crucial role in the activation of the peroxo-vanadium complexes. In this context, it is noteworthy that in the peroxide-bound form of the enzyme a lysine residue (K353) is hydrogen bonded to the equatorial oxygen atom of the peroxo group *(3)* suggesting that this amino acid might play a role similar to H^+ in peroxide activation. To clarify this issue we studied the oxo-transfer step also in model systems in which a CH_3-NH_3^+ group is hydrogen bonded to the equatorial peroxo oxygen atom of the cofactor. Results indicated that a polarizing group such as CH_3-NH_3^+ plays a role, albeit less dramatic than protonation of the peroxo moiety, in the activation of the peroxo form of the cofactor. The barrier to activation for bromide oxidation in this case is 10.2 kcal mol^{-1}, the overall process is exothermic by 7.7 kcal mol^{-1} (Table 1: 8→TS^{Br-} & 8→Product).

Notably, it turned out that when the polarizing power of the R-NH_3^+ group is buffered by counterions, its catalytic role is strongly depressed. Remarkably, the analysis of the VClPO active site *(41)* shows that Lys353 is close to a hydrophobic (phenylalanine) residue and is involved in a single hydrogen bond with the backbone carbonyl of a proline residue in VClPO. On the other hand, in VBrPO *(39)* the phenylalanine residue is replaced by histidine, which forms an additional hydrogen bond with the lysine residue, decreasing its polarizing power. In light of our results this difference might be one of the factors that account for the different reactivity of these enzyme families.

Conclusion

We have demonstrated the complmentary use of computational and synthetic methods to understand catalytic process both in enzymatic and small molecule systems. Our synthetic modeling studies demonstrated the importance of protonation of the oxoperoxovanadium complex in peroxide binding and substrate oxidation. These results were confirmed using computational methods for enzyme models showing that the enzyme undergoes an S_N2-like oxo transfer process. This allows us to define a mechanism for the enzyme which explicity details the binding of hydrogen peroxide and the transitions state geometries. With this information about the enzyme system, we undertook similar computational studies confirming that our synthetic models proceeded through the same transiton state as enzyme models. Additionally these studies demonstrate the importance of complex protonation in catalysis for early transition metal complexes. With this information we can now design ligands to enhance the stereoselectivity of oxidations by limiting the number of isomeric transitions states through hydrogen bonding and steric hindrance.

Acknowledgements

CJS would like to thank both CBI Training Program and the Sokol International Summer Research Fellowship for funding, and Tamas Jakusch for helpful discussions.

References

1. Vilter, H. *Phytochemistry* **1984**, *23*, 1387-1390.
2. Weyand, M.; Hecht, H. J.; Vilter, H.; Schomburg, D. *Acta. Crystallogr., Sec. D: Biol. Crystallogr.* **1996**, *D52*, 864-865.
3. Messerschmidt, A.; Prade, L.; Wever, R. *Biol. Chem.* **1997**, *378*, 309-315.
4. Messerschmidt, A.; Wever, R. *Proc. Natl. Acad. Sci. U. S. A.* **1996**, *93*, 392-396.
5. De Boer, E.; Wever, R. *J. Biol. Chem.* **1988**, *263*, 12326-12332.
6. Arber, J. M.; De Boer, E.; Garner, C. D.; Hasnain, S. S.; Wever, R. *Biochemistry* **1989**, *28*, 7968-7973.
7. Conte, V.; Di Furia, F.; Moro, S.; Rabbolini, S. *J. Mol. Catal.* **1996**, *113*, 175-184.
8. Conte, V.; Bortolini, O.; Carraro, M.; Moro, S. *J. Inorg. Biochem.* **2000**, *80*, 41-49.
9. Santoni, G.; Licini, G.; Rehder, D. *Chem. Eur. J.* **2003**, *9*, 4700-4708.
10. Colpas, G. J.; Hamstra, B. J.; Kampf, J. W.; Pecoraro, V. L. *J. Am. Chem. Soc.* **1996**, *118*, 3469-3478.
11. Zampella, G.; Fantucci, P.; Pecoraro, V. L.; De Gioia, L. *J. Am. Chem. Soc.* **2005**, *127*, 953-960.
12. Zampella, G.; Kravitz, J. Y.; Webster, C. E.; Fantucci, P.; Hall, M. B.; Carlson, H. A.; Pecoraro, V. L.; De Gioia, L. *Inorg. Chem.* **2004**, *43*, 4127-4136.
13. Kravitz, J. Y.; Pecoraro, V. L. *Pure Appl. Chem.* **2005**, *77*, 1595-1605.
14. Kravitz, J. Y.; Pecoraro, V. L.; Carlson, H. A. *J. Chem. Theor. Comp.* **2005**, *1*, 1265-1274.
15. Butler, A. *Coord. Chem. Rev.* **1999**, *187*, 17-35.
16. Everett, R. R.; Kanofsky, J. R.; Butler, A. *J. Biol. Chem.* **1990**, *265*, 4908-4914.
17. Smith, T. S.; Pecoraro, V. L. *Inorg. Chem.* **2002**, *41*, 6754-6760.
18. ten Brink, H. B.; Schoemaker, H. E.; Wever, R. *Eur. J. Biochem.* **2001**, *268*, 132-138.
19. Kimblin, C.; Bu, X.; Butler, A. *Inorg. Chem.* **2002**, *41*, 161-163.
20. Maurya, M. R.; Agarwal, S.; Bader, C.; Rehder, D. *Eur. J. Inorg. Chem.* **2005**, 147-157.

162

21. Plass, W. *Coord. Chem. Rev.* **2003**, *237*, 205-212.
22. Slebodnick, C.; Hamstra, B. J.; Pecoraro, V. L. *Struct. Bonding (Berlin)* **1997**, *89*, 51-108.
23. Butler, A.; Clague, M. J.; Meister, G. E. *Chem. Rev.* **1994**, *94*, 625-638.
24. Bolm, C. *Coord. Chem. Rev.* **2003**, *237*, 245-256.
25. Fernandez, I.; Khiar, N. *Chem. Rev.* **2003**, *103*, 3651-3705.
26. Du, G.; Espenson, J. H. *Inorg. Chem.* **2005**, *44*, 2465-2471.
27. Sun, J.; Zhu, C.; Dai, Z.; Yang, M.; Pan, Y.; Hu, H. *J. Org. Chem.* **2004**, *69*, 8500-8503.
28. Blum, S. A.; Bergman, R. G.; Ellman, J. A. *J. Org. Chem.* **2003**, *68*, 153-155.
29. Ziegler, T.; Autschbach, J. *Chem. Rev.* **2005**, *105*, 2695-2722.
30. Niu, S.; Hall, M. B. *Chem. Rev.* **2000**, *100*, 353-405.
31. Friesner, R. A.; Baik, M.-H.; Gherman, B. F.; Guallar, V.; Wirstam, M.; Murphy, R. B.; Lippard, S. J. *Coord. Chem. Rev.* **2003**, *238-239*, 267-290.
32. Lovell, T.; Himo, F.; Han, W.-G.; Noodleman, L. *Coord. Chem. Rev.* **2003**, *238-239*, 211-232.
33. Siegbahn, P. E. M.; Blomberg, M. R. A. *Chem. Rev.* **2000**, *100*, 421-437.
34. Buhl, M. *J. Comput. Chem.* **1999**, *20*, 1254-1261.
35. Hamstra, B. J.; Colpas, G. J.; Pecoraro, V. L. *Inorg. Chem.* **1998**, *37*, 949-955.
36. Pecoraro, V. L.; Slebodnick, C.; Hamstra, B. *ACS Symp. Ser.* **1998**, *711*, 157-167.
37. Hheida = N-(2-hydroxylethyl) iminodiacetate
38. Schneider, C. J.; Zampella, G.; Pecoraro, V. L.; DeGioia, L. *Submitted* **2006**.
39. Weyand, M.; Hecht, H. J.; Kiesz, M.; Liaud, M. F.; Vilter, H.; Schomburg, D. *J. Mol. Biol.* **1999**, *293*, 595-611.
40. Pooransingh-Margolis, N.; Renirie, R.; Hasan, Z.; Wever, R.; Vega, A. J.; Polenova, T. *J. Am. Chem. Soc.* **2006**, *128*, 5190-5208.
41. Messerschmidt, A.; Prade, L.; Wever, R. *ACS Symp. Ser.* **1998**, *711*, 186-201.
42. Zampella, G.; Fantucci, P.; Pecoraro, V. L.; De Gioia, L. *Inorg. Chem.* **2006**, 45, 7133-7143

Chapter 13

Model Studies of Vanadium-Dependent Haloperoxidation: Structural and Functional Lessons

Synthetic and Computational Models of Supramolecular Interactions and the Formation of Peroxo Species

Winfried Plass, Masroor Bangesh, Simona Nica, and Axel Buchholz

Institut für Anorganische und Analytische Chemie der Friedrich-Schiller-Universität Jena, Carl-Zeiss-Promenade 10, D–07745 Jena, Germany

Vanadate is the prosthetic group of vanadium haloperoxidases and fixed in the active site cavity by just one coordinative bond to a histidine residue and embedded in an environment of extensive hydrogen bonds. Density functional theory has been used to investigate the structure of the resting state of the prosthetic group in the enzyme pocket and to elucidate the mechanism of the formation of its peroxo complex. The role of the protein environment and in particular that of the amino acid residues Ser402 and His404 for the catalytic action of the prosthetic group is discussed and a catalytic mechanism proposed. The relevance of vanadium complexes derived from the versatile tridentate N-salicylidene hydrazide ligand system with a broad variation of the carbonic acid moiety introducing different functional groups in the side chain is presented. This includes supramolecular assemblies with chiral hosts.

© 2007 American Chemical Society

163

Introduction

Vanadium dependent haloperoxidases are enzymes capable of catalyzing the oxidation of halide ions by hydrogen peroxide to the corresponding hypohalous acids and their catalytic properties have been extensively investigated in recent years (1). Crystal structures of vanadium-containing chloroperoxidases (VCPO) (2) and bromoperoxidases (VBPO) (3, 4) have been determined and reveal that an orthovanadate is coordinatively bound to a histidine residue and surrounded by several other amino acid residues forming hydrogen bonds to the anionic oxygen atoms. A schematic drawing of the active site including the prosthetic group of the VCPO from the fungus C. inaequalis is depicted in Figure 1. Nevertheless structural uncertainties concerning the prosthetic group still remain. This is related to the general resolution problem of protein crystallographic data (2.03 Å for the native VCPO and 2.24 Å for the peroxo form) which leads to an estimated mean positional error of the atoms of about ±0.23 Å (5).

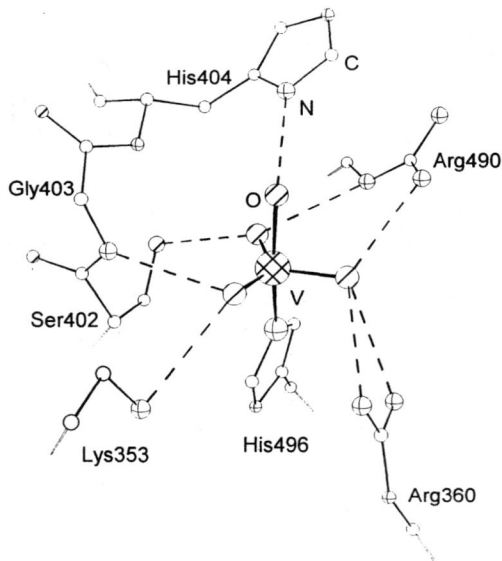

Figure 1. Structure of the active site of vanadium dependent chloroperoxidase from the fungus C. inaequalis.

The active site structure of vanadium haloperoxidase enzymes is the same for VCPO and VBPOs and, moreover, is observed to be rather rigid as the structure of the apo protein is compared with the vanadate and tungstate derivatives of the VCPO (6). An additional striking feature is the observation, that this active site is also observed for a series of acid phosphatases indicating consequences of the chemical similarities of vanadate and phosphate (7).

Although kinetic (8-12) and theoretical studies (13-20) have led to a significant advance in understanding the catalytic mechanism of vanadium haloperoxidase enzymes, there are still important issues that are unresolved, some of which are listed below:

- The protonation state and orientation of the vanadate cofactor in the resting state and the peroxo form.
- The mechanistic pathway from the resting state to the peroxo form.
- The specific role of His404, Lys353 and Ser402 (notation corresponds to Figure 1).
- Supramolecular effects of the protein matrix on structure and mode of action of the vanadate cofactor.

Theoretical Investigations of Haloperoxidases

In the following we will present our results utilizing density functional theory (DFT) to address the above-mentioned key issues related to the mode of action of vanadium haloperoxidase. Since the high catalytic activity of the vanadate cofactor in these enzymes is caused by the interaction with the hydrogen bonding network with a specifically tailored active site pocket which shows a rather high degree of rigidity. This allows for a reasonably sized approximation of the direct protein environment and thereby for investigations on a comparatively high level of theory, in particular as the employed basis sets are concerned. Nevertheless, subtle changes in this system are expected to bring about incisive variations concerning the catalytic properties, as becomes apparent from the comparison of the structure-function relationships of vanadium haloperoxidases and their active site homologous acid phosphatases (6). Therefore any kind of investigation on the mechanism of this system is a difficult task, be it theoretical or experimental.

Resting State Structure

Based on DFT calculations we have recently proposed that the resting state of the haloperoxidase enzymes contains a $[VO_2(OH)_2]^-$ species as prosthetic group with two hydroxyl groups one in equatorial and one in axial position (15, 16). These investigations are based on a protein model matrix build from the amino acids of the first sphere which are directly hydrogen bonded to the vanadium cofactor (cf. Figure 1). Along this search we scanned possible protonation states ranging from $[VO_3(OH)]^{2-}$ to $[VO(OH)_3]$, with some examples

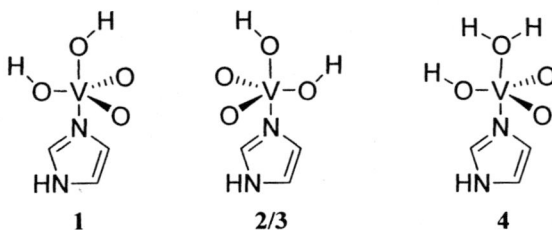

Figure 2. Model structures 1 to 4 of the resting state of the vanadium haloperoxidases, charges not shown; 3 represents the vanadate cofactor with the frozen coordinates taken from the optimized model matrix structure HPO depicted in Figure 3.

shown in Figure 2. Besides the variation in protonation of the vanadate cofactor, we examined also the option of having a protonated His404 residue which is then donating a hydrogen bond to the oxygen atom of the axial hydroxyl group of the vanadate moiety. In Figure 3 the corresponding optimized structure is shown, which we find to be the energetically most favored one. Moreover, our calculations indicate that a mono protonated vanadate cofactor $[VO_3(OH)]^{2-}$ only forms a stable V−N bond to an imidazole within the protein model matrix (*17*).

This resting state model in Figure 3 exhibits distinct differences as compared to other recently reported proposals, which also assume two hydroxyl groups in equatorial and axial position, but are lacking the proton at the His404. Moreover, within the other studies the position of the equatorial hydroxyl group is assigned either to the oxygen atom next to the Arg360 and Arg490 residues (*21, 23*) or to the oxygen atoms in the vicinity of the Ser402 or Lys353 residue (*24*). In this context it is important to note, that the Ser402 is the only residue in the equatorial region of the vanadate cofactor capable of accepting a hydrogen bond.

In addition, a dynamic behavior for the resting state is put forward with a switching of the equatorial proton of the $[VO_2(OH)_2]^-$ moiety toward the axial hydroxyl group forming a $[VO_3(H_2O)]^-$ species with an axial coordinated water molecule (*21, 23*). For this model a first protonation state has been proposed with an equatorial hydroxyl group and an axial water molecule. Although the latter seems to be closely related to our resting state model depicted in Figure 3, attempts to optimize the related structure with an axial water ligand had been unsuccessful. In fact, these calculations always lead to a spontaneous proton transfer from the axial water molecule towards the His404 residue.

At this point it has to be noted, that the successive removal of hydrogen bonding donors interacting with the equatorial oxygen atoms, i.e. Arg360, Arg490 and Lys353, stabilizes a species with an axial water molecule. This

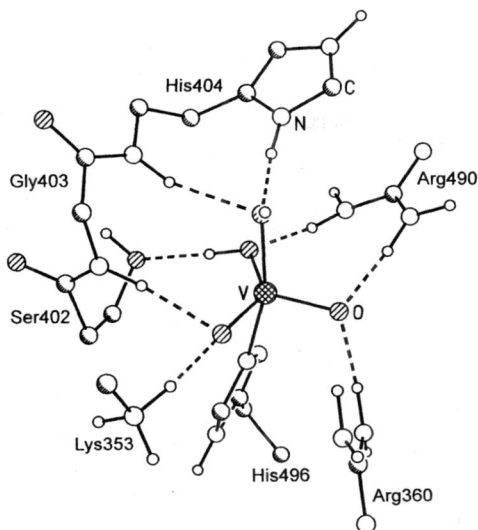

*Figure 3. Optimized structure **HPO** of the resting state with protein matrix.*

shows that the fine tuning of electronic charge on the equatorial oxygen atoms can induce such a proton transfer between a protonated His404 and an axial hydroxyl group. Moreover, reducing the number of equatorial hydrogen bonding donors has also an influence on the stability of the axial V−N bond to the coordinated imidazole of the His496 residue. This clearly indicates that the hydrogen bonding network of the protein matrix plays a significant role for the structure of the vanadate cofactor (*17*).

Electronic Structure of the Resting State and UV/vis spectra

Based on TD-DFT calculations we have investigated the electronic structure of the vanadate cofactor bound to vanadium haloperoxidases (*17*). This is in particular an interesting task, since this can allow for a direct comparison of theoretical and experimental results. In Figure 4 the calculated UV/vis spectra of model structures for $[VO_2(OH)_2]^-$ and $[VO_2(OH)(H_2O)]$ vanadate cofactors not including any hydrogen bonded donors of the protein environment are depicted. Whereas in Figure 5 a comparison of the UV/vis spectra is given for the model structures **3** and **HPO**, where **3** represents the vanadate cofactor with the geometry taken from the optimized model matrix structure **HPO** (see Figure 3), revealing effects related to the hydrogen bonding network of the protein matrix.

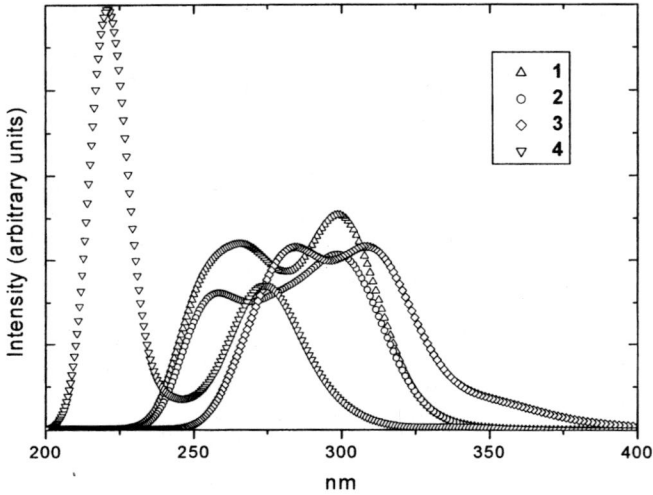

Figure 4. Calculated UV/vis spectra of model structures *1* to *4*.

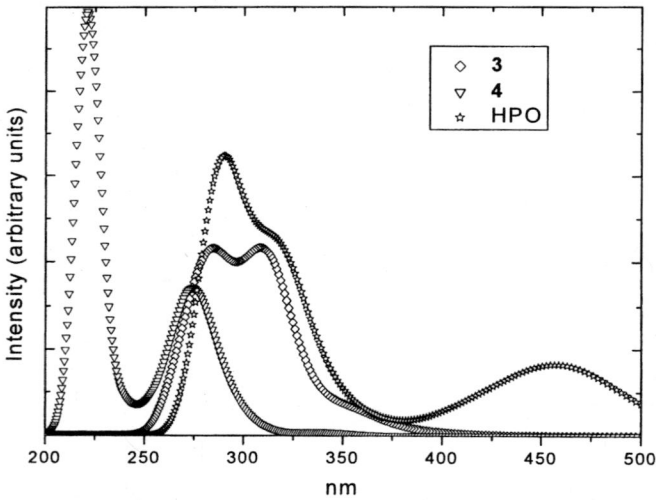

Figure 5. Comparison of the calculated UV/vis spectra of model structures *3* and *4* as well as the structure **HPO** with protein matrix depicted in Figure 3.

From the UV/vis spectra the following conclusions can be drawn:

- High protonation states lead to a blue-shift of the absorption bands and increases the admixture of imidazole orbitals in the HOMO region leading to corresponding LMCT transitions.
- The orientation of the vanadate cofactor with respect to the imidazole ring has only a minor effect on the position of the UV/Vis transitions, but changes their intensities due to variations in the transition dipole moments caused by difference in the charge distribution.
- The comparison of structures **1** or **2** with **3** shows that the structural changes induced by assembling the vanadate cofactor in the protein environment leads to a significant red-shift of the transitions.
- From the comparison of structure **3** with **HPO** it is apparent that inclusion of supramolecular effects of the protein matrix lead to significant changes, manifested by an additional red-shift of the transitions and an increase in intensity as well as a more pronounce contribution of the imidazole (His496) to the lowest energy LMCT transition.
- The comparison of the theoretical and experimental data supports the proposed resting state structure of the vanadate cofactor depicted in Figure 3, as a doubly protonated $[VO_2(OH)_2]^-$ moiety coordinated by His496.

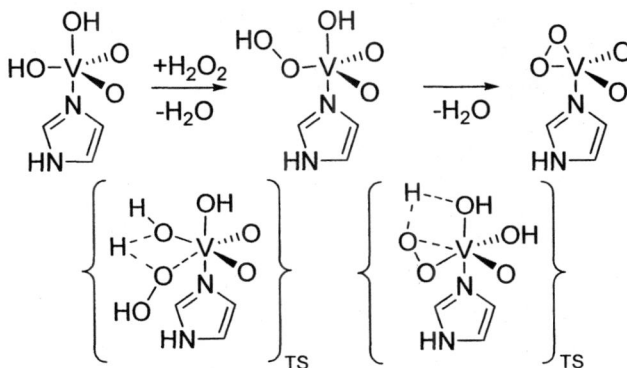

Figure 6. Calculated concerted reaction path for the reaction of $[VO_2(OH)_2(Im)]^-$ with H_2O_2, charges not shown.

Formation of the Peroxo Form

The initial reaction of vanadium haloperoxidases is the conversion from the resting state to the peroxo form of the enzyme. Based on our resting state proposal several scenarios can be envisaged. We sampled the potential reaction pathways starting from our simplified resting state model $[VO_2(OH)_2(Im)]^-$ and its formal first protonation state $[VO_2(OH)(H_2O)(Im)]$. The latter is accessible by proton transfer from the protonated His404 residue to the axial hydroxyl group of the resting state (see Figure 3).

For the $[VO_2(OH)_2(Im)]^-$ model we were able to identify two mechanisms, for which the first transition state refers to a concerted proton transfer and nucleophilic attack of the oxygen atom of the hydrogen peroxide molecule within the equatorial plane of the vanadate moiety. Along this line there are two possible orientations for the hydrogen peroxide molecule to approach. One is between the two oxo groups and the other between one of the oxo groups and the hydroxyl group, where for the latter case two destinations for the transferred proton are possible, which refers to either the adjacent oxo or the hydroxyl group. Only for the latter case with the proton being transferred to the hydroxyl group an energetically feasible reaction path could be calculated (see Figure 6), with an activation energy for the first transition state of about 14 kcal/mol and an overall exothermic reaction profile. Nevertheless, it is interesting to note that the alternative proton transfer path leads to an octahedral intermediate with an end-on coordinated hydroperoxide. Any attempt to initiate a further proton transfer starting from this intermediate leads to the cleavage of the imidazole V−N bond.

Starting from the protonated model structure $[VO_2(OH)(H_2O)(Im)]$ we found four different routes, with two of them leading to high energy intermediates or transition state with octahedral geometry and the peroxide

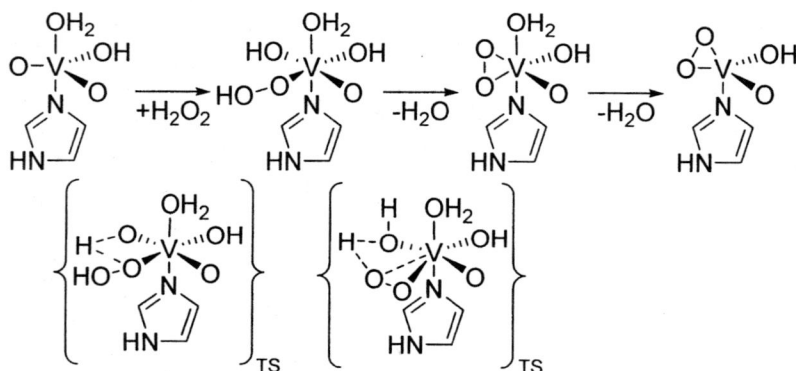

Figure 7. Calculated associative reaction path for the reaction of $[VO_2(OH)(H_2O)(Im)]$ with H_2O_2, charges not shown.

moiety in an equatorial position between two hydroxyl groups. Any search for continuing pathways ends in the cleavage of the imidazole V−N bond, as in the former case with apical hydroxyl group. The two successful reaction pathways an associative and a dissociative are depicted in Figures 7 and 8, respectively.

The first transition state of the associative pathway needs an activation energy of about 20 kcal/mol with the first intermediate structure being endothermic by about 2.5 kcal/mol with respect to the starting encounter complex. All following steps possess activation energies lower than 4 kcal/mol and the overall reaction is exothermic. The situation is somewhat different for the dissociative pathway where the first transition state is only 3 kcal/mol above the initial encounter complex leading to an exothermic intermediate (-5.5 kcal/mol). The highest energy barrier is found for the second step (14.5 kcal/mol) with a slightly exothermic reaction to the next intermediate, which converts to the final peroxo species with a barrier of only about 4 kcal/mol.

The energetically demanding steps in all three presented reaction pathways are proton transfer reactions with a rather strained four-member ring (see Figures 6 to 8). As we did exclude any potential external acid-base catalyst, a mode of action which as a matter of fact is indeed attributed to the His404 residue, these activation barriers are likely to be overestimated.

To address the question, whether or not the final peroxo form of the vanadate cofactor is protonated as found for the pathways starting from $[VO_2(OH)(H_2O)(Im)]$ or unprotonated as obtained for $[VO_2(OH)_2(Im)]^-$ as initial structure, we performed DFT calculations for the cofactor models within the

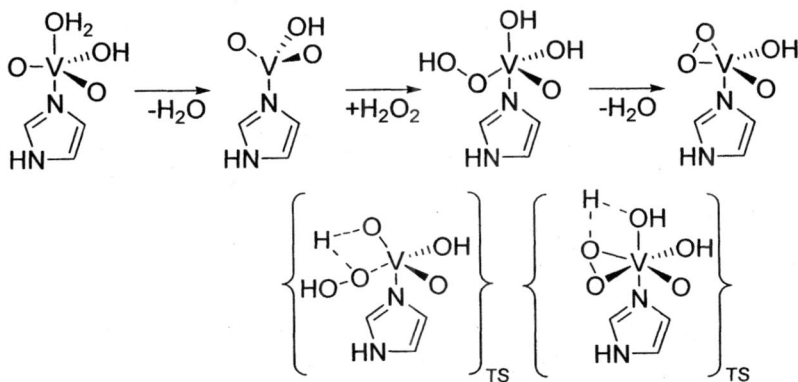

Figure 8. Calculated dissociative reaction path for the reaction of $[VO_2(OH)(H_2O)(Im)]$ with H_2O_2, charges not shown.

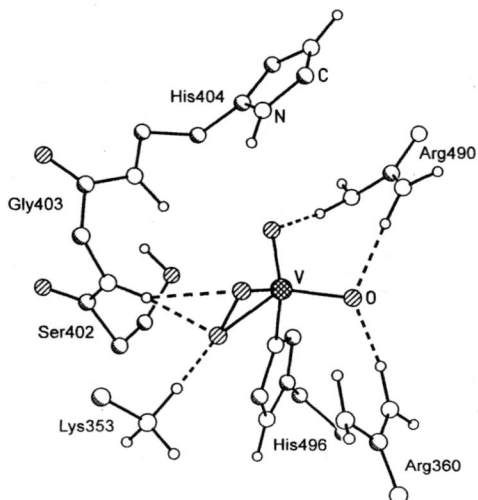

Figure 9. Optimized structure of the peroxo form of the vanadate cofactor with protein matrix.

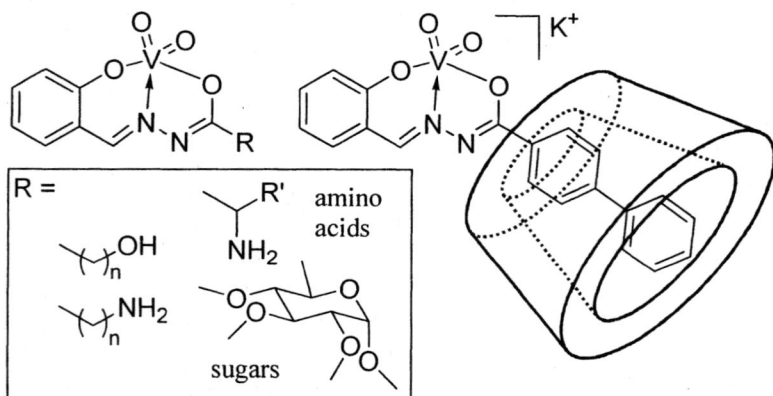

Figure 10. Structure and substitution pattern of N-salicylidene hydrazide ligands (the barrel represents a cyclodextrin molecule).

appropriate protein matrix. The energetically most favored structure is depicted in Figure 9 and represents an unprotonated $[VO_2(O_2)]^-$ with a protonated His404 residue.

Supramolecular Synthetic Model Systems

The chemistry of vanadium model complexes relevant to the haloperoxidase enzyme and its vanadate cofactor is well established (*1, 21-23*). We have recently reported on the versatility of *N*-salicylidene hydrazide ligand systems for the synthesis of vanadium complexes with interesting supramolecular structures and reactivity (*24-30*). It was shown that these tridentate chelate ligands can coordinate the vanadium center either in their mono- or dianionic form. In Figure 10 the schematic structures of corresponding vanadium(V) complexes are shown, which contain a functionalized ligand side chain that is capable of generating supramolecular interactions. This includes also examples of inclusion assemblies with cyclodextrin molecules as supramolecular hosts.

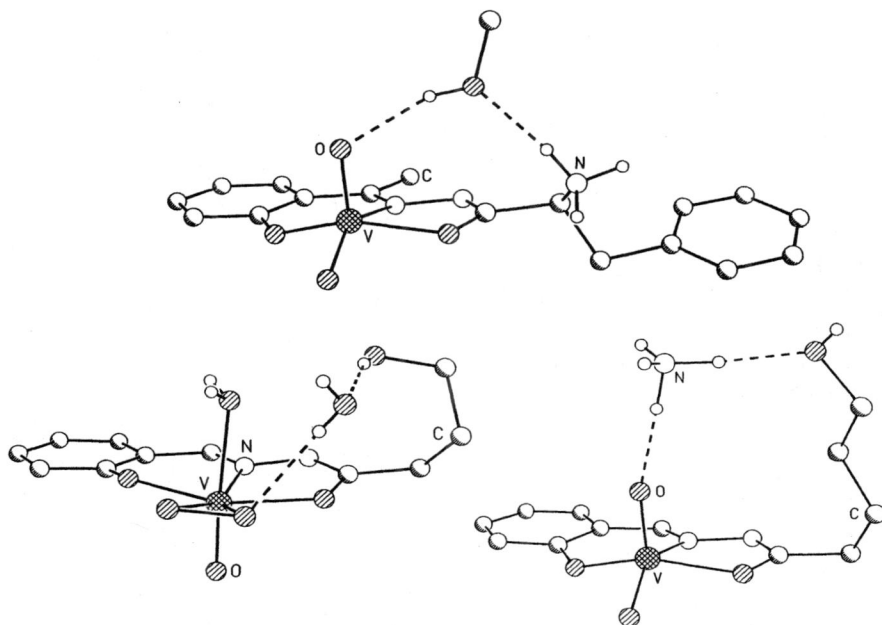

Figure 11. Structures of supramolecular assemblies of vanadium(V) complexes with hydrogen bonding relays.

As has been shown this ligand system can support both oxo and dioxo vanadium(V) species and, moreover, also allows for the variation of the protonation state of the resulting complex (24, 26, 27). An additional interesting feature of complexes with such ligand systems is the feasibility of generating supramolecular assemblies with hydrogen bonding relays of the functionalized side chain and the vanadate moiety. Three particularly interesting examples are shown in Figure 11. In two structures it is shown that protic solvent molecules like alcohol as well as ammonium ions can act as hydrogen bonded linkers to oxo groups of vanadium model complexes (28), whereas the third structure exemplifies the possible interaction with a peroxo group (29).

It should be mentioned here that in the synthesis of peroxo vanadium(V) model complexes generally slightly acidic conditions are required, indicating that a protonation of the starting vanadate species is essential (1, 29, 31).

Of particular interest is the amino acid derivative based on phenylalanine which not only allows a hydrogen bonding relay, but also contains a chiral center next to the vanadium moiety. The latter feature can be also found in model complexes with sugar derived ligands and supramolecular assemblies with chiral hosts like cyclodextrin. It is interesting to note that the phenylalanine based model complex also exhibits efficient catalytic properties with respect to haloperoxidation and sulfoxidation which could be attributed to the influence of the free ammonium group in the vicinity of the vanadium center.

Catalytic Cycle

Based on the results presented and discussed in this contribution, including the relevant cited data from literature, the reaction scheme depicted in Figure 12 is proposed for the catalytic cycle of the halide oxidation by vanadium haloperoxidases. In the following important features are summarized:

- The Ser402 residue functions as a hydrogen bonding acceptor for the equatorial hydroxyl group of the vanadate cofactor, and is not involved in halide binding. It thereby defines the orientation of the $[VO_2(OH)_2]^-$ moiety in the active site pocket.
- The His404 residue is protonated in the resting state, leading to a situation that represents a latent apical water molecule. This masked water molecule can be brought to action by subtle changes in the electronic structure of the vanadate cofactor, which might be induced by an incoming hydrogen peroxide molecule.
- The key steps in the formation of the peroxo form are related to proton transfer reactions and might therefore be facilitated by proton transmitters or acid-base catalysts as is exemplified by the supramolecular assemblies of our model complexes.

- In contrast to model complexes, where a protonation is required for the formation of the peroxo species, the enzyme affinity towards hydrogen peroxide is even increasing at higher pH values. This is consistent with the proposed high protonation state of the resting form of the enzyme, thereby offering all essential prerequisites for the transformation without any additional proton source.

- Based on the pK_a values of hypohalous acids and histidine it can be assumed that the enzyme thus not need to take any particular precaution to facilitate the availability of protons. We therefore formally added a proton and a halide ion X^- in successive manner during the cycle.

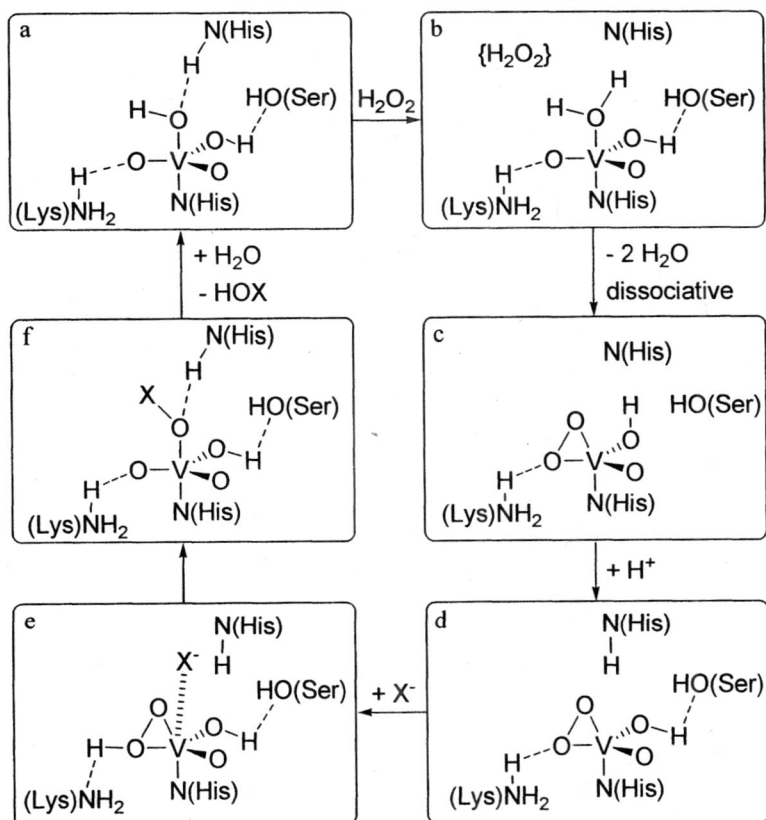

Figure 12. Proposed catalytic cycle for the halide oxidation by vanadium haloperoxidases, charges at the vanadate not schown; emphasizing the specific role of the amino acid residues Lys353, Ser402 and His404 (see Figure 1).

176

Acknowledgment

We thank the Deutsche Forschungsgemeinschaft for financial support by a generous research grant and the John von Neumann Institute for Computing (NIC) at the Forschungszentrum Jülich for the generous supply of computational time on the JUMP supercomputer.

References

1. Crans, D. C.; Smee, J. J.; Gaidamauskas, E.; Yang, L. *Chem. Rev.* **2004**, *104*, 849-902.
2. Messerschmidt, A.; Wever, R. *Proc. Natl. Acad. Sci. USA* **1999**, *93*, 392-396.
3. Isupov, M. N.; Dalby, A. R.; Brindley, A. A.; Izumi, Y.; Tanabe, T.; Murshudov, G. N.; Littelchild, J. A. *J. Mol. Biol.* **2000**, *299*, 1035-1049.
4. Weyand, M.; Hecht. H. J.; Kiess, M.; Liaud, M. F.; Vilter, H.; Schomburg, D. *J. Mol. Biol.* **1999**, *293*, 595-611.
5. Messerschmidt, A.; Prade, L.;Wever, R. *Biol. Chem.* **1997**, *378*, 309-315.
6. Messerschmidt, A.; Wever, R. *Inorg. Chim. Acta* **1998**, *273*, 160-166.
7. Plass, W. *Angew. Chem. Int. Ed.* **1999**, *38*, 909-912.
8. Hemrika, W.; Renirie, R.; Macedo-Ribeiro, S.; Messerschmidt, A.; Wever, R. *J. Biol. Chem.* **1999**, *274*, 23820-23827.
9. Macedo-Ribeiro, S.; Hemrika, W.; Renirie, R.; Wever, R.; Messerschmidt, A. *J. Biol. Inorg. Chem.* **1999**, *4*, 209-219.
10. Renirie, R.; Hemrika, W.; Pierma, S. R.; Wever, R. *Biochemistry* **2000**, *39*, 1133-1141.
11. Tanaka, N.; Hasan, Z.; Wever, R. *Inorg. Chim. Acta* **2003**, *356*, 288-296.
12. Hasan, Z.; Renirie, R.; Kerkman, R.; Ruijssenaars, H. J.; Hartog, A. F., Wever, R. *J. Biol. Chem.* **2006**, *281*, 9738-9744.
13. Borowski, T.; Szczepanik, W.; Chruszcz, M.; Broclawik, E. *Int. J. Quant. Chem.* **2004**, *99*, 864-875.
14. Zampella, G.; Kravitz, J. Y.; Webster, C. E.; Fantucci, P.; Hall, M. B.; Carlson, H. A.; Pecoraro, V. L.; De Gioia,L. *Inorg. Chem.* **2004**, *43*, 4127-4136.
15. Plass, W.; Pohlmann, A.; Yozgatli, H. P. *J. Inorg. Biochem.* **2000**, *80*, 181-183.
16. Plass, W. In *NIC Symposium 2001*; Rollnik, H.; Wolf, D., Eds.; NIC Series; John von Neumann Institut für Computing: Jülich, 2002, Vol. 9, pp 103-110.
17. Bangesh, M.; Plass, W. *J. Mol Struct. Theochem* **2005**, *725*, 163-175.
18. Zampella, G.; Fantucci, P.; Pecoraro, V. L.; De Gioia, L. *J. Am. Chem. Soc.* **2005**, *127*, 953-960.

19. Kravitz, J. Y.; Pecoraro, V. L.; Carlson, H. A. *J. Chem. Theory Comput.* **2005**, *1*, 1265-1274.
20. Raugei, S.; Carloni, P. *J. Phys. Chem. B* **2005**, *110*, 3747.
21. Butler, A.; Clague, M. J.; Meister, G. E. *Chem. Rev.* **1994**, *94*, 625-638.
22. Rehder, D. *Coord. Chem. Rev.* **1999**, *182*, 297-322.
23. Ligtenbarg, A. G. J.; Hage, R.; Feringa, B. L. *Coord. Chem. Rev.* **2003**, *237*, 89-101.
24. Plass, W.; Pohlmann, A.; Yozgatli, H. P. *J. Inorg. Biochem.* **2000**, *80*, 181-183.
25. Pohlmann, A.; Plass, W. *J. Inorg. Biochem.* **2001**, *86*, 381-381.
26. Plass, W.; Yozgatli, H. P. *Z. Anorg. Allg. Chem.* **2003**, *629*, 65-70.
27. Plass, W. *Coord. Chem. Rew.* **2003**, *237*, 205-212.
28. Pohlmann, A.; Nica, S.; Luong, T. K. K.; Plass, W. *Inorg. Chem. Commun.* **2005**, *8*, 289-292.
29. Nica, S.; Pohlmann, A.; Plass, W. *Eur. J. Inorg. Chem.* **2005**, 2032-2036.
30. Becher, J.; Seidel, I.; Plass, W.; Klemm, D. *Tetrahedron* **2006**, 62, 5675-5681.
31. Hamstra, B. J.; Colpas, G. J.; Pecoraro, V. L. *Inorg. Chem.* **1998**, *37*, 949-955.

Chapter 14

^{51}V Solid-State NMR Spectroscopy of Vanadium Haloperoxidases and Bioinorganic Haloperoxidase Mimics

Tatyana Polenova[1], Neela Pooransingh-Margolis[1], Dieter Rehder[2], Rokus Renirie[3], and Ron Wever[3]

[1]Department of Chemistry and Biochemistry, University of Delaware, Newark, DE 19716
[2]Institut für Anorganische und Angewandte Chemie, Universität Hamburg, D–20146 Hamburg, Germany
[3]Van't Hoff Institute for Molecular Sciences, Faculty of Science, University of Amsterdam, Nieuwe Achtergracht 129, 1018 WS Amsterdam, The Netherlands

We present ^{51}V solid-state magic angle spinning NMR spectroscopy as a probe of geometric and electronic environments in vanadium haloperoxidases and in oxovanadium (V) complexes mimicking the active site of these enzymes. In the bioinorganic complexes, ^{51}V MAS spectra are sensitive reporters of the coordination environment and coordination geometry. In vanadium chloro- and bromoperoxidases, the spectra reveal unique electronic environments of the vanadate cofactor in each species. The experimental NMR observables and DFT calculations of the NMR parameters yield the most likely protonation states of the individual oxygen ligands in vanadium chloroperoxidase. A combination of experimental solid-state NMR and quantum mechanical calculations thus offers a powerful strategy for analysis of diamagnetic spectroscopically silent vanadium (V) states in inorganic and biological systems.

© 2007 American Chemical Society

Introduction

Vanadium Haloperoxidases

Vanadium haloperoxidases (VHPO) catalyze a two-electron oxidation of halides to hypohalous acids in the presence of hydrogen peroxide; the native enzymes require diamagnetic V(V) for their activity (*Scheme 1*).

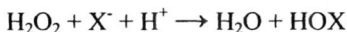

$$H_2O_2 + X^- + H^+ \longrightarrow H_2O + HOX$$

Scheme 1

Haloperoxidases are named after the most electronegative halide they are able to oxidize (i.e., chloroperoxidases oxidize Cl^-, Br^-, and I^-). Vanadium bromoperoxidases (VBPO) are universally present in marine macroalgae; vanadium chloroperoxidases (VCPO) have been found in terrestrial fungi and in lichens (*1-4*). Vanadium bromoperoxidases are thought to be involved in the biosynthesis of halogenated (brominated and iodinated) natural products; the function of the fungal chloroperoxidase is to damage (through oxidation by HOCl) the protective lignocellulose of plant tissues (*1-3, 5, 6*). The kinetics and mechanism of haloperoxidases have been the subject of multiple studies (*7, 8*) due to their potential use as industrial antimicrobial agents and disinfectants (*9, 10*) as well as halogenation catalysts (*11, 12*): these enzymes are the most efficient and stable halide oxidants known to date.

Despite numerous investigations by multiple laboratories addressing various aspects of chemistry and biology of vanadium haloperoxidases, many questions about their function and mechanism remain open preventing their widespread biotechnological use and evaluation of other areas of applications, in part due to the fact that the enzymes are colorless and diamagnetic in their active form, making the spectroscopic studies difficult. Weak UV bands were detected in the region of 300-330 nm that report on vanadate binding to vanadium chloroperoxidase (VCPO), but these do not permit detailed analysis of the vanadium ligands (*13*). The factors determining the substrate specificity, i.e. whether a particular enzyme will or will not display chlorinating activity are not understood (*14-16*). A delicate balance of multiple interactions between the active site residues has been proposed to be responsible for chlorinating activity of the haloperoxidases (*15*). The architectures of the chloro- and bromoperoxidase active sites are found to be very similar, as illustrated in Figure 1, where the active sites of vanadium chloro- and bromoperoxidases are superimposed using the original X-ray coordinates for the two enzymes (*17, 18*). Mutation studies suggest that altering the electrostatic potential distribution either reduces or completely abolishes the chlorinating activity in a series of mutant chloroperoxidases, yet the crystal structures for a number of these mutants with the exception of H496A reveal the active site containing vanadate coordinated to His-496, similar to the native protein

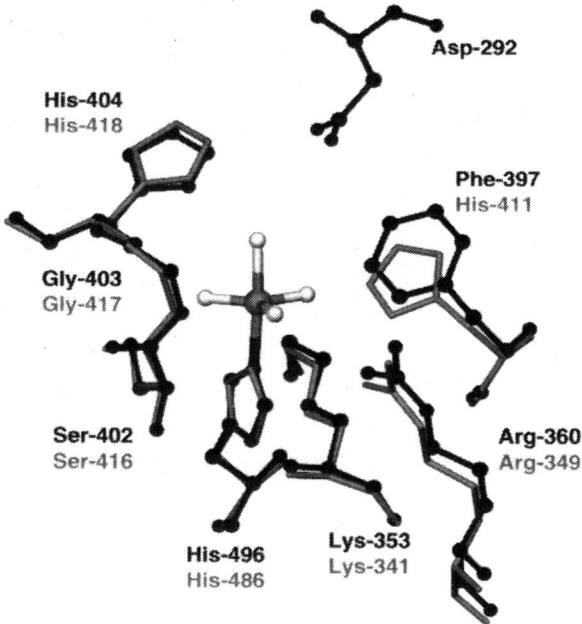

Figure 1. The superimposition of the active sites of VCPO from C. inaequalis (black, ball-and stick) and VBPO from A. nodosum (grey, stick) illustrates their similar geometries. The vanadate cofactor is depicted with larger spheres: vanadium (grey), oxygen (white). Active site residue labels correspond to VCPO (top, black) and VBPO (bottom, grey). The proposed vanadate geometry consists of an axial hydroxo group trans to His-496/His-486 and three equatorial oxo ligands. At pH lower than 8.0 protonation of one of the oxo groups may occur. The figure has been prepared using the original pdb coordinates (pdb codes 1idq and 1qi9) in DSViewer Pro (Accelrys, Inc.)

(*13, 15, 19*). Therefore, these mutagenesis results provide indirect evidence that the electronic structure of the vanadate cofactor is modulated by the protein, thus affecting its chemical reactivity.

The X-ray crystal structure determined at pH 8.0 revealed that in the resting state of VCPO, the vanadate cofactor is covalently bound to the Nε2 atom of a histidine residue (His-496 in the VCPO from *C. inaequalis*) (*17*). The negative charge of the vanadate group is compensated by hydrogen bonds to several positively charged protein sidechains (Lys-353, Arg-360, and Arg-490) (*17*). Ser-402 and Gly-403 form hydrogen bonds with the equatorial oxygens of the vanadate cofactor; an additional hydrogen bond may exist between the axial hydroxo group and His-404. However, the presence and positions of hydrogen atoms could not be unambiguously determined from the crystal structure due to the inherently limited resolution; therefore, the nature of the vanadium first coordination sphere ligands (i.e., whether a particular group is oxo- or hydroxo-) still remains the subject of debate.

Recently, the electronic structure and geometry of the vanadium center in the resting state (*20, 21*) and peroxo forms (*21, 22*) of a series of VHPO active site models were addressed quantum mechanically via Density Functional Theory. Subsequently, hybrid QM/MM calculations were conducted independently by two groups of investigators addressing extended active site models of VCPO treated quantum mechanically and significant portions of the protein treated with classical mechanics (*23, 24*). In parallel, DFT calculations were performed on large models of VCPO active site (*25*). These exciting DFT and QM/MM studies have led to several very interesting conclusions. All of the above calculations suggest that in the resting state, at least one equatorial oxygen needs to be protonated to stabilize the metal cofactor (*20, 23-25*). According to the DFT results, the equatorial hydroxo group is likely to be coordinated to Ser-402 (*25*). As anticipated, QM/MM calculations indicate that the protein environment is crucial for creating the long-range electrostatic field necessary for the stabilization of the resting state (*23*). Interestingly, the QM/MM calculations conducted by Carlson, Pecoraro and Kravitz suggest that the protonated equatorial oxygen is accepting two hydrogen bonds from Arg-360 and Arg-390 residues (*23*). A hybrid resting state consisting of the two lowest-energy minima is likely based on the energetic considerations (*23*). The roles of the individual amino acid residues and of the oxo atoms of the cofactor have been re-examined, and a revised mechanism for VCPO catalyzed halide oxidation reaction has been proposed (*23*). In another QM/MM study, Raugei and Carloni examined the early intermediates and the transition state of the halide oxidation in VCPO, by including a somewhat smaller number of atoms in the QM region of the calculation, and treating the rest of the protein at the MM level (*24*). They concluded that one of the equatorial oxygens is protonated, in agreement with the above studies, but the equatorial hydroxy group is hydrogen bonded to either Ser-402 or to Lys-353 residue (*24*). Thus, the coordination environment of the vanadate cofactor including the protonation states of the individual oxygen atoms still remains an open question.

Based on the above studies, it is clear that there is a need to address experimentally the coordination environment of the vanadium cofactor in vanadium chloro- and bromoperoxidases. Understanding the protonation states of the oxygen atoms is important for elucidating the chlorinating activity in haloperoxidases. The lack of direct site-specific spectroscopic probes has so far prevented gaining further insight on the electronic environment of the vanadium center intimately related to the enzymatic mechanism of vanadium haloperoxidases.

In this work, we introduce ^{51}V solid-state magic angle spinning NMR spectroscopy as a direct spectroscopic probe of the diamagnetic vanadium (V) sites of vanadium haloperoxidases. ^{51}V is a half-integer quadrupolar nucleus (I = 7/2) whose high natural abundance (99.8%), relatively high gyromagnetic ratio (Larmor frequency of 157.6 MHz at 14.1 T), and a relatively small quadrupole moment $(-0.052*10^{-28}$ $V/m^2)(26)$ make it favorable for direct detection in the NMR experiments. The ^{51}V solid-state spectra are typically dominated by the anisotropic quadrupolar and chemical shielding interactions. These interactions in turn report on the geometric and electronic structure of the vanadium site (27-29). As demonstrated for the model bioionorganic oxovanadium complexes mimicking the active site of haloperoxidases, the NMR parameters extracted from numerical simulations of the ^{51}V MAS spectra are sensitive to the coordination environment of the vanadium atom beyond the first coordination sphere. The NMR fine structure constants calculated for the crystallographically characterized complexes using Density Functional Theory, are in good agreement with the experimental results, illustrating that a combination of solid-state NMR experiments and quantum mechanical calculations presents a powerful approach for deriving coordination geometry in these systems.

We present ^{51}V MAS NMR spectra of vanadium chloroperoxidase from *C. inaequalis* and of vanadium bromoperoxidase from *A. nodosum*. The spectra reveal different electronic environments of the vanadate cofactor in each species. In both enzymes, the spectra are dominated by a large quadrupolar interaction, providing the first direct experimental evidence of the asymmetric electronic charge distribution at the vanadium site. The isotropic chemical shifts in VCPO and VBPO are profoundly different, suggesting that the vanadate oxygens are likely not to have the same protonation states. In VCPO, the NMR observables were extracted from the spectra, and the DFT calculations of these observables in the extensive series of the active site models whose electronic structure and energetics were previously addressed by De Gioia, Carlson, Pecoraro and co-workers (20) indicate that one equatorial and one axial oxygen are protonated resulting in an overall anionic vanadate. Our experimental results are in remarkable agreement with the quantum mechanical calculations discussed above (20, 23-25). We anticipate that our approach combining ^{51}V solid-state NMR spectroscopy and quantum mechanical calculations will yield a thorough understanding of the salient features of the vanadium haloperoxidases' active sites.

^{51}V Solid-State NMR Spectroscopy

^{51}V is a half-integer quadrupolar nucleus (I=7/2). Two tensorial interactions dominate the ^{51}V solid-state NMR spectra: the quadrupolar and chemical shielding anisotropies. Due to their different magnitudes and symmetries, both tensors can be extracted from a single NMR spectrum (*27, 28, 30, 31*). It is worth noting that in solution, both interactions are averaged due to molecular tumbling, and only the isotropic component of the chemical shift tensor is observed. Therefore, valuable geometric and electronic information is lost in solution.

In the solid state, the total Hamiltonian can be expressed as:

$$H = H_{Zeeman} + H_{RF} + H_{DIP} + H_Q + H_{CSA} \tag{1}$$

The first three terms are the Zeeman, the radiofrequency field, and the dipolar interactions. The dipolar interaction is typically much smaller than the quadrupolar and chemical shielding anisotropy and will be omitted in the subsequent discussions. The last two terms are the quadrupolar and CSA interactions, which determine the spectral shape. They are commonly expressed in a spherical tensor notation in terms of the spatial (R_{mn}) and spin (T_{mn}) variables (*32*):

$$H_Q^{(1)} = \frac{eQ}{4S(2S-1)} \cdot R_{20}^Q T_{20}^S = \omega_Q \left[3S_z^2 - S(S+1)\right] \tag{2}$$

$$H_Q^{(2)} = \frac{C_Q}{\omega_0} \sum_{m \neq 0} \frac{R_{2m} R_{2-m} \left[T_{2m}, T_{2-m}\right]}{2m} \tag{3}$$

$$H_{CSA} = -\gamma \left(R_{00}^{CS} T_{00}^S + R_{20}^{CS} T_{20}^S\right) = \left(\omega_{CS}^{iso} + \omega_{CS}^{aniso}\right) S_Z \tag{4}$$

$H_Q^{(1)}$ and $H_Q^{(2)}$ are the first- and second-order quadrupolar interactions. The quadrupolar and CSA tensor elements are defined in a spherical harmonics basis set according to the standard notation (*33, 34*):

$$C_Q = \frac{eQV_{zz}}{h}; \qquad \eta_Q = \frac{V_{yy} - V_{xx}}{V_{zz}}; \tag{5}$$

$$\delta_\sigma = \delta_{zz} - \delta_{iso}; \qquad \eta_\sigma = \frac{\delta_{yy} - \delta_{xx}}{\delta_{zz} - \delta_{iso}};$$

$$\delta_{iso} = \tfrac{1}{3}\left(\delta_{xx} + \delta_{yy} + \delta_{zz}\right) \tag{6}$$

where C_Q is the quadrupolar coupling constant (in MHz); V_{xx}, V_{yy}, V_{zz} are the principal components of the electric field gradient (EFG) tensor, with $V_{zz} = eq$ being its largest principal component, and $|V_{zz}| \geq |V_{yy}| \geq |V_{xx}|$. C_Q defines the overall breadth of the spectral envelope. Q is the vanadium quadrupole moment (-0.052*10^{-28} V/m^2) (*26*); e is the electronic charge; h is the Planck constant.

184

The spectral broadening due to the second-rank spatial components R_{20} of tensorial anisotropies of $H_Q^{(1)}$ and H_{CSA} can be averaged into a spinning sideband pattern when the solid sample is rapidly spun at the magic angle (an axis inclined at 54.7° with respect to the static magnetic field). The fourth-rank terms of the $H_Q^{(2)}$ are not completely averaged out, and give rise to characteristic second-order lineshapes. The ^{51}V quadrupolar and the chemical-shielding tensors, and their relative orientations, can be determined from the MAS spectra of the central and the satellite transitions, as demonstrated by Skibsted, Nielsen, Jacobsen and their colleagues as well as by our recent work (27, 28, 30, 31, 35-42). These two interactions can be further correlated with the structure and electronic properties at the vanadium site by classical electrostatic calculations for ionic compounds (30, 31) or more generally via the density functional theory (DFT) (28, 43, 44).

Experimental Protocols

Preparation of Oxovanadium (V) Complexes and Vanadium Haloperoxidases

Preparation of oxovanadium (V) bioinorganic complexes mimicking the active site of vanadium haloperoxidases has been reported previously (28).

Isolation and purification of recombinant vanadium chloroperoxidase from *C. inaequalis* overexpressed in *S. cerevisae* has been described in our recent report (45). The specific activity of the purified protein was approximately 16 units/mg. The vanadate incorporation, calculated by measuring the activity using MCD assay with and without vanadate, was found to be approximately 98%.

For isolation and purification of the wild type vanadium bromoperoxidase from the brown seaweed *A. nodosum*, the procedure reported by Wever and colleagues was followed (46, 47). Quantification of the protein was carried out by Bradford assay (Appendix B). The final protein yield was approximately 187 mg of VBPO (17.80 mg VBPO/kg seaweed). The specific activity, determined by MCD assay was found to be approximately 504 units/mg. Previous studies on VBPO from *A. nodosum* yielded specific activities ranging from 120 to 800 units/mg (48, 49). The vanadate incorporation, calculated by measuring the activity using MCD assay with and without vanadate, was found to be approximately 95%.

In the preparation of VCPO and VBPO, special care was taken to minimize sodium salts during enzyme isolation and purification, in order to eliminate possible sources of ^{23}Na, as the presence of the latter could potentially interfere with the detection of ^{51}V because of the similar gyromagnetic ratios and resonance frequencies for the two isotopes.

Solid-state NMR samples of VCPO and VBPO were prepared by lyophilization of the concentrated protein solutions using a vacuum centrifuge. VCPO (68 mg; the total sample volume of 100 µl) and VBPO (a total of 88 mg, of which 44 mg was pure enzyme; the total sample volume of 100 µl) were packed into 5 mm Doty SiN$_3$ thick-wall rotors and stored at -20 °C for subsequent solid-state NMR experiments. The enzymatic activity of the proteins was recorded before and after the NMR experiments. No detectable change in activity was observed during the measurements.

^{51}V Solid-State NMR Spectroscopy of Oxovanadium (V) Haloperoxidase Mimics

^{51}V solid-state NMR spectra of oxovanadium (V) complexes were acquired on Tecmag Discovery spectrometer at a static magnetic field of 9.4 T (^{51}V Larmor frequency of 105.2 MHz). 5-40 mg of sample was packed into a 4 mm Doty XC4 MAS probe. Spectra were recorded at three MAS frequencies in the range from 10 to 17 kHz for each compound. The spinning speed was controlled to within ± 5 Hz. The magic angle was adjusted using NaNO$_3$, by detecting the ^{23}Na MAS signal. Single pulse excitation spectra were obtained using 1µs-pulses at $\gamma H_1 / 2\pi \approx 80\text{kHz}$; the recycle delays were 0.5 s. The data were processed by linear prediction of the first 66 points to suppress baseline distortions, followed by Fourier transformation and baseline correction. Isotropic chemical shifts are reported with respect to the external reference- neat VOCl$_3$.

^{51}V Solid-State NMR Spectroscopy of Vanadium Chloro- and Bromoperoxidases

^{51}V solid-state NMR spectra of vanadium haloperoxidase were recorded at a static magnetic field of 14.1 T (^{51}V carrier frequency of 157.64539 MHz). A narrow bore Varian InfinityPlus instrument outfitted with a 5 mm Doty single channel MAS probe was used for data collection. The data were acquired the spinning frequencies of 13, 15 and 17 kHz. The temperature was maintained between –25 and –26 °C throughout the measurements, according to the readings on the temperature controller. The magic angle was adjusted using NaNO$_3$ (by detecting the ^{23}Na MAS signal). The radiofrequency field strength was 44.6 kHz (non-selective 90-degree pulse of 5.6 µs), as calibrated using the ^{51}V signal from

186

the neat $VOCl_3$ liquid. A single 1.45-μs excitation pulse (a 23-degree flip angle) was employed with 0.3-s recycle delay between the individual scans. In the VCPO measurements, a total of 1.6 million or 1.5 million transients were added in the final spectra acquired with the spinning frequencies of 15 kHz and 17 kHz, respectively. For VBPO, the number of scans were 2 million, 1.25 million and 1.5 million for spectra acquired with the spinning frequencies of 13 kHz, 15 kHz and 17 kHz respectively. Data were processed by left shifting the spectrum to the first rotor echo to suppress baseline distortions, followed by exponential apodization, with line broadening of 500 Hz. Isotropic chemical shifts are reported with respect to the external reference- neat $VOCl_3$.

Numerical Simulations of ^{51}V MAS NMR Spectra

^{51}V NMR spectra were simulated in SIMPSON (*35*). Quadrupolar interaction to the second order and chemical shift anisotropy were included in the Hamiltonian; the heteronuclear dipolar interaction was omitted as it is expected to be negligible. The seven independent parameters describing the quadrupolar and CSA tensor anisotropies (C_Q, η_Q, δ_σ, and η_σ) and the relative tensor orientations (the Euler angles α, β, and γ) were obtained by the least-squares fitting of the simulated and experimental sideband intensities using a program written in our laboratory under the Mathematica (Wolfram, Inc.) environment.

Quantum Mechanical Calculations of NMR Spectroscopic Parameters

Quantum mechanical calculations of electric field gradient and magnetic shielding anisotropy tensors were conducted under Gaussian03 (*50*). Becke's three-parameter hybrid B3LYP functional was used for all calculations (*51*). The calculations were conducted with three different basis sets for small VCPO active site models, and with two different basis sets for oxovanadium (V) bioinorganic complexes. The TZV basis set(*52*) and effective core potentials for vanadium atoms were used for one set of the calculations; in another set the 6-311+G basis set was employed for VCPO active site models and oxovanadium (V) molecules. Full geometry optimizations were performed for the fifteen smallest models and compounds resulting in the energetically most stable structures, using the starting coordinates utilized by Carlson, Pecoraro, De Gioia, and colleagues and the same basis set utilized in their work (*20*). For V atom, the effective core potential basis set LanL2DZ (*53-55*) was modified to replace the two outermost p functions by a (41) split of the optimized 4p function (341/341/41).(*56*) For N and O atoms, the cc-pVDZ(*57*) basis set was employed, and for H - D95 (*53*). In all cases, the structures converged to the same geometries as reported in the above work. Therefore, for the remaining models the minimized structures obtained by these authors were employed without further geometry optimization for calculations of the NMR parameters. Further details on the calculations for VCPO models and oxovanadium (V) complexes are reported in our recent articles (references (*45*) and (*28*), respectively).

Bioinorganic Vanadium Coordination Complexes Mimicking the Active Site of Vanadium Haloperoxidases

^{51}V Solid-State NMR Spectroscopy

In Figure 2, chemical structures of eleven bioinorganic vanadium(V) complexes mimicking the active sites of haloperoxidases are depicted. The series under investigation contains penta- and hexacoordinate complexes with different geometries and different proximal and distal ligands.

The (ONO) and (ONN) donor sets mimic the nature of the atoms in the first coordination sphere of the haloperoxidase proteins. Compounds **I-VI** mimic the coordination geometry of the resting state of haloperoxidases. Compound **IX** mimics the active site of the peroxide-bound haloperoxidases. Within each series (penta- vs. hexacoordinate compounds), the complexes were selected so as to assess the following: a) how the changes in the proximal substituents affect the NMR spectra (e.g., compounds **II** vs. **I**, vs. **III**, vs. **IV**, or **X** vs. **XI**); b) what the effect of the distal substituents is (e.g., compounds **I** vs. **VI** vs. **VII** vs. **V** vs. **VIII**). Compounds **VIII-XI** have been crystallographically characterized (*58-61*).

In Figure 3, a characteristic ^{51}V MAS NMR spectrum representing the central and the satellite transitions, is displayed for **XI**. The anisotropic lineshapes are determined by the combination of the quadrupolar interaction (the interaction between the electric quadrupole moment of the nucleus and the electric field gradient on the nuclear site), and chemical-shielding anisotropy (CSA) (*26, 27*).

As illustrated in Figure 4, a wide variation in ^{51}V MAS NMR spinning sideband patterns is observed for **I-XI**, indicating that vanadium spectra are very sensitive to the geometric and electronic environment at the nuclear site. Remarkably, it is not only the ligands in the first coordination sphere but also the distal substituents that are an important determinant of the spectral lineshape, as can be inferred from the MAS spectra for **I-VI**.

Numerical simulations of the experimental spectra yielded seven independent parameters describing the quadrupolar and the CSA tensors:

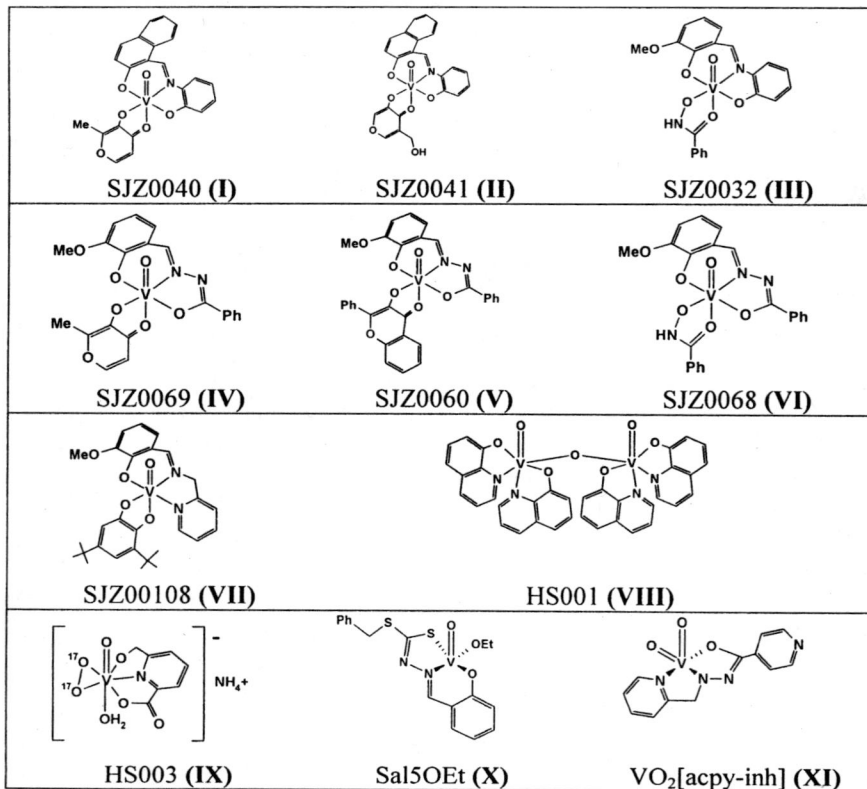

Figure 2. Chemical structures of the oxovanadium(V) compounds mimicking the VHPO active sites.

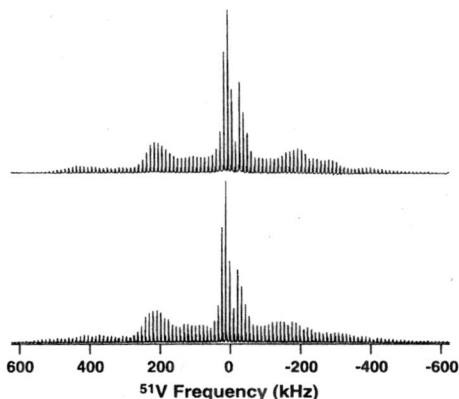

*Figure 3. Top: Experimental ^{51}V solid-state MAS NMR spectrum of **XI** (MAS frequency of 11 kHz). The overall width of the spectral envelope is determined by C_Q, the shape- by η_Q; the width and the shape of the central transition- by δ_σ and η_σ, respectively. Bottom: Best fit simulated spectrum yielding the following fine structure constants: $C_Q = 7.00\pm 0.1$ MHz; $\eta_Q = 0.25\pm0.05$; $\delta_\sigma= 485\pm29$ ppm; $\eta_\sigma = 0.2\pm0.1$; $\alpha = 70°\pm20°$; $\beta= 75°\pm15°$; $\gamma= 0°\pm10°$. Note the excellent agreement between the experimental and simulated spectrum, generally achieved for all eleven compounds.*

quadrupolar coupling constant (C_Q), the isotropic chemical shift (δ_{iso}), the chemical-shielding anisotropy (δ_σ), the asymmetry parameters of the two tensors (η_Q and η_σ), and the Euler angles (α, β, γ) defining their mutual orientations. Remarkably, these parameters display large variations across all eleven compounds (as reported in our work (28), Table 3). Figure 5 summarized the relationships between selected pairs of spectroscopic observables- δ_σ and C_Q, as well as δ_σ and δ_{iso}. Not surprisingly, there is a strong correlation between the parameters describing the chemical shielding anisotropy tensor, δ_σ and δ_{iso} (Figure 5, bottom plot). On the contrary, there appears to be no obvious relationship between the quadrupole coupling constant C_Q and the chemical shielding anisotropy δ_σ, as shown in Figure 5, top plot. This result indicates that δ_σ and C_Q report on different, complementary aspects of the electronic structure in these vanadium coordination complexes. Therefore, it is only via direct measurement of both of these tensorial parameters that most complete understanding of the vanadium electronic environment can be gained.

The experimental results reveal that the peroxo compound **IX** has large fine structure constants (C_Q=6.23 MHz, and δ_σ=728 ppm), which is not surprising in view of its non-symmetric coordination geometry. On the other hand, the large range of the C_Q and δ_σ values for the geometrically similar compounds **I-VI** was unexpected, indicating that ^{51}V NMR spectroscopy can detect with great

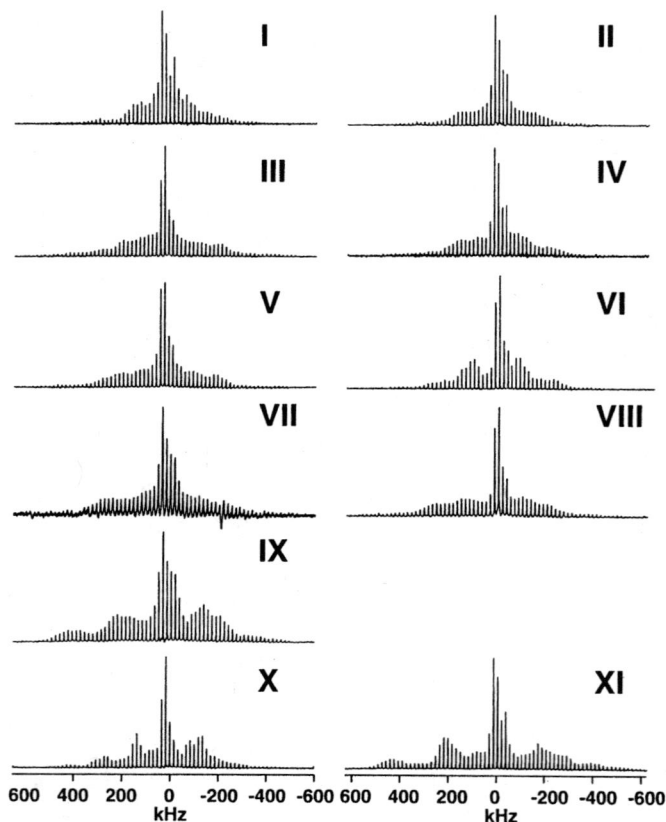

Figure 4. ^{51}V MAS NMR spectra of the oxovanadium(V) complexes acquired at 9.4 T and the spinning speed of 17 kHz. The anisotropic spinning sideband patterns are very sensitive to the details of the molecular environment of the V center, as is manifested by the broad repertoire of the spectral lineshapes for the different compounds in this series. The numbering of the spectra corresponds to the complexes in Figure 2.

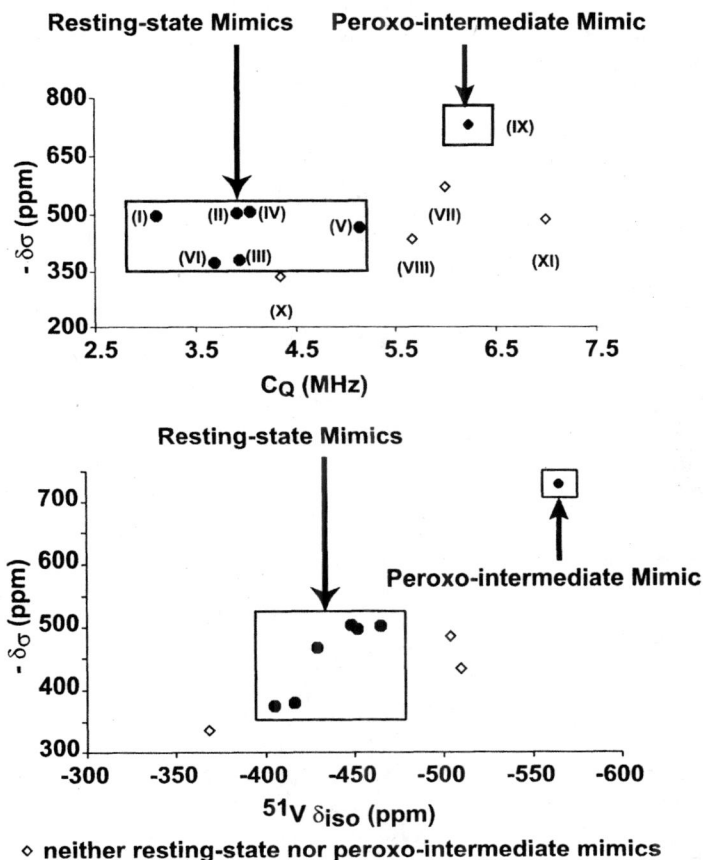

Figure 5. Correlations between selected NMR spectroscopic observables.
Top: quadrupolar coupling constant C_Q the chemical shielding anisotropy δ_σ;
bottom: chemical shielding anisotropy δ_σ and isotropic chemical shift δ_{iso}.
Note the large variation of C_Q and δ_σ values among the geometrically similar
compounds **I-VI**. Note that chemical shielding anisotropy δ_σ is negative.

sensitivity subtle changes in the geometric and electronic structure around the vanadium center.

Moreover, the isotropic chemical shifts in solution and in the solid state display only modest variation in compounds I-VI (28) , in contrast to the anisotropic components of the quadrupolar and CSA tensors. Therefore, the significant differences in the electronic structure in these compounds would be undetectable by solution NMR spectroscopy, which probes isotropic chemical shifts alone.

Density Functional Theory (DFT) Calculations of NMR Spectroscopic Observables

To establish whether the NMR spectroscopic parameters observed experimentally can be interpreted in terms of the molecular structure for the bioinorganic vanadium haloperoxidase mimics, we have conducted quantum mechanical DFT calculations of the CSA and EFG tensors, for the crystallographically characterized compounds VIII-XI (Figure 2).

As illustrated in Figure 6 and in our report (28), the experimental and the theoretically predicted NMR spectroscopic parameters are in good agreement.

The DFT calculations gave additional insights into the electronic structure of the oxovanadium solids. According to both the experiment and calculations, the EFG and CSA tensors are non-coincident. The calculations yielded absolute orientations of these tensors, unavailable from the SSNMR (28). We learned that the most shielded δ_{33} component of the CSA tensor in VIII-XI is roughly perpendicular to the plane formed by the heterocycles, with the other two components approximately lying in this plane. On the contrary, the orientations of the EFG tensor differ among IX-XI, and in each case reflect the unique electronic charge distribution. As was discussed in the preceding sections, the CSA and EFG tensors bear complementary information, and the independent measurements of their sizes, symmetries, and absolute orientations are thus necessary for the complete description of the molecule of interest.

Vanadium haloperoxidases

^{51}V Solid-State MAS NMR Spectroscopy of Vanadium Chloroperoxidase.

In Figure 7, ^{51}V solid-state MAS NMR spectra are shown for the resting state of vanadium chloroperoxidase. This is the first example of ^{51}V solid-state NMR spectroscopy applied to directly probe the vanadium center in a protein. Due to the favorable magnetic properties of the ^{51}V isotope (high natural abundance, high gyromagnetic ratio, small quadrupole moment, and short spin-lattice relaxation times), the signal could be readily detected despite the small concentration of vanadium spins in the sample (one per molecule of 67.5 kDa).

Furthermore, spectra of a signal-to-noise ratio adequate for subsequent numerical simulations can be collected within 20 hours. The quadrupolar and chemical shielding anisotropy tensors have been determined by numerical simulations of the spinning sideband envelopes and the lineshapes of the individual spinning sidebands corresponding to the central transition. The observed quadrupolar coupling constant C_Q of 10.5 ± 1.5 MHz and chemical shielding anisotropy δ_σ of -520 ± 13 ppm are sensitive reporters of the geometric and electronic structure of the vanadium center. As described in the following section, a combination of these anisotropic NMR observables and quantum mechanical DFT calculations of the spectroscopic observables yields important insight regarding the coordination geometry of the vanadate in VCPO, unavailable from any other experimental measurements prior to our work (*45*).

Density Functional Fheory (DFT) Calculations of NMR Spectroscopic Observables in Vanadium Chloroperoxidase.

We addressed the NMR parameters for an extensive series of eighty six VCPO active site models (illustrated in Figure 8), using Density Functional Theory. This approach allowed us to gain insight on the geometric and electronic features of the vanadium site that give rise to the experimental NMR observables. These complexes model the possible coordination environments and protonation states of the vanadium center in the resting state of the protein. The same models were addressed previously by Carlson, Pecoraro, and colleagues, who investigated their energetic and structural parameters to determine the most likely coordination environment of the resting state of VCPO (*20*).

Calculations with two different basis sets indicate that a very limited number of models belonging to only 3 types of coordination environments: **a3**, **c5**, and **c6** (Figure 8) yield computed NMR parameters in agreement with the experimental NMR data. Furthermore, according to the results, the most probable resting state of the protein under our experimental conditions involves an anionic vanadate cofactor with the axial hydroxo ligand, while one hydroxo and two oxo groups are present in the equatorial plane (basic structure termed **a3**). This is in remarkable agreement with the DFT and QM/MM studies by Carlson, Pecoraro and colleagues (*20, 23*), and by Raugei and Carloni (*24*). Two additional cationic geometries with the parent structures termed **c5** and **c6** might be feasible in which the axial ligand is either hydroxo or aqua, while the equatorial plane contains either one hydroxo and two oxo groups or one aqua and two hydroxo groups, respectively. Computationally these were considered

194

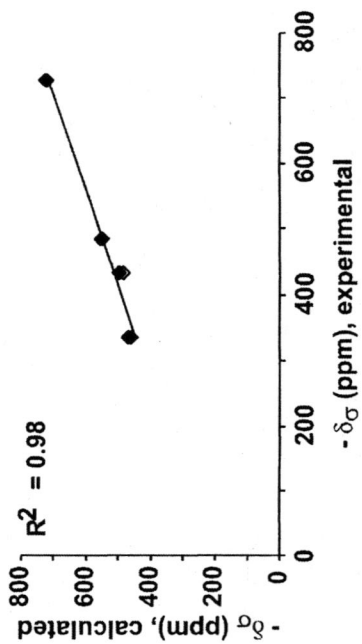

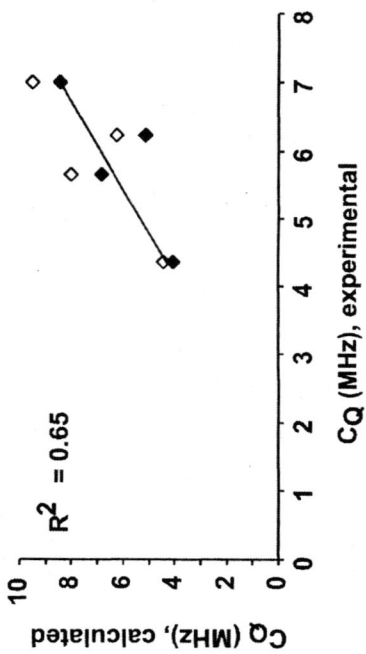

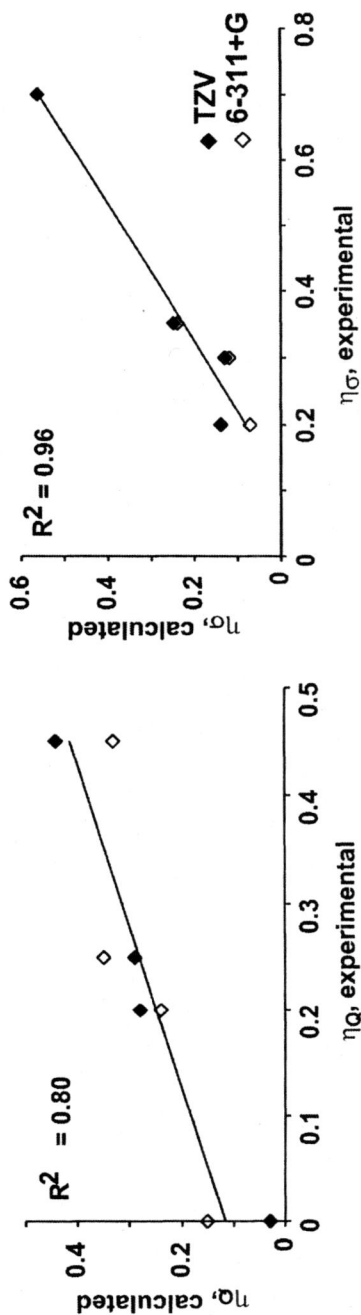

Figure 6. Correlations between experimental NMR parameters in oxovanadium (V) complexes and those calculated quantum mechanically, via Density Functional Theory: C_Q, η_Q; δ_σ and η_σ. Note very good agreement between the experiment and the calculations for two different basis sets (TZV, filled diamonds, 6-311+G, empty diamonds). Note that chemical shielding anisotropy δ_σ is negative.

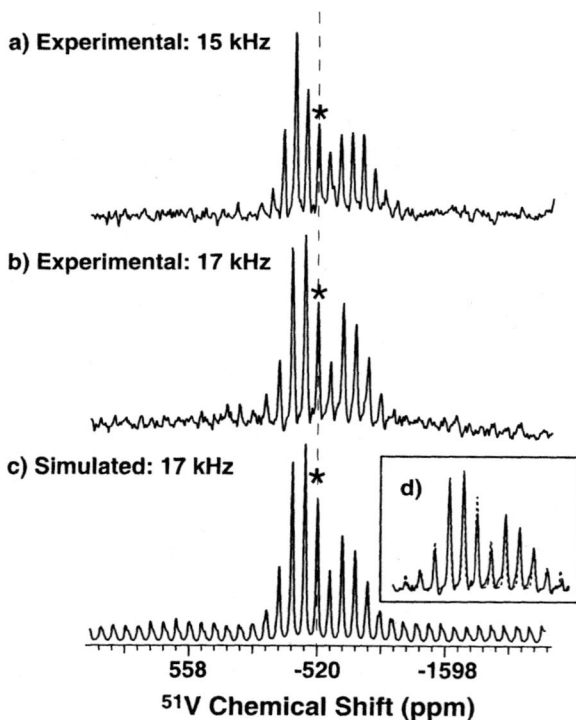

*Figure 7. 14.1 T ^{1}V MAS NMR spectra of vanadium chloroperoxidase: a)
experimental (15 kHz), b) experimental (17 kHz) c) simulated in SIMPSON
using the best fit parameters: C_Q = 10.5 MHz, η_Q = 0.55, δ_σ = -520 ppm, η_σ =
0.4. The isotropic chemical shift corrected for the second-order quadrupolar-
induced shift is -507.5 ppm. The inset d) illustrates the overlay of the
experimental (solid line) and simulated (dashed line) central transition spinning
sidebands, to demonstrate the high quality of the fit. The specific activity of the
purified protein was approximately 16 units/mg. 68 mg of lyophilized VCPO
was used in the measurements; the total sample volume was 100 μl.*

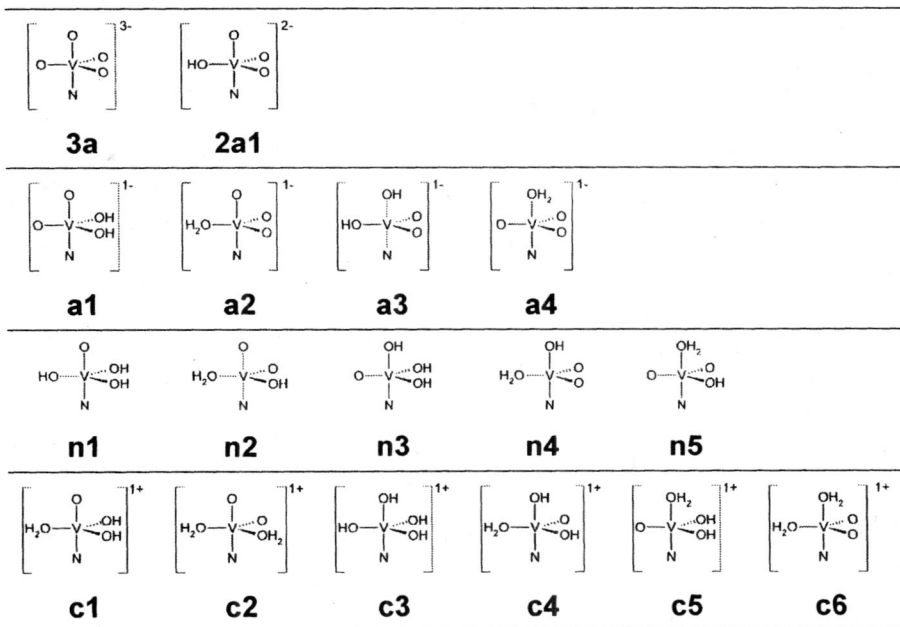

Figure 8. Schematic representations of the smallest size VCPO active site models used in the DFT calculations of the NMR spectroscopic observables.

improbable based on the energetic considerations, but whether they could be stabilized by the protein environment remains to be addressed. The **n3** and **n5** type structures while energetically favorable do not yield NMR observables in any agreement with experiment, and can therefore be ruled out. Exploring the intriguing possibility of the hybrid resting state oscillating between the **a3** and **a4** structures (*23*) entails high-level QM/MM calculations of NMR observables using extended active site models. These will be the subject of a separate study.

[51]V solid-state MAS NMR Spectroscopy of Vanadium Bromoperoxidase.

In Figure 9, the [51]V solid-state MAS NMR spectrum is shown for the resting state of vanadium bromoperoxidase, and contrasted to the spectrum of VCPO.

The outer satellite transition spinning sidebands in VBPO are more intense than in VCPO, indicating different C_Q in VBPO. The observed isotropic chemical shift of -681.3 ppm in VBPO is significantly upfield compared with that of VCPO. That the vanadium site in VBPO is more shielded is consistent with prior solution NMR studies by Rehder and colleagues (*49, 62*). The isotropic solution shift of the resting state of VBPO was reported to be –931 ppm. The observed shielding was hypothesized to be due to the extensive hydrogen bonding network and tightly bound water molecules in the active site. Interestingly, we have not detected the highly shielded peak at –931 ppm in our current studies. We are planning additional experiments, including measurements of ^{51}V solid-state NMR spectra of proteins prepared as hydrated precipitates that are expected to retain the water microenvironment of the active site.

While spectral analysis and simulations are currently under way to extract the numerical values of the NMR observables, these first experimental solid-state NMR results suggest substantially different electronic environments in VCPO and VBPO, and possibly different protonation states of the vanadate cofactor. This hypothesis remains to be corroborated with upcoming NMR measurements.

Conclusions

Our results establish ^{51}V solid-state NMR spectroscopy as a direct and sensitive probe of the coordination environment in vanadium haloperoxidases and oxovanadium (V) complexes mimicking the active site of vanadium haloperoxidases. We have presented the first example of ^{51}V solid-state NMR spectroscopy in proteins, a 67.5 kDa- vanadium chloroperoxidase and a 120.4 kDa- vanadium bromoperoxidase. Our NMR results indicate that despite the strikingly similar geometries, the active sites of VCPO and VBPO have unique electronic environments. The approach involving a combination of experimental solid-state NMR and quantum chemical calculations yields an accurate description of the molecular and electronic structure in the vanadium-containing systems, as was corroborated in our studies of bioinorganic oxovanadium (V) compounds. With this protocol, we have obtained the detailed coordination environment of the vanadate cofactor in vanadium chloroperoxidase unavailable from other experimental measurements. Our strategy is expected to be generally applicable to studies of diamagnetic V(V) sites in vanadium containing proteins.

Acknowledgment

We thank Dr. Alexander J. Vega for very helpful discussions on solid-state NMR spectroscopy of half-integer quadrupoles and of vanadium haloperoxidases in particular; Dr. Zulfiqar Hasan and Mr. Louis Hartog for their assistance in preparation of vanadium haloperoxidases; Dr. Martin Ebel and Dr. Sven Jantzen for preparation of oxovanadium (V) complexes discussed in this work. T.P. acknowledges financial support of the National Science Foundation

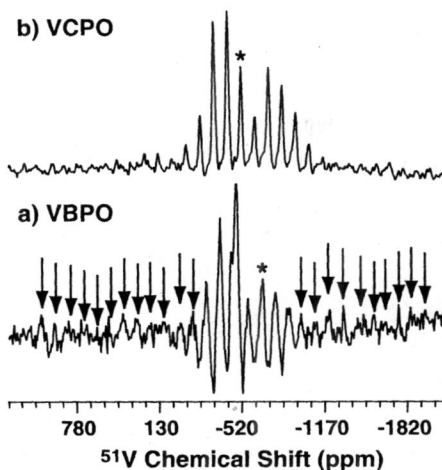

Figure 9. Comparison of ^{51}V NMR spectra of (a) vanadium chloroperoxidase and (b) vanadium bromoperoxidase. Both spectra were acquired at 14.1 T, -25 °C at a spinning frequency of 17 kHz and 1.5 million scans. Exponential apodization was applied, with line broadening of 500 Hz for both spectra. The isotropic chemical shift for each enzyme is indicated by the asterisk. For VBPO, the isotropic shift is the observed value, not corrected for the second-order quadrupolar interaction. The quadrupole-induced shift in VBPO is expected to be in the range of 10-15 ppm, based on the qualitative estimates for the quadrupolar coupling constant. The arrows correspond to spinning sidebands for VBPO, spanning a slightly larger frequency range compared to VCPO. The specific activity of VBPO was 504 units/mg. A total of 88 mg, of which 44 mg was pure VBPO was used in the measurements; the total sample volume was 100 μl.

(NSF-CAREER CHE-0237612) and the National Institutes of Health (P20-17716 under COBRE program). RW and RR acknowledge support of the Netherlands Technology Foundation (STW).

References

1. Vilter, H. *Phytochem.* **1984**, *23* (7), 1387-1390.
2. Vollenbroek, E. G. M.; Simons, L. H.; Van Schijndel, J. W. P. M.; Barnett, P.; Balzar, M.; Dekker, H.; Van Der Linden, C.; Wever, R. *Biochem. Soc. T.* **1995**, *23* (2), 267-271.
3. Butler, A. *Coord. Chem. Rev.* **1999**, *187*, 17-35.

4. Butler, A. *Science.* **1998**, *281* (207-210).
5. Butler, A., Vanadium-Dependent Redox Enzymes: Vanadium Haloperoxidase. In *Comprehensive Biological Catalysis*, ed.; Sinnott, M., Ed., Academic Press: 1998; *Vol.* pp. 1-12.
6. Butler. *Curr. Opin. Chem. Bio.* **1998**, *2*, 279-285.
7. Van Schijndel, J. W. P. M.; Barnett, P.; Roelse, J.; Vollenbroek, E. G. M.; Wever, R. *Eur. J. Biochem.* **1994**, *225* (1), 151-157.
8. Macedo-Ribeiro, S.; Hemrika, W.; Renirie, R.; Wever, R.; Messerschmidt, A. *J. Biol. Inorg. Chem.* **1999**, *4*, 209-219.
9. Svendsen, A.; Jorgensen, L. Haloperoxidases with Altered Ph Profiles. US Patent 6,221,821 B1, 04/24/1002, 2001.
10. Hansen, E. H.; Albertsen, L.; Schafer, T.; Johansen, C.; Frisvad, J. C.; Molin, S.; Gram, L. *Appl. Environ. Microb.* **2003**, *69* (8), 4611-4617.
11. Munir, I. Z.; Hu, S. H.; Dordick, J. S. *Adv. Synth. Catal.* **2002**, *344* (10), 1097-1102.
12. Littlechild, J. *Curr. Opin. Chem. Bio.* **1999**, *3* (1), 28-34.
13. Renirie, R.; Hemrika, W.; Piersma, S. R.; Wever, R. *Biochemistry.* **2000**, *39* (1133-1141).
14. Almeida, M.; Filipe, S.; Humanes, M.; Maia, M. F.; Melo, R.; Severino, N.; da Silva, J. A. L.; da Silva, J. J. R. F.; Wever, R. *Phytochem.* **2001**, *57* (5), 633-642.
15. Renirie, R.; Hemrika, W.; Wever, R. *J. Biol. Chem.* **2000**, *275* (16), 11650-11657.
16. Murphy, C. D. *J. Appl. Microbiol.* **2003**, *94* (539-548).
17. Messerschmidt, A.; Wever, R. *Proc. Natl. Acad. Sci. USA.* **1996**, *93*, 392-396.
18. Weyand, M.; Hecht, H. J.; Kiess, M.; Liaud, M. F.; Vilter, H.; Schomburg, D. *J. Mol. Biol.* **1999**, *293* (3), 595-611.
19. Tanaka, N.; Hasan, Z.; Wever, R. *Inorg. Chim. Acta.* **2003**, *356*, 288-296.
20. Zampella, G.; Kravitz, J. Y.; Webster, C. E.; Fantucci, P.; Hall, M. B.; Carlson, H. A.; Pecoraro, V. L.; De Gioia, L. *Inorg. Chem.* **2004**, *43* (14), 4127-4136.
21. Borowski, T.; Szczepanik, W.; Chruszcz, M.; Broclawik, E. *Int. J. Quantum Chem.* **2004**, *99* (5), 864-875.
22. Zampella, G.; Fantucci, P.; Pecoraro, V. L.; De Gioia, L. *J. Am. Chem. Soc.* **2005**, *127* (3), 953-960.
23. Kravitz, J. Y.; Pecoraro, V. L.; Carlson, H. A. *J. Chem. Theory Comput.* **2005**, *1* (6), 1265-1274.
24. Raugei, S.; Carloni, P. *J. Phys. Chem. B.* **2006**, *110* (8), 3747-3758.
25. Bangesh, M.; Plass, W. *J. Mol. Struc-Theochem.* **2005**, *725* (1-3), 163-175.
26. Smith, M. E.; van Eck, E. R. H. *Prog. NMR Spec.* **1999**, *34*, 159-201.
27. Skibsted, J.; Nielsen, N. C.; Bildsøe, H.; Jacobsen, H. J. *Chem. Phys. Lett.* **1992**, *188*, 405-412.

28. Pooransingh, N.; Pomerantseva, E.; Ebel, M.; Jantzen, S.; Rehder, D.; Polenova, T. *Inorg. Chem.* **2003**, *42*, 1256-1266.
29. Lapina, O. B.; Shubin, A. A.; Khabibulin, D. F.; Terskikh, V. V.; Bodart, P. R.; Amoureux, J. P. *Catal. Today.* **2003**, *78* (1-4), 91-104.
30. Huang, W.; Todaro, L.; Francesconi, L. C.; Polenova, T. *J. Am. Chem. Soc.* **2003**, *125* (19), 5928-5938.
31. Huang, W. L.; Todaro, L.; Yap, G. P. A.; Beer, R.; Francesconi, L. C.; Polenova, T. *J. Am. Chem. Soc.* **2004**, *126* (37), 11564-11573.
32. Frydman, L. *Annu. Rev. Phys. Chem.* **2001**, *52*, 463-498.
33. Cohen, M. H.; Reif, F. *Solid State Phys.* **1957**, *5*, 321-438.
34. Sandstrom, D.; Hong, M.; Schmidt-Rohr, K. *Chem. Phys. Lett.* **1999**, *300* (1-2), 213-220.
35. Bak, M.; Rasmussen, J. T.; Nielsen, N. C. *J. Magn. Reson.* **2000**, *147*, 296-330.
36. Nielsen, U. G.; Jacobsen, H. J.; Skibsted, J. *Inorg. Chem.* **2000**, *39*, 2135-2145.
37. Nielsen, U. G.; Jacobsen, H. J.; Skibsted, J. *J. Phys. Chem. B.* **2001**, *105*, 420-429.
38. Fyfe, C. A.; zu Altenschildesche, H. M.; Skibsted, J. *Inorg. Chem.* **1999**, *38*, 84-92.
39. Jacobsen, H. J.; Skibsted, J.; Bildsøe, H.; Nielsen, N. C. *J. Magn. Reson.* **1989**, *85*, 173-180.
40. Skibsted, J.; Jacobsen, C. J. H.; Jacobsen, H. J. *Inorg. Chem.* **1998**, *37*, 3083-3092.
41. Skibsted, J.; Jakobsen, H. J. *Inorg. Chem.* **1999**, *38*, 1806-1813.
42. Skibsted, J.; Nielsen, N. C.; Bildsøe, H.; Jacobsen, H. J. *J. Magn. Reson.* **1991**, *95*, 88-117.
43. Bryce, D. L.; Wasylishen, R. E. *Phys. Chem. Chem. Phys.* **2002**, *4* (15), 3591-3600.
44. Ooms, K. J.; Wasylishen, R. E. *J. Am. Chem. Soc.* **2004**, *126* (35), 10972-10980.
45. Pooransingh-Margolis, N.; Renirie, R.; Hasan, Z.; Wever, R.; Vega, A. J.; Polenova, T. *J. Am. Chem. Soc.* **2006**, *128* (15), 5190-5208.
46. De Boer, E.; Vankooyk, Y.; Tromp, M. G. M.; Plat, H.; Wever, R. *Biochim. Biophys. Acta.* **1986**, *869* (1), 48-53.
47. Wever, R.; Plat, H.; Deboer, E. *Biochim. Biophys. Acta.* **1985**, *830* (2), 181-186.
48. De Boer, E.; Wever, R. *J. Biol. Chem.* **1988**, *263* (25), 12326-12332.
49. Rehder, D.; Casny, M.; Grosse, R. *Magn. Res. Chem.* **2004**, *42* (9), 745-749.
50. Frisch, M. J., et al. *Gaussian 98*, Revision A.1x; Gaussian, Inc.: Pittsburgh PA, 2001.
51. Becke, A. D. *J. Chem. Phys.* **1993**, *98* (7), 5648-5652.

52. Schafer, A.; Huber, C.; Ahlrichs, R. *J. Chem. Phys.* **1994**, *100* (8), 5829-5835.
53. Dunning, T. H. J.; Hay, P. J., In *Modern Theoretical Chemistry*, ed.; Schafer, H. F. I., Ed., Plenum Press: New York, 1976; *Vol. 3*, pp. 1-28.
54. Hay, P. J.; Wadt, W. R. *J. Chem. Phys.* **1985**, *82* (1), 270-283.
55. Hay, P. J.; Wadt, W. R. *J. Chem. Phys.* **1985**, *82* (1), 299-310.
56. Couty, M.; Hall, M. B. *J. Comput. Chem.* **1996**, *17* (11), 1359-1370.
57. Woon, D. E.; Dunning, T. H. *J. Chem. Phys.* **1993**, *98* (2), 1358-1371.
58. Drew, R. E.; Einstein, F. W. B. *Inorg. Chem.* **1973**, *12* (4), 829-835.
59. Maurya, M. R.; Khurana, S.; Schulzke, C.; Rehder, D. *Eur. J. Inorg. Chem.* **2001**, *3*, 779-788.
60. Wang, D. R.; Ebel, M.; Schulzke, C.; Gruning, C.; Hazari, S. K. S.; Rehder, D. *Eur. J. Inorg. Chem.* **2001**, *4*, 935-942.
61. Yamada, S.; Katayama, C.; Tanaka, J.; Tanaka, M. *Inorg. Chem.* **1984**, *23* (253-255).
62. Vilter, H.; Rehder, D. *Inorg. Chim. Acta.* **1987**, *136* (1), L7-L10.

Enzymology, Toxicology, and Transport

Chapter 15

Vanadium and Bone: Relevance of Vanadium Compounds in Bone Cells

Susana B. Etcheverry[1,2] and Daniel A. Barrio[1]

[1]Cátedra de Bioquímica Patológica, Facultad de Ciencias Exactas, Universidad Nacional de La Plata, La Plata, Argentina
[2]Centro de Química Inorgánica (CEQUINOR), Facultad de Ciencias Exactas, Universidad Nacional de La Plata, La Plata, Argentina

Bone is one of the main tissues for vanadium storage in vertebrates. Vanadium compounds exert interesting biological effects as well as potential pharmacological actions. Herein we review the bioactivity of vanadium with emphasis on results obtained in bone-related cells in culture. Vanadium compounds behaved as antitumoral agents, as growth factor mimetic compounds and as osteogenic drugs. We report for the first time, the effects of a complex of vanadyl(IV) cation with Trehalose on long-term cultures of non transformed osteoblasts. The complex induces cell proliferation, glucose consumption, osteoblast differentiation and the mineralization of the extracellular matrix.

© 2007 American Chemical Society

Bone is a dynamic tissue with a great capacity to remodel and repair throughout life *(1)*. The integrity of the skeleton depends on a series of events that regulate the coordinated activity of osteoblasts (bone-forming cells) and osteoclasts (cells involved in bone resorption). Other cells that play an important role in bone tissue are macrophages *(2)*. All these cells are placed in a mineralized matrix whose main moieties are the mineral phase hydroxyapatite and the organic part with collagen as one of the principal components. Numerous growth factors and cytokines regulate osteogenesis and bone repair *(3)*. Misbalance in skeletal cell activity may cause bone diseases. Vanadium compounds are an interesting group of drugs with potential effects on bone tissue *(4,5)*.

Different studies demonstrate the accumulation of vanadium in bone *(6–8)*. Vanadium accumulation in bone tissue may be due to its analogy with phosphate *(9)*. Vanadium stimulates DNA and collagen synthesis in fibroblasts and promotes bone formation and repair in vivo *(10 - 12)*. Moreover, vanadium derivatives display interesting potential pharmacological actions as insulin- and growth factor-mimetic compounds *(12,13)*, as antitumoral agents *(14 - 16)* and as osteogenic drugs *(17,18)*.

We have previously demonstrated that vanadium(V) and vanadium(IV) species as well as different vanadium complexes regulate the proliferation and differentiation of osteoblast-like cells in culture *(17-19)*. Several mechanisms have been proposed to explain the bioactivity of vanadium *(20,21)*. As a consequence, different cellular cascades can be triggered and translated into biological actions, including the regulation of osteoblast-specific gene expression, proliferation, matrix mineralization, and cell adhesion to the matrix *(20 - 22)*.

Herein, we present new and interesting results of the complex of vanadyl(IV) cation with the disaccharide trehalose (TreVO) on non-transformed osteoblasts that in long-term cultures develop the stages and characteristics of normal bone tissue.

Potential Pharmacological Effects of Vanadium Compounds

As part of a project devoted to the synthesis, characterization and the study of the bioactivity of vanadium derivatives in bone-related cells, we have used Swiss 3T3 fibroblasts, two osteoblast-like cell lines (MC3T3-E1 that are non-transformed osteoblasts derived from mouse calvaria and UMR106 osteoblast-like cells derived from a rat osteosarcoma) *(19, 23, 24)* and the murine macrophage cell line RAW 264.7 *(2)*.

Insulin Mimetic Effects of Vanadium Compounds

Effects of Vanadium on Osteoblast Glucose Consumption

The most interesting pharmacological action of vanadium is its behavior as insulin mimics. Different models of diabetic rats as well as different vanadium compounds were used *(26 - 28)*. On the other hand, the first studies in human beings were developed later *(20-31)*. The activation of phosphorylation cascades has been proposed as one of the most important mechanisms involved in the insulin mimetic actions of vanadium compounds. *In vivo,* insulin signaling transduction pathway is mediated through a complex network of phosphorylation and dephosphorylation reactions. *(32,33)*. *In vivo* and *in vitro* studies demonstrated that vanadium influences different parts of the insulin-signaling pathway.

In vitro studies carried out in our laboratory using a set of inhibitors of different cellular pathways, have allowed us to demonstrate that insulin and a complex of vanadyl(IV) with trehalose stimulate glucose consumption in the osteoblasts in culture through a mechanism independent of PI3-K in contrast with previous results in other cell types *(17)*. Besides, we have recently reported for the first time the effect of TreVO on the phosphorylated and non-phosphorylated forms of GSK-3 (kinase 3 of the glycogen synthase) in osteoblast-like cells. This compound stimulated glucose consumption and the phosphorylation of this protein and a cytosolic protein kinase would be involved in this effect *(17,34)*.

Vanadium Compounds as Growth Factor Mimetic Compounds

Effects of Vanadium on Osteoblast Proliferation

Several research groups have reported that vanadium derivatives behave as growth factor mimetic compounds, resembling EGF, FGF and insulin actions, promoting bone formation and repair *(11,19)*. As an overview, at low concentrations most vanadium compounds behave as weak mitogens in comparison to insulin, while at high doses they exert cytotoxicity *(17)*. Bioactivity of vanadium compounds involves a series of complex events that depends on the cellular type, the vanadium concentration, the oxidation state of vanadium and the nature of the ligands.

The model of two osteoblastic lines in culture has allowed us to demonstrate that, in general, vanadium(V) compounds are more cytotoxic than vanadium(IV) derivatives and MC3T3E1 cells are more sensitive to the cytotoxicity than

UMR106 cells. At present, the signaling mechanisms by which vanadium compounds produced their biological effects are under exhaustive research. One of the pathways that seems to be involved in vanadium bioactivity is the regulation of phosphotyrosine protein levels *(35)*. The insulin mimetic effects of vanadium may be attributed to its analogy with phosphate and to the stimulation of protein tyrosine phosphorylation through the inhibition of protein tyrosine phosphatases (PTPases) *(20,36)*. The activation of the extracellular regulated kinases (ERK) by vanadyl sulphate has been shown to be dependent on PI3-K activation *(21)*. It is assumed that the stimulation of the PI3-K/ras/ERK pathway plays a key role in mediating the insulin-mimetic and growth factor like effects of inorganic vanadium salts. Moreover, it has been shown that several vanadium salts activate ERKs and two ribosomal protein kinases, including p90[rsk], in a maner independent of the kinase of the insulin receptor (IR) *(37,38)*. On the other hand, different authors have pointed to the formation of reactive oxygen species (ROS) as the molecular mechanism by which vanadium exerts its biological effects *(39 - 43)*.

Among the vanadium(V) complexes synthesized and characterized in our group, the complex with trehalose (TreVO) has shown very interesting effects and it may be considered as a promising pharmacological compound. At low doses it induced cell proliferation in MC3T3E1 non-transformed osteoblasts while it caused inhibitory effects on the tumoral cell line. TreVO also stimulates ERK phosphorylation in the MC3T3E1 osteoblast cell line *(17)*. This effect was totally blunted by an inhibitor of MEK (PD98059) and by wortmannin (an inhibitor of PI3K) but not by a mixture of vitamins E and C. Low doses of the complex, which are mitogenic for MC3T3E1 cells, could act though the PI3K-MEK-ERK pathway and by a mechanism independent of free radicals. High concentrations of TreVO strongly increased ERK phosphorylation, an effect that was partially blocked by wortmannin, PD98059 or a mixture of vitamins E and C *(17)*. These results suggest that ERK pathway may be involved in the proliferative effects of TreVO. The activation of the ERK cascade by low doses of the complex seems not to be mediated by ROS. On the contrary, the activation of this pathway by high doses of TreVO would be mediated by ROS.

Effects of Vanadium on Osteoblast Differentiation

Studies in fibroblasts have shown that vanadium stimulates DNA and collagen synthesis, suggesting that it promotes osteoblastic differentiation *(22,23)*. On the other hand, alkaline phosphatase (ALP) is involved in the mineralization process of bone. Vanadium compounds inhibit ALP activity in varying degrees*(17,19,44)*.

Mineralization takes place in the ECM where type I collagen is one of the more important components. Vanadium compounds display opposing actions on these two markers of osteoblastic phenotype: they inhibit ALP activity and some of them enhance the synthesis of collagen. As a consequence between these two aspects, the probability of a vanadium compound acting as a stimulating agent for bone formation would depend on the relationship between the induction of osteoblast proliferation and collagen synthesis with a slight inhibition on ALP activity. The complex of vanadyl(IV) cation with trehalose behaves as a weak inhibitor of ALP *(17)* but has an stimulatory effect on collagen synthesis in UMR106 cells *(45)*, as well as in long-term cultures of MC3T3E1 (see below). These results indicate that this compound can be considered as a good osteogenic agent. As it is well known, diabetic patients with a bad control of the hyperglycemia present important alterations in bone tissue *(46)*. We have also demonstrated that ALP activity is decreased upon a great level of glycosylation *(47)*. The osteogenic action of TreVO together with its effect as a very good promoter of glucose uptake by the osteoblasts, allow us to strongly suggest that this complex is a worthy candidate for clinical trials.

Long-term Studies on the Effects of Vanadium in MC3T3E1 Osteoblasts

The process of maturation from preosteoblasts into osteoblasts requires the expression of several specific markers along bone development *(48)*. This cell line is a good *in vitro* model of preosteoblasts that in long-term cultures differentiate into mature osteoblasts, going through developmental stages similar to the growth of bone *in vivo*. Ascorbic acid (AA) and β-glycerol phosphate (β-GP) are two compounds required for differentiation of MC3T3R1 cell line. The differentiation begins approximately after two weeks of culture and the cells express some proteins such as ALP and collagen.

From previous results obtained in short-term cultures in this cell line and in the UMR106 tumoral osteoblasts, TreVO has shown interesting potential pharmacological actions as insulin mimetics and as an osteogenic compound. It stimulated collagen production and caused a weak inhibition of ALP, a key enzyme for the mineralization process *(45)*. TreVO stimulated MC3T3E1 cell proliferation in the range of 5 - 25 μM in 24 h culture. For the long-term studies (up to 25 days), we used the minor effective dose in the induction of cell proliferation (5 μM). We have also included a control with AA and β-GP as inducers of differentiation. The cells were cultured in 24 well/plates in DMEM medium supplemented with 10 % fetal bovine serum (FBS). When the monolayers reached the confluence, the medium was replaced every two days for a fresh one. The assay conditions were: DMEM plus 10 % FBS (Basal), DMEM plus 10 % FBS plus 5 μM of the complex (TreVO), and DMEM plus 10% FBS plus 140 μM AA and 5 mM β-GP (AA+β-GP). At different culture stages, we determined the cell number, collagen and glucose consumption. In another assay we investigated the effect of 5 μM of TreVO on the process of mineralization induced by AA and β-GP.

Figure 1 shows the effect of the treatment on cell proliferation (Crystal violet assay) *(49)*. As can be seen, the cells grow and proliferate until approximately 10 days of culture. Then the cell number remains constant through 25 days. After 10 days no significant differences in the cell number were observed for each culture condition. The effects on collagen production (Sirius Red staining) *(50)* and on the specific ALP activity *(19)* are shown in Figures 2 and 3, respectively. From figure 2 it can be seen that the mixture of AA and β-GP induced collagen production in this cell line. TreVO also produced an increase in collagen in these cells at 16 - 25 days of culture. Besides, it can be observed that the basal collagen production significantly increased during 25 days. When the results on the ALP activity were analyzed, we can see that the mixture of AA+β-GP stimulated cell differentiation of MC3T3E1 since it induced the collagen production and the activity of ALP. TreVO enhanced collagen synthesis but it partially inhibited the ALP activity. Figure 4 shows the effects on the glucose consumption from MC3T3E1 osteoblasts at different culture times (Glucose oxidase assay) *(17)*. The basal value was very low after 10 days of culture (2 μg/8h/well), but it increased markedly at 16 days (32 μg/8h/well) and then remained constant. The addition of AA+β-GP caused a high increment in glucose consumption (48 μg/8h/well) after 16 days. This effect is maintained during the mineralization stage at 25 days. On the other hand, TreVO also significantly increased the glucose consumption by the osteoblasts at 16 day culture (42 μg/8h/well) and a great increase was detected after 25 days of culture (45 μg/8h/well).

As we can mention above, ALP specific activity and collagen production are markers for the differentiation of the preosteoblastic stage into the mature osteoblast. During the first days of culture, FAL activity was undetectable in this cell line. Nevertheless, after 10 days a little ALP activity could be observed. Between 10 and 16 days, the activity of FAL significantly increased for the three experimental conditions mentioned above. The mixture of AA+ β-GP induced a 400% of increment over basal activity. TreVO (5μM) caused a 50 % of inhibition of FAL activity. Nevertheless, when the cells were incubated with the mixture of AA and β-GP plus TreVO (5μM), the inhibition of ALP was only 30 %, remaining at 70 % of the enzyme activity under this condition. Between 16 - 25 days (mineralization step), FAL activity increased only 20 % for AA+ β-GP mixture while the treatment with TreVO did not produce any change in this period. These results suggest that TreVO could affect the mineralization of MC3T3E1 cells. To evaluate this hypothesis, we incubated the cells for 30 days with AA + β-GP (control) and with AA + β-GP + 5 μM TreVO and analized the mineralization nodule formation by an hystochemical method. The effect of 5

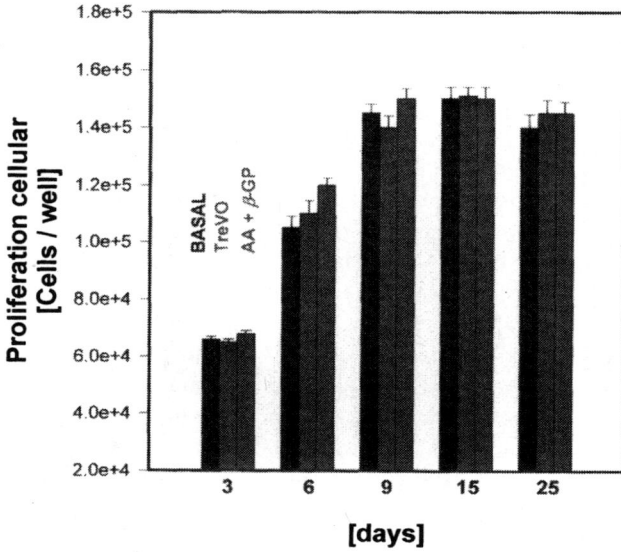

Figure 1. Cell proliferation assay.

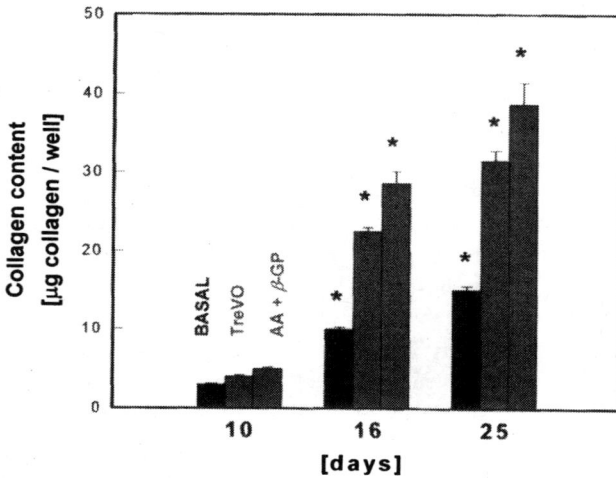

Figure 2. Collagen content. * p<0.001.

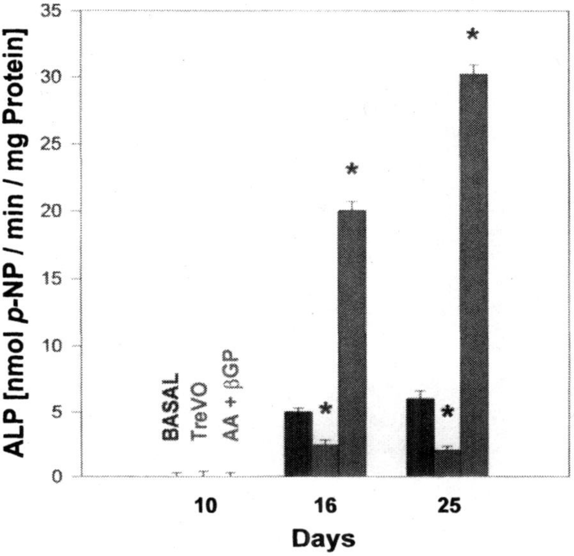

Figure 3. ALP activity. * *p<0.001.*

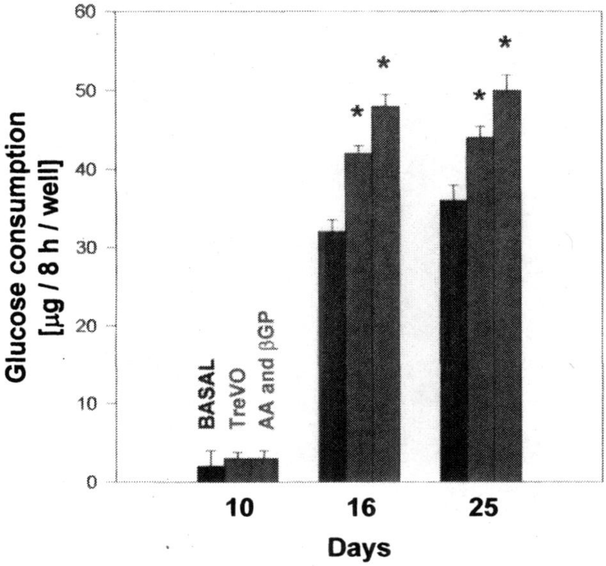

Figure 4. Glucose consumption. * *p<0.001.*

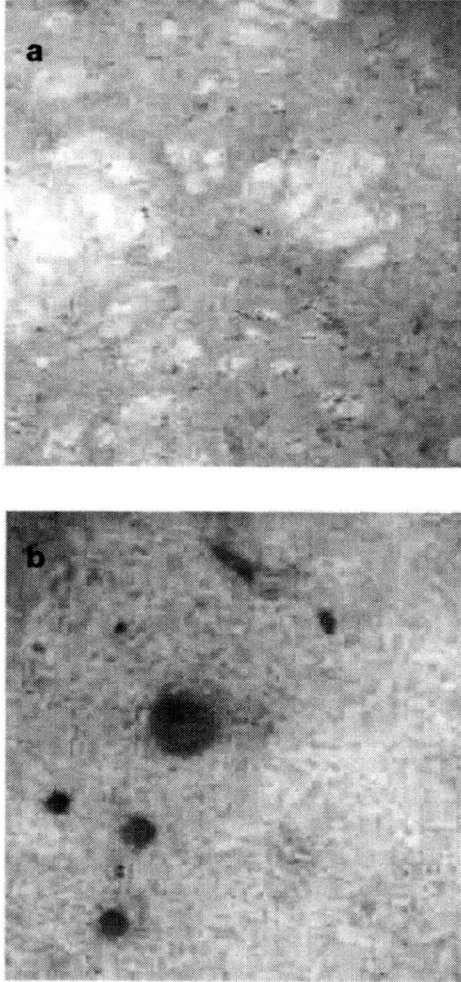

Figure 5. Mineralization nodules. a) AA+β-GP, b) AA + β-GP + 5 μM TreVO.

µM TreVO on the mineralization of MC3T3E1 cells stimulated with AA+β-GP (Alizarine Red staining) *(51)* can be observed in the pictures of figure 5. TreVO induced the formation of a great number of mineralization nodules in presence of AA+β-GP (figure 6a) in comparison with the incubation of the osteoblasts with the mixture alone. In the presence of the mixture, the inhibition of TreVO on ALP specific activity seems to be overcome by its effects on the mineralization process or the 70 % of ALP activity is apparentrly enough for the culture to reach mineralization. Besides, no cytotoxicity was observed after chronic incubation with the complex. Altogether, these results suggest that the complex TreVO caused a positive effect on MC3T3E1 cell differentiation and mineralization and behaves as an osteogenic drug.

In relation to the metabolic effects, it can be seen that the basal glucose consumption is very low for this cell line in an undifferentiated stage. Nevertheless, when the cells differentiated, the glucose consumption increased significantly. Both, TreVO and AA+β-GP promoted the metabolic effect over the basal condition, but the mixture was more effective than the complex. Between 10-16 days the glucose consumption significantly increased and in the period 16-25 days of culture it remained without changes in relation to the previous period. The same is observed for the basal glucose consumption. These results show that at the beginning of the differentiation stage, important changes take place in the glucose metabolism by these cells. One of these changes may be the synthesis of glucose transporters that facilitate the uptake of the metabolite by the osteoblasts.

Moreover, important cellular events such as the synthesis of new proteins occur during the cellular differentiation. TreVO and the mixture can favor these events. The complex showed beneficial effects on the non-transformed osteoblasts. TreVO stimulated cell proliferation, collagen synthesis, the mineralization process and the glucose uptake by the normal osteoblasts. TreVO is a good potential pharmacological compound to be further evaluated in studies with laboratory animal models and also in clinical trials.

Acknowledgements

This work was supported by grants from Universidad Nacional de La Plata, La Plata, Argentina; CONICET (PIP 6366) and ANPCyT (PICT 10968). SBE and DAB are members of the Carrera del Inevstigador, CONICET, Argentina.

References

1. Salgado, A.J.; Coutinho, O.P.; Reis,R.L. *Macromol.Biosci.* **2004**, *4*, 643–765.

214

2. Molinuevo, M.S.; Etcheverry, S.B.; Cortizo, A.M. *Toxicology* **2005**, *210*, 205-212.
3. Tuan, R.S. *Clin. Orthop. Relat. Res.* **2004**, *427*, S105-S117.
4. Nielsen, F. H. In *Vanadium and Its Role in Life*. Sigel, H; Sigel, A., Eds.; Metal Ions in Biological Systems. Marcel Dekker, New York, 1995,Vol. 31, pp 543-573.
5. Anke, M.; Groppel, B.; Krause, U. In *Trace Elements in Man and Animals*. Momciliovic, B. Ed. IMI, Zagreb, 1991, Vol. 7, pp 11.9-11.10.
6. Mongold J.J.; Cros, G.H.; Vian, L.; Tep, A.; Ramanadham, S.; Siou G., Díaz J.; McNeill J.H.; Serrano, J.J. *Pharmacol. Toxicol.* **1990**, *67*, 192-198.
7. Dai, S.; Thompson, K.H.; Vera, E.; McNeill, J.H. *Pharmacol. Toxicol.* **1994**, *75*, 265-273.
8. Setyawati, I.A.; Thompson, K.H.;Yuen, V.G.; Sun,I.; Battell, M.; Lyster, D.M.; Vo, C.; Ruth, T.J.; Teisler, S.; McNeill, J.H., Orvig, C. *J. Appl. Physiol.* **1998**, *84*, 569-575.
9. Etcheverry, S.B., Apella, M.C., Baran, E.J. *J. Inorg. Biochem.* **1984**, *80*, 169-171.
10. Canalis, E. *Endocinology* **1985**, *116*, 855-862.
11. Lau, K.H.; Tanimoto, H.; Baylink, D.J. *Endocrinology* **1988**, *123*, 2858-2867.
12. Thompson, K.H.; Orvig, C. *Met. Ions Biol. Syst.* **2004**, *41*, 221-52.
13. Srivastava, A.K; Mehdi, M.Z. *Diabetes Medicine*. 2005, *22*, 2–13.
14. Djordjevic, C. In *Vanadium and Its Role in Life*. Sigel, H.; Sigel, A. Eds. Metal Ions in Biological Systems. Marcel Dekker, New York, 1995,Vol. 31, pp 595-616.
15. Etcheverry, S.B.; Barrio, D.A.; Cortizo, A.M.; Williams, P.A.M. *J. Inorg. Biochem.* **2002**, *88*, 94-100.
16. Molinuevo, M.S.; Barrio, D.A.; Cortizo, A.M.; Etcheverry, S.B. *Cancer. Chemother. Pharmacol.* **2004**, *53*, 163-172.
17. Barrio, D.A.; Williams, P.A.M.; Cortizo, A.M.; Etcheverry, S.B. *J. Biol. Inorg. Chem.* **2003**, *8*, 459-468
18. Cortizo, A.M.; Molinuevo, M.S.; Barrio, D.A.; Bruzzone, L. *In.t J. Biochem. Cell Biol.* **2006**, *38*, 1171-1180.
19. Cortizo, A.M.; Etcheverry, S.B. *Mol. Cell. Biochem.* **1995**, *145*, 97-102.
20. Tracey, A.S.; Gresser, M.J. *Proc. Natl. Acad. Sci. USA.* **1986**, *83*, 609-613.
21. Pandey, S.K.; Théberge, J.F.; Bernier, M.; Srivastava, A.K. *Biochemistry* **1998**, *38*, 14667-14675.
22. Lai, C.F.; Chaudhary, L.; Fausto, A.; Halstead, L.R.; Ory , D.S.; Avioli, L.V.; Cheng, S *J. Biol. Chem.* **2001**, *276*, 14443–14450.
23. Etcheverry, S.B.; Barrio, D.A.; Zinczuk, J.; Williams, P.A.M.; Baran, E.J. *J Inorg. Biochem.* **2005**, *99*, 2322-2327.
24. Etcheverry, S.B.; Crans, D.C.; Keramidas, A.D.; Cortizo, A.M. *Arch. Biochem. Biophys.* **1997**, *338*, 7-14.

25. Heyliger, C.E.; Tahiliani, A.G.; McNeill, J.H. *Science,* **1985,** *227,* 1474-1477.
26. Meyerovitch, J.; Rothemberg, P.; Shecter, Y.; Bonner-Weir, S.; Kahn, C.R. *J. Clin. Invest.* **1991,** *87,* 1286-1294.
27. Brichard, S.M.; Pottier, A.M.; Henquin, J.C. *Endocrinology,* **1989,** *125,* 2510-2516.
28. Sakurai, H.; Tsuchiya, K.; Nukatsuka, M.; Sofue, M.; Kawada, J. *J. Endocrinol.* **1990,** *126,* 451-459.
29. Goldfine, A.B.; Simonson, D.C.; Folli, F.; Patti, M.E.; Kahn, C.R. *J. Clin. Endocrinol. Metab.* **1995,** *80,* 3311-3320.
30. Cohen, N.,; Halbestam, M.; Shlimovich, P.; Chang, C. J.; Shamoon, H.; Rossetti, L. *J. Clin. Invest.* **1995,** *95,* 2501-2509.
31. Cusi, K.; Cukier, S.; DeFronzo, R.A.; Torres, M.; Puchulu, F.M.; Redondo, J.C. *J. Clin. Endocrinol. Metab.* **2001,** *86,* 1410-1417.
32. Ahmad, F.; Azevedo, J.L.; CortrighT, R.; Do ; hm, G.L.; Goldstein, B.J. *J. Clin. Invest.* **1997,** *100,* 449-458.
33. Cheng, A.; Dube, N.; Gu, F.; Tremblay, M.L. *Eur. J. Biochem.* **2002,** *269,* 1050-1059.
34. Shisheva, A.; Shechter, Y. *FEBS Lett .* **1992,** *23,* 93-96.
35. Stankiewicz , P.J.; Tracey, A.S. In *Vanadium and Its Role in Life.* Sigel, H; Sigel, A., Eds.; Metal Ions in Biological Systems. Marcel Dekker, New York, 1995,Vol. 31, pp 249-286.
36. Swarup, G.; Cohen, S.; Garbers, D.L. *Biochem. Biophys. Res. Commun. .* **1982,** *107,* 1104-1109.
37. D'Onofrio, F.; Le, M.Q.; Chiasson, J.L.; Srivastava, A.K. *FEBSS Lett.* **1994,** *340,* 269, 275.
38. Pandey, S.K.; Chiasson, J-L-; Srivastava, A.K. *Mol. Cell. Biochem.* **1995,** *153,* 69-78.
39. Cortizo, A.M.; Bruzzone, L.; Molinuevo, S.; Etcheverry, S.B. *Toxicology.* **2000,** *147,* 89-99.
40. Wang, Y.Z.; Bonner, J.C. *Am. J. Respir. Cell Mol. Biol.* **2000,** *22,* 590-596.
41. Zhang, Z.; Huang, C.; Li, J.; Leonard, S.S.; Lanciotti, R.; Butterworth, L.; Shi, X. *Arch. Biochem. Biophys.* **2001,** *392,* 311-320.
42. Xia, Z.; Dickens, M.; Raingeaud, J.; Davis, R.J.; Greenberg, M.E. *Science,* **1995,** *270,* 1326-1331.
43. Kitagawa, D.; Tenemura, S.,; Ohata, S.; Shimizu, N.; Seo, J.; Nishitai, G.; Watanabe, T.; Nakagawa, K.; Kishimoto, H.; Wada, T.; Tezuka, T.; Yamamoto, T.; Nishina, H.; Katada, T. *J. Biol. Chem.* **2002,** *277,* 366-371.
44. Crans, D.C.; Tracey, A.S. In: *Vanadium compounds: Chemistry, Biochemistry, and Therapeutic Applications.* Tracey, A.S.; Crans, D.C. Eds. American Chemical Society Symposium Series 711, Washington DC, 1998 pp 2-29.

45. Barrio, D.A.; Cattáneo, E.R.; Apezteguía, M.C.; Eycheverry, S.B. *Can. J. Physiol Pharmacol.* **2006**, *in press.*
46. Mccarthy, A.D.; Etcheverry, S.B.; Bruzzone, L.; Cortizo, A.M. *Mol. Cell. Biochem.* **1997**, *170*, 43-51.
47. Mccarthy, A.D.; Cortizo, A.M.; Giménez Segura, G.; Bruzzone, L.; Etcheverry, S.B. *Mol. Cell. Biochem.* **1998**, *181*, 63-69.
48. Stein, G.S.; Lian, J.B. *Endocrine Rev.* **1993**, *14*, 424-442.
49. Okajima, T.; Nakamura, K.; Zhang, H.; Ling, N.; Tanabe, T.; Yasuda, T.; Rosenfeld, R.G. *Endocrinology,* **1992**, *130*, 2201-2212.
50. Tullberg-Reinert, H.; Jundt, G. *Histochem. Cell. Biol.* **1999**, *112*, 271-276.
51. Ueno, A.; Kitase, Y.; Moiyama, K.; Inoue, H. *Matrix Biol.* **2001**, *20*, 347-355.

Chapter 16

Toxicity of Vanadium Compounds: Pulmonary and Immune System Targets

Mitchell D. Cohen

NYU-NIEHS Center of Excellence, Department of Environmental Medicine, New York University School of Medicine, Tuxedo, NY 10987

Inhalation is the most prevalent route of human exposure to insoluble pentavalent vanadium(V) oxides and soluble salts in urban/occupational settings. While initial pulmonary clearance of both soluble and insoluble forms of V is fairly rapid, complete clearance/degree of absorption of any V agent is ultimately a function of its solubility. Nevertheless, there are still several general toxicologic outcomes that arise from lung deposition of various V agents (as pure compounds or V-contaminated dusts). Workers exposed to V-bearing dusts or fumes display an increased incidence of several lung diseases (e.g., asthma, bronchitis, pneumonia). Similarly, after deposition of urban particulate matter (PM) or residual oil fly ash (ROFA), animals develop states of immunomodulation (inflammation, neutrophilic alveolitis, modified resistance to infection) that correlate with the levels of V in the particles. This presentation focused on how general (and in some cases agent-specific) mechanisms have been formulated to explain how entrained V agents induce toxicity and immunomodulation in the lungs.

© 2007 American Chemical Society

Vanadium is a ubiquitous trace metal in the environ-ment (*1*). Since clays/shales can contain >300 ppm V, coals upwards of 1% V (by weight), and petroleum oils 100-1400 ppm V, combustion of fossil fuels is the most identifiable non-occupational sources for delivering V-bearing particles into the air. Representative air levels of V for several rural and urban sites are presented in Table 1. Values for ambient V levels in some highly polluted cities, i.e., Mexico City (though not listed here) are in publications by Barcelaux, (*2*), Fortoul *et al.* (*3*), the National Toxicology Program (*4*), and the World Health Organization (*5*).

Table 1. Ambient V Levels and Acceptable Workplace V Levels

	Background[a]	Rural[a]	Urban[a]
average air levels (ng V/m^3)	0.001 - 0 002	1 - 40[b] 0.2 - 1.9[c] 0.02-0.80 [d]	3 - 22 [b] 150 - 1400[c]

REGULATIONS AND GUIDELINES[a,f]			
accepted levels (mg V/m^3) [g]	NIOSH	OSHA	ACGIH
V pentoxide as dust	N/A	0.05	0.05
V pentoxide as fume	N/A	0.05	0.05

[a](*1*). [b]United States; [c]Northwest Canada; [d]Eastern Pacific; [e]Northeastern United States. [f]Many states set values for V (in 8 or 24 hr periods) at <<<1 µg/m^3. [g]Eight (8) hour time-weighted averages (TWA).

It is important to note that in all cases, ambient V levels display seasonality - winter urban V levels can be many-fold higher than in summer due to increased combustion of V-bearing coals, oils, and shales for heat and electricity. The majority of V in air is in its pentavalent form (V^V); of these species the most common form is the pentoxide (V_2O_5), with ferrovanadium, vanadium carbide, and various vanadates (VO_3^- and VO_4^{3-}) in lesser amounts. Based on the values shown in Table 1 (level actually used was 50 ng V/m^3), average daily intake by inhalation in a general urban population has been estimated at ≈ 1 µg V/d.

In occupational settings, there are several potential routes of exposure to V; however, inhalation of dusts of V compounds or V-contaminated dusts is most important. Jobs with a high risk for V exposure include mining and milling of V-bearing ores, oil-boiler cleaning, and production of vanadium metal, oxides, and catalysts. In these settings, ambient V levels have been noted to sometimes be >30 mg V/m^3, a value approximating the established value for immediate danger to life or health (IDLH, 70 mg V/m^3) (*5,6*). The most recent National Occupational Exposure Study (NOES) in the 1980s estimated that the total numbers of workers potentially exposed to V was only ≈5,000 (*1*). Compared to other metals, this seems to most likely be an underestimation. Regardless of the NOES value, regulatory standards have been established (Table 1) to minimize the risk of potential worker exposure to V. As with most metals, permissible levels in many states were set substantively lower, i.e., at fractions of a $\mu g/m^3$.

Once inhaled, V is rapidly transported into the systemic circulation. Initial clearance, as either insoluble V_2O_5 or soluble vanadates/vanadyl (VO^{2+}) ions, is fairly rapid, with ≈40% cleared in 1 hr (*7,8*). However, after 24 hr, the two forms diverge in ability to be cleared, with V_2O_5 persisting (*9*). Thus, absorption of V compounds (50-85% of an inhaled dose) vary as a function of solubility; total V clearance is never achieved and commonly 1-3% of an original dose can persist for extended periods (i.e., months→years) (*4, 10, 11, 12*). Though inhalation is the primary means of delivery of V into the lungs, exposure by other routes also leads to increased lung V burden and toxic manifestations (*13, 14*).

Effects of Vanadium on the Lungs and Their Immune System-Associated Components – A Review

Pentavalent vanadates and oxides have long been known to alter pulmonary immunity in exposed hosts (reviewed in *16, 17, 18, 19*). Workers exposed to airborne V display an increased occurrence of prolonged coughing spells, tuberculosis, and general respiratory tract irritation; post-mortems indicated extensive lung damage with the primary cause of death being bacterial infection-induced respiratory failure. Epidemiological studies of workers have shown that acute exposure to high (or chronic exposure to moderate) levels of V-bearing dusts or fumes resulted in a higher incidence of a variety of pulmonary diseases, including: asthma, rhinitis, pharyngitis, ('Boilermakers') bronchitis, and pneumonia (*20, 21, 22, 23, 24, 25, 26*), metal fume fever-like syndrome (*27*), as well as increased localized fibrotic foci (*28, 29*) and lung cancers arising from non-V sources (*30, 31*) or, of increasing speculation, from the V agent itself (*4, 5, 32, 33*). Cytologic studies with cells from exposed workers noted disturbances in the levels and cellularity of neutrophils (PMN) and plasma cells; production of immunoglobulin by the latter were also affected (*29, 34, 35*).

Changes in pulmonary immune function induced by V are reproducible in a variety of animal models. Subchronic and/or acute exposure of various rodent hosts to V^V induced: decreased alveolar macrophage (AM) phagocytosis and lysosomal enzyme activity and release (*36, 37, 38, 39*); altered lung immune cell population numbers and profiles (*40, 41, 42, 43*); modified mast cell histamine release (*44*); increased *in situ* (but not *in vitro*) AM expression/production of

MIP-2 and KC CXC chemokine mRNA (45); and, airway fibrosis (46, 47). This latter study also reported a similar effect when vanadyl sulfate ($VOSO_4$) was used, suggesting that some aspects of the pulmonary immunotoxicity of V may not be valence-dependent.

Several studies implicate the presence of V as a significant factor in the local immunomodulation induced after inhalation or instillation of urban particulate matter (PM) or residual oil fly ash (ROFA) (3, 24, 48, 49, 50, 51, 52, 53, 54, 55, 56, 57). Many effects, including: intense inflammation, eosinophilia, neutrophilic alveolitis; changes in lung compliance/resistance to acetylcholine; and, modified host resistance to lung infection were found to correlate with V levels in the particles and were reproduced by exposures of animals to soluble or insoluble V at amounts equivalent to that of the V in the parent particles. Through *in vitro* studies, it has been shown that V in the ROFA/PM likely undermines pulmonary immunocompetency, in part, by inducing dysregulation of cytokine and/or chemokine production by local immune and epithelial cells (48, 58, 59, 60, 61).

Mechanisms of Immunotoxicity

Studies using non-pulmonary exposure regimens have provided information for determining mechanisms that underlay increased host susceptibility to lung infection after V inhalation. A decreased resistance to, and increased mortaility from, *Listeria monocytogenes* infection in mice that underwent acute/subchronic intraperitoneal (IP) ammonium metavanadate (NH_4VO_3) exposure suggested that cell-mediated immuity was primarily affected (62). Peritoneal macrophages (PEM) in the mice had decreased capacities to phagocytize opsonized *Listeria* and to kill even the few organisms ingested; these defects were attributed to V-induced disturbances in superoxide anion (O_2^-) formation, glutathione redox cycle activity, and hexose-monophosphate shunt activation (63). These findings with peritoneal immune cells suggested then that a major target for the immuno-toxicologic effects of V in the lungs might be the local alveolar macrophage (AM). If so, this raised the possibility of several areas of AM function that might be affected *in situ* by the metal (Figure 1).

While effects on pathways critical to maintaining PEM energy levels alone might underlay these changes in cell function and host resistance, other studies suggested that the decreased phagocytic activity and intracellular killing were related to reduced surface opsonin receptor expression/binding activity and lyso-somal enzyme release and activity (64, 65, 66). Studies with V-treated murine WEHI-3 macrophages also noted that production/release of monokines essential to anti-listeric responses were diminished in conjunction with increases in the spontaneous formation and release of potentially immunoinhibitory prostaglandin E_2 (*67*). Subsequent studies with rats exposed to airborne NH_4VO_3 also demonstrated significant reductions in the ability of AM from these hosts (*in situ* and *ex vivo*)

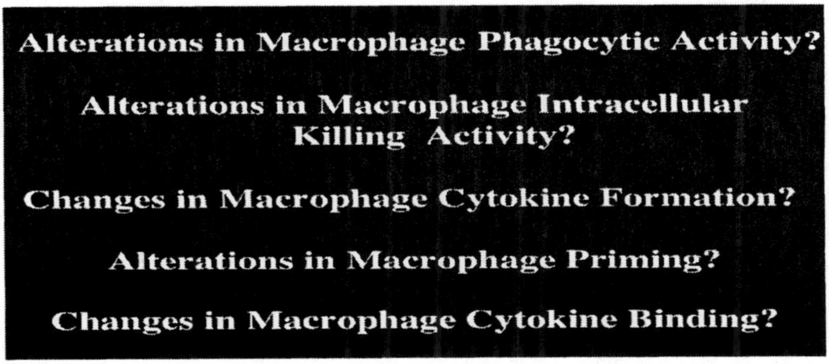

Alterations in Macrophage Phagocytic Activity?

Alterations in Macrophage Intracellular Killing Activity?

Changes in Macrophage Cytokine Formation?

Alterations in Macrophage Priming?

Changes in Macrophage Cytokine Binding?

Figure 1. Potential effects of V on the functions of AM resulting in immunomod-ulation in the lungs.

to produce/induce the formation of these critical cytokines (Figures 2A and B); these included reduced cytokine (e.g., interluekin [IL]-6, tumor necrosis factor [TNF]-α. and interferon [IFN]-γ) and both bactericidal and tumoricidal factor production *in situ* and *ex vivo* (*39, 42*).

An important aspect of immune responses that is very sensitive to V, and may even contribute to the above-described defects in AM function, is their capacity to bind with, and respond to, IFNγ (*39, 42, 68*). Results of *in vitro* exposures of macrophage cell lines indicated that surface levels and binding affinities of two surface IFNγ receptor (IFNγR) classes were greatly modified by V (Figure 2). As a result of these changes in receptor expression and/or affinity for cytokine, IFNγ-inducible responses (such as enhanced: Ca^{2+} influx, Class II antigen expression, and zymosan-induced reactive oxygen intermediate (ROI) formation) in all V-treated cells were diminished. Similar decrements in responsiveness to IFNγ were observed with AM recovered from rats subchronically-exposed to NH_4VO_3 (*42*). These effects on Ca status were subsequently also observed independently by Ishiguro and colleagues (*69*). Effects on surface receptor are apparently a common feature of the overall toxicology of V; lymphocytes (and other non-immune cell types) treated *in vitro* with vanadate also displayed altered affinity for hormones (i.e., epidermal growth factor and insulin) or cytokines (*70, 71, 72*).

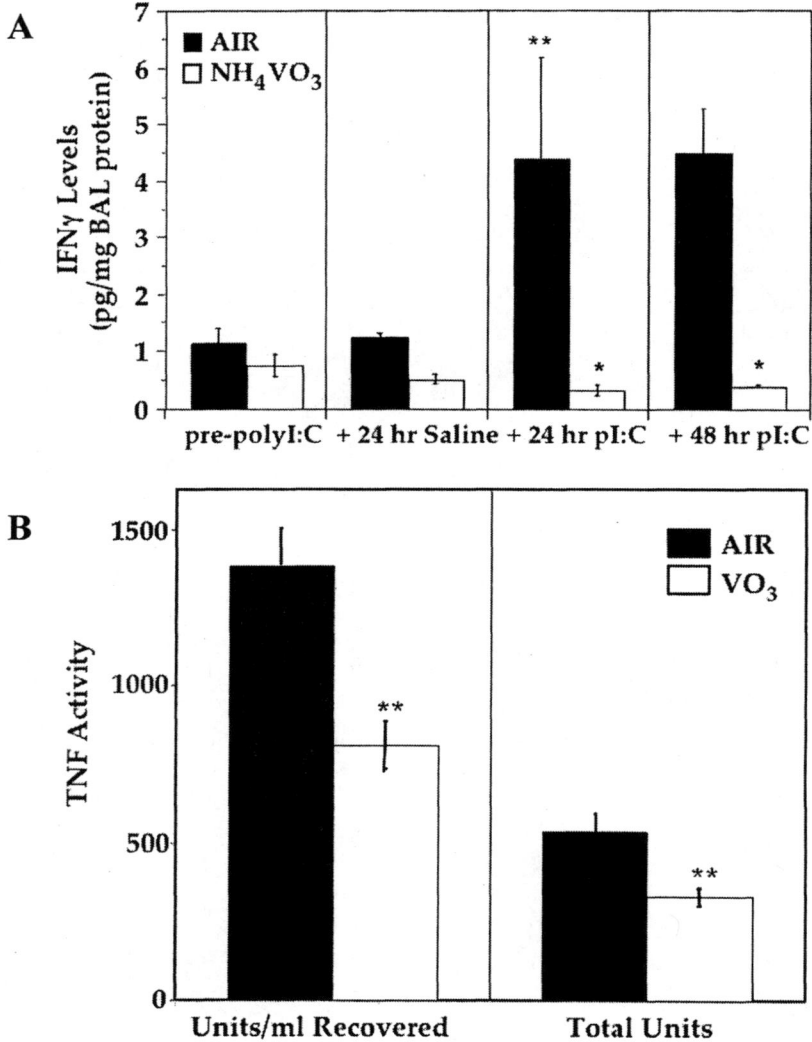

Figure 2. Effects of V on production/induction of critical cytokines. (A) In situ formation of IFNγ after stimulation of rats with poly-I:C. (B) Ex vivo production of TNFα by AM recovered from exposed rats. In both studies, rats were exposed to either filtered air or to 2 mg V/m³ as NH₄VO₃ for 8 hr/d for 4 days prior to pI:C instillation or harvesting of AM (subsequently stimulated with Escherichia coli endotoxin). Asterisks above V-asscociated bars in each figure indicate value significantly different from time-matched controls. Asterisks above air-control rats in (A) indicate significant effect of the pI:C vs. vehicle-instilled counterpart controls. N = 4-5 per each treatment group.

IFNγ binding (22°C)

Control

Class II binding sites: $K_z = 9.858 \times 10^{-9}$
Sites/cell ($\pm$ SE) = 41327 ($\pm$ 2993)

Class I sites: $K_z = 7.722 \times 10^{-9}$
Sites/cell ($\pm$ SE) = 6754 ($\pm$ 1523)

B/F — 0.2, 0.15, 0.1, 0.05, 0
Bound: 0, 10, 20, 30, 40, 50

IFNγ binding (22°C)

V-Exposed

Class II binding sites: $K_z = 4.731 \times 10^{-9}$
Sites/cell ($\pm$ SE) = 54528 ($\pm$ 2365)

Class I sites: $K_z = 1.121 \times 10^{-10}$
Sites/cell ($\pm$ SE) = 3584 ($\pm$ 872)

B/F — 0.2, 0.15, 0.1, 0.05, 0
Bound: 0, 10, 20, 30, 40, 50

Bound (M x 10^{-12})

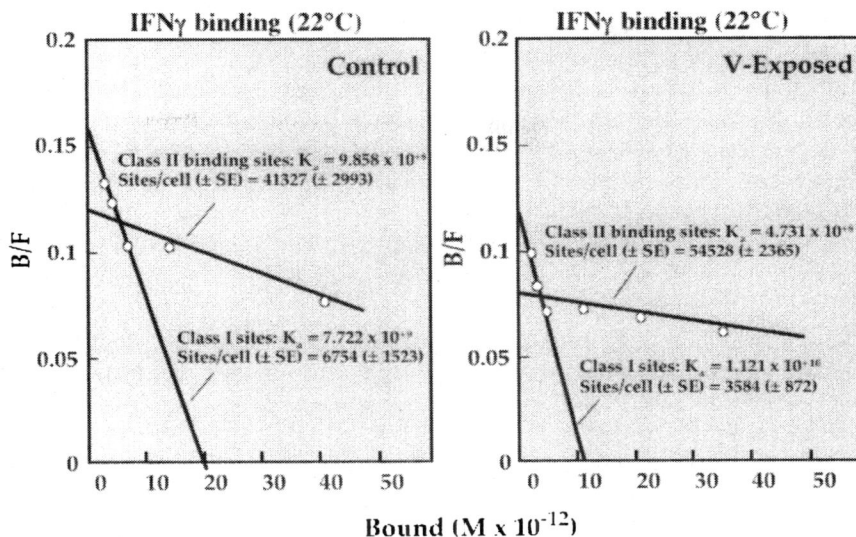

Figure 3. Effects of V on the binding of interferon-γ (IFNγ) and the expression of its specific receptors (IFNγR) on macrophages treated with vanadate in vitro.

Though the underlying causation for the effects on IFNγR is unknown, it may be that V directly modifies proteins that constitute this and other surface cytokine/opsonin receptors on macrophages (*68*). Modified receptor responses might also be related to induced changes in cellular protein kinase/phosphatase activities (*73, 74, 75, 76, 77, 78, 79, 80, 81*). Here, V-induced prolonged phosphorylation of receptor proteins and cytokine-induced secondary messenger proteins might induce prolonged states of cell activation (45, 56, *82, 83, 84, 85, 86, 87*) that modulate the receptor expression. This prolonged phosphorylation of cell proteins could also lead to bypass of normal signal transduction pathways and subsequent activation of cytokine DNA response elements (*88, 89, 90, 91, 92*) that, in turn, lead to down-regulation of cytokine receptor expression and function. This is exemplified in a study in which activation of protein kinases A and C, an event that can result in down-regulated IFNγR expression, was seen in PEM harvested from V-treated mice and in naive cells exposed *in vitro* (*93*).

It is possible that the V effects on receptor expression/functionality might be a result of alterations in various processes involved in cytokine-receptor complex handling. Agents (like V) that are able to disrupt: endocytic delivery of surface receptor-ligand complexes to lysosomes; subsequent complex dissociation;

receptor recycling, and/or *de novo* receptor synthesis, can diminish cytokine-induced responses. In macrophages and other cell types, V has been shown to: disrupt microtubule or microfilament structural integrity (*94, 95, 96*); induce altered local pH due to V polyanion formation (*97*); modify lysosomal enzyme release and activity (*65*); alter secretory vesicle fusion to lysosomes (*98*), and disrupt cell protein metabolism at the levels of both synthesis and catabolism (*99, 100*).

The effect on intracellular phosphorylation may also underlie some observations about general activation states of AM overall as well as induction of local inflammation after V exposure. In studies with AM and lung myofibroblasts, V-induced inhibition of tyrosine phosphatases (in conjunction with subsequently-induced increases in formation of intracellular reactive oxygen species (ROS) via NADPH oxidase (76, 77, 86) - using AM and other types of granulocytes) led to activation of three MAP kinase families (JNK, p38, and ERK (*90, 91, 92, 101, 102, 103, 104*). The precise implication from prolonged activation of JNK and p38 kinases (mediators of signals during response to cytokines) or ERKs (transducers in cell signaling cascades in response to growth factors/hormone signals) on cytokine receptors or responsivity to cytokines remain undefined. However, it is clear that the activation of ERKs is associated with increases in inducibility of macrophage NO (that can act as a suppressor of cell activation) as well as MIP-1α and MIP-2 chemoattractant formation (*105, 106, 107*). ERK activation (as seen in *56*) provides one of the best plausible mechanisms to explain the *in vivo* and *in vitro* findings of Chong *et al.* (*108, 109*) wherein levels of both chemokines were increased by instillation of rats, or treatment of mouse RAW264.7 cells, with soluble sodium orthovanadate (NaVO$_3$).

New Potential Mechanisms of Immunotoxicity: Effects on Macrophage Iron Homeostasis

While many of the above-discussed observations about the means by which V agents might affect the immune system (and its constituent cells) in the lungs imply that it is ultimately the direct effect of the metal that leads to the observed outcomes, studies in this laboratory and in collaboration with investigators at the U.S. Environmental Protection Agency have led to the postulate that vanadium in the lungs might also impact immune function indirectly. Specifically, it was recently hypothesized that, in part, through effects on normal carrier proteins involved in iron (Fe) transport to AM (and local epithelial cells), V can induce disruptions in iron homeostasis in those cells (*110*). As indicated in Figure 4, there are several immunotoxicologic outcomes that could arise from such disruptions, in all cases leading to acute immunomodulation if the disturbance in homeostasis is severe. The case scenarios outlined in the figure reflect how

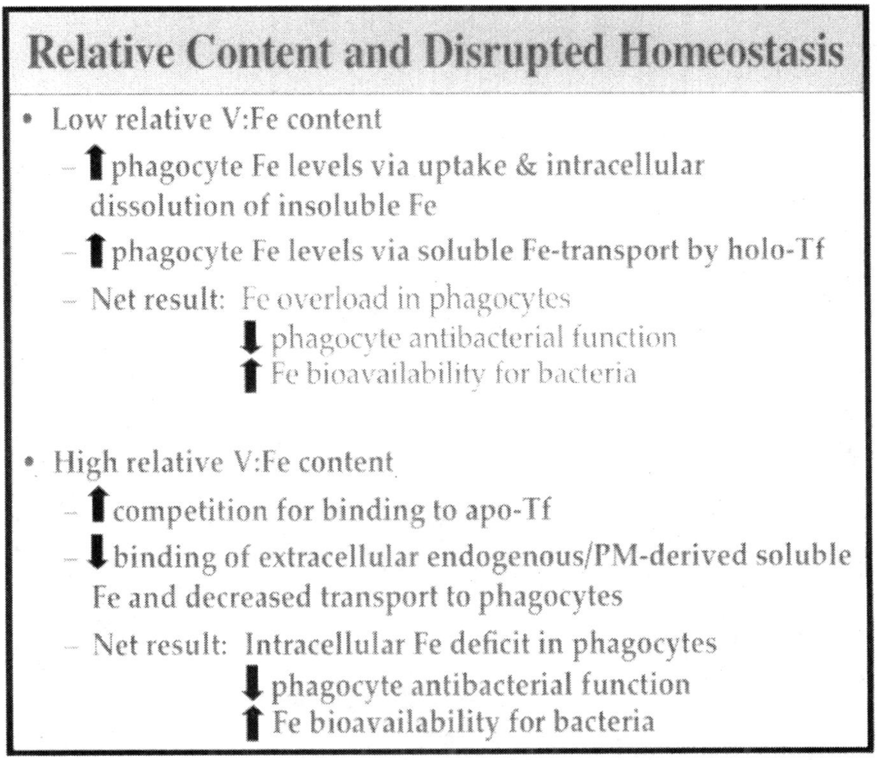

Relative Content and Disrupted Homeostasis

- Low relative V:Fe content
  - ↑ phagocyte Fe levels via uptake & intracellular dissolution of insoluble Fe
  - ↑ phagocyte Fe levels via soluble Fe-transport by holo-Tf
  - Net result: Fe overload in phagocytes
    - ↓ phagocyte antibacterial function
    - ↑ Fe bioavailability for bacteria

- High relative V:Fe content
  - ↑ competition for binding to apo-Tf
  - ↓ binding of extracellular endogenous/PM-derived soluble Fe and decreased transport to phagocytes
  - Net result: Intracellular Fe deficit in phagocytes
    - ↓ phagocyte antibacterial function
    - ↑ Fe bioavailability for bacteria

Figure 4. Indirect mechanisms of immunomodulation in the lungs due to vanadium. In both scenarios, content relationship between the V and Fe in a pollutant particle could then potentially govern the type of alteration that manifests in the AM and thus, the type of immunomodulation that arises.

ultimately it is the mass relationship between the V and Fe contents in ambient pollutant particles that could govern the toxicity that evolves from that particle.

Recent studies in macrophages treated with varying ratios of V:Fe (as found in actual particles in several different metropolitan atmospheres in the United States during Fall, 2001) clearly showed that the presence of increasing amounts of V relative to Fe caused significant shifts in the Fe content of these cells (*56, 57*). In these studies, changes in Fe content were monitored by analyses of the binding activities of iron-response proteins in the cells after a 20 hr treatment of the cells with the various metals-bearing media. When the cells were increaseingly Fe-deficient (see positive control at far right), the binding of these proteins to iron-response elements in mRNA increased (Figure 5).

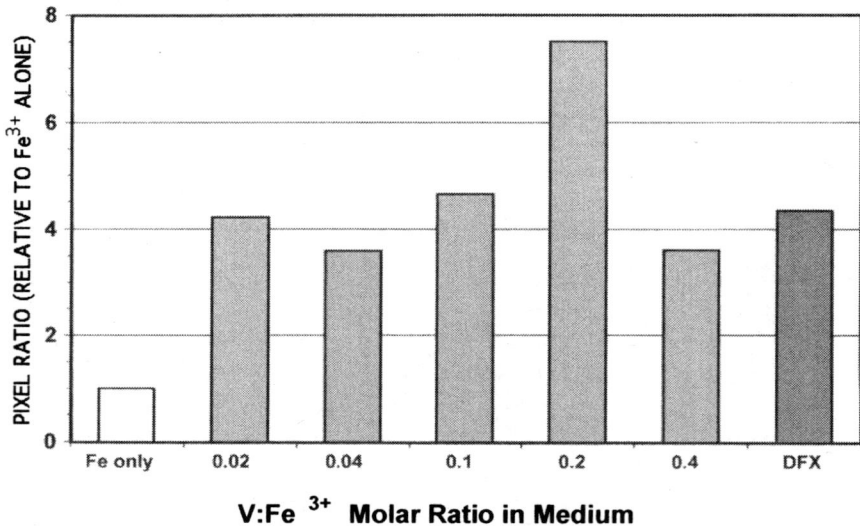

Figure 5. Effects of increasing presence of V relative to ferric iron [Fe^{3+}] on the binding of macrophage iron response proteins to a radiolabeled iron-response element in mRNA. Level of Fe^3 was always 16 μM; all data were normalized to the amount of binding seen in cells that were treated with Fe alone. Both metals were delivered as soluble forms (i.e., $NaVO_3$ and $FeCl_3$). Desferroxamine (DFX, 200 μM), an active Fe chelator, was used as the positive control.

Ongoing studies measuring the Fe content of the airways from rats exposed on five consecutive days to 100 μg V/m³ (as $NaVO_3$ or V_2O_5) have indicated that there are increases in the levels of free Fe - the expected effect when there is a decreased ability of Fe-carrier proteins to bind their normal target for delivery to the AM and local epithelial cells. These studies also showed that levels of transferrin (Tf)– a major carrier protein in the airways – were significantly decreased while those of ferritin (F) – a major Fe storage protein released into airways to prevent oxidative stress that may be induced be an excess of free Fe ions – were significantly increased (*111, 112*). Future studies will clarify more precisely the immunotoxicologic manifestations attributable to these shifts.

Role of Physicochemical Properties in Pulmonary Immunotoxicologic Effects of Vanadium Agents

The presence of V (like several select metals) as free ions or complexes in the air is increasingly believed to be a causative factor in a large number of exposure-associated respiratory diseases (*113*). With each breath, local cell populations are exposed and a variety of biological effects involved in disease pathogeneses, e.g., modifications in immune cell structural, functional, or bio-chemical properties that could impair immunocompetence, are induced. Many *in vivo* and *in vitro* studies have addressed whether, the extent, and the means by which individual metals induce these effects (see *18, 19*). While these studies showed that dose (i.e., amount delivered to lung) was a determinant in the extent of immunomodulation noted, potential effects depended on the agent itself. This suggested that inherent physicochemical properties (i.e., including valency, redox activity, and solubility) of metals (including V) or their compounds were determinative in their *in situ* toxicity.

Solubility of a metal (alone or compound) depends on molecule size, ligand type, charge, and nuclearity. Most inhalation studies with metals generally have investigated only soluble forms; effects of soluble and insoluble forms of a given metal have seldom been compared directly. In some cases, it seems that solubility may be most critical for effects in the lungs as insoluble materials often persist after entrainment. For example, a recent study reported that solubil-ity had a dramatic effect on the *in situ* immunomodulating effects of inhaled chromium compounds (*114*). Valency and redox behavior also likely impact on the immunotoxicity of metals that can exist in several oxidation states. While the fundamental difference between a +4 and +5 state is one electron, the difference is profound and affects how the metal complexes form and their properties (*115*). Coordinated ligands can also affect the valence state that predominates. The inherent ability of a metal to change valence (i.e., redox behavior) generally can occur in shuttling or unidirectional processes; the latter are abundant in biology as most metals have one preferred valence. Valency also determines types of coordinating ligands, complex stereochemistry, solubility, intra- and/or extracellular reactivity, and means of entry into cells.

In recent studies to examine the role of physicochemical properties in the induced immunotoxicities in the lung, effects by each of four chemically distinct V compounds were analyzed (*116*). These studies – using pentavalent insoluble V_2O_5, pentavalent soluble $NaVO_3$, and two novel dipicolinate complexes of tri-

and tetravalent vanadium (i.e., V^{III} dipic and V^{IV} dipic, respectively) - showed that the properties of solubility, redox behavior, and valency likely governed the *in situ* immunotoxic potentials of inhaled V agents. This conclusion was based on changes in host resistance against *Listeria monocytogenes*, an infectivity model suited for evaluation of lung cell-mediated components and their dysregulation by inhaled chemicals (reviewed in [*18*]). As shown in Figure 6, it is clear that although equivalent amounts of V were delivered to each animal (i.e., ≈ 100 μg V/m^3) with each agent, the extent of host resistance/ability to clear viable bacteria from the lung differed widely depending on the V agent employed.

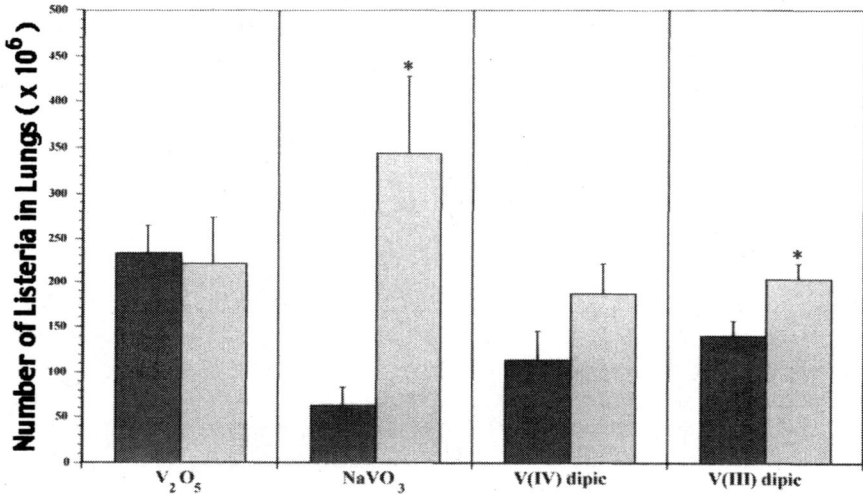

Figure 6. Effects of inhalation exposure to V agents of differing solubility (at a fixed valence state) or redox behaviors/valency (all as soluble forms). Rats were exposed to a given agent for 5 hr/d for 5 d at 100 μg V/m^3 before being having their lungs infected with viable Listeria monocytogenes. Three days later, their lungs were isolated and bacterial burdens determined. Each grey bar represents the mean ($\pm$ SE) from n = 10 animals/V regimen and each hatched bar from n = 5 air-exposed rats (hatched bar). *Value significantly different (p < 0.05) vs. time-matched infected air-exposed control.

In an attempt to begin to prescribe values for "immunotoxic potentials" for each V agent, relative changes in resistance (i.e., bacterial burden compared to levels in time-matched air-exposed counterparts) were examined in the context of the amount of V present in each rat's lungs at the time of infection. From these normalizing analyses (i.e., potential at a per ng V in lung basis; Figure 7), it became clear the most soluble pentavalent form of inhaled V was the strongest immunotoxicants tested, while the equisoluble dipic agents at lower valences had

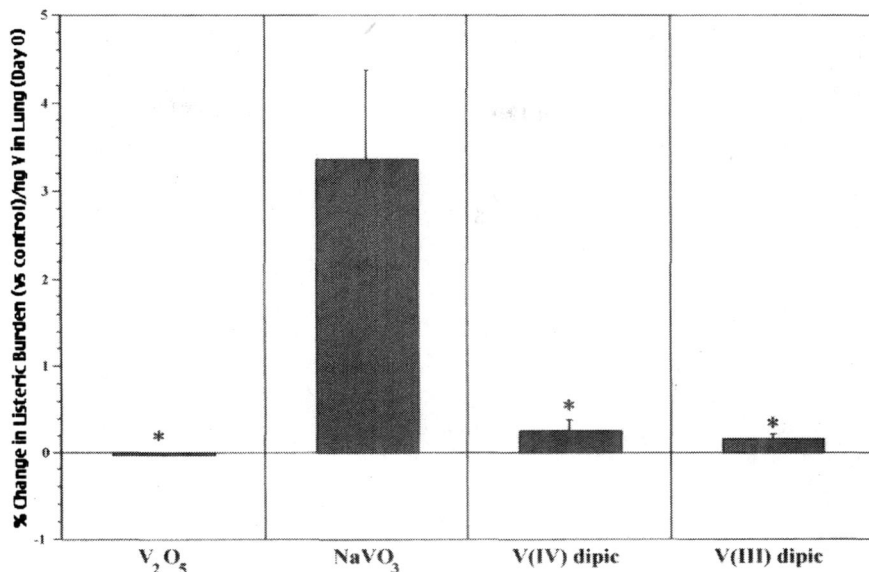

Figure 7. Relative immunotoxicologic potencies of various V agents. Potency is based on percentage change in Listeria burden in the lungs of rats at Day 3 post-infection as a function of Day 0 lung V burdens. Each bar represents the mean (± SE) from n = 10 Day 3 rats/indicated treatment average percentage differences in Listeria levels compared to respective values in air controls, all in the context of ng V in lungs at Day 0. *Value significantly (p < 0.05) different from that in rats in soluble NaVO$_3$ group.

roughly lower – but still immunotoxic - potentials. Whether the effects observed with each dipic agent (i.e., as a chelate form of V) was representative of all soluble V^{IV} and V^{III} agents will require comparison of immunotoxic potentials of the each dipic form vs. other V^{III} or V^{IV} complexes, as well as against their free salts.

To see if the differing solubility-related effects among the V compounds were not simply due to differences in how they were generated and administered, comparative analyses of data obtained here with that from studies with other metal agents (i.e., chromium, lead, and zinc; [114] and unpublished data) of varied solubility was performed. As netiher all soluble nor all insoluble agents induced equivalent immunomodulating effects, this meant that the means of atmosphere generation here was not a factor dictating differences in effects from each V compound. Nevertheless, how these three properties governed the *in situ* immunotoxic potentials of the tested V agents – and possible reflected general potentials of all inhalable V agents – required a more global examination of results from several earlier studies from this and other laboratories.

Differences in how V particles are handled in the lung are best reflected by clearance or retention patterns (see *11, 12, 13*). Inhaled insoluble V_2O_5 particles localize in AM phagosomes after ingestion and undergo a slow dissolution to vanadate ions. As such, rapid diffusion through the lung epithelia is limited and clearance would rely on mucociliary transport of uningested particles and V_2O_5-bearing AM. In comparison, soluble V^V and V^{IV} agents - if not complexed with nascent lining fluid constituents - are more readily deposited in epithelia and AM, with the limiting step being entry via membrane phosphate (PO_4^{3-}) transport systems. The results here reflected differences in post-entrainment processing as rats that received V_2O_5 had the greatest lung V burdens and those exposed to $NaVO_3$ the least. Soluble V^{III} dipic and V^{IV} dipic exposures led to comparable retentions and both were closer to that of $NaVO_3$ than V_2O_5.

Beyond affecting their deposition in, retention by, and distribution among lung cells, solubility as a factor in immunotoxicity of the two V^V agents was likely reflected in their differing effects on AM release of cytokines critical to resistance to *Listeria*. As noted earlier, these likely included differential effects on production of IL-1, TNFα, and IFNγ (*39, 42, 67*), as well as of select chemokines (i.e., MIP-2 and KC [*45*]) critical to inflammatory cell recruitment during the antibacterial response. Solubility-based differential effects on phagocytic function (*4, 38, 68, 117*) also likely affected how rats responded to *Listeria*. If AM only displayed reduced phagocytic function, increased amounts of *Listeria* would stay extracellular and need to be ingested and killed by any still functional AM. If AM had normal phagocytic activity but compromised killing, they would be less able to resist any ingested bacteria and would quickly attain high burdens; on cell death, progeny bacteria would be released to infect neighboring cells. Here, V_2O_5-exposed rat AM likely bore V but their phagocytic and killing activities would be depressed less than in the AM in $NaVO_3$-exposed rats. While the situation in V_2O_5-exposed rats would thus be more in line with the first scenario above, that in $NaVO_3$-exposed rats would be even worse than in the second scenario, i.e., there would be a delayed, but substantive, release of *Listeria* into a lung bearing few uncompromised AM to process the progeny.

Results obtained with the three soluble valence forms suggested that redox behavior also governed *in situ* immunotoxic potentials. Among the forms, V^V is the strongest oxidant under physiological conditions (depending on parent ion, $E°(V^V/V^{IV})$ = +1.02 to +1.31 V [*97*]). Extrapolating what is known about the possible chemistry of V from isolated systems, it can be inferred that on entry into cells, V^V will oxidize/complex with glutathione [GSH] and NAD(P)H (*74, 80, 81*). NAD(P)H levels may also be affected by activation of vanadate-dependent membrane-associated NAD(P)H oxidation (*118, 119*). This ongoing loss of nicotinamide equivalents would be expected to impair AM capacity to form adequate amounts of ROS for use in intraphagosomal killing, due to potential effects on NADPH oxidase activity (*62*). Similarly, loss of GSH would affect GSH redox cycle use for protection against peroxida-tive damage arising during intracellular killing or any increased presence of reactive species derived from V^V/V^{IV} reactions with cellular oxygen (O_2)/ROS (*63*). Due to the critical need to repair peroxidative damage (to maintain viability), V^V-induced increases in GSH oxidation would force the AM to consume most remaining available NAD(P)H equivalents for reduction of oxidized GSH (i.e., GSSG) rather than for ROS formation. Based on these scenarios, it could be assumed that a presence of oxidizing V^V in AM of $NaVO_3$-exposed rats may have decreased killing of ingested *Listeria*, in part, via both direct (self-perpetuating ROS$\leftrightarrow V^V/V^{IV}$ reactions) and indirect diminution of levels of ROS needed for killing.

Unlike V^V, V^{III} is a poor oxidant ($E°(V^{III}/V^{II})$ = -0.24 to -0.27 V) and will act as a reductant with cellular O_2 or ROS. It would then be expected that airway interactions with O_2 should result in V^{III} conversion to V^{IV} and then to V^V. Since levels of V delivered in each atmosphere did not differ significantly, and since V^{III} dipic exposure caused higher pre-infection V burden but a smaller impact on resistance than V^V (and marginally more than V^{IV}), it was inferred that most inhaled V^{III} was not converted to V^V in the airways. Though this conversion was expected based on its chemical properties, a recent study (*120*) showed that orally administered V^{III} dipic was more stable than expected. In an absence of significant conversion to the potent V^V form, it would seem one way V^{III} dipic caused declined resistance to *Listeria* was via direct reductive effects on ROS. It is also possible V^{III} indirectly perturbed AM function. With the highest affinity for Tf among the valences tested, any *in situ* V^{III} complexation with Tf (*56, 57, 121*) could have altered Fe delivery to AM and its capacity to generate ROS.

That V^{IV} dipic appeared to be retained more than its V^V counterpart yet had less immunomodulating effect was a finding of interest related to differences in valency and redox potential. The ability of V^{IV} (as vanadyl unit chelated to dipic ligand) to complex with PO_4^{3-} ions is many-fold that of V^V (*74*) and reflects a difference in each compound's valency. Rat lung lining fluid contains $\approx$1 mM PO_4^{3-} (*122*), and the total V in the 90 μl lining fluid of V^{IV} dipic-exposed rats is only $\approx$50 μM on Day 0. Unlike for V^V, it is likely that most V^{IV} entrained may not have been able to enter AM or epithelia but instead remained extracellular. An additional important factor to consider is that vanadate was not chelated. Previous studies have suggested that V salts were more toxic than their chelated counterparts (*81, 123, 124*). Since the $[VO_2$dipic]$^-$ system is unstable in a pH neutral reducing environment (*125*) like the lung and likely to dissociate (*126*), and V^V ions are converted intracellularly to V^{IV}, similar effects were expected to arise from the vanadate and V^{IV} dipic. Instead, the lack of any substantive effect on host resistance suggests that V^{IV} (in dipic form) is really not a potent direct toxicant. That V^{IV} had any effect at all in the rats may be attributed to its ability to bind Tf and possibly affect AM Fe homeostasis.

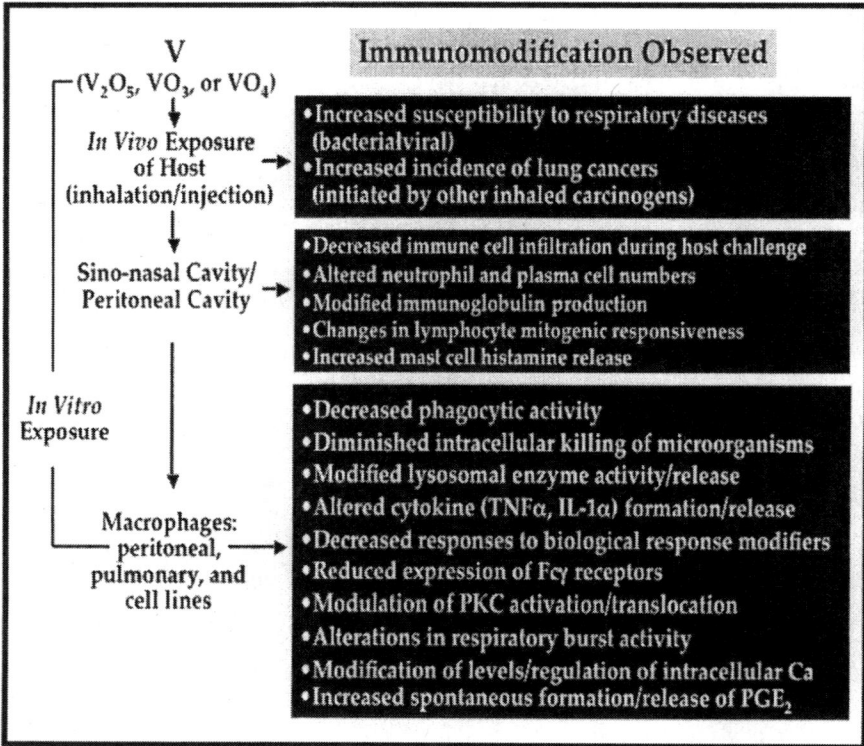

Figure 8. Review of the known immunomodulating effects of V compounds.

Summary

As outlined in Figure 8, there have been a large number of studies that have demonstrated the immunotoxic effects from exposure to vanadium. It is clear that there are a multitude of effects on overall immunologic function in the lungs of hosts exposed to various V compounds, and that many of these can be traced back to the direct effects of the metal on the local macrophages. However, this is not to say that effects on other critical cell types (e.g., PMN, mast cells, lymphocytes, interstitial macrophages) do not occur – rather, that the majority of studies have rightfully focused on the AM as they are critical to orchestrating the overall immune response in the lungs. The recent studies clarifying the potential for V agents to alter pulmonary iron homeostasis have opened up a new venue for many Investigators to better define potential mechanisms by which airborne pollutants (containing V as well as other metals) might be contributing to the increased incidence of several debilitating lung diseases and acute infections, as well as cardiovascular changes, increasingly documented among urban and suburban dwellers. Similarly, the studies to define the roles of physicochemical properties in the toxicity of V agents will allow Investigators in the pharmaceutical sciences to better pre-select/-design V-containing agents for potential use in medical treatment regimens.

In conclusion, that V can affect so many aspects of AM function – at both the cellular and genetic level - is testimony to the long-held criticism that V is mostly overlooked (as compared to other metals) as a truly potent (immuno)tox-icant. It is hoped that this Chapter has provided a concise, but extensive, over-view of the (immuno)toxicologic properties of vanadium agents to scientists who have only previously focused on its chemistry-related aspects or to those who have not given this widely-used and widely-encountered metal its due as a potential health threat to the indiviudals exposed to it daily.

Acknowledgment

Much of the more recent data reported in this chapter arose from studies originally supported primarily by funds from NIGMS/NIH Grants GM065458 and GM40525, NCI/NIH Grant 1U19CA105010, as the USEPA/PM Center Grant R82735101. The Author is grateful to the services and assistance provided, in part, by Center Programs in the NYU Department of Environmental Medicine supported by NIEHS (Grant ES00260). The Author also would like to acknowledge the support provided from USEPA/PM Center Grant R82735501 at the Northwest Center for Particulate Matter & Health in Seattle, and by USEPA Grants R82735201 and CR8280260-01-0 at the Southern California Particle Center and Supersite in Los Angeles.

References

1. ATSDR. Agency for Toxic Substances and Disease Registry. *Toxicological Profile for Vanadium*; United States Department of Public Health and Human Services: Atlanta, GA, 1991.
2. Barcelaux, G. D. *Clin. Toxicol.* **1999**, *37*, 265-278.
3. Fortoul, T. I.; Quan-Torres, A.; Sanchez, I.; Lopez, I. E.; Bizarro, P.; Mendoza, M. L.; Osorio, L.; Espejel-Maya, G.; Avila-Casado, C.; Avila-Costa, M.; Colin-Barenque, L.; Villanueva, D. N.; Olaiz-Fernandez, G. *Arch. Environ. Health* **2002**, *57*, 446-449.
4. NTP. National Toxicology Program. *NTP Technical Report on Toxicology and Carcinogenesis Studies of Vanadium Pentoxide (CAS No. 1314-62-1)*

in F344/N Rats and $B_6C_3F_1$ Mice (Inhalation Studies); NTP TR-507, NIH Publication #03-4441, United States Department of Health and Human Services: Research Triangle Park, NC, 2002.

5. IARC. International Agency for Research on Cancer. *Working Group on the Evaluation of Carcinogenic Risks to Humans. Cobalt in hard metals and cobalt sulfate, gallium arsenide, indium phosphide and vanadium pentoxide. No. 86*; World Health Organization: Lyon, 2006.

6. NIOSH. National Institute for Occupational Safety and Health. *Pocket Guide to Chemical Hazards: Fifth Edition*; United States Department of Health and Human Services: Washington, DC, 1985, pp. 234-235.

7. Conklin, A. W.; Skinner, C. S.; Felten, T. L.; Sanders, C. L. *Toxicol. Lett.* **1982**, *11*, 199-203.

8. Sharma, R. P.; Flora, S. J.; Brown, D. B.; Oberg, S. G. *Toxicol. Indust. Health* **1987**, *3*, 321-329.

9. Edel, J.; Sabbioni, E. *J. Trace Elem. Electrolytes Health Dis.* **1988**, *2*, 23-30.

10. Oberg, S. G.; Parker, R. D.; Sharma, R. P. *Toxicology* **1978**, *11*, 315-323.

11. Rhoads, K.; Sanders, C. L *Environ. Res.* **1985**, *36*, 359-378.

12. Paschoa, A. S.; Wrenn, M. E.; Singh, M. P.; Bruenger, F. W.; Miller, S. C.; Cholewa, M.; Jones, K. W. *Biol. Trace Elem. Res.* **1987**, *13*, 275-282.

13. Dill, J. A.; Lee, K. M.; Mellinger, K. H.; Bates, D. J.; Burka, L. T.; Roycroft, J. H. *Toxicol. Sci.* **2004**, *77*, 6-18.

14. Hopkins, L. L.; Tilton, B. E. *Am. J. Physiol.* **1966**, *211*, 169-172.

15. Kacew, S.; Parulekar, M. R.; Merali, Z. *Toxicol. Lett.* **1982**, *11*, 119-124.

16. Zelikoff, J. T.; Cohen, M. D. In *Experimental Immunotoxicology*; Smialowicz, R. J.; Holsapple, M. P., Eds., CRC Press: Boca Raton, FL, 1995, pp. 189-228,.

17. Cohen, M. D. In *Experimental Immunotoxicology*; Zelikoff, J. T.; Thomas, P. T.; Eds.; Taylor and Francis: London, 1998, pp. 207-229.

18. Cohen, M. D. In *Toxicology of the Lung, 4th Edition*; Gardner, D., Ed., Taylor and Francis/CRC Press: Boca Raton, FL, 2006, pp. 351-420.

19. Cohen, M. D. *J. Immunotoxicol.* 2004, *1*, 39-70.

20. Lees, R. E. *Br. J. Ind. Med.* **1980**, *37*, 253-256.

21. Musk, A. W.; Tees, J. G. *Med. J. Aust.* **1982**, *1*, 183-184.

22. Levy, B. S.; Hoffman, L.; Gottsegen, S. *J. Occup. Med.* **1984**, *26*, 567-570.

23. Pistelli, R.; Pupp, N.; Forastiere, F.; Agabit, N.; Corbo, G. M.; Tidel, F.; Perucci, C. A. *Med. Lav.* **1991**, *82*, 270-275.

24. Kielkowski, D.; Rees, D. *Report on Exposure and Health Assessment of Vanadium in Workers*. National Centre for Occupational Health, Project No. 925242/96, Report No. 3, 1997.

25. Irsigler, G. B.; Visseer, P. J.; Spangenberg, P. *Am. J. Ind. Med.* **1999**, *35*, 336-374.

26. Woodin, M. A.; Liu, Y.; Neuberg, D.; Hauser, R.; Smith, T.; Christiani, D. C. *Am. J. Ind. Med.* **2000**, *37*, 353-363.
27. Vandenplas, O.; Binard-van Cangh, F.; Gregoire, J.; Brumagne, A.; Larbanois, A. *Occup. Environ. Med.* **2002**, *59*, 785-787.
28. Kivuoloto, M. *Br. J. Ind. Med.* **1980**, *37*, 363-366.
29. Kivuoloto, M.; Rasanen, O.; Rinne, A.; Rissanen, M. *Scand. J. Work Environ. Health* **1979**, *5*, 50-58.
30. Stocks, P. *Br. J. Cancer* **1960**, *14*, 397-418.
31. Hickey, R. J.; Schoff, E. P.; Clelland, R. C. *Arch. Environ. Health* **1967**, *15*, 728-739.
32. Ress, N. B.; Chou, B. J.; Renne, R. A.; Dill, J. A.; Miller, R. A.; Roycroft, J. H.; Hailey, J. R.; Haseman, J. K.; Bucher, J. R. *Toxicol. Sci.* **2003**, *74*, 287–296.
33. Desoize, B. *In Vivo* **2003**, *17*, 529-539.
34. Kivuoloto, M.; Pakarinen, A.; Pyy, L. *Arch. Environ. Health* **1980**, *36*, 109-1130.
35. Kivuoloto, M.; Rasanen, O.; Rinne, A.; Rissanen, A. *Annt. Anz. Jena* **1981**, *149*, 446-450.
36. Waters, M. D.; Gardner, D. E.; Coffin, D. L. *Toxicol. Appl. Pharmacol.* **1974**, *28*, 253-263.
37. Fisher, G. L.; McNeill, K. L.; Whaley, C. B.; Fong, J. *J. Reticuloendothel. Soc.* **1978**, *24*, 243-252.
38. Labedzka, M.; Gulyas, H.; Schmidt, N.; Gercken, G. *Environ. Res.* **1989**, *48*, 255-274.
39. Cohen, M. D.; Becker, S.; Devlin, R.; Schlesinger, R. B.; Zelikoff, J. T. *J. Toxicol. Environ. Health* **1997**, *51*, 591-608.
40. Knecht, E. A.; Moorman, W. J.; Clark, J. C.; Hull, R. D.; Biagini, R. E.; Lynch, D. W.; Boyle, T. J.; Simon, S. D. *J. Appl. Toxicol.* **1992**, *12*, 427-434.
41. Knecht, E. A.; Moorman, W. J.; Clark, J. C.; Lynch, D. W.; Lewis, T. R. *Am. Rev. Respir. Dis.* **1985**, *132*, 1181-1185.
42. Cohen, M. D.; Yang, Z.; Zelikoff, J. T.; Schlesinger, R. B. *Fundam. Appl. Toxicol.* **1996b**, *33*, 254-263.
43. Toya, T.; Fukuda, K.; Takaya, M.; Arito, H. *Ind. Health* **2001**, *39*, 8-15.
44. Al-Laith, M.; Pearce, F. L. *Agents Action* **1989**, *27*, 65-67.
45. Pierce, L. M.; Alessandrini, F.; Godleski, J. J.; Paulauskis, J. D. *Toxicol. Appl. Pharmacol.* **1996**, *138*, 1-11.
46. Bonner J. C.; Rice A. B.; Ingram J. L.; Moonmaw C. R.; Nyska A.; Bradbury A.; Sessoms A. R.; Chulada P. C.; Morgan D. L.; Zeldin D. C.; Langenbach R. *Am. J. Pathol.* **2002**, *161*, 459-470.
47. Bonner, J. C.; Rice, A. B.; Moonmaw, C. R.; Morgan, D. L. *Am. J. Physiol.* **2000**, *278*, L209-L216.

236

48. Schiff, L. J.; Graham, J. A. *Environ. Res.* **1984**, *34*, 390-402.
49. Pritchard, R. J.; Ghio, A. J.; Lehmann, J. R.; Winsett, D. W.; Tepper, J. S.; Park, P.; Gilmour, M. I.; Dreher, K. L.; Costa, D. L. *Inhal. Toxicol.* **1996**, *8*, 457-477.
50. Dreher, K. L.; Jaskot, R. H.; Lehmann, J. R.; Richards, J. H.; McGee, J. K.; Ghio, A. J.; Costa, D. L. *J. Toxicol. Environ. Health* **1997**, *50*, 285-305.
51. Gavett, S. H.; Madison, S. L.; Dreher, K. L.; Winsett, D. W.; McGee, J. K.; Costa, D. L. *Environ. Res.* **1997**, *72*, 162-172.
52. Kodavanti, U. P.; Hauser, R.; Christiani, D. C.; Meng, Z.; McGee, J.; Ledbetter, A.; Richards, J.; Costa, D. L. *Toxicol. Sci.* **1998**, *43*, 204-212.
53. Saldiva, P.; Clarke, R. W.; Coull, B. A.; Stearns, R. C.; Lawrence, J.; Krishna Murthy, G. G.; Diaz, E.; Koutrakis, P.; Suh, H.; Tsuda, A.; Godleski, J. J. *Am. J. Respir. Crit. Care Med.* **2002**, *165*, 1610-1617.
54. Nadadur, S. S.; Kodavanti, U. P. *J. Toxicol. Environ. Health* **2002**, 65, 1333-1350.
55. Huang, Y. C.; Soukup, J.; Harder, S.; Becker, S. *Am. J. Physiol. Cell Physiol.* **2003**, *284*, C24-C32.
56. Prophete, C.; Maciejczyk, P.; Salnikow, K.; Gould, T.; Larson, T.; Koenig, J.; Jaques, P.; Sioutas, C.; Lippmann, M.; Cohen, M. D. *J. Toxicol. Environ. Health* **2006**, *69*, 935-951.
57. Doherty, S. P.; Prophete, C.; Maciejczyk, P.; Salnikow, K.; Gould, T.; Larson, T.; Koenig, J.; Jaques, P.; Sioutas, C.; Lippmann, M.; Cohen, M. D. *Inhal. Toxicol.* **2007**, In Press.
58. Carter, J. D.; Ghio, A. J.; Samet, J. M.; Devlin, R. B. *Toxicol. Appl. Pharmacol.* **1997**, *146*, 180-188.
59. Dye, J. A.; Adler, K. B.; Richards, J. H.; Dreher, K. L. Am. J. Physiol. 1999, 277, L498-L510.
60. Riley, M. R.; Boesewetter, D. E.; Kim, A. M.; Sirvent, F. P. *Toxicology* **2003**, *190*, 171-184.
61. Klein-Patel, M. E.; Diamond, G.; Boniotto, M.; Saad, S.; Ryan, L. K. *Toxicol. Sci.* **2006**, *92*, 115-125.
62. Cohen, M. D.; Chen, C. M.; Wei, C. I. *Int. J. Immunopharmacol.* **1989**, *11*, 285-292.
63. Cohen, M. D.; Wei, C. I. *J. Leukocyte Biol.* **1988**, *44*, 122-129.
64. Cohen, M. D.; Wei, C. I.; Tan, H.; Kao, K. J. *J. Toxicol. Environ. Health* **1986**, *19*, 279-298.
65. Vaddi, K.; Wei, C. I. *Int. J. Immunopharmacol.* **1991b**, *13*, 1167-1176.
66. Vaddi, K.; Wei, C. I. *J. Toxicol. Environ. Health* **1991a**, *33*, 65-78.
67. Cohen, M. D.; Parsons, E.; Schlesinger, R. B.; Zelikoff, J. T. *Int. J. Immunopharmacol.* **1993**, *15*, 437-446.
68. Cohen, M. D.; Zhang, Z.; Qu, Q.; Schelsinger, R. B.; Zelikoff, J. T. *Toxicol. Appl. Pharmacol.* **1996a**, *138*, 110-120.

69. Ishiguro S.; Miyamoto A.; Tokushima T.; Ueda A.; Nishio A. *Magnesium Res.* **2000**, *13*, 11-18.
70. Kadota, S.; Fantus, I. G.; Deragon, G.; Guyda, H. J.; Posner, B. I. *J. Biol. Chem.* **1987**, *262*, 8252-8256.
71. Torossian, K.; Freedman, D.; Fantus, I. G.. *J. Biol. Chem.* **1988**, *263*, 9353-9359.
72. Evans, G. A.; Garcia, G. G.; Erwin, R.; Howard, O. M.; Farrar, W. L. *J. Biol. Chem.* **1994**, *269*, 23407-23412.
73. Swarup, G.; Cohen, S.; Garbers, D. L. *Biochem. Biophys. Res. Commun.* **1982**, *107*, 1104-1109.
74. Nechay, B. R.; Nanninga, L. B.; Nechay, P. E.; Post, R. L.; Grantham, J. J.; Macara, I. G.; Kubena, L. F.; Phillips, T. D.; Nielsen, F. H. *Fed. Proc.* **1986**, *45*, 123-132.
75. Klarlund, J. K.; Latini, S.; Forchhammer, J. *Biochim. Biophys. Acta* **1988**, *971*, 112-120.
76. Grinstein, S.; Furuya, W.; Lu, D. J.; Mills, G. B. *J. Biol. Chem.* **1990**, *265*, 318-327.
77. Trudel, S.; Paquet, M. R.; Grinstein, S. *Biochem. J.* **1991**, *276*, 611-619.
78. Crans, D. C.; Willging, E. M.; Butler, S. R. *J. Amer. Chem. Soc.* **1990**, *112*, 427-432.
79. Slebodnick, C.; Hamstra, B. J.; Pecoraro, V. L. *Struct. Bond.* **1997**, *89*, 51-108.
80. Baran, E. J. *J. Inorg. Biochem.* **2000**, *80*, 1-10.
81. Crans, D. C.; Smee, J. J.; Gaidamauskas, E.; Yang L. *Chem. Rev.* **2004**, *104*, 849-902.
82. Pumiglia, K. M.; Lau, L.; Huang, C.; Burroughs, S.; Feinstein, M. B. *Biochem. J.* **1992**, *286*, 441-449.
83. Stern, A.; Yin, X.; Tsang, S. S.; Davison, A.; Moon, J. Biochem. Cell Biol. 1993, 71, 103-112.
84. Imbert, V.; Peyron, J. F.; Far, D. F.; Mari, B.; Auberger, P.; Rossi, B. *Biochem. J.* **1994**, *297*, 163-173.
85. Radloff, M.; Delling, M.; Gercken, G. Toxicol. Lett. 1998, 96, 69-75.
86. Grabowski, G. M.; Paulauskis, J. D.; Godleski, J. J. Toxicol. Appl. Pharmacol. 1999, 156, 170-178.
87. Huang, Y. C.; Wu, W.; Ghio, A. J.; Carter, J. D.; Silbajoris, R.; Devlin, R. B.; Samet, J. M. Exp. Lung Res. 2002, 28, 19-38.
88. Igarishi, K.; David, M.; Larner, A. C.; Finbloom, D. S. *Mol. Cell. Biol.* **1993**, *13*, 3984-3989.
89. Chen, F.; Demers, L. M.; Vallyathan, V.; Ding, M.; Lu, Y.; Castranova, V.; Shi, X. *J. Biol. Chem.* **1999**, *274*, 20307-20312.
90. Silbajoris, R.; Ghio, A. J.; Samet, J. M.; Jaskot, R.; Dreher, K. L.; Brighton, L. E. *Inhal. Toxicol.* **2000**, *12*, 453-468.

238

91. Jaspers, I.; Samet, J. M.; Erzurum, S.; Reed, W. *Am. J. Resp. Cell Mol. Biol.* **2000**, *23*, 95-102.

92. Wang, Y. Z.; Ingram, J. L.; Walters, D. M.; Rice, A. B.; Santos, J. H.; Van Houten, B.; Bonner, J. C. *Free Rad. Biol. Med.* **2003**, *35*, 845-855.

93. Vaddi, K.; Wei, C. I. *J. Toxicol. Environ. Health* **1996**, *49*, 631-645.

94. Wang, E.; Choppin, P. W. *Proc. Natl. Acad. Sci. USA* **1981**, *78*, 2363-2367.

95. Bennett, P. A.; Dixon, R. J.; Kellie, S. *J. Cell Sci.* **1993**, *106*, 891-901.

96. Ramirez, P.; Eastmond, D. A.; Laclette, J. P. *Mutat. Res.* **1997**, *386*, 291-298.

97. Rehder, D. In *Vanadium and Its Role in Life; Metal Ions in Biological Systems, Vol. 31*; Sigel, H.; Sigel, A., Eds., Marcel Dekker, Inc:.: New York, 1992, pp. 1-44.

98. Goren, M. B.; Swendsen, S. L.; Fiscus, J.; Miranti, C. *J. Leukocyte Biol.* **1984**, *36*, 273-282.

99. Montero, M. R.; Guerri, C.; Ribelles, M.; Grisoia, S. *Physiol. Chem. Phys.* **1981**, *13*, 281-287.

100. Seglen, P. O.; Gordon, P. B. 1981, *J. Biol. Chem.* **1981**, *256*, 7699-7703.

101. Zhao, Z. T.; Zhongjia, T.; Diltz, C.; You, M.; Fischer, E. *J. Biol. Chem.* **1996**, *271*, 22251-22255.

102. Samet, J. M.; Graves, L. M.; Quay, L.; Dailey, L. A.; Devlin, R. B.; Ghio, A. J.; Wu, W.; Bromberg, P. A.; Reed, W. 1998. *Am. J. Physiol.* **1998**, *275*, L551-L558.

103. Wang, Y.; Bonner, J. *Am. J. Respir. Cell Mol. Biol.* **2000**, *22*, 590-596.

104. Torres, M.; Forman, H. J. *Ann. N. Y. Acad. Sci.* **2002**, *973*, 345-348.

105. Chan, E. D.; Riches, D. *Am. J. Physiol.* **2001**, *280*, C441-C450.

106. Jaramillo M.; Olivier M. *J. Immunol.* **2002**, *169*, 7026-7038.

107. Kurosaka K.; Takahashi M.; Kobayashi Y. *Biochem. Biophys. Res. Commun.* **2003**, *306*, 1070-1074.

108. Chong, I.; Lin, S.; Hwang, J.; Huang, M.; Wang, T.; Tsai, M.; You, J.; Paulauskis, J. D. *Inflammation* **2000a**, *24*, 127-139.

109. Chong, I.; Shi, M. M.; Love, J. A.; Christiani, D. C.; Paulauskis, J. D. *Inflammation* **2000b**, *24*, 505-517.

110. Ghio A. J.; Cohen, M. D. *Inhal. Toxicol.* **2005**, *17*, 709-716.

111. Bowser, D.; Prophete, C.; Gould, T.; Larson, T.; Koenig, J.; Jaques, P.; Sioutas, C.; Lippmann, M.; Cohen, M. D. *Am. J. Resp. Cell Mol. Biol.* (In Preparation).

112. Cohen, M. D.; Sisco, M.; Prophete, C.; Chen, L.; Holder, A.; Stonehuerner, J. D.; Crans, D. C.; Ghio, A. J. *Am. J. Physiol.* (In Preparation)

113. EPA: United States Environmental Protection Agency. 2004. *Air Quality Criteria for PM, 2004*; United States Environmental Protection Agency, EPA/600/P-99/002aF, bF, Washington, D.C., 2004.

114. Cohen, M. D.; Prophete, C.; Sisco, M.; Chen, L. C.; Zelikoff, J. T.; Smee, J. J.; Holder, A. A.; Crans, D. C. *J. Immunotoxicol.* **2006**, *3*, 69-81.
115. McCleverty, J. A.; Meyer, T. J. (Eds.). *Comprehensive Coordination Chemistry II – From Biology and Nanotechnology, Volume 4 Transition Metal Groups 3-6*; Elsevier Pergamon: San Diego, CA, 2004.
116. Cohen, M. D.; Sisco, M.; Prophete, C.; Chen, L.; Zelikoff, J. T.; Smee, J. J.; Holder, A. A.; Ghio, A. J.; Stonehuerner, J. D.; Crans, D. C. *J. Immunotoxicol.* **2007**, In Press.
117. Graham, J. A.; Gardner, D. E.; Waters, M. D.; Coffin, D. C. *Infect. Immun.* **1975**, 11, 1278-1283.
118. Liochev, S. I.; Fridovich, I. *Arch. Biochem. Biophys.* **1990**, *279*, 1-7.
119. Minasi, L. A.; Willsky, G. R. *J. Bacteriol.* 1991, 173, 834-841.
120. Buglyo, P.; Crans, D. C.; Nagy, E. M.; Lindo, R. L.; Yang, L.; Smee, J. J.; Jin, W.; Chi, L. H.; Godzala, M. E.; Willsky, G. R. *Inorg. Chem.* **2005**, *44*, 416-5427.
121. Nagaoka, M. H.; Yamazaki, T.; Maitani, T. *Biochem. Biophys. Res. Commun.* **2002**, *296*, 1207-1214.
122. Cowley, E. A.; Govindaraju, K.; Lloyd, D. K.; Eidelman, D. H. *Am. J. Physiol.* **1997**, *273*, L895-899.
123. Willsky, G. R.; Goldfine, A. B.; Kostyniak, P. J.; McNeill, J. H.; Yang, L.; Khan, A. R.; Crans, D. C. *J. Inorg. Biochem.* **2001**, *85*, 33-42.
124. Thompson, K. H.; Barta, C. A.; Orvig, C. *Chem. Soc. Rev.* **2006**, *35*, 545-556.
125. Crans, D. C.; Yang, L.; Jakusch, T.; Kiss, T. *Inorg. Chem.* **2000**, *39*, 4409-4416.
126. Jakusch, T.; Jin, W.; Yang, L.; Kiss, T.; Crans, D. C. *J. Inorg. Biochem.* **2003**, *95*, 1-13.

Chapter 17

Biological Effects of Vanadium in the Lung

Andrew J. Ghio and James M. Samet

Human Studies Facility, National Health and Environmental Effects Research Laboratory, Environmental Protection Agency, Chapel Hill, NC 27599–7315

Exposures of the lung to vanadium can affect a biological response in cells and an injury in tissues. Investigation supports the postulate that vanadium either produces an oxidative stress or disrupts phosphotyrosine metabolism resulting in cell changes in signaling, transcription factor activation, and induction of mediator expression. These culminate in inflammatory and fibrotic lung injuries.

While vanadium is an essential trace element for humans and certain animals, it occurs rarely in living systems (1). Certain plants can have higher levels of vanadium (e.g. sugar beets, vines, and beech and oak trees) but the greatest concentrations are found in lower marine animals (e.g. tunicates) (1). Since oil is derived from fossilized marine organisms, vanadium can be found in this fuel at high concentrations and, subsequently, in its fly ash. Higher contents of the metal occur in the heavy oils which are left (i.e. the residual) after the more volatile fractions such as petrol, paraffin, and diesel oil have been distilled, hence the term "residual oil" fly ash (1). Fugitive fly ash from the combustion of oil and residual fuel oil contributed 76,000 and 49,000 tons, respectively, to the national ambient particle burden in 1992. This is a major source of vanadium since, in the general atmosphere, concentrations of vanadium in the atmosphere are usually quite low. In a rural setting, vanadium in ambient air can vary between 25 to 75 ng/m^3 (2). In other environments where combustion of oil products is frequent, levels of vanadium can be significantly greater than this ranging up to 300 ng/m^3 (3).

© 2007 American Chemical Society

Specific occupational environments can similarly expose individuals to vanadium. Among these are included the worksites of steel workers and boilermakers.

Lung Injury after Exposure to Vanadium Compounds

The first report of a toxic human exposure to vanadium was in 1911 (*4*). "Vanadiumism" was defined to be a chronic intoxication with the principal evidence of injury observed in the lungs, kidneys, and gastrointestinal tract (*5*). Such significant exposures occurred during the mining, separation, and use of V_2O_5 in the steel and chemical industries. Vanadium exposure was mostly via inhalation, and excretion followed via the urine with a smaller amount in the feces. Vanadium dust caused symptoms of respiratory tract irritation with conjunctivitis, sneezing, rhinorrhea, sore throat, and chest tightness (*4-6*). The cough was prominent and characteristically dry and paroxysmal (*4*). Examination of vanadium-exposed individuals revealed a greenish discoloration of the tongue, wheezing, rhonchi, and rales (*5,6*). An increase in the inflammatory cells in nasal smears and biopsies from the nasal mucosa accompanied symptoms of respiratory tract irritation. There could be concurrent changes in pulmonary function indices associated with vanadium exposure (*5,6*). Vanadium workers were observed to be more susceptible to tuberculosis and could rapidly succumb to this disease (*4*). At high exposure levels, the lungs became highly congested and show a marked destruction of the alveolar epithelium (*4*). At high vanadium exposures, hemorrhages were frequent and severe, even causing death (*5,6*). Workers who died from vanadium exposure showed congested lungs with destruction of the alveolar epithelium (*5,6*).

Human lung injury after exposure to oil fly ash has been reported predominantly after occupational exposures of workers engaged in the maintenance of oil-fired boilers in power generating stations (*7-13*). Vanadium is accepted as that component of oil fly ash responsible for toxicity of this dust. The clinical presentation of these workers has been termed "boilermakers' bronchitis" or "vanadium bronchitis". Individuals exposed to high concentrations of oil fly ash provide a history of eye irritation, sore throat, hoarseness, cough, dyspnea, wheezing, and, infrequently, symptoms consistent with pneumonitis. Physical examination can demonstrate rhinitis, conjunctivitis, and wheezing. Within 24 hours of exposure, dose-dependent losses in pulmonary function have been observed, including diminished forced vital capacity, forced expiratory volume in one second, and forced expiratory flows (*12,13*). Bronchoscopic examination shows a bronchitis with erythema and discharge in oil fly ash exposed individuals. Symptoms and signs subside, and pulmonary function decrements can resolve, within a few days or weeks of cessation of the exposure (*11*).

Mechanism of Lung Injury Following Exposure to Vanadium Compounds

The mechanism of biological effect and injury after exposure to vanadium has been postulated to be mediated by metal-catalyzed oxidant generation, metal ion dysregulation of phosphotyrosine metabolism, or possibly elements of both (Figure 1). These events are then proposed to result in phosphorylation-dependent cell signaling, an activation of specific transcription factors, an increased expression of pro-inflammatory proteins whose genes have binding sites for these transcription factors in their promoter regions, and inflammatory and fibrotic injuries to the lung.

Vanadium generates oxygen-based free radicals to present an oxidative stress to the cell (14) comparable to acellular systems (15). This production of oxidants can be inhibited by either the metal chelator deferoxamine or the antioxidants N-acetylcysteine and dimethylthiourea (DMTU). *In vivo* oxidative stress in the lungs of animals instilled with both vanadium and oil fly ash has been verified by electron spin resonance (Figure 2) (16).

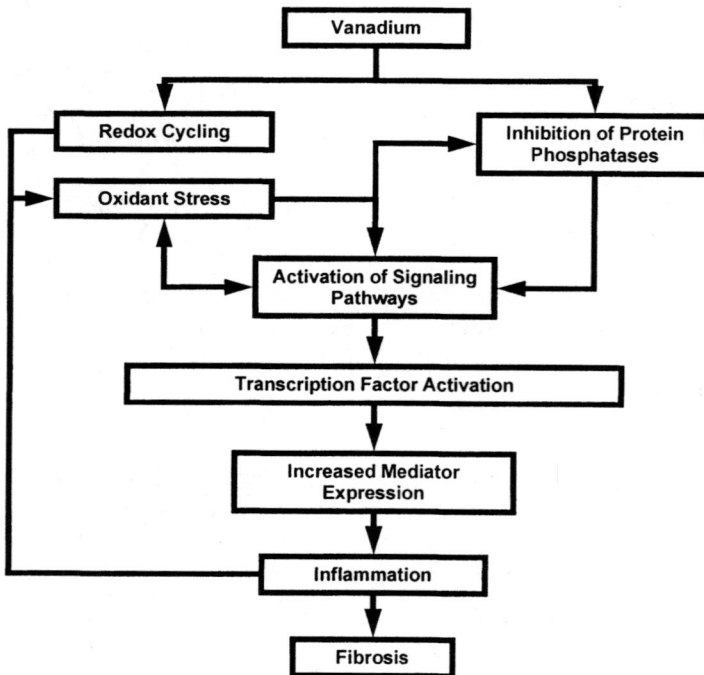

Figure 1. Proposed mechanism for the biological effect of vanadium on respiratory cells.

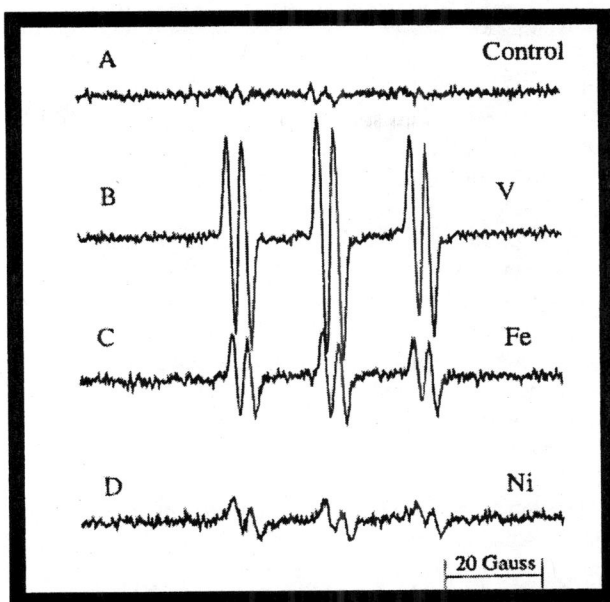

Figure 2. Electron spin resonance in the lipid soluble fraction of rat lung following intratracheal exposure to vanadium, nickel, and iron compounds in concentrations equivalent to that found in an oil fly ash (previously published, 16). Vanadium is that metal associated with the greatest signal reflecting a significant in vivo oxidative stress. Vanadium in oil fly ash accounts for the oxidative stress presented both in vitro and in vivo by this specific particle.

Soluble vanadium salts are also known inducers of protein phosphorylation and, as such, they have been used as an experimental stimulus for many years. As a result, there are numerous studies showing the effects of vanadium on a broad array of signal transduction processes. In addition, the use of oil fly ash as a model particle has produced a substantial literature that is relevant to these specific effects of vanadium exposure (17).

The relationship between oxidative stress and protein phosphorylation is only now beginning to be understood. The pivotal initiating event of vanadium on signaling is believed to be the inhibition of protein tyrosine phosphatases (18). Inhibition of protein tyrosine phosphatases by vanadium leads to unopposed kinase activity and a net accumulation of tyrosine phosphoproteins, which would be expected to result in a simultaneous activation of multiple signaling cascades in the cell (19). Multiple mechanisms have been proposed for the inhibitory effect of vanadium on protein tyrosine phosphatases. The vanadate ion has been proposed as a phosphate analog that acts as a competitive (and, therefore, reversible) inhibitor of protein tyrosine phosphatases (18). Vanadyl and vanadate species occur in equilibrium in solution and solutions of either form appear equally potent in inducing phosphotyrosine accumulation in respiratory epithelial cells (18,19).

Oxidized forms of vanadium such as pervanadate, produced by the reaction of vanadate with H_2O_2, act as potent irreversible inhibitors of protein tyrosine phosphatases by oxidizing the catalytic cysteine in the active site (20-22). A third potential mechanism of vanadium-induced inhibition of protein tyrosine phosphatase activity is catalyzed formation of reactive oxidant species through redox cycling. H_2O_2 is a direct inhibitor of protein tyrosine phosphatase activity that functions by oxidizing the reactive cysteine to form the sulfenyl derivative of the cysteinyl thiol (23). Oxidation of protein tyrosine phosphatases is now recognized as an essential event in physiological signaling (23-30).

Activation of kinases involved in cell signaling by vanadium ions has been shown in a human airway epithelial cell line (31,32) (Figure 3). Similarly, it has been reported that vanadium pentoxide (V_2O_5) induces kinase activation in human lung fibroblasts and rat lung myofibroblasts (33,34).

Separate studies have reported that exposure to vanadium ions increases the activation of a transcription factor and protein kinase C dependent signaling. Vanadium induced transcriptional activity in mouse epidermal cells through a protein kinase C-dependent process, and this effect was inhibited by superoxide dismutase, catalase and N-acetyl cysteine (35). Another report using the same cells confirmed vanadate-induced activation of protein kinase C (36). Also, it was demonstrated that vanadate induces transactivation of the tumor suppressor protein p53 in cells that could be blocked by pre-treatment with the antioxidants N-acetyl cysteine, catalase and deferoxamine, and enhanced by superoxide dismutase, implicating H_2O_2 as the reactive oxygen species involved (37). In a subsequent study, the same authors described activation of another transcription factor in vanadate exposed fibroblasts and epidermal cells (38). Again, N-acetyl cysteine, catalase and deferoxamine blocked transcription factor activation but superoxide dismutase synergized its activation by vanadate.

As previously noted, incubation of respiratory epithelial cells with oil fly ash is associated with the initiation of phosphorylation-dependent signaling reactions that may be modulated by specific redox changes (19). Redox active vanadium compounds can reproduce these events while catalytically active iron and nickel compounds have no effect (Figure 4) (19). One transcription factor that is known to be associated with oxidant responses is nuclear factor kappa B (NFkB). NFkB is normally sequestered in the cytoplasm as an inactive multiunit complex bound by an inhibitory protein (IkB). In the nucleus, NFκB binds to promoter and enhancer regions of a multitude of genes involved in the inflammatory response, including cytokines, chemokines, and growth factors. It is postulated that these genes then function to initiate, amplify, and coordinate the inflammatory response. Oil fly ash induces phosphorylation and degradation of IkB, with a resulting translocation of the active dimer into the nucleus in respiratory epithelial cells (39). Oil fly ash-induced activation of NFκB is

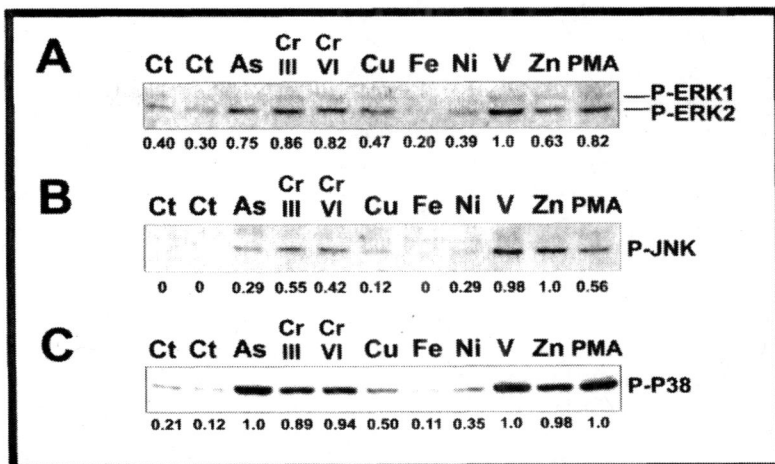

Figure 3. MAP kinase signaling following exposure of respiratory epithelial cells to equivalent concentrations of several metals (previously published, 31). Vanadium exposure is associated with significant elevations in the phosphorylation of the three MAP kinases assayed (ERK, JNK, and P38).

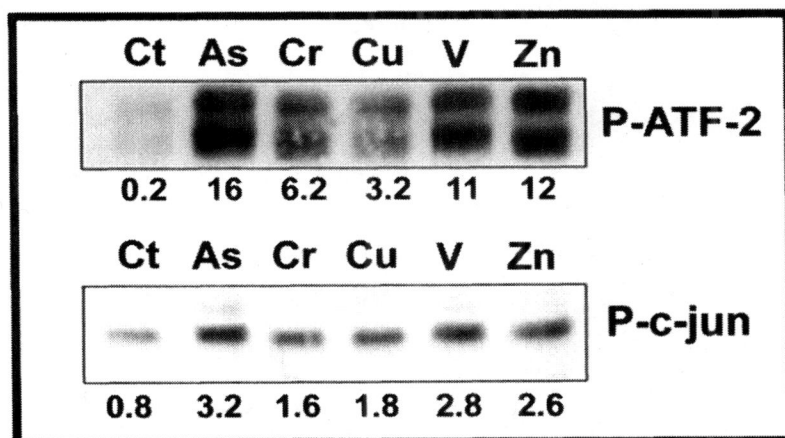

Figure 4. Transcription factor activation following exposure of respiratory epithelial cells to equivalent concentrations of several different metals (previously published, 31). Vanadium exposure is associated with significant activation of the transcription factors assayed (ATF-2 and c-jun).

blocked by metal chelators and free radical scavengers, suggesting that this activation is dependent on the generation of oxidants (39). Effects of oil fly ash on the translocation of NFkB and phosphorylation of several transcription factors have been demonstrated in the rat lung (17).

Respiratory epithelial cells exposed to either oil fly ash or vanadium, but not iron or nickel, showed increased mRNA and protein expression of numerous cytokines, including IL-6, IL-8, and TNF (40). In addition, prostaglandin H synthase 2 expression is induced, and there is concomitant enhanced secretion of prostaglandins E_2 and $F_{2\alpha}$ from normal human airway epithelial cells exposed to oil fly ash (41). As with NFkB activation, deferoxamine and an antioxidant diminishes the release of inflammatory mediators induced by oil fly ash in these cells.

mRNA and protein expression of mediators of inflammation and fibrosis are also elevated in tissues following instillation of oil fly ash and vanadium (42). Finally, exposure to oil fly ash and vanadium results in a dose-dependent influx of inflammatory cells (Figure 5) (43). Almost always, this is neutrophilic but occasional eosinophilic infiltration into the lower respiratory tract has been noted (43). The peak of this influx occurs 18 to 24 hours following exposure. Detachment of ciliated and mucus cells from the epithelial lining of the terminal bronchioles and hemorrhage can also be observed at 24 hours following oil fly ash and vanadium exposure. While incursion of inflammatory cells appears to best correlate with vanadium exposure, injury assessed as protein concentrations in the lavage fluid correlated best with the nickel content in oil fly ash (44). The cellular influx persists 96 hours later and resolution occurs slowly.

Pulmonary inflammatory injury induced by oil fly ash is reproducible by instillation of a mixture of soluble forms of vanadium, nickel, and iron in the proportions found in a saline leachate (45). Pretreatment with DMTU significantly decreased the number of neutrophils present in bronchoalveolar lavage fluid, further supporting metal-catalyzed oxidative stress as a factor determining inflammatory injury after oil fly ash instillation in animals (46).

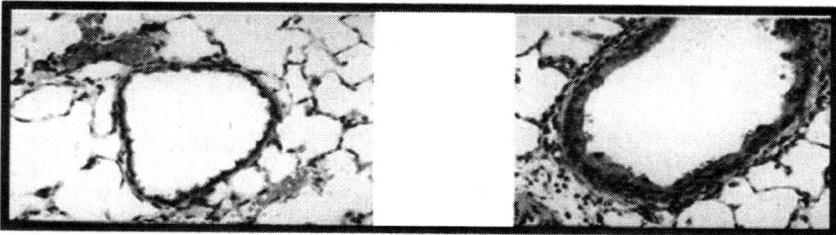

Figure 5. Airways in an animal model following exposure to saline (left) and oil fly ash (right). Twenty four hours after exposure of a Sprague Dawley rat to oil fly ash (intratracheal), there is hyperplasia of the airway epithelial cells, perivascular edema, inflammatory influx, and cell debris in the airway.

Finally, animals exposed to vanadium-containing compounds demonstrate neutrophilic inflammatory injury of the bronchi and the distal lung accompanied by a significant air-flow limitation, confirming that it is a significant determinant of injury presented by oil fly ash in the respiratory system (*47*).

References

1. Byrne, A.R.; Kosta, L. *Sci. Total Environ.* **1978**, *10*, 17-30.
2. Zelikoff, J.T.; Cohen, M.D. *Experimental Immunotoxicology*; CRC Press: New York, **1996**, pages 189-228.
3. Schroeder, W.H.; Dobson, M.; Kane, D.M.; Johnson, N.D. *J. Air Pollut. Control Assoc.* **1987**, *37*, 1267-1285.
4. Dutton, W.F. *JAMA* **1911**, *1*, 1648-1652.
5. Kiviluoto, M.; Rasanen, O.; Rinne, A.; Rissanen, M. *Scand. J. Work Environ. Health* **1979**, *5*, 50-58.
6. Kiviluoto, M. *Br. J. Ind. Med.* **1980**, *37*, 363-366.
7. Wyers, H. *Br. J. Ind. Med.* **1946**, *3*, 177-182.
8. Williams, N. *Br. J. Ind. Med.* **1952**, *9*, 50-55.
9. Browne, R.C. *Br. J. Ind. Med.* **1955**, *12*, 57-59.
10. Sjoberg, S.-G. *AMA Arch. Ind. Health* **1955**, *11*, 505-512.
11. Lees, R.E.M. *Br. J. Ind. Med.* **1980**, 37, 253-256.
12. Hauser, R.; Elreedy, S.; Hoppin, J.; Christiani, D.C. *Am. J. Respir. Crit. Care Med.* **1995**, *152*, 1478-1484.
13. Hauser, R.; Daskalakis, C.; Christiani, D.C. *Am. J. Respir. Crit. Care Med.* **1996**, *154*, 974-980.
14. Jiang, N.; Dreher, K.L.; Dye, J.A.; Li, Y.; Richards, J.H.; Martin, L.D.; Adler, K.B. *Toxicol. Appl. Pharmacol.* **2000**, *163*, 221-230.
15. Pritchard, R.J.; Ghio, A.J.; Lehmann, J.R.; Winsett, D.W.; Tepper, J.S.; Park, P.; Gilmour, M.I.; Dreher, K.L.; Costa, D.L. *Inh. Tox.* **1996**, *8*, 457-477.
16. Kadiiska, M.B.; Mason, R.P.; Dreher, K.L.; Costa, D.L.; Ghio, A.J. *Chem. Res. Toxicol.* **1997**, *10*, 1104-1108.
17. Ghio, A.J.; Silbajoris, R.; Carson, J.L.; Samet, J.M. *Environ. Health Perspect.* **2002**, *110* (Supple. 1), 89-94.
18. Gordon, J. A. *Methods Enzymol.* **1991**, *201*, 477-482.
19. Samet, J. M.; Stonehuerner, J.; Reed, W.; Devlin, R. B.; Dailey, L. A.; Kennedy, T. P.; Bromberg, P. A.; Ghio, A. J. *Am. J. Physiol.* **1997**, *272*, L426-L432.
20. Krejsa, C. M.; Nadler, S. G.; Esselstyn, J. M.; Kavanagh, T. J.; Ledbetter, J. A.; Schieven, G. L. *J. Biol. Chem.* **1997**, *272*, 11541-11549.
21. Krejsa, C. M.; Schieven, G. L. *Environ. Health Perspect.* **1998**, *106* (Supple. 5), 1179-1184.
22. Huyer, G.; Liu, S.; Kelly, J.; Moffat, J.; Payette, P.; Kennedy, B.; Tsaprailis, G.; Gresser, M. J.; Ramachandran, C. *J. Biol. Chem.* **1997**, *272*, 843-851.
23. Groen, A.; Lemeer, S.; van der Wijk, T.; Overvoorde, J.; Heck, A. J.; Ostman, A.; Barford, D.; Slijper, M.; den Hertog, J. *J. Biol. Chem.* **2005**, *280*, 10298-10304.

248

24. Persson, C.; Sjoblom, T. ; Groen, A.; Kappert, K.; Engstrom, U.; Hellman, U.; Heldin, C. H.; den Hertog, J.; Ostman, A. *Proc. Natl. Acad. Sci. U S A* **2004**, *101*, 1886-1891.

25. Meng, T. C.; Fukada, T.; Tonks, N. K. *Mol. Cell* **2002**, *9*, 387-399.

26. Kim, J. H.; Cho, H.; Ryu, S. E.; Choi, M. U. *Arch. Biochem. Biophys.* **2000**, *382*, 72-80.

27. Kamata, H.; Shibukawa, Y.; Oka, S. I.; Hirata, H. *Eur. J. Biochem.* **2000**, *267*, 1933-1944.

28. Meng, T. C.; Buckley, D. A.; Galic, S.; Tiganis, T.; Tonks, N. K. *J. Biol. Chem.* **2003**, *279*, 37716-37725

29. Salmeen, A.; Andersen, J. N.; Myers, M. P.; Meng, T. C.; Hinks, J. A.; Tonks, N. K.; Barford, D. *Nature* **2003**, *423*, 769-773

30. Meng, T. C.; Tonks, N. K. *Methods Enzymol.* **2003**, *366*, 304-318

31. Samet, J. M.; Graves, L. M.; Quay, J.; Dailey, L. A.; Devlin, R. B.; Ghio, A. J.; Wu, W.; Bromberg, P. A.; Reed, W. *Am. J. Physiol.* **1998**, *275*, L551-558

32. Wu, W.; Graves, L. M.; Jaspers, I.; Devlin, R. B.; Reed, W.; Samet, J. M. *Am. J. Physiol.* **1999**, *277*, L924-931.

33. Ingram, J. L.; Rice, A. B.; Santos, J.; Van Houten, B.: Bonner, J. C. *Am. J. Physiol .* **2003**, *284*, L774-782.

34. Wang, Y. Z.; Ingram, J. L.; Walters, D. M.; Rice, A. B.; Santos, J. H.; Van Houten, B.; Bonner, J. C. *Free Radic. Biol. Med.* **2003**, *35*, 845-855.

35. Ding, M.; Li, J. J.; Leonard, S. S.; Ye, J. P.; Shi, X.; Colburn, N. H.; Castranova, V.; Vallyathan, V. *Carcinogenesis* **1999**, *20*, 663-668.

36. Li, J.; Dokka, S.; Wang, L.; Shi, X.; Castranova, V.; Yan, Y.; Costa, M.; Huang, C. *Mol. Cell. Biochem.* **2004**, *255*, 217-225.

37. Huang, C.; Zhang, Z.; Ding, M.; Li, J.; Ye, J.; Leonard, S. S.; Shen, H. M.; Butterworth, L.; Lu, Y.; Costa, M.; Rojanasakul, Y.; Castranova, V.; Vallyathan, V.; Shi, X. *J. Biol. Chem.* **2000**, *275*, 32516-32522.

38. Huang, C.; Ding, M.; Li, J.; Leonard, S. S.; Rojanasakul, Y.; Castranova, V.; Vallyathan, V.; Ju, G.; Shi, X. *J. Biol. Chem.* **2001**, *276*, 22397-22403.

39. Quay, J.L.; Reed, W.; Samet, J.; Devlin, R.B. *Am. J. Respir. Cell Mol. Biol.* **1998**, *19*, 98-106.

40. Carter, J.D.; Ghio, A.J.; Samet, J.M.; Devlin, R.B. *Toxicol. Appl. Pharmacol.* **1997**, *146*, 180-188.

41. Samet, J.M.; Ghio, A.J.; Madden, M.C. *Exp. Lung Res.* **2000**, *26*, 57-69.

42. Su, W.-Y.; Kodavanti, U.P.; Jaskot, R.H.; Costa, D.L.; Dreher, K.L. *J. Environ. Pathol. Toxicol. Oncology* **1995**, *14*, 215-225.

43. Dreher, K.L.; Jaskot, R.H.; Lehmann, J.R.; Richards, J.H.; McGee, J.K.; Ghio, A.J.; Costa, D.L. *J. Toxicol. Environ. Health* **1997**, *50*, 285-305.

44. Kodavanti, U.P.; Hauser, R.; Christiani, D.C.; Meng, Z.H.; McGee, J.; Ledbetter, A.; Richards, J.; Costa, D.L. *Toxicol. Sci.* **1998**; *43*, 204-212.

45. Dreher, K.: Jaskot, R.; Kodavanti, U.; Lehmann, J.; Winsett, D.; Costa, D. *Chest* **1996**, *109* (Supple. 3), 33s-34s.

46. Dye, J.A.; Adler, K.B.; Richards, J.H.; Dreher, K.L. *Am. J. Respir. Cell Mol. Biol.* **1997**, *17*, 625-633.

47. Knecht, E.A.; Moorman, W.J.; Clark, J.C.; Lynch, D.W.; Lewis, T.R. *Am. Rev. Respir. Dis.* **1985**; *132*, 1181-1185.

Chapter 18

Biological Effects of Decavanadate: Muscle Contraction, In Vivo Oxidative Stress, and Mitochondrial Toxicity

Manuel Aureliano[1,2,*], Sandra S. Soares[1,3], Teresa Tiago[1,2], Susana Ramos[1,2], and Carlos Gutiérrez-Merino[4]

[1]CCMAR and [2]Departamento de Química e Bioquímica, Faculdade de Ciências e Tecnologia, Universidade do Algarve, Campus de Gambelas, 8005–139 Faro, Portugal
[3]Faculdade de Ciências do Mar e do Ambiente, Universidade do Algarve, Campus de Gambelas, 8005–139 Faro, Portugal
[4]Departamento de Bioquímica y Biología Molecular, Facultad de Ciencias, Universidad de Extremadura, Av. Elvas s/n, 06071 Badajoz, Spain
[*]Corresponding author: University of Algarvbe, 8005–139 Faro, Portugal
(maalves@ualg.pt)

Decameric vanadate species (V10) can be formed at physiological pH values in vanadate solutions presumably containing only monomeric vanadate species (V1). Sarcoplasmic reticulum Ca^{2+}-ATPase and myosin are known to interact with decameric vanadate species. V10 interaction with myosin is favored by conformational changes that take place in myosin during the catalytic cycle. Apparently, V10 operates at a different protein state in comparison with monomeric vanadate (V1) that mimics the protein at the hydrolysis transition state. V10 also clearly differs from V1, by inhibiting sarcoplasmic reticulum calcium accumulation in non-damage native vesicles, besides affecting calcium efflux associated with ATP synthesis and proton ejection associated with ATP hydrolysis. Recently reported studies referred that V10 is stabilized by actin during the process of the protein polymerization since the decomposition half-life time increases from 5 to 27 hours, suggesting that the interaction is also supported by a protein conformation induced during ATP hydrolysis followed by the formation of protein filaments. Besides affecting muscle contraction and its regulation, V10, as low as 100 nM, inhibits 50% of oxygen consumption in mitochondria, pointing that this organelle is a potential cellular target for V10, while a 100-fold higher concentration of V1 (10 µM) is needed to induce the same effect. Furthermore, *in vivo* studies have shown that following an acute exposure, decavanadate induced different changes, when compared with vanadate, on oxidative stress markers, vanadium intracellular accumulation as well as in lipid peroxidation. Putting it all together, it is suggested that the biological effects of decameric vanadate species contribute, at least in part, to the understanding of the versatility of vanadium biochemistry.

Vanadium and skeletal muscle are strongly connected to each other since almost thirty years ago when vanadium was found in commercial ATP obtained from horse skeletal muscle. Vanadium is currently used as inhibitor of E1-E2 ion transport ATPases, e.g. the sarcoplasmic reticulum (SR) Ca^{2+}- ATPase, besides being considered as a tool for the comprehension of several biochemical processes. However, the complex chemistry of vanadium difficult the interpretation of the effects promoted in biological systems and concomitantly the understanding of the role of this element in life sciences (*1*). This is not surprising if we consider that among the chemical properties of vanadium, besides the multiple oxidation states, is the capacity of vanadium(V) species to condense, forming vanadate oligomers, and with many compounds with biological interest. Certainly , serendipity combined with increasing interest from young researchers worldwide allowed to recognize motivating aims for this versatile element. This review focuses the biological effects of decavanadate. We apologize to include mainly references from our research group of "BioVanadium". In this sense, we now transmit some studies, data, ideas and future perspectives performed by our group regarding the biological effects of a vanadate oligomer, decavanadate, on: (i) contractile system and SR calcium pump, (ii) mitochondria and iii) vanadium accumulation, oxidative stress markers and lipid peroxidation following *in vivo* administration.

Formation of Decavanadate in Vanadate Solutions

Different vanadate species can occur simultaneously in vanadium(V) solutions, e.g. monomeric (V1), dimeric (V2), tetrameric (V4) and pentameric

(V5) in some cases, with different states of protonation and structures, depending on several factors such as vanadate concentration, pH and ionic strength (2). Besides these vanadate species, decameric vanadate can also occur in solution, in particular at acidic pH values. To our knowledge, it is not possible to prevent the formation of decameric vanadate upon acidification of vanadate solutions. In fact, many researchers working in different areas ranging from life sciences to chemical engineering, using vanadate in their studies, frequently observed the orange/yellow color in their solutions. This is due to the occurrence of decameric vanadate, for instance in a cell culture, after an acidification procedure during protein purification, after the adjustment of the pH value of the reaction medium or even in methods for vanadate quantification. Thus, if after the preparation of a vanadate solution (10 mM) an acidification occurs, it is possible to observe instantaneously the appearance of a yellow colour due to the occurrence of decameric vanadate species, even if the global pH value of the solution does not changes significantly. The formation of decameric species can be easily confirmed by NMR analysis. After a small pH acidification, from 6.8 to 6.6, of a metavanadate solution containing V1, V2, V4 and V5 vanadate species (Figure 1A), as evaluated by NMR spectroscopy, it is possible to observe NMR signals ascribed to decavanadate species (Figure 1B). Further acidification with HCl up to pH near 4, it is obtained the nominated decavanadate solution containing mainly decameric vanadate species (Figure 1C). Therefore, even at physiological pH values an eventual local acidification of a vanadate solution will induce the formation of V10 species. Once formed, decameric vanadate disintegration is in general slow enough to allow the study of its effects even in the micromolar range. Besides, it may become inaccessible to decomposition due to their stabilization upon binding to target proteins (3). In fact, we recently verified that in the presence of sarcoplasmic reticulum vesicles or actin, the half-life time of decameric vanadate as low as 10 μM decameric species (100 μM total vanadate), increases from 5 hours to 17 or 27 hours, respectively, at room temperature and pH 7.0, as appraised by UV/vis at 400 nm. The disintegration of decameric vanadate, follows first order kinetics, independent of concentration, and is prevented by some proteins known to interact with V10 (3).

Muscle Contraction

In our laboratory there is a long tradition in the study of the effects of decavanadate on molecular mechanisms involved in muscle contraction, since original findings describing that decameric species affect the myosin and actomyosin ATPase activity were reported since 1987 (4-8). Nevertheless, and almost twenty years after, the oligomeric vanadate species that can be present in vanadate solutions are often not accounted for consideration in most biological

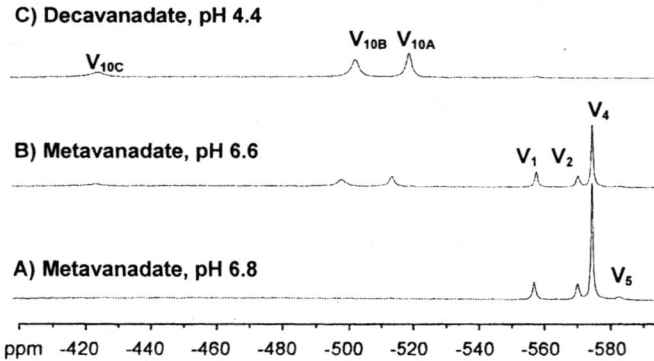

Figure 1. 105.2 MHz ^{51}V NMR spectra, at room temperature, of 10 mM metavanadate solution, pH 6.8, before (A) and after addition of HCl up to pH 6.6 (B) and to pH 4.4 (C). Decavanadate solutions correspond to spectra C. VI, monomeric; V2, dimeric; V4, tetrameric; V5, tetrameric and V10, decameric vanadate.

studies, although it is recognized that the individual species may influence enzyme activities differentially (9). More recently, it was suggested that the presence of vanadate oligomers, such as tetrameric ($V_4O_{12}^{4-}$) or decameric ($V_{10}O_{28}^{6-}$) species, prevent the stimulation of myosin ATPase activity by actin suggesting a different reactivity due, eventually, to different inhibition mechanisms in comparison to monomeric vanadate (10-12). In fact, it was shown that the ATPase activity of the actomyosin complex is inhibited (Ki = 0.27 ± 0.05 µM) by decameric but not by the monomeric form of vanadate (13). The results were consistent with binding of decavanadate to the conserved regions of the myosin phosphate binding-loop, adjacent to the ATP binding site, as it was subsequently showed by means of protein photocleavage studies in the presence of decavanadate (14). This V10 high-affinity binding site produces non-competitive inhibition of the actin-stimulated myosin ATPase activity, without causing dissociation of the ATP-free rigor actomyosin complex. Moreover, the affinity of myosin for V10 is modulated by the conformational changes that take place in myosin during the catalytic cycle, as indicated by the two to three-fold increase of the dissociation constant produced in the presence of ATP analogues such as ADP·Vi and ADP·AlF₄, which induce a conformational state close to the metastable myosin.ADP·P$_i$ intermediate state generated during the contractile cycle (13). Apparently, in the mechanism of myosin ATP hydrolysis inhibition, V10 blocks the protein at a state different from the one described for monomeric vanadate. Using an uncomplicated model for the myosin ATP hydrolysis cycle, it is suggested that V10 interaction is favour by a myosin step prior to energy transduction input, probably in the pre-hydrolysis state or upon a "back-stroke" of the myosin head, but at several myosin conformational states before ATP hydrolysis (Figure 2). Putting it all together, it is suggested that decameric vanadate species would populate myosin states with different properties in comparison to vanadate; therefore it can be used as a tool for the understanding of muscle contraction processes.

Actin Cytoskeleton

Actin is one of the two major proteins in muscle. Monomeric actin, G-actin, by itself does not stimulate myosin ATPase activity. G-actin polymerizes to F-actin, the only form known to have biological function. In muscle cells, the polymerization process is very important to maintain the thin filaments required for contraction. F-actin is the major component of muscle thin filaments and also of the microfilaments of the multifunctional cytoskeletal systems of nonmuscle cells. Besides being responsible for muscle contraction, along with myosin, F-actin filaments play very important roles in all eukaryotic cells such as locomotion, cytokinesis, structural functions and phagocytosis. In non-muscle cells, polymerization/depolymerization equilibrium regulates the process of actin filaments formation needed for specific functions such as the acrossomal process of sperm, besides others structural functions. For all eukaryotic cells if this property of actin is affected then several physiological processes would be blocked.

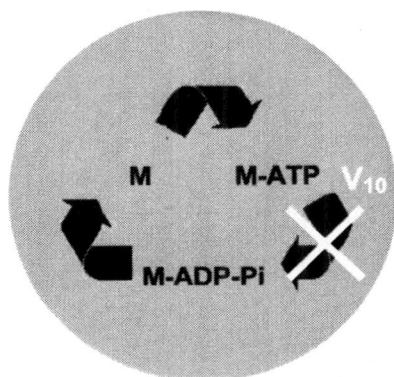

Figure 2. Scheme of the proposed decavanadate interaction with myosin during the mechanism of ATP hydrolysis. M, myosin; V10, decameric vanadate species.

It has been described that vanadate stabilizes F-actin filaments through the formation of F-actin-ADP·V1 complexes and it induces actin polymerization by inhibiting specific tyrosine phosphatases. However, the putative effects of other vanadate oligomers in actin structure and function were inexistent. Using actin from rabbit skeletal muscle we report the effects of a decavanadate solution on the capacity of actin to polymerize. Our results suggest that decameric vanadate interactions with actin inhibit G-actin polymerization and stabilize decameric vanadate species. In fact, decameric vanadate species inhibit the rate and the extent of G-actin polymerization with an IC_{50} of 68 ± 22 μM and 17 ± 2 μM, respectively, whilst they induce F-actin depolymerization at a lower extent (3). On contrary, no effect on actin polymerization and depolymerization was detected for 2 mM concentration of metavanadate solution that contains monomeric and metavanadate species, as observed by combining kinetic with ^{51}V-NMR spectroscopy studies. In those studies, it was also described that, at 25°C, decameric vanadate (10 μM) half-life time increases 5-fold (from 5 to 27 h) in the presence of G-actin only at experimental conditions favoring protein polymerization (Figure 3), whereas no effects were observed in the presence of phosphatidylcholine liposomes, myosin or G-actin alone. It was proposed that the decavanadate interaction with G-actin, favored by the G-actin polymerization processes, stabilizes decameric vanadate species and induces inhibition of G-actin polymerization (3).

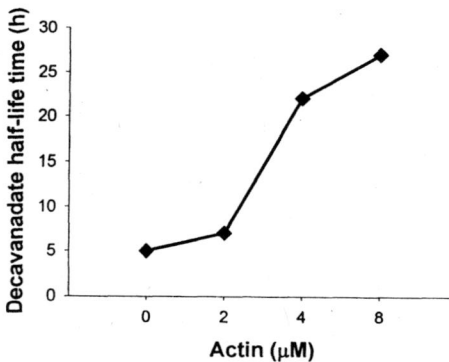

Figure 3. Effect of actin on the decameric vanadate half-life time. Decavanadate concentration of 0.1 mM (10 μM decameric species) was used in the studies (n=3).

Sarcoplasmic Reticulum Calcium ATPase

Sarcoplasmic reticulum (SR) Ca^{2+}-ATPase is a transmembrane transport system, which accumulates Ca^{2+} at expense of ATP splitting during the process of muscle relaxation. Therefore, there is an important physiological relationship between the contractile system constituted also by the proteins myosin and actin described above and the SR in the process of muscle contraction/relaxation. In our country there is a tradition in the study of the molecular mechanisms of the Ca^{2+} translocation by the SR calcium pump. More recently, significant additional data were obtained on the interaction of species present in vanadate solutions with ionic pumps (*15-17*). In these studies, it was shown that in non damage native vesicles, for instance, without the presence of calcium ionophores, decameric vanadate clearly differs from other oligomeric species in inhibiting Ca^{2+} uptake by SR coupled to ATP hydrolysis, Ca^{2+} efflux coupled with ATPase reversed activity (ATP synthesis) and H^+ ejection promoted by the SR ATPase (*9, 15-16*).

In native vesicles, the measurements of Ca^{2+} accumulation by the SR calcium pump reflect simultaneously the uptake of Ca^{2+} through the pump and the Ca^{2+} efflux (Figure 4). At this condition, where a gradient of calcium modulated the calcium pump activity, only decameric vanadate (V10) inhibits the calcium pump (Figure 4A, coupled uptake). The Ca^{2+} efflux could be passive when not associated with the pump activity or active when directly associated with intrinsic reactions to the SR Ca^{2+} pump mechanisms. In a passive efflux of calcium the ATPase works as a Ca^{2+} channel and vanadate might behave as natural ligands of the enzyme. For an active efflux of calcium, coupled to ATP synthesis, it was demonstrated that in SR vesicles loaded with radioactive Ca^{2+}, sediment by centrifugation and diluted into media containing EGTA, ADP and phosphate, Ca^{2+} active efflux from the vesicles was strongly depressed (85%) by 40 µM decameric species while milimolar concentrations of V1 has no effect (Figure 4B, right side, active efflux). In another different experimental condition, it was observed that when the gradient of calcium is destroy, meaning using phosphate or oxalate to reduce the calcium concentration inside the vesicles to almost zero, and see only accumulation, the calcium ATPase is inhibited by both V10 and V1 solutions (Figure 4C, left side).

When the vesicles are damaged, for example by the presence of a calcium ionophore, the ATP hydrolysis is not associated with calcium translocation, the calcium leaks, and in this condition the ATPase activity is also inhibited by both vanadate solutions (*9,15,18*). Therefore, "decavanadate" exert noticeable effects, on comparison to metavanadate, on "physiological" calcium accumulation, e.g. coupled with ATP hydrolysis besides on the efflux of Ca^{2+} in particular when coupled with ATP synthesis (Figure 4).

Vanadium ions and complexes are known to affect the activity of various enzymes. Sarcoplasmic reticulum E1-E2 Ca^{2+}-ATPase (SR calcium pump) is one of the proteins known to interact with vanadate (Figure 5). As a transmembranar transport system, it accumulates Ca^{2+} at expense of ATP during the process of muscle relaxation. The catalytic mechanism includes a covalent phosphorylated enzyme intermediate, which is formed by transfer of the ATP terminal

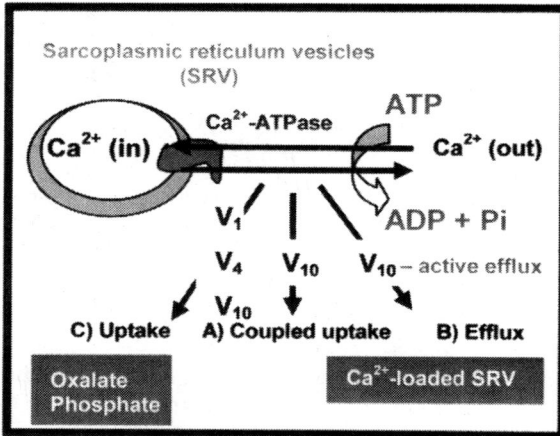

Figure 4. Schematic representation of calcium translocation by sarcoplasmic reticulum at 3 different experimental conditions: (A) coupled uptake (centre), (B) active efflux (right side) and (C) uncoupled accumulation (left side) as affected by different vanadate oligomers.

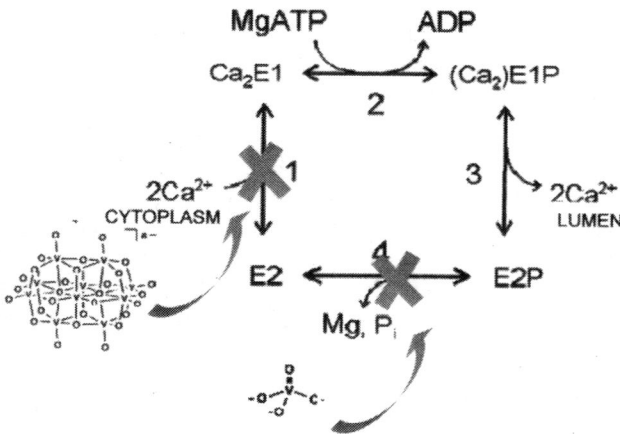

Figure 5. Effects of decavanadate on the mechanism of calcium translocation by the SR calcium ATPase. V10 block calcium efflux and calcium accumulation associate with ATP hydrolysis whereas monomeric vanadate interaction is favoured by E2 conformation.

phosphate to an aspartyl residue at the catalytic site. Several kinetic studies have suggested that, in the absence of ATP, orthovanadate can bind to the SR-ATPase and form a transition state analogue of the phosphorylated intermediate blocking the E2 conformation of the protein (Figure 5). In addition to monovanadate (V1), it has been reported that other vanadate oligomers, such as decameric vanadate, interact with the SR calcium pump (9) at distinct sites from the phosphorylation site. Some of these interactions, e.g. decavanadate (V10), as described above strongly inhibit calcium accumulation. Besides, decameric vanadate also stabilized the protein conformation at conformation E2, although others conformation may be also favourable, as described by [51]V- NMR spectroscopy (9). Being clearly different in inhibiting the SR calcium pump it is proposed a different mode of action for decavanadate in the mechanism of calcium accumulation associated with ATP hydrolysis (Figure 5).

In vivo Studies Following Decavanadate Administration

Several biological studies associate vanadium with the ability to produce reactive oxygen species (ROS), resulting in antioxidant enzymes alterations and leading to lipid peroxidation. In order to explore the hypothesis that the vanadate effects on antioxidante stress markers and on cellular responses are dependent on the oligomeric species, several kinetic studies following decavanadate administration were performed combining NMR and UV/vis spectroscopy analysis. To our knowledge, only our group has performed *in vivo* administration of decavanadate in order to understand the contribution of V10 to the toxic effects of vanadate (*19-25*). In these studies, a metavanadate solution not containing V10 was also administered as a comparison group, besides a placebo group. Until now, following *in vivo* administration of V10, several parameters were analysed such us: subcellular vanadium distribution; lipid peroxidation; antioxidants enzymes activities besides several oxidative stress markers. Among the different experimental conditions described, it was included different: mode of V10 administration (intraperitoneal, i.p. versus intravenous, i.v.); animal species (*Halobatrachus didactylus* and *Sparus aurata*); vanadate concentration (1 and 5 mM); tissues (cardiac, hepatic, renal, blood); subcellular fractions (cytosol, mitochondria, red blood cells, blood plasma); exposure time (1, 6, 12, 24 hours, 2 and 7 days). Although decameric vanadate is unstable in the assay medium, it decomposes with a half-life time from 5 to 16 hours (*19-25*), depending on the experimental conditions, allowing studying its effects not only *in vitro* but also *in vivo*. Moreover, besides the interest of piscine models to oxidative stress studies, being more sensitive to heavy metals toxicity than mammals, it has been shown to be very useful to study decavanadate toxicity, since at the fish physiological temperature decameric vanadate species is stable enough to induce different effects than vanadate itself.

By analysing the vanadium subcellular distribution following *in vivo* administration, it was described that the amount of vanadium in *Sparus aurata* cardiac tissue (46 ± 11 ng/g dry tissue) and blood (231 ± 45 ng/g dry tissue) depends on total vanadium concentration administration (25). After 1 and 6 hours of 1 mM vanadate i.v. administration, individuals intoxicated with metavanadate exhibits a higher amount of vanadium in heart (114 ± 28 ppb and 94 ± 16 ppb) relatively to those injected with decavanadate (31 ± 7 ppb and 41 ± 7 ppb), respectively, whereas after 12 hours a similar value was obtained for both solutions, approximately 80 ppb (Figure 6A). However, upon administration of a higher vanadate concentration (5 mM), the amount of vanadium detected in cardiac tissue does not depend on the type of vanadate solution administrate (Figure 6B). Therefore, decavanadate does only affect vanadium distribution at the earlier times and at lower vanadate concentrations in comparison to metavanadate (*19-25*).

Vanadate toxicity is known to induce oxidative stress that leads to lipid peroxidation. Recent unpublished results demonstrated that for higher vanadate concentration (5 mM), different longer exposure times, fish species (*H. didactylus*) but for the same tissue (heart), a significant increase (P <0.05) in cardiac tissue lipid peroxidation propagation was observed after 1 (+123%) and 7 (+64%) days after decavanadate exposure, whereas no effects where observed for metavanadate (Figure 7). In liver tissue from a same fish species, the same vanadate concentration (5 mM) and mode of administration (intravenous, i.v.) an

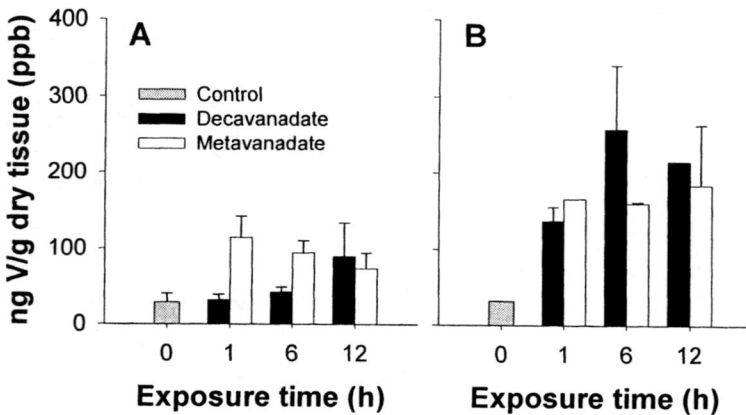

Figure 6. Total vanadium amount in cardiac tissue of Sparus aurata individuals (n=4) 1, 6 and 12 hours after 1 mM (A) and 5 mM (B) total vanadate concentration of decavanadate and metavanadate exposure.
*Significantly different from Control (P <0.05).

80% increase (P <0.05) in lipid peroxidation was observed 24 hours after i.v. administration of both vanadate solutions (23).

Therefore, the reactivities of decavanadate and metavanadate leading to lipid peroxidation are different according to the different organs. Putting it all together, at specific experimental conditions the administration of decameric vanadate clearly induces different biological responses such as vanadium accumulation and lipid peroxidation than other labile oxovanadates pointing out the importance of taking in account the V10 species in the evaluation of vanadate effects.

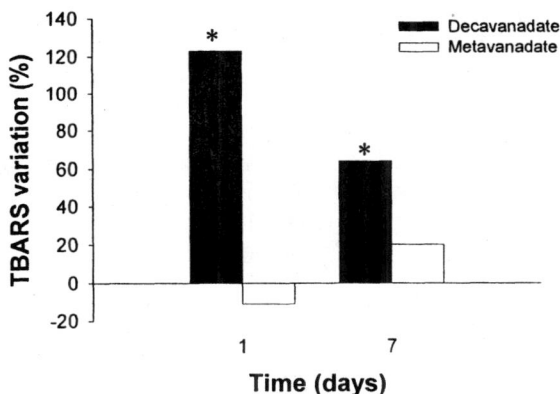

Figure 7. Lipid peroxidation propagation variation in H. didactylus heart, 1 and 7 days following decavanadate and metavanadate (5 mM total vanadate) in vivo administration. Variation is calculated based on basal values. Values are present as means ± SD (n=6). * Significantly different from Control (P <0.05).

Decavanadate in vivo administration also differs from metavanadate in not inducing cardiac mitochondrial ROS production and superoxide dismutase (SOD) activity besides decreasing catalase (CAT) activity (25). Apparently, more pronounced prooxidant effects occur in cardiac mitochondria following i.v. metavanadate exposure whereas decavanadate administration seems to prevent this effect. Decavanadate in vivo exposure also induce a decrease in cardiac mitochondrial CAT activity (-60%), 7 days following 5 mM decavanadate i.p. administration in H. didactylus (18). In liver, it was described that decavanadate and metavanadate administration clearly induce different changes in oxidative stress markers (23). Therefore, the antioxidants responses induced by vanadate may depend on the total vanadium concentration administered, on the way of exposure and/or vary between fish species, besides the vanadate species composition of vanadate solutions.

Mitochondria: a Target for Decavanadate Toxicity

Mitochondria tend to be concentrated in tissues in which the energy demand is high. Thus, a compound that inhibits mitochondrial oxidative phosphorylation can therefore have a profound effect on the metabolism of important organs like the heart, kidney, liver and brain. As described above, decavanadate *in vivo* administration point out to specific effects in mitochondrial activity besides affecting mitochondrial anti-oxidants enzyme activities *(25)*. Apparently, the mitochondria seem to be a target for decavanadate. This hypothesis was further explored and recent results are now described. Mitochondria isolated from rat liver have the advantage to be a well-characterized model for toxicological studies, very easily obtainable and the preparation produces large quantities of functionally intact mitochondria. Thus, it was show that V10 inhibits mitochondrial respiration and induces mitochondrial membrane depolarization in a larger extent that metavanadate, in both hepatic and cardiac mitochondria (Sandra *et al.*, unpublished results). For instance, decavanadate concentration as low as 100 nM, inhibits 50% of oxygen consumption in mitochondria, while a 100-fold higher concentration of V1 (10 µM) is needed to induce the same effect pointing out mitochondria as a potential cellular target for V10 toxicity (Figure 8).

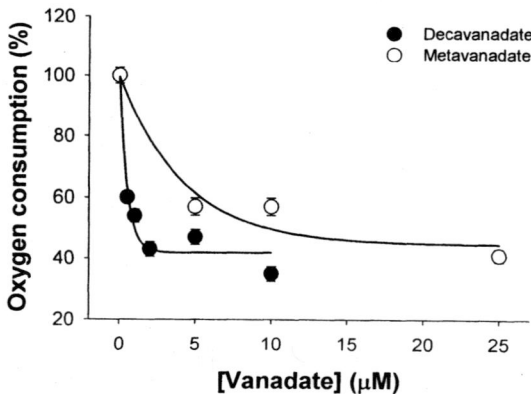

Figure 8. Effect of decavanadate and metavanadate solutions on rat liver mitochondrial oxygen consumption (n=4).

Concluding Remarks and Future Perspectives

This review focuses biological effects of decameric vanadate species on phosphohydrolases and ATPases such as the Ca^{2+}-pump of sarcoplasmic reticulum (SR) and myosin from skeletal muscle. Current studies are in progress in order to evaluate decavanadate toxic effects on neonatal rat cardiac myocytes and in the modulation of muscle contraction since it was proposed that decavanadate inhibits actomyosin ATPase and SR ATPase activity by a different mechanism than the one described for monomeric vanadate, besides affecting actin polymerization. Recently results suggesting that mitochondria are a potential target for decameric toxicity also associate the effects of decavanadate with bioenergetics processes. Once decavanadate solutions contain decameric vanadate species, whose dissociation is slow enough to study its *in vivo* effects in piscine models, it was observed that antioxidant stress markers, lipid peroxidation and vanadium subcellular distribution is dependent on the administration of solutions containing or not decameric vanadate species. In fact, decavanadate *in vivo* intoxication clearly induces different changes on oxidative stress markers and lipid peroxidation than others oligomers. Upon intoxication with decavanadate the metabolism of vanadium is affected, being mitochondria a potential toxicity target.

These studies point to the importance of taking into account decavanadate, which once formed may not completely fall apart, in the evaluation of vanadium biological effects. Questions that remain to be addressed include for instance: i) How can it be prevented the formation of decavanadate? ii) Once formed, can it be prevented the disintegration of decavanadate? iii) How can it be formed in cells an *in vivo*? iv) Does decavanadate interaction with myosin induce different *in vivo* populated states from the ones induce by monomeric vanadate? v) Can be those myosin conformation states induced by V10 relevant for the understanding of muscular contraction processes? vi) Is mitochondria a preferential target for decavanadate toxicity? vii) How can decavanadate be stabilized by actin polymerization? viii) Does V10 enter into SR vesicles or does it operate through the outside? ix) Are the *in vivo* effects induced by V10 due to interactions with membrane proteins? x) Does V10 induce the *in vivo* effects due to specific citoplasmic protein targets?

Acknowledgment

This work has been supported by Joint Spanish-Portuguese Grant HP2004-0080 (to C.G.-M. and M.A.), by POCTI program funded through FEDER for the research project POCTI/38191/QUI/2001 (to M.A.), by Grant 3PR05A078 of the Junta de Extremadura (to C.G.-M.) and CRUP project E-106/05. T. Tiago and S.S. Soares are recipients of a post-doctoral fellowship (SFRH/BPD/20777/2004) and a PhD grant (SFRH/BD/8615/2002), respectively, from the Portuguese Foundation for Science and Technology (FCT). The authors gratefully acknowledge Dr.ª Maria do Rosário Caras Altas the excellent technical assistance provided at the Laboratório de Ressonância Magnética Nuclear, Departamento de Química, Universidade Nova de Lisboa.

References

1. Nriagu, J.O., Vanadium in the environment (Vol. 1 and 2), J.O. Nriagu (Ed.), John Wiley & Sons, New York, **1998**.
2. Amado, A.M.; Aureliano, M.; Ribeiro-Claro, P.J.; Teixeira-Dias, J.J.C., *J. Raman Spect.* **1993**, *24*, 669-703.
3. Ramos, S.; Manuel, M.; Tiago, T.; Gândara, R.M.C.; Duarte, R.O.; Martins, J.; Moura, J.J.G; Aureliano. M., *J. Inorg. Biochem.* **2006**, *100*, 1734-1743.
4. Aureliano, M.; Silva, P.C.; Pires E.M.V., **1987**, Abstract, X Chemistry Portuguese Society Meeting, Porto, Portugal.
5. Aureliano, M.; Lima, M.C.P.; Pires E.M.V., **1988**, Abstract, III Portuguese-Spanish Biochemistry Congress, Santiago de Compostela, Spain.
6. Aureliano, M.; Geraldes, C.F.G.C.; Pires E.M.V., **1991**, Abstract, IV Portuguese-Spanish Biochemistry Congress, Póvoa de Varzim, Portugal.
7. Teixeira-Dias, J.J.C.; Pires, E.M.V.; Ribeiro-Claro, P.J.A.; Batista de Carvalho; L.A.E., Aureliano, M.; Amado, A.M., Cellular Regulation by Protein Phosphorylation, **1991**. NATO-ASI Series, H56, Pp. 29-33, Springer Verlag, Heidelberg.
8. Aureliano, M., MSc thesis, University of Coimbra, Coimbra, **1991**.
9. Aureliano, M.; Madeira, V.M.C., Wiley Series in *Adv. Environ. Sci. Tecnol.*, **1998**, *30*, 333-357.
10. Aureliano, M., *J. Inorg. Biochem*, **2000**, *80*, 141-143.
11. Tiago, T.; Aureliano, M.; Duarte, R.O.; Moura, J.J.G., *Inorg. Chim. Acta*, **2002**, *339*, 317-321.
12. Tiago, T.; Aureliano, M.; Gutiérrez-Merino, C., *J. Fluorescence* **2002**, *12*, 87-90.
13. Tiago, T.; Aureliano, M.; Gutiérrez-Merino C., *Biochemistry*, **2004**, *43*, 5551-5561.
14. Tiago, T.; Aureliano, M.; Moura, J.J.G., *J. Inorg. Biochem*, **2004**, *98*, 1902-1910.
15. Aureliano, M.; Madeira, V.M.C., *Biochim. Biophys. Acta*, **1994**, *1221*, 259-271.
16. Aureliano, M.; Madeira, V.M.C., *Biochem. Biophys. Res. Commun.*, **1994**, *205*, 161-167.
17. Aureliano, M., PhD thesis, University of Coimbra, **1996**.
18. Aureliano, M., *J. Inorg. Biochem*, **2000**, *80*, 144-146.
19. Aureliano, M.; Joaquim, N.; Sousa, A.; Martins, H.; Coucelo, *J. Inorg. Biochem*, **2002**, *90*, 159-165.

20. Soares, S.S.; Aureliano, M.; Joaquim, N.; Coucelo, J.M., *J. Inorg. Biochem*, **2003**, *94*, 285-290.
21. Borges, G.; Mendonça, P.; Joaquim, N.; Aureliano, M.; Coucelo, J., *Arch. Environ. Contam. Toxicol.*, **2003**, *45*, 415-422.
22. Aureliano, M.; Gândara, R.M.C., *J. Inorg. Biochem.*, **2005**, *99*, 979-985.
23. Gândara, R.M.C.; Soares S.S.; Martins, H.; Gutiérrez-Merino, C.; Aureliano M., *J. Inorg. Biochem.*, **2005**, *99*, 1238-1244.
24. Soares, S.S.; Martins, H.; Aureliano, M., *Arch. Environ. Contam. Toxicol.*, **2006**, *50*, 60-64.
25. Soares, S.S.; Coucelo, J.M., Gutiérrez-Merino, C.; Aureliano, M., *J. Inorg. Biochem.*, **2006**, in press available online.

Chapter 19

Genes and Proteins Involved in Vanadium Accumulation by Ascidians

Hitoshi Michibata[1], Masafumi Yoshinaga[1], Masao Yoshihara[1], Norifumi Kawakami[1], Nobuo Yamaguchi[2], and Tatsuya Ueki[1]

[1]Department of Biological Science, Graduate School of Science, Hiroshima University, Higashi-Hiroshima 1–3–1, Hiroshima 739–8526, Japan
[2]Marine Biological Laboratory, Graduate School of Science, Hiroshima University, Mukaishima 2445, Hiroshima 722–0073, Japan

About a hundred years ago, Henze discovered high levels of vanadium in the blood (coelomic) cells of an ascidian collected from the Bay of Naples. The intracellular vanadium concentration of some species in the family Ascidiidae can be as high as 350 mM, which is 10^7 times the concentration in seawater. Vanadium ions, thought to be present in the +5 oxidation state in seawater, are reduced to the +3 oxidation state via the +4 oxidation state and are stored in the vacuole of vanadocytes, the vanadium-containing cells, where high levels of protons and sulfate are also contained. To investigate this unusual phenomenon, we have isolated many proteins and genes that might be involved in the accumulation and reduction of vanadium. To date, more than five types of vanadium-binding protein, designated as Vanabin family, have been isolated from vanadocytes. In addition, four types of enzyme related to the pentose phosphate pathway that produces NADPH were revealed to be located in vanadocytes, and NADPH produced by the pentose phosphate pathway participates in the reduction of vanadium(V) to vanadium(IV). Vacuolar-type H^+-ATPase (V-ATPase) maintains the low pH in the vacuole and is thought to provide the energy for vanadium accumulation. Using an immobilized metal-affinity

© 2007 American Chemical Society

chromatography (IMAC) and gene homology cloning, we have further obtained a Vanabin homologue in blood plasma, metal-ATPase, glutathione-S-transferase and SO_4^{-2} transporter. Now, it becomes important to elucidate not only physiological roles of these proteins but also to resolve how these proteins share their roles to accumulate vanadium in ascidians.

Introduction

About a hundred years ago, a German physiological chemist, Martin Henze, discovered high levels of vanadium in the blood (coelomic) cells of an ascidian, known as the sea squirt, collected from the Bay of Naples (1). His discovery attracted the interdisciplinary attention of chemists, physiologists, and biochemists, in part because of considerable interest in the possible role of vanadium in oxygen transport as a third possible prosthetic group in respiratory pigments in addition to iron and copper, later shown not to be such a role for vanadium, and in part because of the strong interest in the extraordinarily high levels of vanadium never before reported in other organisms (2-9). Much of the interest developed because vanadium was found in ascidians, which phylogenetically belong to the Chordata. After Henze's finding (1), many chemists looked for vanadium in other species of ascidians.

About 3 decades ago, we started to collect many species of ascidians, belonging to the Phlebobranchia and Stolidobranchia, two of the three suborders, from the Mediterranean and from the waters around Japan, and quantified the vanadium levels in several tissues definitively using neutron-activation analysis, which is an extremely sensitive method for quantifying vanadium (10-12). The data obtained are summarized in Table I. Although vanadium was detected in samples from almost every species examined, the ascidians belonging to the suborder Phlebobranchia appeared to contain higher levels of vanadium than those belonging to the Stolidobranchia. Levels of iron and manganese, determined simultaneously, did not vary much among the members of the two suborders.

Of the tissues examined, we confirmed that blood cells contain the highest amounts of vanadium. The highest concentration of vanadium (350 mM) was found in the blood cells of *Ascidia gemmata* belonging to the suborder Phlebobranchia (12). This concentration is 10^7 times higher than that in seawater (14-15). The mechanism of vanadium accumulation and reduction by ascidians revealed up to date is schematically represented in Figure 1.

Table I. Concentrations of Vanadium in the Tissues of Several Ascidians (mM)

Species	Tunic	Mantle	Branchial Sac	Serum	Blood Cells
Phlebobranchia					
Ascidia gemmata	N.D.	N.D.	N.D.	N.D.	347.2
A. ahodori	2.4	11.2	12.9	1.0	59.9
A. sydneiensis	0.06	0.7	1.4	0.05	12.8
Phallusia mammillata	0.03	0.9	2.9	N.D.	19.3
Ciona intestinalis	0.003	0.7	0.7	0.008	0.6
Stolidobranchia					
Styela plicata	0.005	0.001	0.001	0.003	0.003
Halocynthia roretzi	0.01	0.001	0.004	0.001	0.007
H. aurantium	0.002	0.002	0.002	N.D.	0.004

NOTE: N.D.: not determined. Vanadium contents in each tissue were quantitatively determined by a neutron activation analysis (11) and that in A. gemmata was determined by an ESR (electron spin resonance) spectrometry (12).

SOURCE: Reproduced from Reference 9. Copyright 2003 Elsevier.

Vanadium-Accumulating Blood Cells, Vanadocytes

Ascidian blood cells can be classified into nine to eleven different types, which are grouped into six categories on the basis of their morphology: hemoblasts, lymphocytes, leukocytes, vacuolated cells, pigment cells, and nephrocytes (16). The vacuolated cells can be further divided into at least four different types: morula cells, signet ring cells, compartment cells, and small compartment cells.

Identification of the true vanadocytes became a matter of the highest priority for those concerned with the mechanism of accumulation of vanadium by ascidians. Using density-gradient centrifugation to isolate specific types of blood cell, and thermal neutron-activation analysis to quantify vanadium in isolated subpopulations of blood cells, we showed that vanadium is accumulated in signet ring cells in Ascidia ahodori (17). The same experiment was repeated with three different ascidian species and signet ring cells were found to be the true vanadocytes in all three species (12, 18-19).

In addition, more convincing evidence was required to clarify whether any other cell type(s) accumulate vanadium, and where the vanadium is localized. What could provide direct evidence for the location of vanadium was the

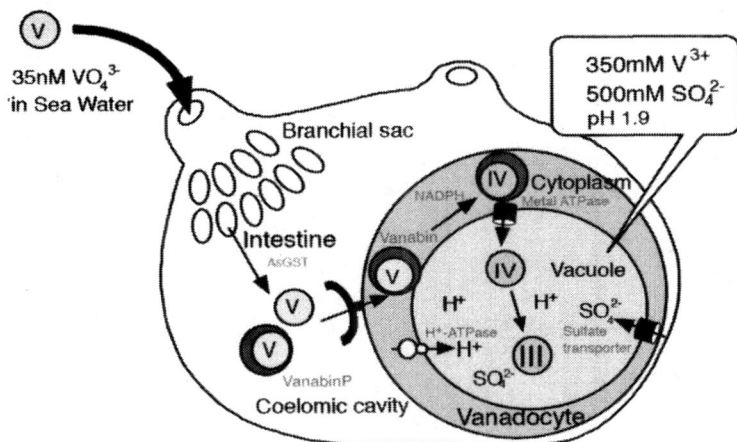

Figure 1. Schematic representation of vanadium accumulation and reduction by ascidians. The concentration of vanadium dissolved in sea water is only 35 nM in the +5 oxidation state. While, the highest concentration of vanadium in ascidian blood cells attains up to 350 mM. In addition 500 mM of sulfate is contained. The contents of vacuoles are maintained in an extremely low pH of 1.9 by H^+-ATPases. Under the environment, almost all vanadium accumulated is reduced to V(III) via V(IV). The first step of vanadium uptake may occur at a branchial sac or digestive organ (intestine), where glutathione-S-transferase was identified as a major vanadium carrier protein. We have found out vanadium binding proteins, designated Vanabins in the blood plasma and the cytoplasm of vanadocytes. The pentose phosphate pathway, which produces NADPH, has been disclosed to localize in the cytoplasm and by in vitro experiments NADPH has been revealed to reduce V(V) to V(IV). A Metal-ATPase might be involved in vanadium transport has been found in vacuolar membrane.

scanning x-ray microscope installed at the European Synchrotron Radiation Facility (ESRF) (20). The beam energy was set to 5.500 keV to ensure good fluorescence yield for vanadium (K-edge energy at 5.470 keV). To obtain images without subjecting cells to freezing or fixation, *Phallusia mammillata* blood cells were suspended in a liquid or gel medium, sealed between two plastic films with a thin spacer film and observed by x-ray microscopy (Figure 2). Signet ring cells, morula cells, compartment cells and a vacuolated amoebocyte were clearly identified by the x-ray transmission detector. The vanadium image obtained by simultaneously integrating the fluorescence signal in only the vanadium window clearly showed that the signet ring cells and vacuolated amoebocytes contained vanadium, but the morula cells and compartment cells did not. Signet ring cells of *Ascidia sydneiensis samea* also gave a clear image of vanadium accumulation (20). X-ray fluorescence energy spectra covering vanadium, potassium, chloride, and argon showed an intense, 4.952 keV vanadium signal that was emitted only from signet ring cells (20). Consequently, we concluded that vanadium accumulates in signet ring cells and vacuolated amoebocytes. We did not examine the granular amoebocytes and type-II compartment cells, reported by Scippa and colleagues (21) to contain vanadium, because they are relatively rare to be found by x-ray microscope.

Vanadium Reducing Agents

In an early attempt to identify genes specific for vanadocytes, we prepared several monoclonal antibodies which reacted specifically with signet ring cells. Afterward, it was revealed that the antigen of the S4D5 monoclonal antibody specific to vanadocytes, is 6-phosphogluconate dehydrogenase (6-PGDH: EC1.1.1.44) localized in the cytoplasm of vanadocytes (22). 6-PGDH is the third enzyme in the pentose phosphate pathway. Glucose-6-phosphate dehydrogenase (G6PDH: EC1.1.1.49), the first enzyme in the pentose phosphate pathway, was also localized immunocytologically and enzymatic activity in the cytoplasm of vanadocytes was confirmed (23). These two enzymes are known to produce 2 mols of NADPH in the pentose phosphate pathway. In addition, transketolase (TKT: EC2.2.1.1), a rate-limiting enzyme in the non-oxidative pathway, is exclusively expressed in the vanadocytes (24).

It has been reported that V(V) stimulates the oxidation of NAD(P)H; specifically, V(V) is reduced to V(IV) in the presence of NAD(P)H *in vitro* (25-28). These observations suggest that NADPH conjugates the reduction of V(V) to V(IV) in the vanadocytes of ascidians. We have, in fact, found that V(V) species are reduced to V(IV) directly by NADPH in the presence of EDTA (29). Moreover, we discovered that cysteine methyl ester can reduce vanadium(IV) to V(III) in the presence of aminopolycarboxylate in water (30), although it should be confirmed whether these phenomena occur in vanadocytes.

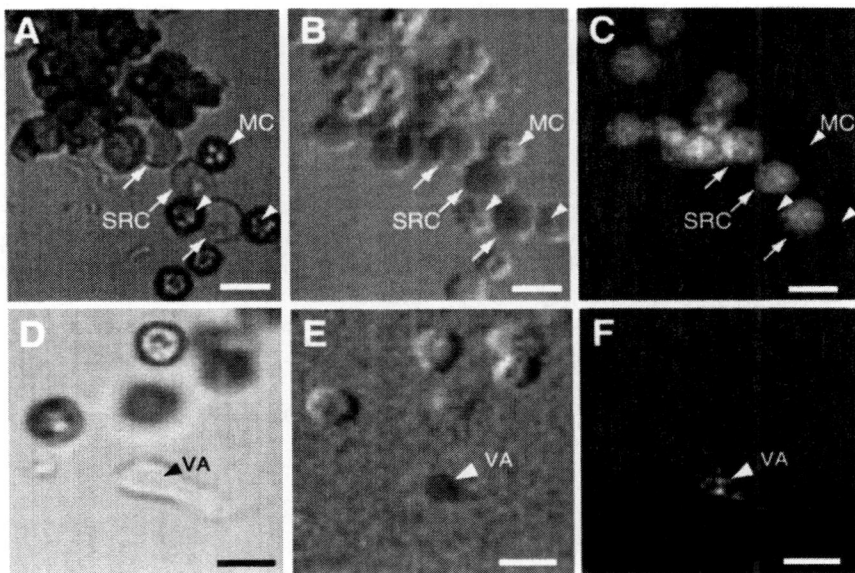

Figure 2. Phallusia mammillata blood cells observed by differential interference contrast optical microscopy (A, D), by x-ray microscopy in transmission mode (B, E), and in the fluorescence mode for vanadium (C, F). Photographs A to C are from the same field of view. Photographs D to F are from another field. Transmission and fluorescence images were taken by scanning cells with a 5.500 keV x-ray at a 1 μm x 1 μm resolution for 500 ms per pixel. Vanadium is accumulated in signet ring cells (SRC, shown by arrows) and in a vacuolated amoebocyte (VA, shown by arrowheads in D-F), but not in morula cells (MC, shown by arrowheads in A-C). Each scale bar = 10 μm. Reproduced from Reference 20. Copyright 2002 Zoological Society of Japan.

Reducing agents must, therefore, participate in the accumulation of vanadium in vanadocytes. Several candidates for the reduction of vanadium in ascidian blood cells have been proposed: tunichromes, a class of hydroxy-Dopa containing tripeptides (*31*), glutathione, H_2S, NADPH, dithiothreitol (*32*), and thiols such as cysteine (*33*).

Vanabin, Vanadium-Binding Protein

Many vanadium-binding proteins and vanadium-transporting proteins must be involved in the accumulation and reduction of vanadium in the vanadocytes. Recently, we have identified a novel family of V(IV)-binding proteins, designated as Vanabins, from *Ascidia sydneiensis samea*. In this species, the Vanabin family consists of at least five closely related proteins. Among them, Vanabin1 and Vanabin2 were first extracted from the cytoplasmic fraction of vanadocytes as major vanadium-binding proteins (*34*) and subsequently cDNAs were cloned (*35*). Vanabin3 and Vanabin4 were then identified by an expressed sequence tag (EST) database analysis of vanadocytes (*36-37*), and VanabinP was isolated from the ceolomic fluid by immobilized metal ion affinity chromatography with V(IV) (*38*). All of these five vanabins possess eighteen cysteine residues, and the intervals between cysteines are conserved very well (Figure 3). Vanabin1, Vanabin2 and vanabin3 are exclusively localized in cytoplasm of vanadocytes, while Vanabin4 is loosely associated with cytoplasmic membrane of signet ring cells (*39*).

A homology search with both the Vanabin1 and Vanabin2 sequences using BLASTP and public protein databases did not reveal any other proteins with striking similarities. On the other hand, we identified five new Vanabins (CiVanabin1 to CiVanabin5) by an EST database search of another vanadium-accumulating ascidian species, *Ciona intestinalis*, the whole genome of which has been sequenced, and have reported that these CiVanabins also bind V(IV) ions (*40*). We have also found five similar genes from the related species *C. savignyi*. Vanabins, therefore, seem to be common among the vanadium-accumulating ascidians and to hold the key toward resolving the mechanism underlying the highly selective accumulation of vanadium ions.

Using recombinant Vanabin1 and Vanabin2, we revealed that they bind up to 10 and 20 V(IV) ions, with dissociation constants of 2.1×10^{-5} M and 2.3×10^{-5} M, respectively (*35*). Recently, an EPR study revealed that Vanabin2 can bind up to ~23.9 V(IV) ions per molecule, and most of the vanadium ions are in a mononuclear state and coordinated by amine nitrogens (*41*). The EPR study also suggested that no allosteric effects are involved in the process of binding multiple vanadium ions. These data suggested that Vanabins are a new class of metal binding proteins.

Judging from their exclusive localization in the cytoplasm of vanadocytes and the metal binding properties, Vanabin1 and Vanabin2 are considered to be

```
Vanabin1  1  - - M V S K F T I L L G V V V L M A L S V N A Y E S E F D D D E T F E K - - - - G P G  36
Vanabin2  1  - - - M S K V I F A L V L V V V L V A C I N A T Y V E F E E A Y - - - - - - A P V D  33
Vanabin3  1  M A S K L F L L L F L G M F V L I A A S D E S F D E E E D F E D E V M A Q S Y Y P E  42
Vanabin4  1  M V T K S H I I F F L G M V V V I V G C P A F E K F V S K I E E S V I V D S - - - -  38
VanabinP  1  - - - - - - - M R V T I V V L V V V A C L L V A A E A - - - - - - - - R K K K K K M  27

            1       2      3        4        5       6                7        8      9
37  C K - - C Q S V C G E V K K C G V K C F R S C N G D R D - - - C T K D C A K A K C G  73
34  C K G Q C T T P C E P L T A C K E K C A E S C E T S A D K K T C R R N C K K A D C E  75
43  C D - - C R Q E C G T F R N C R A T C R A N C G D G R - - - - C R R E C K R T K C I  78
39  C K T N C S T E C L P L K N C T E N C T E H C E G L S D K K A C H Q N C R K V T C K  80
28  C R V A C K S D C K A A K A C M K P C K R G C R T S T N K K A C K K G C R T S - C K  68

              10      11       12          13          14        15
74  K V - - - - P N A G D C G H C M L S - C E G - - - - K C R A D H C A S A C P G - - - -  103
76  - - - - - P Q D K V C D A C R M K - C H K - - - - A C R A A N C A S E C P K - - - -  103
79  N - - - - - M K S Q C R N C N G D - C R E - - - - R C R S K Y C S K P C Y - - - - -  105
81  - - - - - A E D G Q C R A C K K K - C K D - - - - E C K K A N C K S S C E E - - - -  108
69  Q S V G A S I I Q Q C R K C V T D N C P L G D V K M C V A N N C V R P C L S T R N R  110

                              16      17       18
104  - - - - - - - - - - K V S K A P A C L D C M K L N C V - - - - - - -  120
104  - - - - - - - - - - H E H K S D T C R A C M K T N C K - - - - - - - - -  120
106  - - - - - - - - - - K S L K V R K C V R C M V V S C H L R F - - - - - -  125
109  - - - - - - - - - - K A M K S P A C K S C M E K N C H - - - - - - - - -  125
111  P I P G E K P N S A K M F D N P F C L E C M K E N C E E Q F E A L F E G  146
```

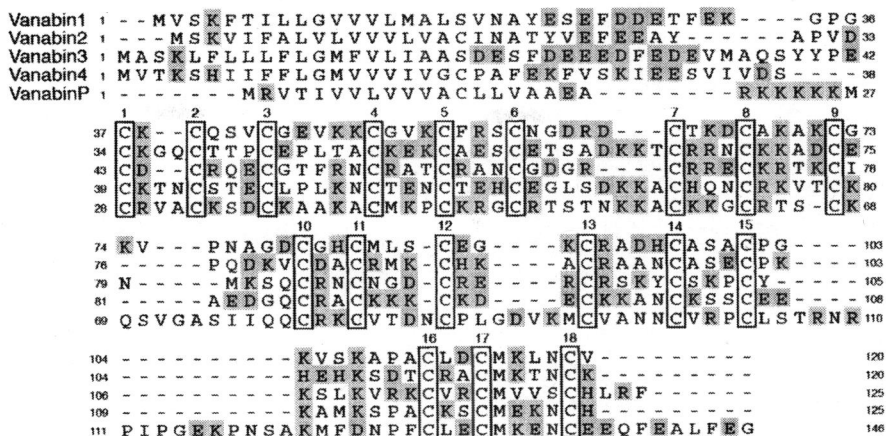

Figure 3. Amino acid sequences of Vanabins isolated from Ascidia sydneiensis samea. Conserved cysteines were boxed. Note that all the five amino acid sequences were derived from corresponding cDNAs, and the N-termini of Vanabin1 (G34), Vanabin2 (A29) and VanabinP (R21) are experimentally determined. Positively and negatively charged residues are shown in gray. See (35, 37, 38)

vanadium metallochaperone proteins that transport vanadium ions from the cytoplasm to the vacuole, as illustrated in Figure 1. A better understanding of the functions of Vanabins and the mechanism of vanadium accumulation in ascidians requires high-quality 3D structures of the proteins in the presence and absence of vanadium ions. We reported the solution structure of Vanabin2, determined by multidimensional NMR experiments (42). The structure provides, to our knowledge, the first 3D picture of a protein with a novel protein fold that binds to multiple V(IV) ions (Figure 4). There are no structural homologues reported so far. Moreover, the NMR titration experiments, showing the ^{1}H-^{15}N heteronuclear single-quantum coherence (^{15}N HSQC) spectra of Vanabin2 upon the gradual addition of VO^{2+} revealed the putative vanadium-binding sites on Vanabin2. The elucidation of this structure provides a key to solving a riddle that has attracted interdisciplinary interest for a century.

On the other hand, the mechanism of metal-selectivity of Vanabin2 has not yet been determined. The effects of acidic pH on selective metal binding to Vanabin2 and on the secondary structure of Vanabin2 were examined. Using an immobilized metal-affinity chromatography (IMAC), Vanabin2 was revealed to selectively bind to V(IV), Fe(III), and Cu(II) ions under acidic conditions (Table 2). In contrast, Co(II), Ni(II), and Zn(II) ions were bound at pH 6.5 but not at pH 4.5. Changes in pH had no detectable effect on the secondary structure of

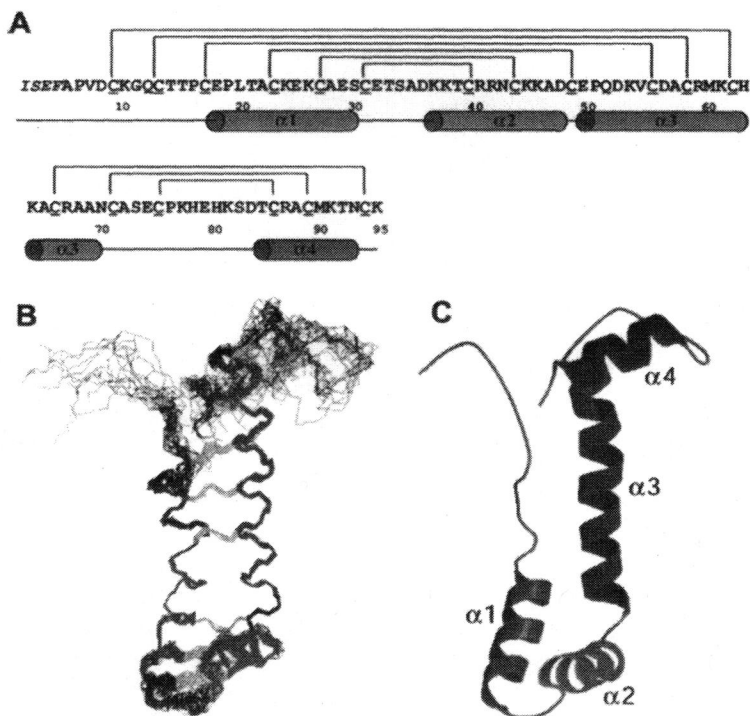

Figure 4. Structure of Vanabin2. (A) Amino acid sequence of Vanabin2. The amino-terminal tag is italicized. The disulfide bond pairings, determined by the CYANA calculation, are indicated at the top of the sequence. The secondary structure elements of Vanabin2 are indicated at the bottom of the sequence and are colored correspondingly in all panels. (B) Final 10 structures superposed over the backbone heavy atoms of residues 18-70. The side chains of the half-cystine residues are shown as yellow lines. (C) Ribbon representation of a single structure, in the same orientation as in panel B. Reproduced from Reference 42. Copyright 2005 American Chemical Society.
(See page 2 of color inserts.)

Vanabin2 under acidic conditions, as determined by circular dichroism spectroscopy, and little variation in the dissociation constant for V(IV) ions was observed in the pH range 4.5-7.5, suggesting that the binding state of the ligands is not affected by acidification (Table II). Taken together, these results suggest that the reason for metal ion dissociation upon acidification is attributable not to a change in secondary structure but, rather, that it is caused by protonation of the amino acid ligands that complex with V(IV) ions (*43*). Site-directed mutagenesis of Vanabin2 may elucidate the contribution of each amino acid residue coordinating V(IV) or cysteine residues that are responsible for maintaining the overall structure of Vanabin2.

Table II. Metal ion binding ability of Vanabin2

	Maximum binding number	*Kd[M]*
pH4.5 VO^{2+}	4.9	9.3×10^{-5}
pH7.5 VO^{2+}	20.2	2.3×10^{-5}
pH4.5 Cu^{2+}	4.5	7.0×10^{-5}
pH7.5 Cu^{2+}	3.2	6.4×10^{-5}

SOURCE: Reproduced from Reference *43*. Copyright 2006 Elsevier.

Vanabin Homologue and Interactive Proteins

The process of vanadium accumulation in ascidians has not yet been elucidated. As mentioned above, we have isolated and cloned cDNA of a novel vanadium-binding protein, designated as VanabinP, from the blood plasma of *A. sydneiensis samea* (*38*). The predicted amino acid sequence of VanabinP was highly conserved and similar to those of other Vanabins. The N-terminus of the mature form of VanabinP was rich in basic amino acid residues. VanabinP cDNA was originally isolated from blood cells, as were the other four Vanabins. However, Western blot analysis revealed that the VanabinP protein was localized to the blood plasma and was not detectable in blood cells. RT-PCR analysis and *in situ* hybridization indicated that the VanabinP gene was transcribed in some cell types localized to peripheral connective tissues of the alimentary canal, muscle, blood cells, and a portion of the branchial sac. Recombinant VanabinP bound a maximum of 13 V(IV) ions per molecule with a *Kd* of 2.8×10^{-5} M. These results suggest that VanabinP is produced in several types of cell, including blood cells, and is immediately secreted into the blood plasma where it functions as a V(IV) carrier.

Metal binding proteins must interact with other proteins such as metal binding proteins, membrane metal transporters, membrane anchor proteins, or metal reducing/oxidizing proteins. Using two-hybrid analysis, some proteins that interact with Vanabins have obtained. Among them, VIP1, Vanabin-interacting protein 1, is a novel protein localized in vanadocytes having no homology with reported protein sequences and is under functional assay (Shintaku et al., unpublished data). We are also analyzing ceolomic proteins that were absorbed by V(IV)-IMAC column together with VanabinP. Those proteins are expected to be V(IV)-binding proteins or VanabinP-interacting proteins. The analysis of protein interaction network should illustrate the overall image of vanadium transport pathway.

Membrane Metal Transporters

Vanadium ions must enter the cell or vacuole by the help of some membrane metal transporters, anyhow as V(V) or V(IV). As candidates to transport V(IV) in vanadocytes, we focus on two common types of metal transporters, metal-ATPase and Nramp.

Metal-ATPase is a P1-type ATPase that transfers divalent cations across the cytoplasmic or organelle membranes. We have isolated a homologue of metal-ATPase from blood cells of *A. sydneiensis samea*. The ascidian metal-ATPase homologue possesses six metal binding motifs which is a common characteristics of mammalian Wilson and Menkes proteins. The six metal binding motifs have been revealed to bind vanadium and copper ions (unpublished data). Nramp (natural resistance-associated macrophage protein) family is known to transport several heavy metals including Fe^{2+}, Zn^{2+}, Mn^{2+}, Co^{2+}, Cd^{2+}, Cu^{2+}, Ni^{2+}, and Pb^{2+} and is highly conserved among mammals, nematodes, yeast, and bacteria (*44*). We isolated an ascidian Nramp homologue, named AsNramp, from the blood cells of the same ascidian species and showed that AsNramp was expressed in the vanadocytes exclusively by *in situ* hybridization. To examine whether AsNramp acts as a proton-coupled vanadium transporter in ascidians, we constructed a plasmid expressing a fusion protein of AsNramp and a green fluorescent protein (GFP) under the control of cytomegalovirus (CMV) promoter. We have preliminarily found that the cultured CHO-K1 cells overexpressing the fusion protein specifically uptake vanadium (unpublished data).

Low pH and Energetics of the Accumulation of Vanadium

Henze also reported that the homogenate of the blood cells was extremely acidic (*1, 45-47*). This unusual phenomenon has also attracted the interest of investigators because of the possible role of the highly acidic environment in changing or maintaining the redox potential. From microelectrode measurements of blood cell lysate, and non-invasive ESR measurements on intact cells under anaerobic conditions, we found a correlation between the concentration of V(III) ions and the pH within the vacuole (*12*). In *Ascidia gemmata*, which contains the highest concentration of vanadium (350 mM), the vacuoles have the lowest pH (1.86). Vacuoles of *A. ahodori* containing 60 mM vanadium have a pH of 2.67, and those of *A. sydneiensis samea* containing 13 mM vanadium have a pH of 4.20 (*12*). Comparison of pH values and levels of vanadium in the signet ring cells of three different species suggested that there might be a close correlation between a higher level of vanadium and lower pH, namely, a higher concentration of protons.

It is well known that vacuolar-type H^+-ATPases (V-ATPases) play a role in pH homeostasis in various intracellular organelles, including clathrin-coated vesicles, endosomes, lysosomes, Golgi-derived vesicles, multivesicular bodies and chromaffin granules that belong to the central vacuolar system (*48-50*). Immunocytological studies, using antibodies against subunits *A* and *B* of bovine V-ATPase show that V-ATPases are localized in the vacuolar membranes of vanadocytes (*51*). A specific V-ATPase inhibitor, bafilomycin A_1 (*52*), inhibits the proton pump in the vanadocyte vacuoles, neutralizing the vacuoles' contents (*51*). Therefore, V-ATPase functions to accumulate protons in the vanadocytes. One possible mechanism involves the extremely tight coupling of ATP hydrolysis and proton pumping by V-ATPase in the vanadocytes. As a first step to assess the possibility, we isolated and analyzed cDNA of subunits *A*, *B* and *C* of V-ATPase from the blood cells of *Ascidia sydneiensis samea* (*53-54*). By expressing the ascidian cDNA for subunit *C*, the pH-sensitive phenotype of the corresponding *vma5* mutant of a budding yeast was successfully rescued (*54*). Functional assay of more subunits and whole enzyme are necessary to bring out the actual function of V-ATPase in vanadium accumulation.

Sulfate in Vanadocytes

A considerable amount of sulfate has always been found in association with vanadium in ascidian blood cells (*21, 47, 55-68*), suggesting that sulfate might be involved in the biological function and/or the accumulation and reduction of vanadium. Frank *et al.* (*33*) suggested the existence of a non-sulfate sulfur compound, such as an aliphatic sulfonic acid, in ascidian blood cells.

As the first step towards an analysis of the possible correlation between the accumulation and/or reduction of vanadium and sulfate. Raman spectroscopy can be used to detect sulfate ion selectively in ascidian blood cells because sulfate ion gives a very intense Raman band at the diagnostic position, 983 cm^{-1}. We observed fairly good Raman spectrum of the blood cell lysate from *Ascidia gemmata*, which has the highest concentration of V(III) among ascidians (*69*). V(III) ions in the blood cells were converted to V(IV) ions by air-oxidation prior to Raman measurements so as to facilitate detection based on V=O stretching vibration. From analysis of the band intensities due to $V=O^{2+}$ and SO_4^{2-} ions, we estimated the content ratio of sulfate to vanadium to be approximately 1.5, as would be predicted if sulfate ions were present as the counter ions of V(III). We also found evidence that an aliphatic sulfonic acid was present in the blood cells (*69*). Carlson (*70*) reported a similar value of the content ratio for *Ascidia ceratodes*, but lower values were obtained by Bell *et al.* (*63*) and Frank *et al.* (*59*).

Recently, we have isolated ascidian homologues of sulfate transporter from ascidian blood cells. We introduced ascidian sulfate transporter gene in yeast mutant strain that lacks corresponding genes. The mutant strain which cannot grow on sulfur-limited medium became able to grow by expressing ascidian sulfate transporter (unpublished data). The growth and transport activity as well as its expression pattern are being studied in detail.

Glutathione-S-transferase

Few studies have examined the pathway of vanadium accumulation from seawater into the ascidian coelom. Recently, we isolated novel proteins with a striking homology to glutathione-S-transferases (GSTs), designated *As*GST-I and *As*GST-II, from the digestive system of the vanadium-accumulating ascidian *Ascidia sydneiensis samea*, in which the digestive system is thought to be involved in vanadium uptake (*71*). Analysis of recombinant *As*GST-I confirmed that *As*GST-I has GST activity and forms a dimer, as do other GSTs. In addition, *As*GST-I was revealed to have vanadium-binding activity, which has never been reported for GSTs isolated from other organisms. *As*GST-I bound about 16 vanadium atoms as either V(IV) or V(V) per dimer, and the apparent dissociation constants for V(IV) and V(V) were 1.8×10^{-4} M and 1.2×10^{-4} M, respectively. Western blot analysis revealed that *As*GSTs were expressed in the digestive system at exceptionally high levels, although they were localized in almost all organs and tissues examined (Figure 5).

GSTs are a superfamily of enzymes that utilize glutathione (GSH) in reactions contributing to the detoxification of a wide range of toxic compounds. GSTs are found in all eukaryotes and have various functions. Notwithstanding their versatility, GSTs have never been considered to be involved in metal homeostasis. Considering these results, we postulate that AsGSTs play important roles in vanadium accumulation in the ascidian digestive system. Based on the results of our experiments, it is likely that the main function of AsGSTs in the vanadium-accumulating process is vanadium-carrier in the digestive system. Vanadium ions taken up by ascidians via the digestive system might be captured

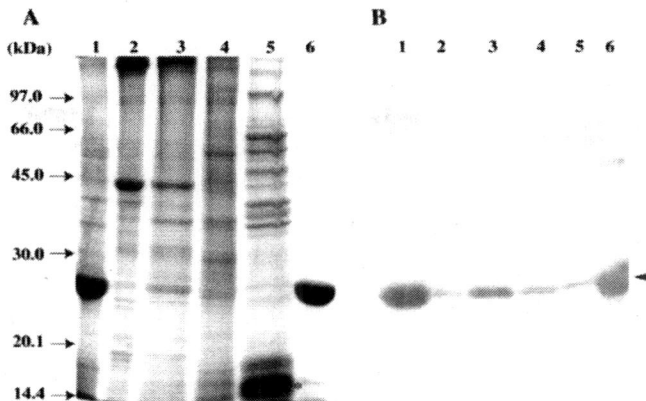

Figure 5. Localization of AsGSTs using antiserum to AsGST-I. Homogenates prepared from major organs or tissues of the ascidian A. sydneiensis samea were analyzed using (A) SDS-PAGE and (B) Western blotting using antiserum to AsGST-I. Purified AsGST-I recombinant protein was used as a positive control. Lane 1, digestive system; lane 2, body wall; lane 3, endostyle; lane 4, branchial sac; lane 5, blood cells; lane 6, purified AsGST-I recombinant protein. AsGSTs were detected in all specimens (arrowhead). The expression in the digestive system was exceptionally high and rivaled the level of Vanabins in blood cells (asterisk). Reproduced from Reference 71. Copyright 2006 Elsevier.

by AsGSTs, vanadium-binding proteins that are locally abundant. Since AsGSTs might be one of the first molecules involved in the influx of vanadium ions through the digestive system, these seem to be a quite important clue to elucidate the first step of the vanadium-accumulating process in ascidians.

Conclusion

Many proteins and genes that might be involved in the accumulation and reduction of vanadium have been now found out. Vanabin family, four types of enzyme related to the pentose phosphate pathway that produces NADPH, vacuolar-type H^+-ATPase (V-ATPase), metal-ATPase, Nramp, glutathione S-transferase and SO_4^{-2} transporter. Now, it is the time to consider how these proteins share their roles to accumulate vanadium in ascidians, to resolve the unusual phenomenon whereby some ascidians accumulate vanadium to levels more than ten million times higher than those in seawater. Attempts to characterize this phenomenon can be expected to promote more information about the unusual accumulation of vanadium by one class of marine organisms.

278

However, we have not yet obtained any clue to resolve the physiological roles of vanadium in ascidians.

References

1. Henze, M. *Hoppe-Seyler's Z. Physiol. Chem.* **1911**, *72*, 494-501.
2. Kustin, K. Perspectives on vanadium biochemistry. *Vanadium compounds, chemistry, biochemistry, and therapeutic applications*, Tracey, A. S.; Crans, D. C., Eds., Am. Chem. Soc. Syp. Ser., 711,1998; p 170-185.
3. Michibata, H. *Adv. Biophys.* **1993**, *29*, 103-131.
4. Michibata, H. *Zool. Sci.* **1996**, *13*, 489-502.
5. Michibata, H.; Kanamori, K. Selective accumulation of vanadium by ascidians from seawater. *Vanadium in Environment. Part 1: Chemistry and Biochemistry*, Nriagu, J. O., Ed., John Wiley & Sons, Inc., New York, 1998; p 217-249.
6. Michibata, H.; Sakurai, H.; Vanadium in ascidians. *Vanadium in biological systems*, Chasteen, N. D., Ed., Kluwer Acad. Publ., Dortrecht, 1990, p 153-171.
7. Michibata, H.; Uyama, T.; Kanamori, K. *Am. Chem. Soc. Symp. Ser.* **1998**, 248-258.
8. Michibata, H.; Uyama, T.; Ueki, T.; Kanamori, K. *Microsc. Res. Tech.* **2002**, *56*, 421-434.
9. Michibata, H.; Yamaguchi, N.; Uyama, T.; Ueki, T. *Coord. Chem. Rev.* **2003**, *237*, 41-51.
10. Michibata, H. *Comp. Biochem. Physiol.* **1984**, *78A*, 285-288.
11. Michibata, H.; Terada, T.; Anada, N.; Yamakawa, K.; Numakunai, T. *Biol. Bull.* **1986**, *171*, 672-681.
12. Michibata, H.; Iwata, Y.; Hirata, J. *J. Exp. Zool.* **1991**, 257, 306-313.
13. Webb, D. A. *J. Exp. Biol.* **1939**, *16*, 499-523.
14. Cole, P. C.; Eckert, J. M.; Williams, K. L. *Anal. Chim. Acta.* **1983**, *153*, 61-67.
15. Collier, R. W. *Nature* **1984**, *309*, 441-444.
16. Wright, R. K. Urochordata. *Invertebrate blood cells. vol. 2*, Ratcliffe, N. A.; Rowley, A. F., Ed., Acad. Press, London, 1981, p 565-626.
17. Michibata, H.; Hirata, J.; Uesaka, M.; Numakunai, T.; Sakurai, H. *J. Exp. Zool.* **1987**, *244*, 33-38.
18. Michibata, H.; Uyama, T.; Hirata, J. *Zool. Sci.* **1990**, *7*, 55-61.
19. Hirata, J.; Michibata, H. *J. Exp. Zool.* **1991**, *257*, 160-165.
20. Ueki, T.; Takemoto, K.; Fayard, B.; Salomé, M.; Yamamoto, A.; Kihara, H.; Susini, J.; Scippa, S.; Uyama, T.; Michibata, H. *Zool. Sci.* **2002**, *19*, 27-35.
21. Scippa, S.; Zierold, K.; M. de Vincentiis, M. *J. Submicroscop. Cytol. Pathol.* **1988**, *20*, 719-730.

22. Uyama, T.; Kinoshita, T.; Takahashi, H.; Satoh, N.; Kanamori, K.; Michibata, H. *J. Biochem.* **1998**, *124*, 377-382.
23. Uyama, T.; Yamamoto, K.; Kanamori, K.; Michibata, H. *Zool. Sci.* **1998**, *15*, 441-446.
24. Ueki, T.; Uyama, T.; Yamamoto, K.; Kanamori, K.; Michibata, H. *Biochim. Biophys. Acta* **2000**, *1494*, 83-90.
25. Erdmann, E.; Kraweitz, W.; Philipp, G.; Hackbarth, I.; Schmitz, W.; Scholtz, H.; Crane, F. L. *Nature* **1979**, *282*, 335-336.
26. Liochev, S. I.; Fridovich, I. *Arch. Biochem. Biophys.* **1990**, *279*, 1-7.
27. Shi, X.; Dalal, N. S. *Arch. Biochem. Biophys.* **1991**, *289*, 355-361.
28. Shi, X.; Dalal, N. S. *Arch. Biochem. Biophys.* **1993**, *302*, 300-303.
29. Kanamori, K.; Sakurai, M.; Kinoshita, T.; Uyama, T.; Ueki, T.; Michibata, H. *J. Inorg. Biochem.* **1999**, *77*, 157-161.
30. Kanamori, K.; Kinebuchi, Y.; Michibata, H. *Chem. Lett.* **1997**, 423-424.
31. Bruening, R. C.; Oltz, E. M.; Furukawa, J.; Nakanishi, K.; Kustin, K. *J. Am. Chem. Soc.* **1985**, *107*, 5298-5300.
32. Ryan, D. E.; Ghatlia, N. D.; McDermott, A. E.; Turro, N. J.; Nakanishi, K. *J. Am. Chem. Soc.* **1992**, *114*, 9659-9660.
33. Frank, P.; Hedman, B.; Carlson, R. K.; Tyson, T. A.; Row, A. L.; Hodgson, K. O. *Biochemistry* **1987**, *26*, 4975-4979.
34. Kanda, T.; Nose, Y.; Wuchiyama, J.; Uyama, T.; Moriyama, Y.; Michibata, H. *Zool. Sci.* **1997**, *14*, 37-42.
35. Ueki, T.; Adachi, T.; Kawano, S.; Aoshima, M.; Yamaguchi, N.; Kanamori, K.; Michibata, H. *Biochim. Biophys. Acta* **2003**, *1626*, 43-50.
36. Yamaguchi, N.; Togi, A.; Ueki, T.; Uyama, T.; Michibata, H. *Zool. Sci.* **2002**, *19*, 1001-1008.
37. Yamaguchi, N.; Kamino, K.; Ueki, T.; Michibata, H. *Mar. Biotechnol.* **2004**, *6*, 165-174.
38. Yoshihara, M.; Ueki, T.; Watanabe, T.; Yamaguchi, N.; Kamino, K.; Michibata, H. *Biochim. Biophys. Acta* **2005**, *1730*, 206-214.
39. Yamaguchi, N.; Amakawa, Y.; Yamada, H.; Ueki, T.; Michibata, H. *Zool. Sci.* **2006**, 23, in press.
40. Trivedi, S.; Ueki, T.; Yamaguchi, N.; Michibata, H. *Biochim. Biophys. Acta* **2003**, *1630*, 64-70.
41. Fukui, K.; Ueki, T.; Ohya, H.; Michibata, H. *J. Am. Chem. Soc.* **2003**, *125*, 6352-6353.
42. Hamada, T.; Asanuma, M.; Ueki, T.; Hayashi, F.; Kobayashi, N.; Yokoyama, S.; Michibata, H.; Hirota, H. *J. Am. Chem. Soc.* **2005**, *127*, 4216-4222.
43. Kawakami, N.; Ueki, T.; Matsuo, K.; Gekko, K.; Michibata, H. *Biochim. Biophys. Acta* **2006**, *1760*, 1096-1101.
44. Gunshin, H.; Mackenzie, B.; Berger, U. V.; Gunshin, Y.; Romero, M. F.; Boron, W. F.; Nussberger, S.; Gollan, J. L.; Hediger, M. A. *Nature* **1997**, *388*, 482-488.

45. Henze, M. *Hoppe-Seyler's Z. Physiol. Chem.* **1912**, *79*, 215-228.
46. Henze, M. *Hoppe-Seyler's Z. Physiol. Chem.* **1913**, *88*, 345-346.
47. Henze, M. *Hoppe-Seyler's Z. Physiol. Chem.* **1932**, *213*, 125-135.
48. Forgac, M. *Physiol. Rev.* **1989**, *69*, 765-796.
49. Forgac, M. *J. Exp. Biol.* **1992**, *172*, 155-169.
50. Nelson, N. *J. Exp. Biol.* **1992**, *172*, 19-27.
51. Uyama, T.; Moriyama, Y.; Futai, M.; Michibata, H. *J. Exp. Zool.* **1994**, *270*, 148-154.
52. Bowman, E . J.; Siebers, A.; Altendorf, K. *Proc. Natl. Acad. Sci. U.S.A.* **1988**, *85*, 7972-7976.
53. Ueki, T.; Uyama, T.; Kanamori, K.; Michibata, H. *Zool. Sci.* **1999**, *15*, 823-829.
54. Ueki, T.; Uyama, T.; Kanamori, K.; Michibata, H. *Mar. Biotechnol.* **2001**, *3*, 316-321.
55. Botte, L. S.; Scippa, S.; de Vincentiis, M. *Experientia* **1979**, *35*, 1228-1230.
56. Scippa, S.; Botte, L.; de Vincentiis, M. *Acta. Zool. (Stockh)* **1982**, *63* , 121-131.
57. Scippa, S.; Botte, L.; de Vincentiis, M. *Acta Embryol. Morphol. Exper. ns.* **1982**, *3*, 22-23.
58. Scippa, S.; Botte, L.; Zierold, K.; M. de Vincentiis, M. *Cell Tissue Res.* **1985**, *239*, 459-461.
59. Frank, P.; Carlson, R. M. K.; Hodgson, K. O. *Inorg. Chem.* **1986**, *25*, 470-478.
60. Califano, L.; Boeri, E. *J. Exp. Zool.* **1950**, *27*, 253-256.
61. Bielig, H.-J.; Bayer, E.; Califano, L.; Wirth, L. *Publ. Staz. Zool. Napoli* **1954**, *25*, 26-66.
62. Botte, L.; Scippa, S.; de Vincentiis, M. *Dev. Growth Differ.* **1979**, *21*, 483-491.
63. Bell, M. V.; Pirie, B. J. S.; McPhail, D. B.; Goodman, B. A.; Falk-Petersen, I.-B.; Sargent, J. R. *J. Mar. Biol. Ass. U.K.* **1982**, *62*, 709-716.
64. Pirie, B. J. S.; Bell, M. V. *J. Exp. Mar. Biol. Ecol.* **1984**, *74*, 187-194.
65. Lane, D. J. W.; Wilkes, S. L. *Acta Zool. (Stockh)* **1988**, *69*, 135-145.
66. Frank, P.; Hedman, B.; Carlson, R. M. K.; Hodgson, K. O. *Inorg. Chem.* **1994**, *33*, 3794-3803.
67. Frank, P.; Kustin, K.; Robinson, W. E.; Linebaugh, L.; Hodgson, K. O. *Inorg. Chem.* **1995**, *34*, 5942-5949.
68. Anderson, D. H.; Swinehart, J. H. *Comp. Biochem. Physiol.* **1991**, *99A*, 585-592.
69. Kanamori, K.; Michibata, H. *J. Mar. Biol. Ass. U.K.* **1994**, *74*, 279-286.
70. Carlson, R. M. K. *Proc. Natl. Acad. Sci. U.S.A.* **1975**, *72*, 2217-2221.
71. Yoshinaga, M.; Ueki, T.; Yamaguchi, N.; Kamino, K.; Michibata, H. *Biochim. Biophys. Acta* **2006**, *1760*, 499-503.

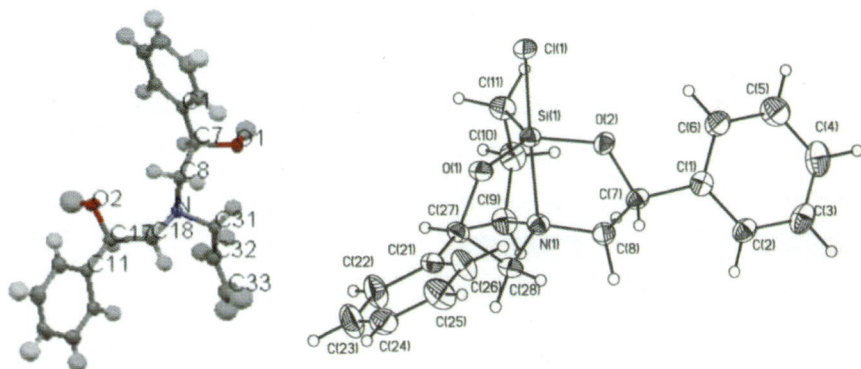

Figure 5.2. Molecular structures of R,R-H₂L_A (left) and R,R-ClSiL_A (right).

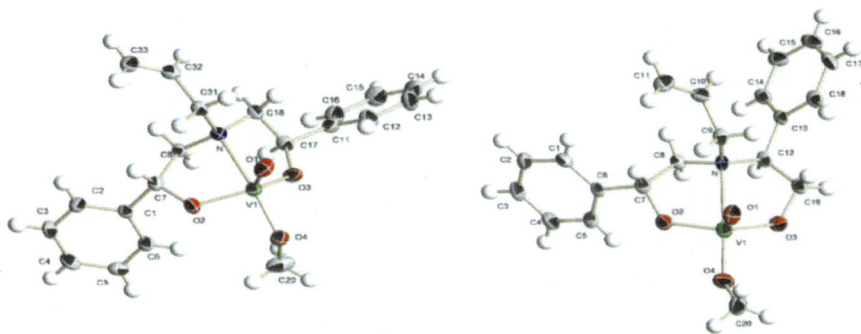

*Figure 5.3. Molecular structures of the complexes
[VO(OMe)L_A] (1, left) and [VO(OMe)L_B] (2).*

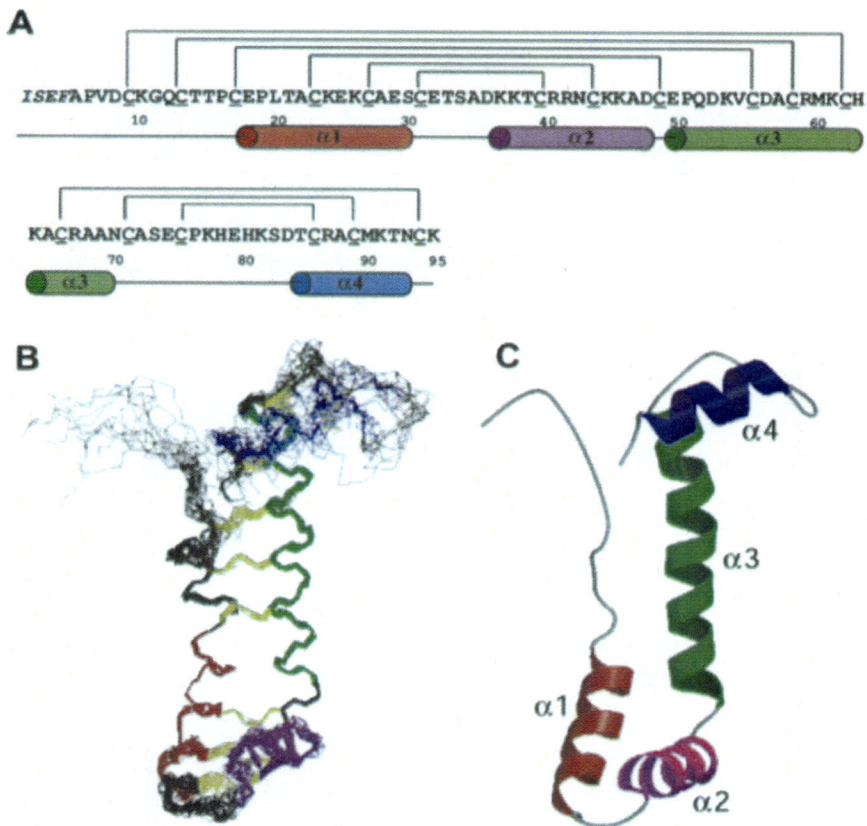

*Figure 19.4. Structure of Vanabin2. Reproduced from Reference 42.
Copyright 2005 American Chemical Society.
(See page 272 for the full caption.)*

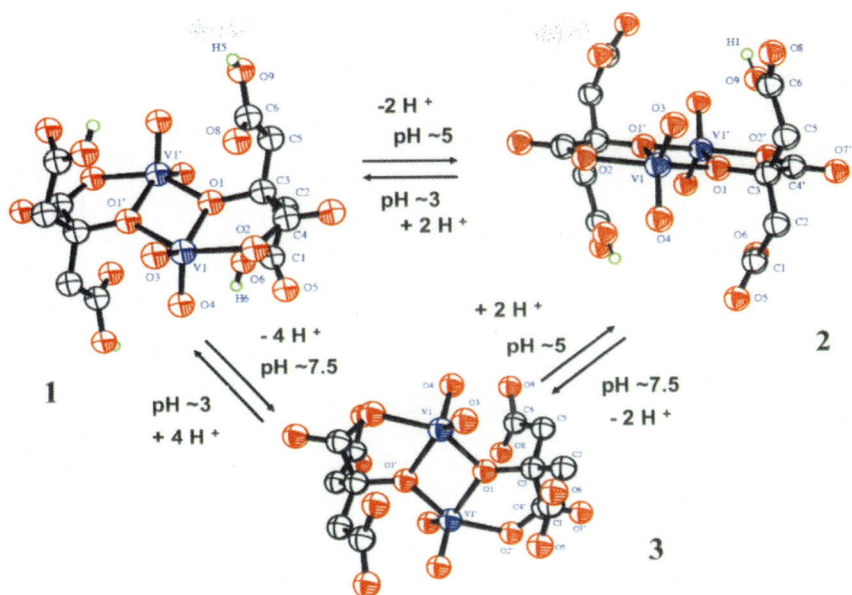

Figure 27.1. pH-structural variants of binary V(V)-citrate species and their pH-dependent acid-base chemistry.

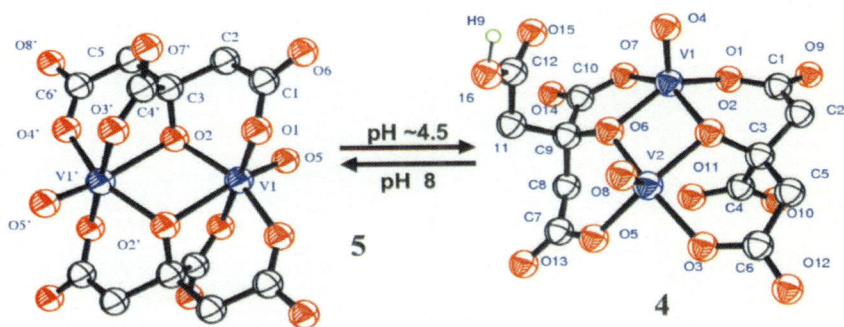

Figure 27.2. pH-Dependent acid-base chemistry between pH-structural variants of binary V(IV)-citrate species.

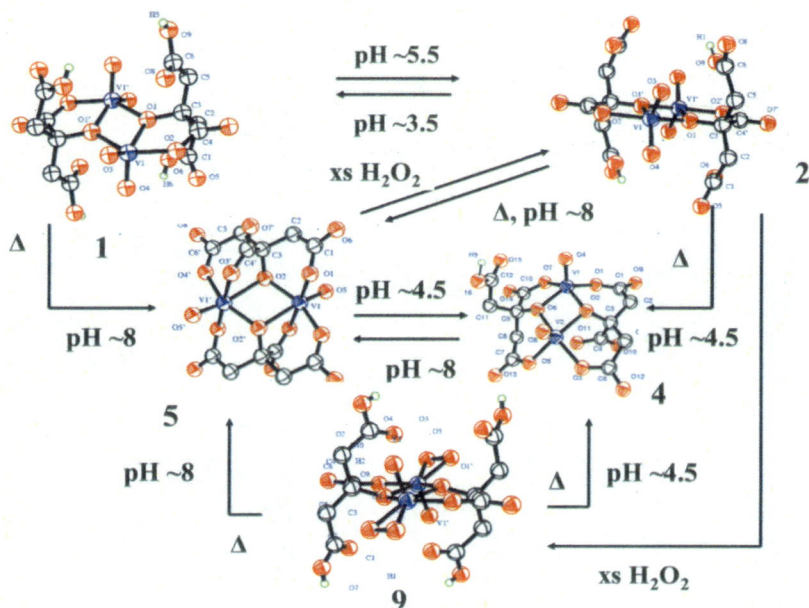

Figure 27.3. pH-Dependent acid-base and (non)-thermal transformations among pH-structural variants of binary and ternary V-(peroxo)-citrate species.

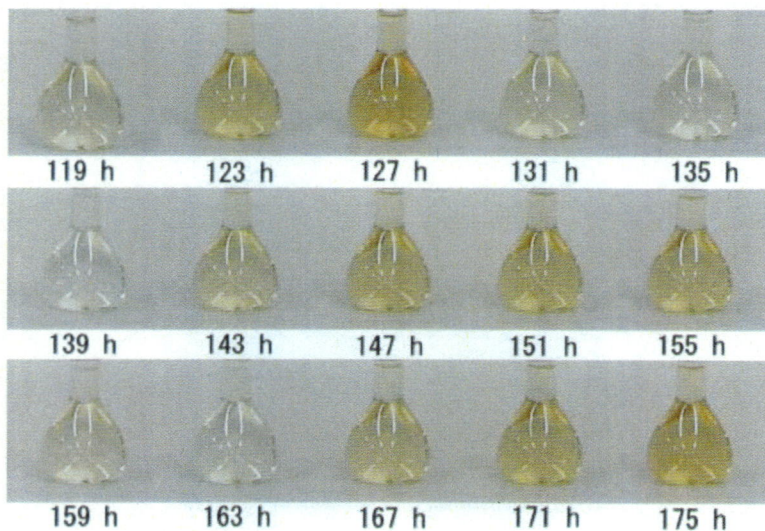

Figure 30.1. Video frames of the oscillating reaction: [V(IV)OCl₂(bpy)] (0.34 mM) in 20 ml of dichloromethane; stirring rate = 900 rpm.

Chapter 20

Toward the Biological Reduction Mechanism of Vanadyl Ion in the Blood Cells of Vanadium-Sequestering Tunicates

Patrick Frank[1,2,*], Elaine J. Carlson[3,5], Robert M. K. Carlson[4], Britt Hedman[1], and Keith O. Hodgson[1,2]

[1]Stanford Synchrotron Radiation Laboratory, SLAC, Stanford University, Stanford, CA 94309
[2]The Department of Chemistry, Stanford University, Stanford, CA 94305–5080
[3]The Buck Institute, Novato, CA 94945
[4]The Chevron Energy Technology Company, P.O. Box 1627, Richmond, CA 94802
[5]Current address: Department of Biochemistry and Biophysics, University of California, San Francisco, CA 94143–0984
[*]Corresponding author: telephone: 1-650-723-2479; fax: 1-650-723-4817; email: frank@ssrl.slac.stanford.edu

Nearly one hundred years after Henze reported high concentrations of vanadium and acid in some ascidians, the mechanism for the reduction of ambient V^{5+} to cellular V^{3+} remains unknown. We will report the results of x-ray absorption spectroscopic (XAS) measurements that queried the fate of vanadyl ion following uptake by living blood cells from the tunicate *Ascidia ceratodes*. These new results, in addition to previous results from XAS experiments and insights from the known inorganic chemistry of vanadium, will form the basis of a proposed mechanism for the biological reduction of vanadyl ion. The new field of vanadium redox-enzymology, long suspected but virtually undetected until now, has thus achieved infancy and awaits growth.

Introduction

The fascinating problems of the endogenesis and utility of the acid and reduced vanadium found in blood cells of Phlebobranch and Aplousobranch ascidians has been under study for almost 100 years (*1-11*). While much has been learned concerning the cellular locale and chemical status of endogenous vanadium (*4,11-16*), less is known about vanadium uptake (*17-19*) and subsequent intra-organismal transport, and virtually nothing is known about the mechanism of reduction of ingested marine vanadate. We describe here preliminary results from experiments potentially bearing on the biological vanadate-vanadic transformation in ascidians.

Materials and Methods

Ascidia ceratodes were collected from beneath the floating docks of the Spud Point Marina, Bodega Bay, California, USA. Whole blood was obtained by cardiac puncture as described previously (*11*). The whole blood cell incubation experiments were carried out in triplicate, with each series including an untreated control. The blood cells were kept on ice during incubation, and all the incubation reagents were prepared as solutions in 0.5 M NaCl with 25 mM phosphate pH 6.6 buffer. Following incubation, the whole blood cells were cleared of reagents using centrifugation (100×g, 5 min). To prepare samples for XAS analysis, treated and control whole blood cell pellets were suspended in the minimum volume of ice-cold 0.5 M NaCl in 30% aqueous ethylene glycol. Lexan XAS sample cells were then filled with the suspended blood cells, and the filled and capped Lexan cells were frozen in liquid nitrogen. Control experiments including cell counts showed that the incubation conditions did not materially affect cell viability. Photomicrographs taken of all nine blood cell samples did not reveal any obvious morphological changes from the incubation conditions. These results will be reported in detail elsewhere.

Blood cell vanadium K-edge XAS spectra were measured in fluorescence mode on SSRL SPEAR2 beam line 7-3 under dedicated operating conditions of 3 GeV, a wiggler field of 17 kG, and 70-98 mA of current, detuned 50% at 5861 eV. All XAS samples were held at 10 K using an Oxford Instruments CF1208 continuous flow liquid helium cryostat. Whole blood cell XAS spectra were calibrated against the first rising edge inflection of a vanadium foil standard (5464.0 eV), measured concurrently, and were processed as described in detail previously (*20,21*).

Fits to XAS spectral pre-edges were carried out using the program EDG_FIT, which is part of the EXAFSPAK data analysis program, written by Prof. Graham George, Department of Geological Sciences, University of Saskatchewan, and available free-of-charge at: http://www.ssrl.slac.Stanford.

edu/exafspak.html. The fits utilized pseudo-Voigt lines, fixed at a 1:1 numerical sum of a Gaussian and a Lorentzian. The half-widths of the Voigt lines were refined as a group and were not allowed to become larger than the core-hole lifetime of vanadium (1.01 eV; *22*) convolved with the spectrometer resolution (1.1 eV).

Results and Discussion

Vanadium in blood cells from *Phallusia nigra,* is divided among several biologically-bounded environments and two oxidation states. Fits to the vanadium K-edge XAS of *P. nigra* whole blood cells in terms of model complexes (*23*), discerned a V(III) to $V^{IV}O^{2+}$ ratio of about 2:1. The V(III), in turn, was, on average, divided almost equally between the free aqua ion, $[V(H_2O)_6]^{3+}$ and complexed forms of V(III) (see below). The aqua ion clearly cannot exist as such in the absence of strong acid. In *P. nigra*, the acid present did not produce any intracellular $[V(SO_4)_n]^{(3-2n)}$ complex ions, which represents almost a categorical distinction from the V(III) present in the acidic vacuole within blood cells of *Ascidia ceratodes* (*11*). The complexed V(III), also unique thus far to *P. nigra*, was found to be similar to the air-sensitive $K_3[V(catecholate)_3]$ and $[V(acac)_3]$ complexes, further indicating an unexpected biological ability to frustrate a strong oxidative gradient against atmospheric oxygen.

In the course of studying blood-cell vanadium in *P. nigra*, it was found that treating the blood cells with dithiothreitol (DTT) induced the appearance of a unique endogenous vanadium complex, notably different from those described above. During fits to whole blood cell vanadium XAS spectra, this complex could be modeled as $Na[V(edta)(H_2O)]$. Systematic tests in the course of these fits showed that the 7-coordinate edta complex was distinctly favored over otherwise similar 6-coordinate complexes, such as $Na[V(trdta)]$ (*24*). The bar-graph in Figure 1 compares the results of these fits in terms of component percents. The need for a $Na[V(edta)(H_2O)]$ model in the fits was restricted to those cells that were treated with DTT, which, in the context of metabolic theory, implies the induction of a new endogenous complex.

These results become more significant in light of aqueous inorganic chemistry of vanadium reported in 1997 (*25*), namely, that the methyl ester of cysteine will reduce $[VO(edta)]^{2-}$ to produce the 7-coordinate $[V(edta)(H_2O)]^-$ complex, but will not react with the structural homologue $[VO(trdta)]^{2-}$. In addition, NADPH will not reduce $[VO(edta)]^{2-}$ to the V(III) complex (*26*) despite providing a somewhat stronger electromotive driving force than cysteine (*27,28*).

The specificity of the vanadyl-to-vanadic reduction to the edta ligand can be explained at least in part by the 10^8-fold increase in the stability constant

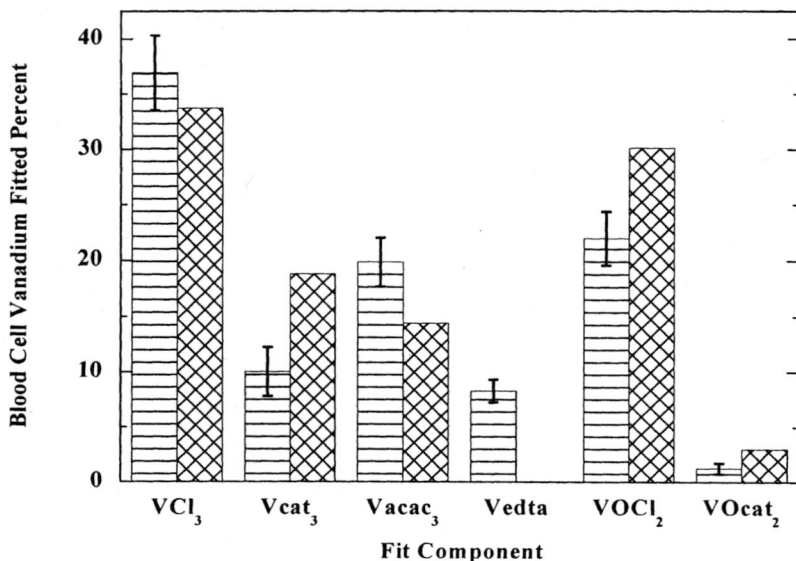

Figure 1. Bar-graph representing fits to the K-edge XAS spectra of whole blood cells from Phallusia nigra. Lined bars: Component percents averaging four fits to blood cell samples treated with DTT, weighted according to number of specimens contributing to each sample. The error bars show the standard variation calculated as $\sqrt{[\Sigma(avg.-sample)^2]/3}$. Cross-hatched bars: Component percents for the fit representing the control blood cell sample without DTT-treatment. The components are: VCl₃, 0.1 M VCl₃ in 1 M HCl; Vcat₃, K₃[V(catecholate)₃]; Vacac₃, V(acetylacetonate)₃; Vedta, Na[V(edta)(H₂O)]; VOCl₂, 0.1 M VOCl₂ in 1 M HCl; VOcat₃, K₂[VO(catecholate)₂]

governing the reaction of V(III) with edta relative to that associated with trdta (29). We looked for the source of this difference in terms of 3d-level electronic structures, and the vanadium K-edge XAS spectra of solid Na[V(edta)(H₂O)], Na[V(trdta)], and [V(picolinate)₃] are shown in Figure 2. The latter complex was chosen as a pseudo-octahedral reference (30).

These data show that the rising edge of the edta complex is significantly higher in energy than those of the two 6-coordinate complexes. The maximum of the first major XAS absorption band, marked in the Figure, is at 5481.8 eV for the edta complex, compared to 5481.1 eV for the trdta complex. This difference likely indicates a higher effective nuclear charge for V(III) in the edta complex. That is, the energies of the unoccupied p-symmetry virtual orbitals of 3d transition metal ions are relatively insensitive to minor changes in bonding. Shifts in the 1s→(n>3)p transition energies in the XAS spectra of the 3d

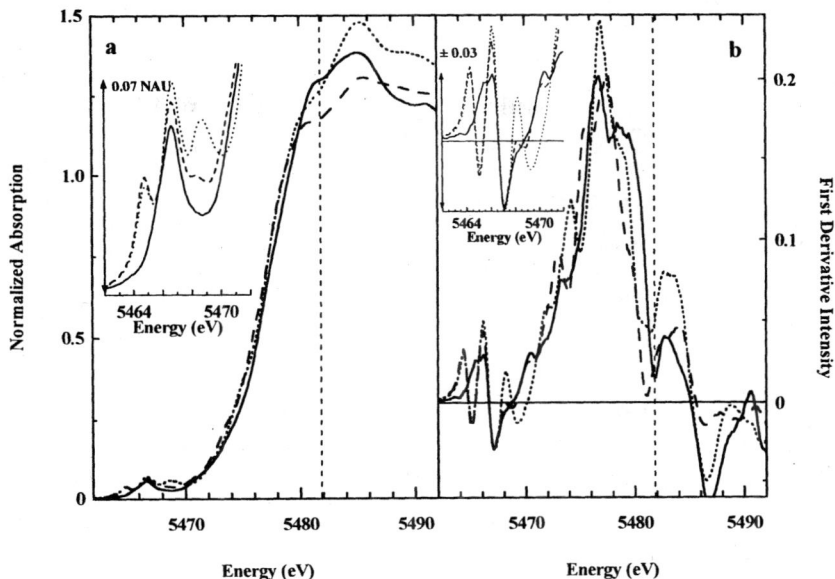

Figure 2. Vanadium K-edge XAS spectra of: a. (—) Na[V(edta)(H₂O)]; (− −),
Na[V(trdta), and; (···), V(picolinate)₃ measured as the solids finely ground in
BN. Inset: Expansion of the pre-edge energy region. Part b: First derivatives of
the K-edge XAS spectra of part a. Inset b: Expansion of the pre-edge energy
region. The dashed vertical line marks the first rising edge transition maximum
of the edta complex.

transition metals are then primarily indicative of changes in the 1s level energy
due to greater or lesser nuclear shielding by valence electrons.

Figure 2, inset shows the remarkable change that occurs in the pre-edge
energy region when the V(III) coordination number changes from 6 to 7. This
energy region of the XAS spectrum reflects the 1s→3d electronic transitions.
For the pseudo-octahedral complex [V(picolinate)]3, three clearly resolved
transitions are seen. These reflect the valence-level $3d^3$ excited states $^4A_2$, $^2E_2$,
and $^4T_2$ that are consistent with near-octahedral symmetry. The same three
transitions are observed for the [V(trdta)]⁻ complex although the intensity of the
$^4T_2$ transition near 5468.4 eV has significantly diminished. For the 7-coordinate
[V(edta)(H₂O)]⁻ complex, however, the three 1s→3d transitions have apparently
coalesced into a single feature at 5466.6 eV.

In order to pursue these effects, the pre-edge absorption envelopes for the
three complexes were fit with pseudo-Voigt lines. The results of these fits are
gathered in Table 1, and shown in Figure 3 for the two diaminetetraacetate
complexes.

Table 1. Final State Energies of Pre-Edge XAS Transitions[a]

Final State[b]	V(pic)₃)	Area[c]	Na[V(trdta)]	Area[c]	Final State[b]	Na[V(edta)(H₂O)]	Area[c]
$^4A_2$	5464.8	0.0523	5464.7	0.0635	$^2E_1$	5465.5	0.0227
$^2E_2$	5466.7	0.1187	5466.7	0.1136	$^4A_1$	5466.7	0.0949
$^4T_2$	5468.5	0.0700	5468.4	0.0446	$^4E_2$	5468.3	0.0225
Center[d]	5466.8		5466.5			5466.8	

[a] XAS spectra are shown in Figure 2. [b] The final state assignments are taken from the d^3 Tanabe-Sugano diagram assuming a moderate ligand field. [c] Integrated pseudo-Voigt areas in normalized units. [d] The center of gravity of the transition energies was calculated as the intensity-ratioed sum of the individual transition energies.

All three complexes show the three XAS valence transitions predicted for V(III), Table 1, and the energy positions of the transitions vary only slightly. However, the relative intensities show significant variation. In particular, the two flanking transitions in the pre-edge XAS of the 7-coordinate [V(edta)(H₂O)]⁻ complex exhibit severely diminished intensities, with no compensating intensity increase in the transition at 5466.7 eV. For the 6-coordinate [V(trdta)]⁻ complex, a reduced intensity relative to V(picolinate)₃ was found at 5468.4 eV, only.

The ground state configuration of 7-coordinate $3d^2$ V(III) is likely analogous to that depicted for the frontier β-magnetic orbitals of 7-coordinate Co(II) (31), which include a set of two unpaired $3d^1_{xz}$, $3d^1_{yz}$ spins embedded in a spherical shell of 5 electrons of α-spin. The 1s→3d transitions in the XAS pre-edge energy region of all three V(III) complexes include $3d^3$ excited states. The spin-pairing energy in $3d^3$ transition metals ($\sim$7000 cm⁻¹, $\sim$0.9 eV; 32) is much smaller than the 10Dq splitting of octahedral V(III) ($\sim$18,000 cm⁻¹; 33,34). Thus, the first 1s→3d XAS excited state of octahedral or trigonal prismatic V(III) should be the $^4A_2$ state, having a $3d^1_{xz}$, $3d^1_{yz}$, $3d^1_{xy}$ electronic configuration. In pentagonal bipyramidal 7-coordination, the $^3t_{2g}$-^0e_g octahedral ground state splits further into e_1, a_1, and e_2 states (35). Therefore, in the 7-coordinate edta complex the $^3E_1$ ground state orbitals consist of only the $3d^1_{xz}$, $3d^1_{yz}$ pair. The d_{z^2} orbital is about 1 eV higher in energy than the this pair (34), i.e., more than the spin-pairing energy. The highest energy E_2 state is primarily a mixture of the d_{xy} and $d_{x^2-y^2}$ orbitals. The first 1s→3d XAS excited state of 7-coordinate V(III) should then be the $^2E_1$ state, with a $3d^3_{xz,yz}$ electronic configuration. If this tentative assignment is correct, then the three pre-edge transitions represent $^4A_2$, $^2E_2$, $^4T_2$ states for the 6-coordinate complexes, but $^2E_1$, $^4A_1$, $^4E_2$ states for the 7-coordinate edta complex (Table 1). Thus, the more intense transition at 5466.7 eV represents the $^2E_2$ ($d^3_{xz,yz}$) excited state in the 6-coordinate complexes, but the $^4A_1$ (d^1_{xz}, d^1_{yz}, $d^1_{z^2}$) excited state in the 7-coordinate edta complex (34). The only pre-edge transition that maintains its intensity following 7-coordination then uniformly reflects the ^4A excited state in which the 3d-electronic configuration is both high-spin and π-symmetry.

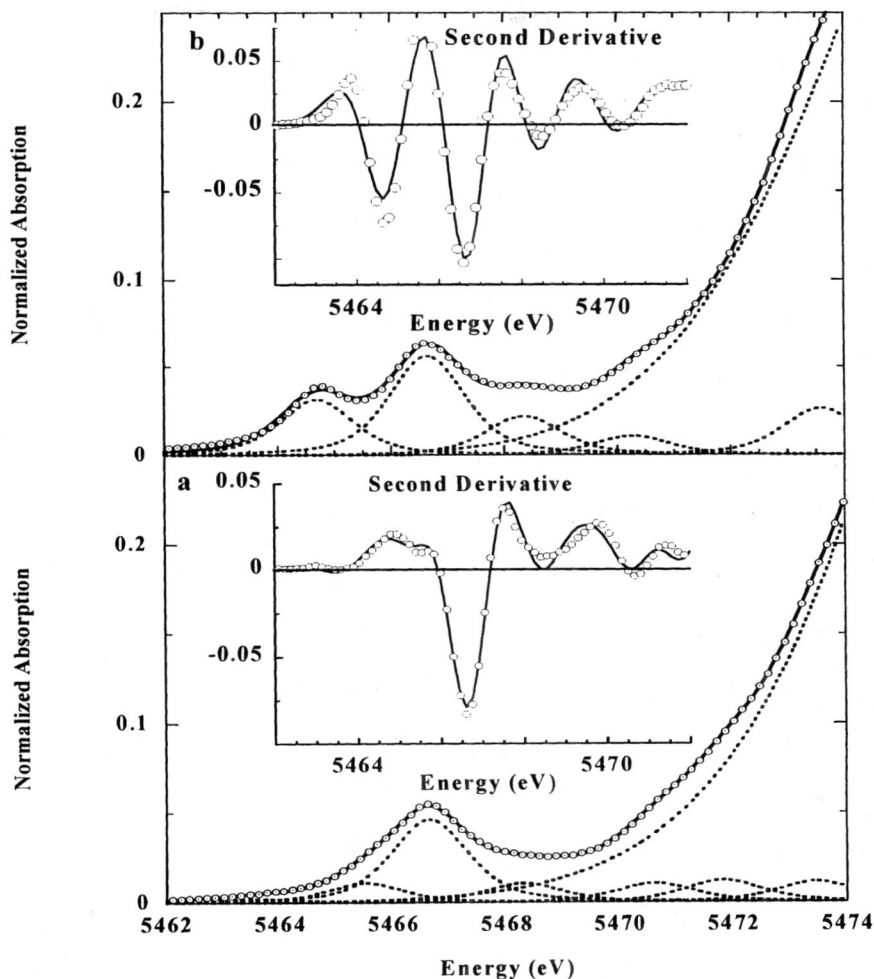

Figure 3. Pseudo-Voigt fit to the vanadium pre-K-edge energy region of the XAS spectrum of: a. Na[V(edta)(H$_2$O)], and; b. Na[V(trdta)]. (o), data points; (—), the fit; (···), the pseudo-Voigt fit components. Insets: the second derivatives of the data and the fits.

The crystal structures of all three complexes have been solved (*24, 30, 36, 37*), and the first shell bond distances are reproduced in Table 2. These data show that the equatorial bond lengths in the 7-coordinate edta complex are, with one exception, ~0.06-0.10 Å longer than those in the two 6-coordinate complexes.

Table 2. First–Shell Bond Lengths (Å) for V(pic)$_3$^{a}, Na[V(trdta)]^b and Na(V(edta)(H$_2$O)]^c

First Shell Atom	V(pic)$_3$	Na[V(trdta)]	Na[V(edta)(H$_2$O)]
N$_{eq}$	2.122	2.145	2.225
N$_{eq}$	2.112	2.141	2.217
O$_{eq}$	1.966	2.011	2.070
O$_{eq}$	1.944	1.986	2.013
(N,O)$_{ax}$	1.936	1.963	2.069
O$_{ax}$	2.153	1.951	2.057
(H$_2$O)$_{eq}$	---	---	2.066

^{a} reference 30; ^{b} reference 24 ^{c} reference 37

Further, the data in Table 1 show that the ^{2}E excited state of the edta complex is 1.2 eV lower in energy than the same excited states of the two 6-coordinate complexes. Likewise the ^{4}A state is 1.9 eV higher in energy in the 7-coordinate edta complex than it is in the 6-coordinate complexes. As noted above, the ~0.8 eV shift to higher energy for the rising K-edge XAS maximum of the edta complex relative to the trdta complex implies a relatively less-shielded V(III) 1s orbital in the former, consistent with the longer bond-lengths. That is, the V(III) 1s orbital in the edta complex is about 0.8 eV lower in energy than the 1s orbital of V(III) in the trdta complex. If the d-orbital manifolds were unaffected by the ligation differences, the difference in 1s orbital energy predicts that the 1s→3d transitions of the edta complex should occur about 0.8eV higher in energy than those of the trdta complex. However, the transition center-of-gravity energy difference of the 3d orbitals is only about +0.3 eV (Table 1). If the energy difference is maintained in the ground state, this result implies that the 3d-orbitals of the edta complex are, on average, ~0.5 eV *lower* in energy than those of the 6-coordinate trdta complex; again consistent with the longer bond lengths in the former complex. This energy difference predicts that the 7-coordinate edta complex should be 11.5 kcal more stable than the 6-coordinate trdta complex. If this energy difference appears in the free energy of complexation, it would produce an increase in binding constant of $\Delta K=2.2\times10^8$ M^{-1} favoring the edta complex over the trdta complex. This predicted difference is almost identical to the observed difference (*38*), noted above. This result has bearing on blood cell vanadium biochemistry, as will be shown below.

Figure 4 shows the XAS spectrum of *A. ceratodes* blood cells prior to incubation with vanadyl ion and dithiothreitol (see Materials and Methods section), compared with a numerical composite that includes 5% vanadyl sulfate plus 95% V(III) sulfate in pH 1.8 sulfuric acid. The energy positions and intensities of the small pre-edge features are known to be sensitive to pH, to sulfate ligation, and to the presence of vanadyl ion. The first derivative spectra in Figure 4b, inset, show that the intensity maximums of the pre-edge features of the numerical spectrum are well within 0.1 eV of those of the blood cell spectrum. The position and shape of the first derivative inflection feature at 2476 eV follows the presence of the $[VSO_4]^+$ complex ion (21). This complex is consistently found endogenous to *A. ceratodes* blood cells, and the numerically derived spectrum again properly reproduces this feature. The numerical XAS spectrum thus is a reasonable facsimile of the blood cell spectrum, and reflects a composition found to typify *A. ceratodes* blood cell vanadium (11).

The *A. ceratodes* blood cells represented by the spectrum of Figure 4 were then incubated for 30 minutes with vanadyl ion and DTT. These experiments were carried out to test the generality of the effect noted to occur on incubation of *P. nigra* blood cells with DTT, as discussed above (cf. Figure 1). The vanadium K-edge XAS spectrum of the blood cells washed free of the incubation reagents was then measured and is shown in Figure 5. Comparison with the blood cell spectrum of Figure 4 immediately reveals the presence of endocytosed vanadyl ion, visible in the pre-edge energy region near 5469 eV. Charged aqua ions cannot diffuse through cell membranes, and so some specific cellular transport process must have taken up the vanadyl ion newly found in the blood cells. The remainder of the spectrum shows little further change, reflecting the continued intracellular primacy of V(III) in sulfuric acid solution, as also indicated, e.g., by the first derivative inflection at 5476 eV in Figure 5b.

A close examination of Figure 5a inset, however, shows that in contrast to the case illustrated in Figure 4a, aqua vanadyl ion does not reproduce the features attending the newly uptaken vanadyl ion. The intensity maximum is shifted 0.4 eV to lower energy, which is consistent with Complexation (15, 23). The first derivative spectra, Figure 5b, inset, show a similar energy shift. Here, the blood cell vanadyl ion feature also shows evidence of an increased line-width relative to the aqua ion in the numerical spectrum, possibly indicating multiple endogenous vanadyl species.

As noted in Figure 1 and the associated discussion above, *P. nigra* blood cells are rich in aqua vanadyl ion, in contrast with the case for *A. ceratodes*. Incubating *P. nigra* blood cells with DTT alone produced a new form of endogenous vanadium. Fits to the K-edge XAS spectrum indicated a new blood cell component similar to $Na[V(edta)(H_2O)]$ (23). In Figure 6, a comparison is made between the XAS spectra of *P. nigra* whole blood cells that were treated with DTT, and that of *A. ceratodes* after treatment with DTT and VO^{2+}. The total blood cell VO^{2+} in the given *P. nigra* sample proved to be 19% (23), which is comparable with the estimated ~16% VO^{2+} in the *A. ceratodes* blood cell sample, following incubation. The overall shapes of the two spectra are similar (Figure 6a) and reflect the intracellular predominance of V(III). Nevertheless, comparison of the first derivative XAS spectra at 5476 eV reveals the absence of the endogenous $[VSO_4]^+$

290

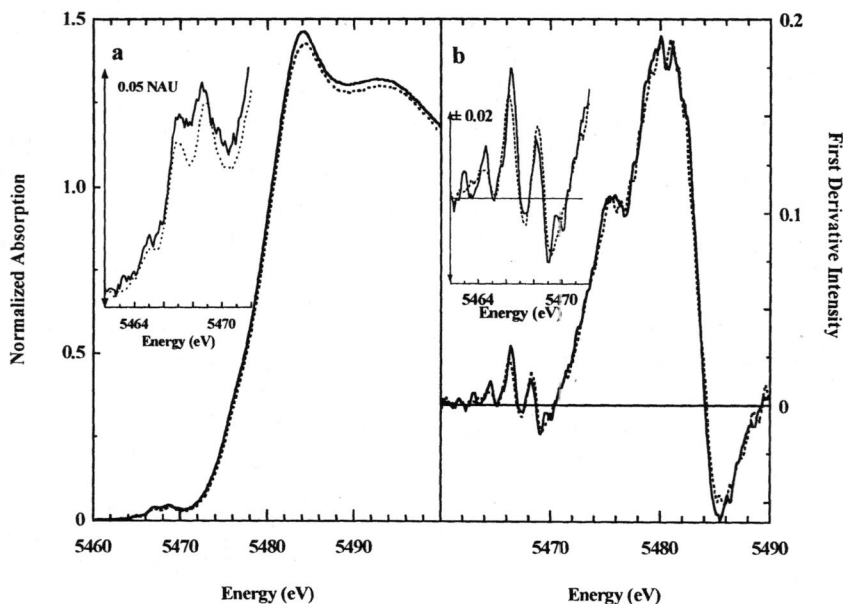

Figure 4. Vanadium K-edge XAS spectra of: a. (—), untreated whole blood cells of A. ceratodes, and; (···), a numerical sum consisting of 0.95×the vanadium K-edge spectrum of 0.1 M $V_2(SO_4)_3$ in pH 1.8 solution plus 0.05×the vanadium K-edge spectrum of 0.1 M $VOSO_4$ in 0.1 M H_2SO_4 solution. Inset: Expansion of the pre-edge energy region. Part b shows the first derivatives of the K-edge XAS spectra of part a.

complex ion from *P. nigra* blood cells is as complete as its presence is obvious in blood cells from *A. ceratodes*. This result has been invariably reproduced and can serve to differentiate these ascidian species.

Comparison of the features in the pre-edge energy region is particularly revealing. The coincidence in shape and energy position of the main pre-edge features is now almost exact. This is in contrast with the results from comparison in Figure 5 with the XAS spectrum reflecting aqua vanadyl ion plus aqua V(III)-sulfate. This similarity is especially evident in the first derivatives of the pre-edge energy region shown in Figure 6b, inset. The coincidence of the absorption maxima is evident where the first derivatives pass through the zero line. Likewise the similarity in line-widths is evident in the peak-to-trough separations of the major features. These results indicate that uptake of vanadyl ion, in the presence of DTT, has produced a new vanadium environment in *A. ceratodes* blood cells. This environment must be similar in many respects to that of the new vanadium environment in *P. nigra* blood cells, as also induced by DTT. A

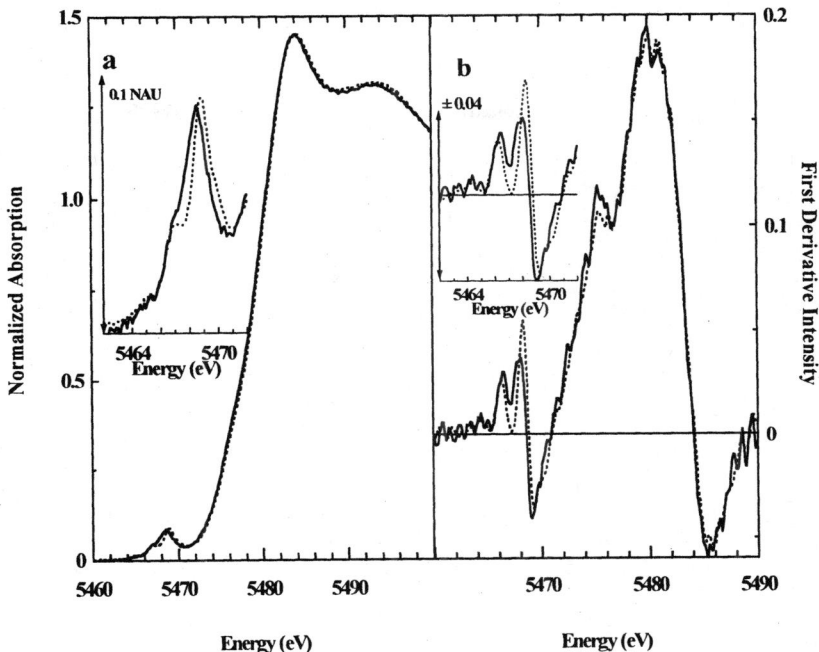

Figure 5. Vanadium K-edge XAS spectra of: a. (—), whole blood cells of A. ceratodes after treatment with DTT and VO^{2+} ion (see text for details), and; (···), a numerical sum consisting of 0.84×the vanadium K-edge spectrum of 0.1 M $V_2(SO_4)_3$ in pH 1.8 H_2SO_4 solution plus 0.16×the vanadium K-edge spectrum of 0.1 M $VOSO_4$ in 0.1 M H_2SO_4 solution. Inset: Expansion of the pre-edge energy region. Part b shows the first derivatives of the K-edge XAS spectra of part a.

qualitative judgment is thus warranted, namely that DTT plus VO^{2+} produced a vanadium environment in *A. ceratodes* blood cells similar to that of an edta-like ligand.

The likely recurrent induction of edta-like vanadium environments in the blood cells of at least two ascidians by DTT is suggestive of a connection between the biological chemistry of vanadium reduction and relevant inorganic chemistry of vanadyl ion. That is, at neutral pH the edta complex of vanadyl ion can be reduced to $[V(edta)(H_2O)]^-$ by the methyl ester of cysteine (*25*), but not by NADPH (*26*). This connection is reinforced by the known high concentration of sulfur-containing metabolites in *A. ceratodes* whole blood cells, including disulfide (*39*). The distinction with NADPH implies the necessity of an inner-sphere mechanism of reduction, in view of the relative E_o's of the two reagents (-0.245 V and –0.324 V, resp.; *40, 41*).

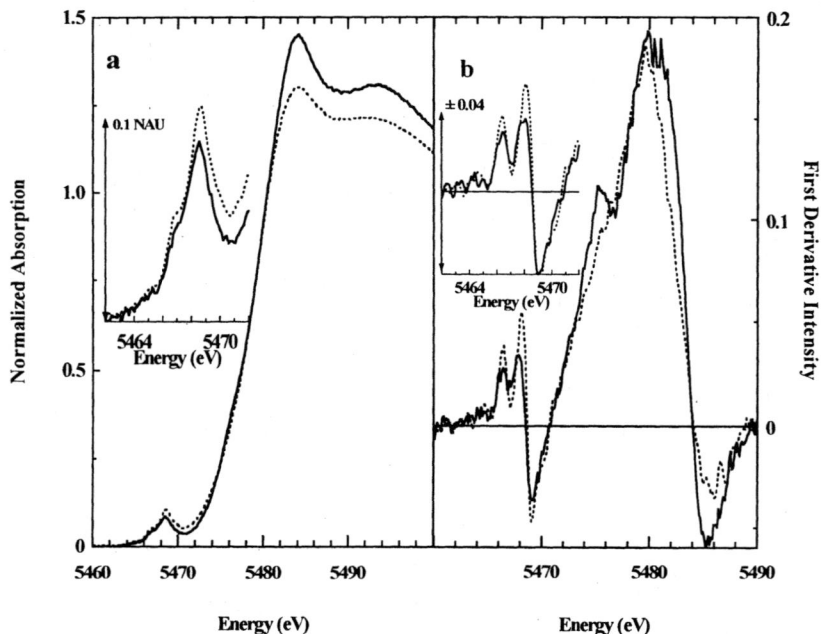

Figure 6. a. Vanadium K-edge XAS spectra of: (—), A. ceratodes whole blood cells following treatment with vanadyl ion and DTT, as also shown in Figure 5, and; (···), P. nigra whole blood cells following treatment with DTT alone. b. First derivatives of the XAS spectra shown in part a. The insets show an expansion of the pre-edge energy region of the spectra.

The appearance of an edta-like vanadium environment when thiol is endocytosed is suggestive of an induced enzymatic redox system. This enzyme may have an active site that reproduces, in some explicitly molecular fashion, the 7-coordinate chelating array of edta. A specific advantage of this array derives from the 10^8 differential in binding constants for V(III) that favors edta over trdta (29). That is, a site that accommodates 7-coordination, as opposed to 6-coordination, translates to a 470 mV differential in EMF driving the reduction of VO^{2+} to V(III) (29, 38, 42). The results described here, therefore, are sufficient to sustain a hypothesis that the blood cells of Phlebobranch ascidians contain a vanadium reductase with an aminocarboxylate-like active site. This site will provide five ligands to VO^{2+} but a total of seven ligands to V(III), one of which will likely be water. This water may derive from the oxyl-ligand of vanadyl ion. The VO^{2+} to V(III) transformation will be accompanied by minimal amino acid side-chain motions. The enzyme active site will also include, or have nearby, a thiol binding site. The isolation of this enzyme will open a whole new field of redox metallobiochemistry; that of vanadium.

Acknowledgments

PF thanks Ms. Ritimukta Sarangi for helpful conversations. This work was supported by grant NIH RR-01209 (to KOH). XAS data were collected at SSRL, which is supported by the Department of Energy, Office of Basic Energy Sciences, Divisions of Chemical and Materials Sciences. The SSRL Structural Molecular Biology Program is supported by the National Institutes of Health, National Center for Research Resources, Biomedical Technology Program and by the Department of Energy, Office of Biological and Environmental Research.

References

1. Henze, M. *Hoppe-Seyler's Z. Physiol. Chem.* **1911**, *72*, 494-501.
2. Henze, M.; Stohr, R.; Muller, R. *Hoppe-Seyler's Z. Physiol. Chem.* **1932**, *213*, 125-135.
3. Boeri, E.; Ehrenberg, A. *Arch. Biochem. Biophys.* **1954**, *50*, 404-416.
4. Carlson, R. M. K. *Proc. Natl. Acad. Sci. USA* **1975**, *72*, 2217-2221.
5. Macara, I. G.; McLeod, G. C.; Kustin, K. *Comp. Biochem. Physiol.* **1979**, *63B*, 299-302.
6. Tullius, T. D.; Gillum, W. O.; Carlson, R. M. K.; Hodgson, K. O. *J. Am. Chem. Soc.* **1980**, *102*, 5670-5676.
7. Frank, P.; Carlson, R. M. K.; Hodgson, K. O. *Inorg. Chem.* **1986**, *25*, 470-478.
8. Brand, S. G.; Hawkins, C. J.; Parry, D. L. *Inorg. Chem.* **1987**, *26*, 627-629.
9. Kustin, K.; Robinson, W. E.; Smith, M. J. *Invert. Reproduc. Devel.* **1990**, *17*, 129-139.
10. Michibata, H.; Uyama, T.; Ueki, T.; Kanamori, K. *Microsc. Res. Tech.* **2002**, *56*, 421-434.
11. Frank, P.; Carlson, R. M. K.; Carlson, E. J.; Hodgson, K. O. *J. Inorg. Biochem.* **2003**, *94*, 59-71.
12. Frank, P.; Hedman, B.; Carlson, R. M. K.; Tyson, T. A.; Roe, A. L.; Hodgson, K. O. *Biochemistry* **1987**, *26*, 4975-4979.
13. Oltz, E. M.; Pollack, S.; Delohery, T.; Smith, M. J.; Ojika, M.; Lee, S.; Kustin, K.; Nakanishi, K. *Experientia* **1989**, *45*, 186-190, and references therein.
14. Michibata, H.; Iwata, Y.; Hirata, J. *J. Exp. Zool.* **1991**, *257*, 306-313.
15. Frank, P.; Hodgson, K. O.; Kustin, K.; Robinson, W. E. *J. Biol. Chem.* **1998**, *38*, 24498-24503.
16. Takemoto, K.; Ueki, T.; Fayard, B.; Yamamoto, A.; Sasaki, H.; Salomé, M.; Susini, J.; Michibata, H.; Kihara, H. In *Portable Synchrotron Light Sources and Advanced Applications*; Yamada, H., Mochizuki-Oda, N.,

294

Sasaki, M., Eds.; American Institute of Physics: Shiga, Japan, 2004; Vol. 716, pp 65-68.

17. Rummel, W.; Bielig, H.-J.; Forth, W.; Pfleger, K.; Rüdiger, W.; Seifen, E. *Protides Biol. Fluid* **1966**, *14*, 205-210.

18. Dingley, A. L.; Kustin, K.; Macara, I. G.; McLeod, G. C. *Biochim. Biophys. Acta* **1981**, *649*, 493-502.

19. Sakurai, H.; Hirata, J.; Michibata, H. *Inorg. Chim. Acta* **1988**, *152*, 177-180.

20. Frank, P.; Kustin, K.; Robinson, W. E.; Linebaugh, L.; Hodgson, K. O. *Inorg. Chem.* **1995**, *34*, 5942-5949.

21. Frank, P.; Hodgson, K. O. *Inorg. Chem.* **2000**, *39*, 6018-6027.

22. Krause, M. O.; Oliver, H. H. *J. Phys. Chem. Ref. Data* **1979**, *8*, 329-338.

23. Frank, P.; Robinson, W. E.; Kustin, K.; Hodgson, K. O. *J. Inorg. Biochem.* **2001**, *86*, 635-648.

24. Robles, J. C.; Mitsuzaka, Y.; Inomata, S.; Shimoi, M.; Mori, W.; Ogino, H. *Inorg. Chem.* **1993**, *32*, 13-17.

25. Kanamori, K.; Kinebuchi, Y.; Michibata, H. *Chem. Lett.* **1997**, 423-424.

26. Kanamori, K.; Sakurai, M.; Kinoshita, T.; Uyama, T.; Ueki, T.; Michibata, H. *J. Inorg. Biochem.* **1999**, *77*, 157-161.

27. Eldjarn, L.; Pihl, A. *J. Am. Chem. Soc.* **1957**, *79*, 4589-4593.

28. Schafer, F. Q.; Buettner, G. R. *Free Red. Biol. & Med.* **2001**, *30*, 1191-1212.

29. Meier, R.; Boddin, M.; Mitzenheim, S. In *Bioinorganic Chemistry: transition metals in biology and their coordination chemistry*; Trautwein, A. X., Ed.; Bonn: Wiley-VCH, 1997, pp 69-97.

30. Chatterjee, M.; Ghosh, S.; Nanda, A. K. *Polyhedron* **1997**, *16*, 2917-2923.

31. Platas-Iglesias, C.; Vaiana, L.; Esteban-Gomez, D.; Avecilla, F.; Real, J. A.; De Blas, A.; Rodriguez-Blas, T. *Inorg. Chem.* **2005**, *44*, 9704-9713.

32. Vanquickenborne, L. G.; Haspeslagh, L. *Inorg. Chem.* **1982**, *21*, 2448-2454.

33. Rahman, H. U.; Runciman, W. A. *J. Phys. C,* **1971**, *4*, 1576-1590.

34. Meier, R.; Boddin, M.; Mitzenheim, S.; Schmid, V.; Schönherr, T. *J. Inorg. Biochem.* **1998**, *69*, 249-252.

35. Hoffmann, R.; Beier, B. F.; Muetterties, E. L.; Rossi, A. R. *Inorg. Chem.* **1977**, *16*, 511-522.

36. Ogino, H.; Shimoi, M.; Saito, Y. *Inorg. Chem.* **1989**, *28*, 3596-3600.

37. Shimoi, M.; Saito, Y.; Ogino, H. *Bull. Chem. Soc. Jpn.* **1991**, *64*, 2629-2634.

38. Meier, R.; Bodin, M.; Mitzenheim, S.; Kanamori, K. In *Metal Ions in Biological Systems: Vanadium and Its Rôle for Life*; Sigel, H., Sigel, A., Eds.; New York: Marcel Dekker, 1995; Vol. 31, pp 45-88.

39. Frank, P.; Hedman, B.; Hodgson, K. O. *Inorg. Chem.* **1999**, *38*, 260-270.
40. Millis, K. K.; Weaver, K. H.; Rabenstein, D. L. *J. Org. Chem.* **1993**, *58*, 4144-4146.
41. Hager, G.; Brolo, A. G. *J. Electroanal. Chem.* **2003**, *550-551*, 291-301.
42. Meier, R.; Werner, G.; Otto, M. *Inorg. Chim. Acta* **1990**, *236*, 315-324.

Chapter 21

Chemical and Biochemical Insights on Preventing Cancer with Vanadium and Selenium

Elizabeth E. Hamilton, Jessica M. Fautch, Sarah M. Gentry, and Jonathan J. Wilker*

Department of Chemistry, Purdue University, West Lafayette, IN 47907

A growing body of studies is showing that vanadium and selenium have significant abilities to prevent many types of cancer. At this time, however, few mechanistic insights are available to explain this promising avenue of health research. Here we summarize our related chemical and biochemical studies. In particular, we focus on a "carcinogen interception" mechanism in which nucleophilic inorganic oxo species react with electrophilic alkylating carcinogens, thereby preventing DNA damage. Working with small molecules, vanadates and selenates are shown to detoxify carcinogens. Biochemical studies on DNA alkylation show that the damage can be prevented with vanadates or selenates. In this review, we summarize the most salient points regarding "carcinogen interception" and provide some perspectives for future developments.

†We dedicate this paper to Dieter Rehder in honor of his receipt of the Second Vanadis Award.

© 2007 American Chemical Society

Cancer Prevention with Selenium and Vanadium

Exciting results in cancer prevention include repeated demonstrations of the chemoprotective effects of dietary selenium, an essential trace nutrient (reviewed (*1-4*)). There is more evidence for the cancer preventing properties of selenium than any other normal component of the human diet (*1*). In various studies, selenium levels have been shown to display inverse correlations with cancer mortality (*5-9*). High selenium blood levels were also inversely correlated with cancer mortality rates (*7-9*). In a comprehensive and widely cited study, patients taking selenium supplemented brewer's yeast experienced a significant reduction of colorectal, lung, and prostate cancer incidences as well as cancer-related mortality (*10*). Chemoprotective effects have also been seen in rats fed diets with high selenium broccoli and garlic (*11,12*). Selenium is an essential nutrient, but can become toxic above ~1 mg/kg body weight (*13*).

The potential for dietary inorganics to have significant impact on cancer prevention is illustrated by the Selenium and Vitamin E Chemoprevention Trial ("SELECT") sponsored by the National Cancer Institute (*14*). In this ongoing trial, over 30,000 males at risk for development of prostate cancer are having their diets supplemented with selenium alone, vitamin E alone, both selenium and vitamin E, or a placebo (*14*). These test subjects will be followed for 7 to 12 years. The Prevention of Cancer by Intervention with Selenium ("PRECISE") trial is of a similar scale and ongoing in Europe (*15*).

The chemoprotective effects of dietary vanadium have also come to light in recent rat studies (*16-21*). Supplementing drinking water with 0.5 ppm vanadium resulted in lower incidences of cancer, fewer visible tumors, smaller tumor size, and longer life in the test rats when compared to controls (*16-21*). Interestingly, the chemoprotective effects of dietary vanadium were most significant when this inorganic component was added to the diet weeks prior to injection of the carcinogen. Already present in Western diets and beneficial at low levels, vanadium can be toxic above ~50 mg/kg body weight (*22*).

Possible Mechanisms of Cancer Prevention by Vanadium and Selenium

Although intriguing, the derivation of health benefits from vanadium and selenium is poorly understood. The biochemical implications of elevated selenium, and possible explanations for the cancer preventing properties, are varied (*1-3*). A few of these include possible antioxidant roles and subsequent inhibition of DNA damage (*1*), induction of detoxifying enzymes (*3*), and influences on carcinogen metabolism (*1*).

Like the case with selenium, a mechanism for the cancer preventing properties of vanadium remains to be found. Among the known biochemical effects of elevated vanadium are increased glutathione levels and increased expression of cytochromes P450 but diminished superoxide dismutase activity (*19*). With these and other (*16-18,20*) effects possibly influencing the beneficial properties of dietary selenium and vanadium, we can see that a coherent picture remains to be formed.

Many studies provide evidence suggesting that the chemoprotective effects of selenium and vanadium are a result of minimized DNA damage. Two research groups showed that when rats were assaulted with the alkylating carcinogen 7,12-dimethylbenzo[*a*]anthracene (DMBA), fewer DMBA-DNA adducts were formed if the diet was supplemented with Na_2SeO_3 (*12,23*). Bacterial studies also found fewer mutations formed in *Escherichia coli* cells subjected to the alkylating agent *N*-methyl-*N*'-nitro-*N*-nitrosoguanidine when the growth medium was supplemented with Na_2SeO_3 (*24*). For analogous studies with vanadium, fewer strand breaks (*16*) and chromosomal aberrations (*25*) were observed when rats exposed to diethylnitrosamine had their diets supplemented with NH_4VO_3. Although various cancer prevention mechanisms have been suggested, none fully account for decreased DNA alkylation. A mechanism explaining cancer prevention by selenium and vanadium needs to take into account lower levels of DNA damage.

Environmental Exposure to Carcinogens

We are constantly exposed to chemical carcinogens such as nitrosamines and polycyclic aromatic hydrocarbons (PAHs). These compounds are prevalent in cooked food, tobacco smoke, and the combustion of fossil fuels. The toxicity is derived from enzymatic activation into highly reactive alkylating agents which attack the genome, as shown in Figure 1. Metabolic processing of the procarcinogens yields alkyl diazoniums or epoxides which are able to react with nucleophilic positions of DNA (Figure 1). Upon replication, the resulting covalently modified adducts of DNA bring about base mismatching. Such mutations can then result in cancerous growth.

Inorganic Oxo Species and Aqueous Speciation

In water, selenium and vanadium equilibrate to various species, depending upon solution pH and dissolved oxygen (*22*). Administered selenium and vanadium salts thus convert according to environmental conditions. Indeed, three different vanadium salts ($VOSO_4$, NH_4VO_3, and Na_3VO_4) produced identical tumor inhibition profiles (*26*). The predominant forms of soluble, aqueous selenium and vanadium are almost exclusively anionic oxo species. In cells, vanadium exists predominantly as V^{5+} and V^{4+} species (*27,28*), with $(HVO_4)^{2-}$, $(H_2VO_4)^-$, $[VO(OH_2)_5]^{2+}$, and $[VO(OH)(OH_2)]^+$ most prevalent (*29*).

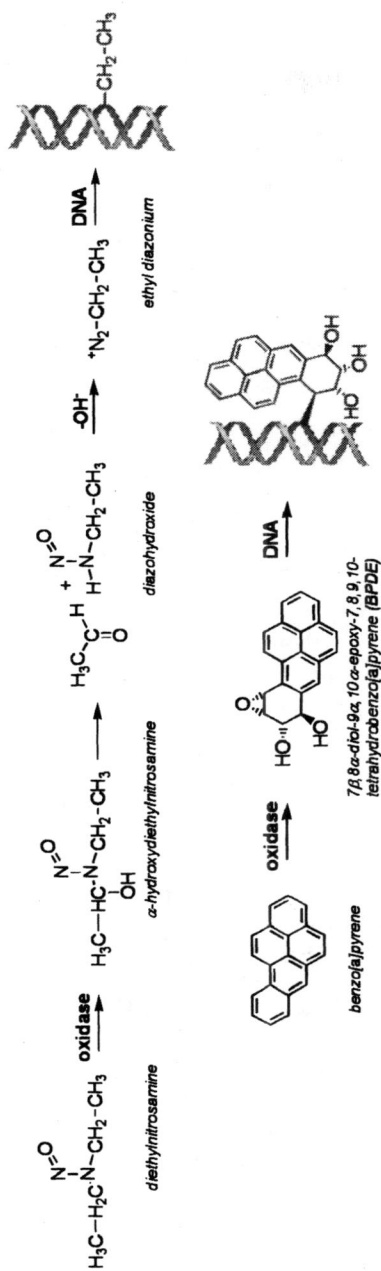

Figure 1. Metabolism of nitrosamines and polycyclic aromatic hydrocarbons yields DNA alkylation damage.

Cellular metabolism of selenium follows a path from $(SeO_4)^{2-}$ to $(SeO_3)^{2-}$ and then Se^{2-}. Subsequent processing yields H_3CSeH followed by $(H_3C)_2Se$ and, finally, $(CH_3)_3Se^+$ (30).

Parallels Between Vanadium and Selenium

Interesting similarities can be found when considering the cancer preventing properties of selenium and vanadium. Both elements are present in Western diets: Selenium at 150 – 200 µg per day (31) and vanadium at tens of micrograms per day (32). Evidence has been presented to suggest that the cancer preventing properties of these nutrients are maximized during carcinogen exposure or the initiation phase of cancer (16-18,20). Multiple forms of each nutrient can be administered to bring about appreciable cancer preventing effects (26,33). The aqueous chemistry of both selenium and vanadium salts is predominated by oxo anion species.

"Carcinogen Interception"

In trying to understand the mechanism of cancer prevention by the micronutrients selenium and vanadium, we have been considering a hypothesis of reactivity between inorganic oxo species and carcinogens. Toxic, electrophilic alkylating agents may react directly with nucleophilic metal oxo compounds, thereby preventing DNA damage. Thus, the dietary inorganics could detoxify carcinogens. Figure 2 depicts this concept in which an inorganic compound may "intercept" a toxin prior to any reaction with DNA. In particular, anionic, aqueous forms of vanadium and selenium could be potent nucleophiles for reactivity with the electrophilic alkylating toxins.

Preventing DNA damage by direct detoxification of carcinogens is a new avenue that has not yet been studied extensively (34-37). We are curious to see if such processes may be relevant to the known cancer preventing effects of vanadium and selenium. Currently our laboratory is exploring this mechanism at chemical, biochemical, and cellular levels. As will be presented below, we have begun our studies by examining potential reactions (34,36) between inorganic oxo species (37) and alkylating toxins. Progressing to a larger perspective, we are exploring alkylation damage of DNA in the presence or absence of vanadium and selenium compounds (35,36). Discovery of a new mechanism for preventing cancer may guide the development of compounds with enhanced activity and reduced toxicity. We could then make an impact on the 65% of United States cancer deaths caused by diet and tobacco (38).

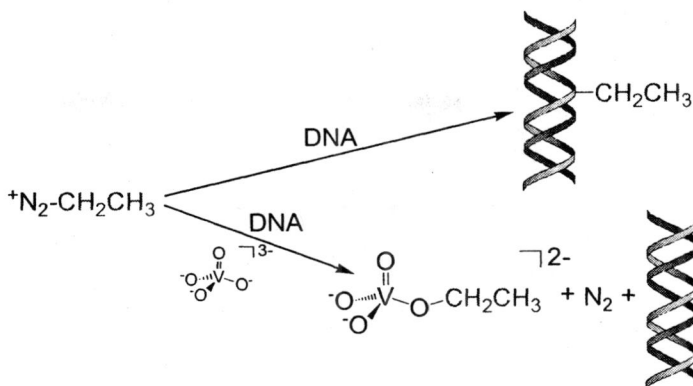

Figure 2. Representation of carcinogen interception in which a nucleophilic inorganic compound may be capable of reacting with an alkylating toxin, thereby preventing DNA damage.

Isolation, Characterization, and Solution Speciation of Vanadates

The aqueous chemistry of vanadium is quite complex (*22*). Multiple oxo species exist under different conditions of pH and solution oxygenation such as $(VO_4)^{3-}$, $(HVO_4)^{2-}$, $(H_2VO_4)^-$, $(H_2V_2O_7)^{2-}$, and $(V_4O_{12})^{4-}$ (*22*). With so many species present in one solution, mechanistic understanding of reaction chemistry can become difficult. To work with discrete vanadium oxo complexes, we turned our efforts toward the use of organic solvents. Without the effects of pH (e.g., hydroxide reactions with metal centers) or hydrolysis, we reasoned that organic solvents may provide inroads to individual vanadates for detailed reactivity studies. We prepared oxo anions with bulky counterions such as tetrabutylammonium, $[(C4H9)4N]+$, in order to permit solubility in organic solvents. Synthesis of the tetrabutylammonium salt of the $(V3O9)3-$ anion, along with characterization by single crystal X-ray diffraction methods (Figure 3), provided the first isolation of this vanadate (*37*). Proposed to exist in water (*39,40*), the $(V_3O_9)^{3-}$ anion had not been isolated previously.

Reactivity Studies of Vanadates and Alkylating Agents

After describing the detailed solution speciation of $[(C_4H_9)_4N]_3(V_3O_9)$ in acetonitrile, we progressed to study reactions of vanadates with alkylating

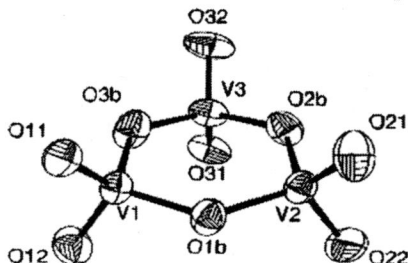

Figure 3. X-ray crystal structure of the $(V_3O_9)^{3-}$ anion (43).

carcinogens. The model carcinogen of choice was the classic ethylating agent diethyl sulfate (*41*). This carcinogen is present in food (*42*), is easier to monitor by ^1H NMR spectroscopy than any methylating agent, and provides a reactivity profile with DNA similar to ethyl diazonium (*41*).

The reactivity found for the trimer $[(C_4H_9)_4N]_3(V_3O_9)$ with diethyl sulfate is typical of most vanadates we have examined. When combined in a 1:1 ratio (10 mM each), diethyl sulfate was converted to ethanol and monoethyl sulfate, quantitatively (Figure 4). This assignment of ethanol was confirmed by gas chromatography mass spectrometry. The product remains colorless and diamagnetic indicating that all vanadium remains in the +5 oxidation state. We used ^{51}V NMR spectrometry to follow the course of reaction and identify the vanadium-containing products (*34*).

The reaction of $(V_3O_9)^{3-}$ and diethyl sulfate describes a general trend in reactivity: The vanadates $(HV_4O_{12})^{3-}$, $(V_5O_{14})^{3-}$, $(V_{10}O_{26})^{4-}$, and $(H_3V_{10}O_{28})^{3-}$, prepared from literature procedures, also reacted with diethyl sulfate to yield ethanol. Other alkylating agents such as ethyl trifluoromethanesulfonate yielded similar results. The $(V_3O_9)^{3-}$ anion also reacts with *N*-ethyl-*N*-nitrosourea (ENU) rapidly, to yield ethanol.

Alkylation Reactivity of Selenium Oxo Species

The vanadate literature is extensive. By contrast, significantly less selenium oxo chemistry has been studied in detail. In order to carry out analogous alkylation studies to those described above with vanadates, we attempted the preparation of anionic selenium oxo species with large counterions for solubility in organic solvents. We isolated and characterized the tetraphenylphosphonium salt of the $(O_3SeOCH_2OSeO_3)^{2-}$ anion (*34*). When reacted with diethyl sulfate or $CH_3CH_2OSO_2CF_3$, this selenate consumed the alkylating toxins. According to

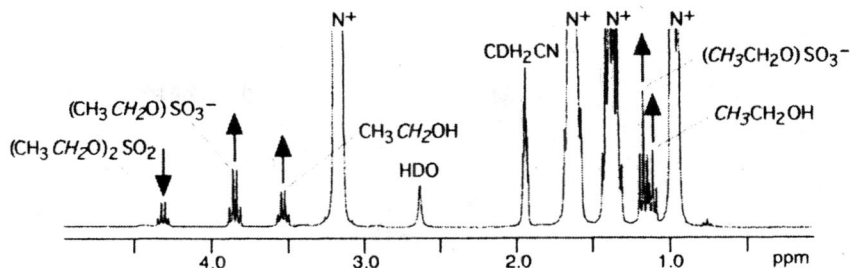

Figure 4. ^1H NMR spectrum showing a 1:1 reaction of $[(C_4H_9)_4N]_3(V_3O_9)$ + $(C_2H_5O)_2SO_2$ in progress. Arrows indicate consumption of $(C_2H_5O)_2SO_2$ and formation of C_2H_5OH. N^+ designates the $[(C_4H_9)_4N]^+$ counterion. (Reproduced with permission from reference 34. Copyright 2004 Wiley-VCH Verlag.)

^1H and ^{77}Se NMR spectrometries, reaction products included ethyl selenate $((CH_3CH_2O)SeO_3^-)$ and diethyl selenate $((CH_3CH_2O)_2SeO_2)$. Hydrolysis of $(CH_3CH_2O)_2SeO_2$ generated ethanol.

Testing the Ability of Inorganic Compounds to Prevent DNA Damage

After showing that select inorganic oxo compounds can detoxify carcinogens, we wondered if such processes may be relevant to DNA damage. In particular, we reacted plasmid DNA with alkylating agents in the presence or absence of inorganic compounds. For observation of DNA alkylation, we used an enzyme-based assay (*35*). This assay provides easy visualization of the starting, undamaged DNA (supercoiled) versus damaged forms (nicked, linear) resulting from alkylation (c.f., Figure 5) (*35*).

Figure 5 (left) shows a typical gel in which the alkyl intercepting capabilities of Na_2SeO_4 were examined in pH = 7.0 unbuffered solutions. Lane 1 shows a control of the starting DNA. Lane 2 is a reaction of the plasmid DNA with the alkylating agent. Some streaking can be seen, indicating alkylation followed by hydrolysis and strand cleavage. Although a difference can be seen between DNA with and without the alkylating agent, the changes are subtle. Lane 3 shows that Na_2SeO_4, alone, brings no major change to the DNA. Similarly, lane 4 shows minimal change when DNA, diethyl sulfate, and Na_2SeO_4 are combined. With the endonuclease enzyme Exo III added to DNA, alone, (lane 5), little change is observed.

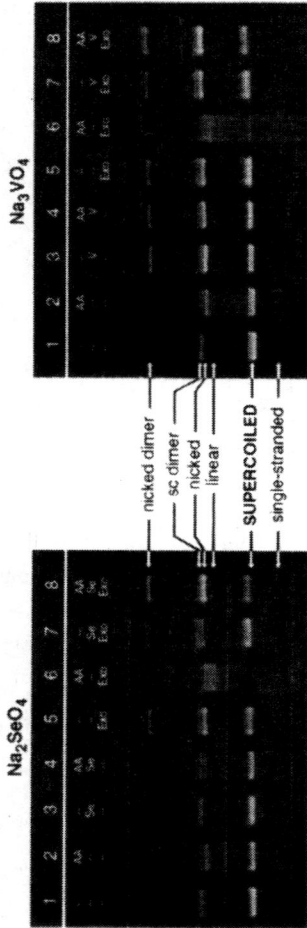

Figure 5. Gels showing the ability of Na₂SeO₄ (left) and Na₃VO₄ (right) to prevent DNA alkylation. "AA" indicates the alkylating agent was added to that lane, "Se" is Na₂SeO₄, "V" is Na₃VO₄, and "Exo" is exonuclease III enzyme. (Reproduced with permission from reference 33. Copyright 2004 Springer-Verlag.)

In lane 6 is a reaction of DNA, diethyl sulfate, and Exo III. This reaction shows a stark contrast to the aforementioned controls. The DNA has been alkylated, hydrolyzed to yield abasic sites, and cleaved by Exo III at these abasic sites. Such cleavage consumes almost all supercoiled plasmid, yielding nicked and linear fragments of DNA. Lane 7 shows a reaction of DNA, Na_2SeO_4, Exo III, and the resulting lack of cleavage. In lane 8 is the reaction of DNA, diethyl sulfate, Na_2SeO_4, and Exo III with some supercoiled form maintained. These results indicate that Na_2SeO_4 consumed the alkylating agent, thereby preventing alkylation of the DNA, subsequent hydrolysis, and the final Exo III cleavage. Control experiments demonstrate that neither Na_2SeO_4 nor Na_3VO_4 inhibit activity of the Exo III enzyme (data not shown). From these results, selenium appears to prevent DNA alkylation damage (35).

Nearly identical results were found in experiments with Na_3VO_4 replacing Na_2SeO_4. Figure 5 (right) also shows a representative gel for these reactions with lane assignments analogous to those described above for the selenium reactions. Similar to selenium, vanadium appears capable of preventing DNA alkylation damage (35).

We wanted to explore the origin of the protective effect afforded by selenium and vanadium. Undamaged DNA could be a result of preventing alkylation or repair of alkylation lesions. To distinguish between these two cases, alkylated DNA was incubated with 1 and 10 equivalents of selenium or vanadium relative to the alkylating agent. Gel electrophoresis showed no reversal of DNA alkylation. We conclude that selenium and vanadium protect DNA by preventing alkylation rather than reversing damage (35,36).

Conclusions and Future Perspectives

Cancer prevention with vanadium and selenium has shown great promise in animal models and human trials. Of course, we are curious to understand the mechanism of this simple means of avoiding disease. Our laboratory has shown that vanadates and selenates consume alkylating toxins. In the case of vanadates, these toxic carcinogens are transformed into relatively harmless alcohols. We have also found that DNA alkylation can be inhibited with select inorganic species, including vanadates and selenates. These insights provide support for the idea that a "carcinogen interception" mechanism may, at least in part, be at play when carcinogenesis is inhibited by vanadium and selenium. Of course, we are in need of "second-generation" compounds for cancer prevention. Although both vanadium and selenium are present in Western diets, toxicity is found with particularly high doses. Thus we are seeking compounds with decreased toxicity relative to simple metal salts. Additionally, we are exploring the design of new compounds with enhanced reactivity toward carcinogens. We may be able to use metal-ligand complexes to both decrease toxicity and enhance reactivity with toxins. Already, the trends found in reactivity between inorganics and carcinogens suggest design parameters for these second-generation compounds. Further exploration of this mechanism will also require insights on a cellular level and, eventually, in animal models. Simple diet supplementation may provide a way in which vanadium and selenium chemistry can have a great impact upon human health.

Acknowledgments

We thank Debbie Crans and João Costa Pessoa for all their work in putting together the excellent Fifth International Symposium on the Chemistry and Biological Chemistry of Vanadium in San Francisco, September 2006. Throughout the course of our work in this area, we have benefited greatly from conversations with and assistance from Kevin Kelley, John Grutzner, John Harwood, Ian Henry, and John Protasiewicz. We also are greatly appreciative of financial support provided by the Cancer Research and Prevention Foundation and an Alfred P. Sloan Foundation Research Fellowship.

References

1. Combs, G. F.; Gray, W. P.; "Chemopreventive Agents: Selenium" *Pharmacol. Ther.* **1998**, *79*, 179-192.
2. Ganther, H. E.; "Selenium Metabolism, Selenoproteins, and Mechanisms of Cancer Prevention: Complexities with Thioredoxin Reductase" *Carcinogenesis* **1999**, *20*, 1657-1666.
3. Ip, C.; "Lessons from Basic Research in Selenium and Cancer Prevention" *J. Nutrition* **1998**, *128*, 1845-1854.
4. Kim, Y. S.; Milner, J.; "Molecular Targets for Selenium in Cancer Prevention" *Nutr. Cancer* **2001**, *40*, 50-54.
5. Shamberger, R. J.; Frost, D. V.; "Possible Protective Effect of Selenium Against Human Cancer" *Can. Med. Assoc. J.* **1969**, *100*, 682.
6. Shamberger, R. J.; Willis, C. E.; "Selenium Distribution and Human Cancer Mortality" *CRC Crit. Rev. Clin. Lab. Sci.* **1971**, *2*, 211-221.
7. Schrauzer, G. N.; "Selenium and Cancer: A Review" *Bioinorg. Chem.* **1976**, *5*, 275-281.
8. Schrauzer, G. N.; White, D. A.; Schneider, C. J.; "Cancer Mortality Correlation Studies- III. Statistical Associations with Dietary Selenium Intakes" *Bioinorg. Chem.* **1977**, *7*, 23-34.
9. Schrauzer, G. N.; White, D. A.; Schneider, C. J.; "Cancer Mortality Correlation Studies- IV. Associations with Dietary Intakes and Blood Levels of Certain Trace Elements, Notably Se-Antagonists" *Bioinorg. Chem.* **1977**, *7*, 35-56.

10. Clark, L. C.; Combs, G. F.; Turnbull, B. W.; Slate, E. H.; Chalker, D. K.; Chow, J.; Davis, L. S.; Glover, R. A.; Graham, G. F.; Gross, E. G.; Krongrad, A.; Lesher, J. L.; Park, H. K.; Sanders, B. B.; Smith, C. L.; Taylor, J. R.; "Effects of Selenium Supplementation for Cancer Prevention in Patients with Carcinoma of the Skin" *J. Am. Medical Assoc.* **1996**, *276*, 1957-1963.

11. Finley, J. W.; Ip, C.; Lisk, D. J.; Davis, C. D.; Hintze, K. J.; Whanger, P. D.; "Cancer-Protective Properties of High-Selenium Broccoli" *J. Agric. Food Chem.* **2001**, *49*, 2679-2683.

12. Ip, C.; Lisk, D. J.; "Efficacy of Cancer Prevention by High-Selenium Garlic Is Primarily Dependent on the Action of Selenium" *Carcinogenesis* **1995**, *16*, 2649-2652.

13. Subcommittee on Selenium, Committee on Animal Nutrition, Board on Agriculture, National Research Council *Selenium in Nutrition*; National Academy Press: Washington, D.C., 1983.

14. Klein, E. A.; Lippman, S. M.; Thompson, I. M.; Goodman, P. J.; Albanes, D.; Taylor, P. R.; Coltman, C.; "The Selenium and Vitamin E Cancer Prevention Trial" *World J. Urol.* **2003**, *21*, 21-27.

15. Rayman, M. P.; "The Importance of Selenium to Human Health" *Lancet* **2000**, *356*, 233-241.

16. Basak, R.; Saha, B. K.; Chatterjee, M.; "Inhibition of Diethylnitrosamine-Induced Rat Liver Chromosomal Aberrations and DNA-Strand Breaks by Synergistic Supplementation of Vanadium and 1α,25-Dihydroxyvitamin D_3" *Biochim. Biophys. Acta* **2000**, *1502*, 273-282.

17. Bishayee, A.; Chatterjee, M.; "Inhibitory Effects of Vanadium on Rat Liver Carcinogenesis Initiated with Diethylnitrosamine and Promoted by Phenobarbital" *British J. Cancer* **1995**, *71*, 1214-1220.

18. Bishayee, A.; Chatterjee, M.; "Inhibition of Altered Liver Cell Foci and Persistent Nodule Growth by Vanadium During Diethylnitrosamine-Induced Hepatocarcinogenesis" *Anticancer Res.* **1995**, *15*, 455-462.

19. Bishayee, A.; Oinam, S.; Basu, M.; Chatterjee, M.; "Vanadium Chemoprevention of 7,12-Dimethylbenz(a)anthracene-Induced Rat Mammary Carcinogenesis: Probable Involvement of Representative Hepatic Phase I and II Xenobiotic Metabolizing Enzymes" *Breast Cancer Res. Treatment* **2000**, *63*, 133-145.

20. Chakraborty, A.; Selvaraj, S.; "Differential Modulation of Xenobiotic Metabolizing Enzymes by Vanadium During Diethylnitrosamine-Induced Hepatocarcinogenesis in Sprague-Dawley Rats" *Neoplasma* **2000**, *47*, 81-89.

21. Thompson, H. J.; Chasteen, N. D.; Meeker, L. D.; "Dietary Vanadyl(IV) Sulfate Inhibits Chemically-Induced Mammary Carcinogenesis" *Carcinogenesis* **1984**, *5*, 849-851.

22. Rehder, D.; "The Bioinorganic Chemistry of Vanadium" *Angew. Chem. Int. Ed. Eng.* **1991**, *30*, 148-167.

23. Liu, J.; Gilbert, K.; Parker, H.; Haschek, W.; Milner, J. A.; "Inhibition of 7,12-Dimethylbenz(*a*)anthracene-Induced Mammary Tumors and DNA Adducts by Dietary Selenite" *Cancer Res.* **1991**, *51*, 4613-4617.

24. Sato, M.; Nunoshiba, T.; Nishioka, H.; Yagi, T.; Takebe, H.; "Protective Effects of Sodium Selenite on Killing and Mutation by *N*-Methyl-*N'*-nitro-*N*-nitrosoguanidine in *E. coli*" *Mutat. Res.* **1991**, *250*, 73-77.

25. Bishayee, A.; Banik, S.; Mandal, A.; Marimuthu, P.; Chatterjee, M.; "Vanadium-Mediated Suppression of Diethylnitrosamine-Induced Chromosomal Aberrations in Rat Hepatocytes and its Correlation with Induction of Hepatic Glutathione and Glutathione S-Transferase" *Int. J. Oncology* **1997**, *10*, 413-423.

26. Hanauske, U.; Hanauske, A.-R.; Marshall, M. H.; Muggia, V. A.; Hoff, D. D. V.; "Biphasic Effect of Vanadium Salts on In Vitro Tumor Colony Growth" *Int. J. Cell Cloning* **1987**, *5*, 170-178.

27. Baran, E. J.; "Oxovanadium(IV) and Oxovanadium(V) Complexes Relevant to Biological Systems" *J. Inorg. Biochem.* **2000**, *80*, 1-10.

28. Chasteen, N. D.; Grady, J. K.; Holloway, C. E.; "Characterization of the Binding, Kinetics, and Redox Stability of Vanadium(IV) and Vanadium(V) Protein Complexes in Serum" *Inorg. Chem.* **1986**, *25*, 2754-2760.

29. Rehder, D.; "Biological and Medicinal Aspects of Vanadium" *Inorg. Chem. Comm.* **2003**, *6*, 604-617.

30. Birringer, M.; Pilawa, S.; Flohe, L.; "Trends in Selenium Biochemistry" *Nat. Prod. Rep.* **2002**, *19*, 693-718.

31. Combs, G. F.; Gray, W. P.; "Chemoprotective Agents: Selenium" *Pharmacol. Ther.* **1998**, *79*, 179-192.

32. Byrne, A. R.; Kosta, L.; "Vanadium in Foods and in Human Body Fluids and Tissues" *Sci. Total Env.* **1978**, *10*, 17-30.

33. Ip, C.; Birringer, M.; Block, E.; Kotrebai, M.; Tyson, J. F.; Uden, P. C.; Lisk, D. J.; "Chemical Speciation Influences Comparative Activity of Selenium-Enriched Garlic and Yeast in Mammary Cancer Prevention" *J. Agric. Food Chem.* **2000**, *48*, 2062-2070.

34. Hamilton, E. E.; Fanwick, P. E.; Wilker, J. J.; "Alkylation of Inorganic Oxo Compounds and Insights on Preventing DNA Damage" *J. Am. Chem. Soc.* **2006**, *128*, 3388-3395.

35. Hamilton, E. E.; Wilker, J. J.; "Inhibition of DNA Alkylation Damage with Inorganic Salts" *J. Biol. Inorg. Chem.* **2004**, *9*, 894-902.

36. Hamilton, E. E.; Wilker, J. J.; "Inorganic Oxo Compounds React with Alkylating Agents: Implications for DNA Damage" *Angew. Chem. Int. Ed.* **2004**, *43*, 3290-3292.

37. Hamilton, E. E.; Fanwick, P. E.; Wilker, J. J.; "The Elusive Vanadate $(V_3O_9)^{3-}$: Isolation, Crystal Structure, and Nonaqueous Solution Behavior" *J. Am. Chem. Soc.* **2002**, *124*, 78-82.

38. Doll, R.; Peto, R. *The Causes of Cancer: Quantitative Estimates of Avoidable Risks of Cancer in the United States Today*; Oxford: New York, 1981.

39. Brito, F.; Ingri, N.; Sillen, L. G.; "Are Aqueous Metavanadate Species Trinuclear, Tetranuclear, or Both? Preliminary LETAGROP Recalculation of EMF Data" *Acta Chem. Scand.* **1964**, *18*, 1557-1558.

40. Heath, E.; Howarth, O. W.; "Vanadium-51 and Oxygen-17 Nuclear Magnetic Resonance Study of Vanadate(V) Equilibria and Kinetics" *J. Chem. Soc. Dalton* **1981**, 1105-1110.

41. Singer, B.; "N-Nitroso Alkylating Agents: Formation and Persistence of Alkyl Derivatives in Mammalian Nucleic Acids as Contributing Factors in Carcinogenesis" *J. Nat. Cancer Inst.* **1979**, *62*, 1329-1339.

42. National Research Council, *Carcinogens and Anticarcinogens in the Human Diet: A Comparison of Naturally Ocurring and Synthetic Substances*; National Academy Press: Washington, DC, 1996.

Coordination Chemistry:
Speciation and Structure

Chapter 22

Speciation of Peroxovanadium(V) Complexes Studied by First-Principles Molecular Dynamics Simulations and ^{51}V NMR Chemical Shift Computations

Michael Bühl

Max-Planck-Institut für Kohlenforschung, Kaiser-Wilhelm-Platz 1, D–45470 Mülheim an der Ruhr, Germany

Structure and speciation of selected peroxovanadium(V) complexes are studied with Car-Parrinello molecular dynamics (CPMD) simulations in aqueous solution. Structures are assessed by comparing experimental ^{51}V NMR chemical shifts to those computed and averaged over the MD trajectories. Using this approach, two high-field resonances observed in vanadate solutions with excess H_2O_2 and at high pH can be assigned to $[V(O_2)_4]^{3-}$ and $[VO(O_2)_2(OOH)]^{2-}$, the latter of which has not been described so far.

Introduction

Vanadate complexes display a very rich aqueous chemistry, often characterized by several species being present simultaneously, or in rapid equilibrium. A valuable tool to study the speciation of vanadium(V) complexes is ^{51}V NMR spectroscopy (*1*), because each stable constituent of the solution is

© 2007 American Chemical Society

identified by a characteristic ^{51}V chemical shift. Assignment of a particular signal to a specific structure is a difficult task, relying for the most part on additional information such as $\delta(^{51}V)$ data for related species from the literature, solid-state structures determined by X-ray crystallography, or other spectroscopic sources. Such assignments can be particularly difficult when substances of poor stability are investigated. One such example comprises the blue $[V(O_2)_4]^{3-}$ ion (**1**), which is stable as dodecahedral, side-on peroxo complex in the solid (*2*), but tends to decompose rapidly in aqueous solution. In two studies investigating the speciation of the vanadyl/H_2O_2 system over the entire pH range (*3*), strongly shielded resonances at $\delta = -734$ (*3a*) or $\delta = -742$ (*3b*) appearing above pH 13, have been ascribed to this species. An even more shielded signal at $\delta = -845$ (or $\delta = -838$) has been assigned to $[VO(O_2)_3]^{3-}$ (**2**). As the pH is lowered, these signals are replaced by others that are successively shifted to higher frequencies (lower field). In neutral or slightly basic solutions, the predominant species can probably be formulated as $[VO(O_2)_2(H_2O)]^-$ (**3**), $\delta = -697$ (*3b*), whereas at pH below 1, a cationic monoperoxo complex is the most likely candidate, viz. $[VO(O_2)(H_2O)_x]^+$ (**4**, with $x = 3$ or 4).

We have been interested in the past years in applying the modern tools of density functional theory (DFT) to compute chemical shifts of transition metal nuclei, with a recent focus on a proper description of thermal and solvation effects on this property (*4,5*). DFT-based molecular dynamics (MD) simulations with explicit description of the bulk solution and ensemble-averaged NMR quantities derived thereof have proven to be quite successful, in particular for highly charged anions in aqueous solution (*5*). For the $\delta(^{51}V)$ values of vanadate and peroxovanadate complexes studied so far, the combined thermal and solvent effects obtained with this approach (i.e. the difference between static equilibrium values in the gas phase and the dynamic average in solution) turned out to be relatively small, from a few ppm up to ca. 100 ppm (*4,6*). Much larger effects exceeding 1000 ppm had been noted for highly charged Fe and Co complexes. Because the target complexes **1** and **2** are also highly charged (−3), larger solvent effects than for the vanadates studied so far could be possible.

DFT-based MD simulations have the additional advantage that they can reveal spontaneous processes such as ligand dissociation or proton transfers, which are of immediate relevance for the speciation of the complexes under scrutiny. For a vanadate-peptide complex, for instance, Car-Parrinello MD (CPMD) (*7*) simulations have shown that a neutral, six-coordinate species VO(OH)(glygly')(H_2O) (glygly' = H_2N-CH_2-C(O)-N-CH_2-CO_2) is unstable in water and affords an anionic five-coordinate complex, $[VO_2(glygly')]^-$, within a few picoseconds (*6*). Analoguous simulations are now reported for the peroxo complexes **1** and **2**. As it turns out, **2** is indicated to be unstable under the experimental conditions and can rearrange to other species, one of which has not been described before.

Computational Details

The same methods and basis sets as in the previous MD and NMR studies of vanadium complexes were employed (4,6); see these papers for additional details and references. Specifically, geometries were optimized in the gas phase using the CPMD program (8) (denoted CP-opt), employing the BP86 functional, norm-conserving Troullier-Martins pseudopotentials [semi-core for vanadium (4b)], and a basis of plane waves expanded at the Γ-point up to a kinetic energy cutoff of 80 Ry. Periodic boundary conditions were imposed using a cubic supercell with a box size of 13.0 Å. Long-range electrostatic interactions were treated with the Ewald method with a compensating background charge. For the aqueous solutions, the boxes were filled with 63 additional water molecules, rendering a density of 1. Car-Parrinello molecular dynamics simulations were performed using a fictitious electronic mass of 600 a.u. and a time step of 0.121 fs. Unconstrained simulations (NVE ensemble) were performed at ca. 300 K. The systems were equilibrated for 0.5 ps maintaing a temperature of $300(\pm 50)$ K via velocity rescaling, and were then propagated without constraints for up to 4 ps (denoted CPMD). For the singly charged systems smaller box sizes were employed, namely 9.8692 Å and 11.5 Å for **3** and **4**, respectively, together with 24 and 42 additional water molecules, respectively, in the aqueous solution. The larger complex **8** was placed in a 13.0 Å box with 60 water molecules.

Magnetic shieldings were computed for equilibrium structures and for snapshots along the trajectories employing gauge-including atomic orbitals (GIAOs) and the B3LYP hybrid functional, together with basis AE1+, which consists of the augmented all-electron Wachters' basis (9) on V (contracted to 8s7p4d), and 6-31+G* basis on O and H. Snapshots were taken every 20 fs over the last $1-2$ ps. No periodic boundary conditions were employed and solvent water molecules were not included specifically, but in form of point charges employing values of -0.9313 and $+0.4656$ for O and H atoms, respectively, as obtained for a single water molecule from natural population analysis (water molecules from the six adjacent boxes were also included as point charges, cf. the procedure in reference 4). These computations were carried out with the Gaussian 03 suite of programs. Chemical shifts are reported relative to $VOCl_3$, optimized at the same level ($\sigma = -2264$ for the CP-opt geometry) or averaged over a 1 ps CPMD simulation in the gas phase (mean σ value -2292 ppm) (4b).

Results and Discussion

Speciation in the Vanadate/H₂O₂ System at High pH

Tris-anionic complex **1** remained stable in CPMD simulations, both in the gas phase and in water (for at least 1.5 ps and 2.5 ps, respectively). Geometrical parameters and ^{51}V chemical shifts are summarized in Tables I and II, respectively. The combined thermal and solvent effect on both properties is only moderate (compare CP-opt and CPMD entries), despite the high negative charge of the complex, which makes the peroxo atoms strong hydrogen-bond acceptors. On average, 13.4 water molecules are H-bonded to **1** in aqueous solution, as judged from geometrical criteria (*10*). The CPMD bond distances compare favorably to those observed in a solid containing **1** (*2b*), and the B3LYP $\delta(^{51}V)$ value is strongly shielded, $\delta = -818$ in water, even more shielded than the experimental signal assigned to **1**, $\delta = -742$ (*3b*).

Table I. Selected Equilibrium (Gas Phase) and Averaged (in Solution) Geometrical Parameters for Peroxovanadium Complexes

Compound	Level	parameters[a]			
1 (³⁻)		$r(V-O^1)$	$r(V-O^2)$	$r(O^1-O^2)$	
	CP-opt	1.932	1.947	1.456	
	CPMD	1.94(6)	1.95(6)	1.49(3)	
	X-ray[b]	*1.913*	*1.923*	*1.470*	
5 (²⁻)		$r(V-O^1)$	$r(V-O^2)$	$r(V-O^3)$	$r(O^2-O^3)$
	CP-opt	1.981	1.876	1.935	1.488
	CPMD	1.97(5)	1.86(5)	1.94(5)	1.48(3)
6 (²⁻)		$r(V-O^1)$	$r(V-O^2)$	$r(V-O^3)$	$r(V-O^4)$
	CP-opt	1.625	1.922	1.923	1.969
	CPMD	1.64(2)	1.90(5)	1.92(5)	1.94(5)
		$r(O^2-O^3)$	$r(O^4-O^5)$	$a(V-O^4-O^5)$	
	CP-opt	1.492	1.513	110.7	
	CPMD	1.49(4)	1.48(3)	119(60)	

[a]Distances in Å, angle in degrees; CP-opt: equilibrium values in the gas phase, CPMD: mean values in aqueous solution (in parentheses: standard deviation). Where applicable, distances have been averaged to the idealized symmetry shown.
[b]From reference (*2b*).

At high pH, a signal in a similarly shielded range as computed for **1** can be observed, at $\delta = -838$ (*3b*), but this has been attributed to **2** rather than to **1**, because its larger line width compared to that of the signal at $\delta = -742$ suggested a less symmetrical environment about vanadium than in the latter. The tris-peroxo species **2** could arise from **1** by oxygen-atom transfer, e.g. to the solvent, a reaction typical for peroxo complexes. In a CPMD simulation starting from a "paddlewheel" structure with idealized C_3 symmetry in water, **2** immediately (within less than 100 fs) abstracted a proton from a water molecule initially H-bonded to the terminal oxo ligand, affording $[V(OH)(O_2)_3]^{2-}$ (**5**) and aqueous OH^-. A new simulation was subsequently started for pure aqueous **5** in water, which remained stable for at least 2.5 ps. One of the high-field resonances in the vanadate/H_2O_2 system, at $\delta = -733$, had been assigned (*3a*) to a structure with the same composition as **5**. However, the $\delta(^{51}V)$ value computed for **5** ($\delta = -644$, Table II) is outside this strongly shielded region and does not support this assignment (see also discussion of energies below).

Table II. Computed (B3LYP level) and Experimental $\delta(^{51}V)$ Values of Peroxovanadium Complexes

Species	CP-opt[a]	CPMD[b]	Expt[c]
1	−853	−818	−742
5	−690	−644	−733
6	−756	−758	−838[d]

[a]Equilibrium values in the gas phase.

[b]Average in water (solvent included as point charges)

[c]From reference (*3*).

[d]Assigned to **2**, which rearranges to **6** with CPMD.

The signal ascribed to **2** appears only at very high pH. Free **2** could thus be a strong base, populated to a significant extent only at high OH^- concentration. In order to model such a situation, another simulation was started for **2** with an explicit hydroxide ion present in the water box. With one OH^- and 62 water molecules, the resulting OH^- concentration is ca. 0.9 M, corresponding to a pH of ca. 14, essentially the same as in the experiments. Specifically, the above-mentioned simulation for **2** in water was started again from the same coordinates, but without the proton that would eventually be transferred. This starting point thus had an OH^- ion placed near the terminal oxo ligand of **2** (and an overall charge of −4). In the subsequent CPMD run, this OH^- ion very rapidly abstracted a proton from a nearby water molecule, commencing the well-known relay mechanism for hydroxide transport in water (*11*). For the remainder of the simulation the terminal oxo ligand was surrounded by water molecules, which, however, were not deprotonated. Instead, one peroxo ligand spontaneously (within 200 fs) opened up from the normal side-on bonding mode to a η^1-coordination. The resulting $[VO(\eta^2-O_2)_2(\eta^1-O_2)]^{3-}$ species (**5a**) remained stable for almost 3 ps, when the terminal peroxo ligand abstracted a proton from one of the two nearby, H-bonded water molecules, affording $[VO(O_2)_2(\eta^1\text{-}OOH)]^{2-}$ (**6**). These processes are illustrated in Figure 1 by monitoring salient interatomic distances.

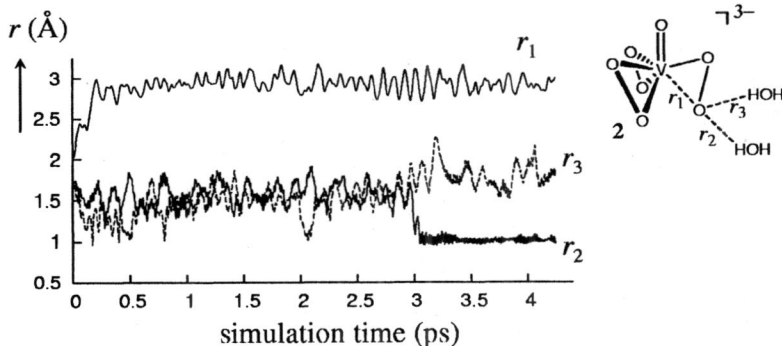

*Figure 1. Evolution of selected distances in a CPMD simulation starting from **2** + OH⁻. Note the immediate increase of r_1, the occasional, reversible transfer of a proton (r_3), and the final transfer at ca. 3 ps (r_2).*

The hydroperoxy complex **6** was not only stable in that simulation in the presence of 2 OH⁻ ions (at least for another picosecond), but also in a subsequent simulation in pure water (for at least 2.5 ps). It is the latter "neutral" simulation of **6** where the snapshots for the NMR computations in Table II were taken from. Here the ^{51}V nucleus is again highly shielded, at $\delta = -758$, albeit not as much as in the resonance initially assigned to **2**, $\delta = -838$. When NMR snapshots were sampled from the first simulation during the lifetime of **5a** (i.e. from ca. 1 – 3 ps, cf. Figure 1), $\delta = -679$ was obtained (*12*), that is, significantly deshielded with respect to **2** and outside the characteristic high-field region. This finding argues against significant population of such a species. Taken together, the present results thus strongly suggest that **2** would not be stable under the experimental conditions, but would rather rearrange to **6**.

Unfortunately, no further quantitative information concerning the relative stability of various isomers can be extracted from these unconstrained MD simulations. More elaborated techniques would be needed for this purpose (such as thermodynamic integration along predefined reaction coordinates or the so-called metadynamics approach), which are beyond the scope of the present paper. It is interesting to note, however, that in the gas phase, pristine **6** is more stable than isomeric **5** by 10.3 kcal/mol at the CP-opt level (15.0 kcal/mol at B3LYP/AE1+ for CP-opt geometries). Assuming that both ions do not differ radically in their hydration energies, the occurrence of **5** appears to be quite unlikely also on energetic grounds. In summary, the hydroperoxo complex **6** presents itself as plausible candidate for one of the constituent species of the vanadate/H_2O_2 system at high pH. To the best of this author's knowledge, such a species has not been proposed before. There is experimental evidence that in oxidation reactions with peroxovanadium complexes, the hydroperoxide anion, HOO^-, is the active oxidant (*13*). Hydroperoxo complexes such as **6** may well be viable precursors for the formation of this species.

Other Peroxovanadium Complexes

When looking at the $\delta(^{51}V)$ data in Table II, it appears that the value computed for **6** fits better to that assigned to **1** and vice versa (assuming that the originally proposed constitution **2** should be formulated as **6**). Reversal of this assignment would thus improve the accord with experiment considerably. In order to gauge the reliability of the computed chemical shifts, a few additional peroxo complexes with well established structures and ^{51}V NMR characteristics have been included in the present study. These comprise the bis-peroxo complexes **3** and $[VO(O_2)_2(Im)]^-$ (**7**, Im = imidazole), which had been computed with the same methodology before (*14,4b*), as well as the mono-peroxo complexes **4** and $[VO(O_2)(glygly')]^-$ (**8**), which have now been studied for the first time. Simulated and observed ^{51}V chemical shifts of these are collected in Table III.

Table III. Computed (B3LYP level) and Experimental $\delta(^{51}V)$ Values of Peroxovanadium Complexes

Species	CP-opt[a]	CPMD[b]	Expt (Ref.)
$[VO(O_2)_2(H_2O)]^-$ (**3**)	−588[c]	−719[e]	−697 (*3b*)
$[VO(O_2)_2(Im)]^-$ (**7**)	−722[d]	−833[d]	−748 (*15*)
$[VO(O_2)(H_2O)_3]^+$ (**4**)	−484[e]	−560[e]	−543 (*3b*)
$[VO(O_2)(glygly')]^-$ (**8**)	−616[e]	−656[e]	−649 (*16*)

[a,b]See corresponding footnotes in Table II.
[c]From reference (*14*).
[d]From reference (*7b*).
[e]This work.

The cationic species **4** is predominant in the vanadate/H_2O_2 system at low pH. According to the CPMD and NMR calculations, details of which will be published elsewhere, **4** is indicated to bind three rather than four water molecules in the first hydration sphere (*17*), in agrement with previous DFT results (*18*). The $\delta(^{51}V)$ value computed for aqueous **4** is in very good agreement with experiment; the same is true for anionic **3** and **8**, which are all found within ca. 20 ppm from experiment (Table III). A larger deviation is apparent for the imidazole complex **7**, which is computed too strongly shielded by 85 ppm.

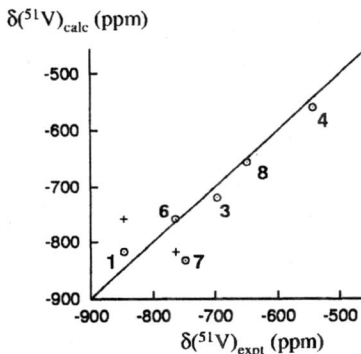

Figure 2. Plot of computed (B3LYP averaged for CPMD in water) vs. exper-imental $\delta(^{51}V)$ data of peroxovanadates, together with the ideal line. Assignment for 1 and 6 reversed (see text); +: original assignment as in Table II.

The overall performance of the theoretical data is illustrated in Figure 2, a plot of computed vs. experimental δ values of **1, 3, 4, 6-8**. When the B3LYP values for **1** and **6** are matched to the experimental data according to best fit (i.e. reversed from that in Table II), a slope of 0.92 is obtained for the linear regression line (*y*-intercept −73 ppm), together with a mean absolute deviation of 28 ppm over all points. The original assignment of **1** and **6** (i.e. that given in Table II, cf. the crosses in Figure 2), would afford a slope of 0.83 (intercept −137 ppm) and a mean absolute error of 46 ppm. Clearly, the former assignment results in a much better accord with experiment than the latter.

The present results thus provide some evidence that the experimental assignment of certain high-field ^{51}V NMR signals in the vanadate/H_2O_2 system may need revision. The accuracy of the present theoretical level would have to be further increased, however, in order to bestow a very high confidence on that proposal. The case of **7** indicates that errors approaching 100 ppm are possible. Even though small compared to the total ^{51}V chemical-shift range of several thousand ppm, such an error is still sizeable compared to the much smaller variations in the resonances of aqueous vanadium(V) complexes, which span but a few hundred ppm.[1] Within this limitation, however, it appears that the present MD-based methodology can be used to reproduce (and predict) $\delta(^{51}V)$ values of peroxovanadium comomplexes with reasonable accuracy.

On the other hand, the same approach has furnished systematic errors of 100 ppm and more for $\delta(^{51}V)$ of simple vanadate complexes such as $[H_2VO_4]^-$, $[VO_2(H_2O)_3]^+$, or $[VO_2(glygly')]^-$ (*4,6*), which all were computed too strongly shielded. Further work is needed to identify theoretical methods and levels that allow computation of ^{51}V NMR properties with a uniform, high accuracy.

Conclusion

Car-Parrinello MD simulations have been performed for aqueous solutions of highly charged vanadium complexes containing three or four peroxo ligands, which are believed to be the key species in the vanadate/H_2O_2 system at high pH. While the known dodecahedral $[V(O_2)_4]^{3-}$ (**1**) remained stable in such a simulation for several picoseconds, the proposed $[VO(O_2)_3]^{3-}$ (**2**) rapidly formed a hydroperoxy complex, $[VO(O_2)_2(OOH)]^{2-}$ (**6**), after rearrangement and capture of a proton from the solvent. Both structures were validated by computation of their ^{51}V chemical shifts, obtained as averages over the dynamic ensemble of the solution. These shifts appeared in the same highly shielded range between ca. −740 ppm and −840 ppm as the observed resonances ascribed to **1** and **2**. If the latter is identified with **6**, as suggested by the present CPMD simulations, and the assignment reversed, excellent accord between theoretical and experimental $\delta(^{51}V)$ values is achieved. This degree of agreement also extends to other prototypical peroxovanadium complexes, a singular deviation of ca. 90 ppm for a peroxo-imidazole complex notwithstanding.

The quest for theoretical methods capable of predicting metal chemical shifts with a high and uniform accuracy continues, but the combination of CPMD and NMR calculations that are feasible at present, already make for a potent tool to study structure and speciation of transition metal complexes in aqueous solution. Even when no assignments can be strictly proven with this approach, they can be strongly disfavored in case of unusually large discrepancies between calculated and observed chemical shifts or, as in the present case, when a proposed structure proves to be intrinsically unstable during the simulation in solution.

Acknowledgment

The author wishes to thank Prof. Walter Thiel and the Max-Planck society for support, and Prof. Valeria Conte for helpful discussion. Computations were performed on a local compute cluster of Intel Xeon and Opteron PCs at the MPI Mülheim, and on an IBM regatta supercomputer at the Rechenzentrum Garching of the Max-Planck society.

References

1. See for instance Rehder, D.; in: Pregosin, P. S. (Ed.) *Transition Metal Nuclear Magnetic Resonance*, Elsevier, Amsterdam, **1991**, p. 1.
2. (a) K_3VO_8: Fergusson, J. E.; Wilkins, C. J.; Young, J. F. *J. Chem. Soc.* **1962**, 2136; (b) $Na_3[V(O_2)_4]\cdot 14H_2O$: Won, T.-J.; Barnes, C. L.; Schlemper, E. O.; Thompson, R. C. *Inorg. Chem.* **1995**, *34*, 449.
3. (a) Howarth, O. W.; Hunt, J. R. *Dalton Trans.* **1979**, 1388; (b) Campbell, N. J.; Dengel, A. C.; Griffith, W. P. *Polyhedron.* **1989**, *11*, 1379.
4. See for instance (a) Bühl, M.; Parrinello, M., *Chem. Eur. J.* **2001**, *7*, 4487; (b) Bühl, M.; Schurhammer, R.; Imhof, P. *J. Am. Chem. Soc.* **2004**, *126*, 3310.
5. See for instance (a) Bühl, M.; Mauschick, F. T.; Terstegen, F.; Wrackmeyer, B. *Angew. Chem. Int. Ed.* **2002**, *41*, 2312; (b) Bühl, M.; Grigoleit, S.; Kabrede, H.; Mauschick, F. T. *Chem. Eur. J.* **2006**, *12*, 477.
6. Bühl, M. *Inorg. Chem.* **2005**, *44*, 6277.
7. Car, R.; Parrinello, M. *Phys. Rev. Lett.* **1985**, *55*, 2471.
8. CPMD Version 3.7.0, Copyright IBM Corp. 1990-2001, Copyright MPI für Festkörperforschung Stuttgart 1997 - 2001.
9. (a) Wachters, A. J. H. *J. Chem. Phys.* **1970**, *52*, 1033-1036. (b) Hay, P. J. *J. Chem. Phys.* **1977**, *66*, 4377-4384
10. O-H···O moieties were counted as H-bonded when the O···O distance is smaller than 3.5 Å and the O-H-O angle is larger than 140 degrees, cf.: Schwegler, E.; Galli, G.; Gygi, F. *Phys. Rev. Lett.* **2000**, *84*, 2429.
11. Tuckerman, M. E.; Chandra, A.; Marx, D. *Acc. Chem. Res.* **2006**, *39*, 151.
12. Because during the lifetime of **5a**, the two water molecules shown in Figure 2 were bonded quite closely to the terminal peroxo atom, they were included explicitly in the NMR computations, together with the point charges representing the remaining solvent molecules.
13. Bonchio, M.; Bortolini, O.; Conte, V.; Moro, S. *Eur. J. Inorg. Chem.* **2001**, 2913.
14. Anionic **3** has been studied in reference 4a using the BLYP functional in conjunction with CPMD and a slightly higher density for the aqueous solution; for the corresponding BP86 results see: Bühl, M.; Mauschick, F. T.; Schurhammer, R. in: *High Performance Computing in Science and Engineering, Munich 2002*, Wagner, S.; Hanke, W.; Bode, A.; Durst, F. (Eds.), Springer Verlag, Berlin, 2003, pp.189-199.

15. Crans, D. C.; Keramidas, A. D.; Hoover-Litty, H.; Anderson, O. P.; Miller, M. M.; Lemoine, L. M.; Pleasic-Williams, S.; Vandenberg, M.; Rossomando, A. J.; Sweet, L. J. *J. Am. Chem. Soc.* **1997**, *23*, 5447.

16. Einstein, F. W. B.; Batchelor, R. J.; Angus-Dunne, S. J.; Tracey, A. S. *Inorg. Chem.* **1996**, *35*, 1680.

17. Because in pure water, **4** rapidly lost a proton to the solvent, the NMR snapshots were taken from a simulation that contained an additional proton (in form of H_3O^+) in the box.

18. Bagno, A.; Conte, V.; Di Furia, F.; Moro, S. *J. Phys. Chem. A* **1997**, *101*, 4637.

Chapter 23

Biospeciation of Insulin–Mimetic VO(IV) Complexes

Tamás Kiss[1,2], Tamás Jakusch[2], Dominik Hollender[2],
and Ágnes Dörnyei[1]

[1]Department of Inorganic and Analytical Chemistry, University of Szeged,
P.O. Box 440, Szeged, H–6701, Hungary
[2]Bioinorganic Chemistry Research Group of the Hungarian Academy
of Sciences, University of Szeged, P.O. Box 440, Szeged, H–6701, Hungary

The possible transformation reactions of several insulin-mimetic vanadium(IV) complexes taking place in the organism after oral administration are discussed in the chapter. These reactions involve the (i) absorption processes in the gastro-intestinal tract, (ii) their transport in the blood stream and (iii) interactions with endogenous binding molecules in the glucose metabolizing cells. Modeling studies are used mostly to determine the actual chemical form of the vanadium(IV) complexes in the various biological environments. In some cases *in vitro* and *in vivo* biological results confirm the basic findings obtained by the modeling.

The experimental conditions of preparation of metal complexes with potential biological activity usually differ considerably from the milieu in the living systems, where they exert the biological effects. The solvent, the pH of the biological fluids, cells and tissues might be significantly different. Moreover, various molecules may also be present, having high affinity to the metal ion and accordingly, these molecules may partly or fully displace the original metal binding ligand(s) and thus the original complex may undergo transformations during the (i) absorption processes in the gastro-intestinal (GI) tract in the case of complexes administered orally, (ii) their transport processes in the blood

© 2007 American Chemical Society

stream with the serum/plasma components, and (iii) in the cell with its endogenous binding molecules. Accordingly, the original carrier ligand(s) might be fully lost in these processes and the real biological/physiological activity is bound to an entirely different chemical entity.

In the past few years we studied the potential transformation reactions of several insulin-mimetic (IM) vanadium(IV) complexes (1–3) after oral administration. The structures of the complexes are shown in Figure 1. The complexes are all neutral bis complexes of bidentate ligands having (O,O), (O,N) or (O,S) binding mode (4).

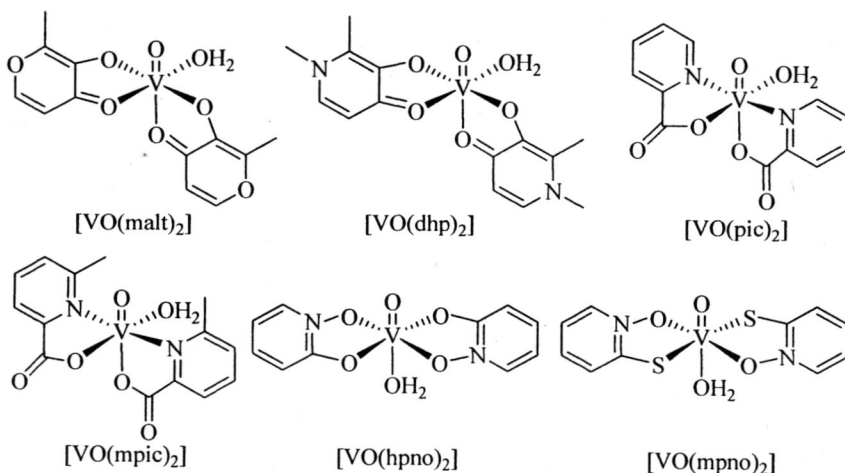

[VO(malt)₂]   [VO(dhp)₂]   [VO(pic)₂]

[VO(mpic)₂]   [VO(hpno)₂]   [VO(mpno)₂]

Figure 1. Structural formulae of the studied IM vanadium(IV) complexes (malt=maltolate, dhp=1,2-dimethyl-3-hydroxy-4-pyridinone, pic=picolinate, mpic=6-methyl-picolinate, hpno=2-hydroxy-pyridine-N-oxide, mpno=2-mercapto-pyridine-N-oxide)

Their possible transformation reactions were followed taking place during the absorption, serum transport and cell processes. In the knowledge of the speciation description of these IM complexes with the constituents of the various biological fluids, modeling calculations were made to identify the actual chemical forms existing under biological conditions. pH-potentiometric technique was used to describe the reactions quantitatively with the low molecular mass components, and spectroscopic methods to characterize these interactions with the high molecular mass protein components. Ultrafiltration served to separate the high and low molecular mass fraction bound metal ion content.

Absorption Processes of Insulin-Mimetic Vanadium Complexes in the GI Tract

Due to parallel protonation processes of the metal-binding sites of the coordinating ligands, these neutral bis complexes (allowing good passive membrane transport) will certainly decompose in the acidic pH range, e.g. at the pH (~2) of the gastric juice. This is demonstrated in Figure 2, in which the species distribution of two well-studied systems, the VO(IV)–malt and the VO(IV)–dhp systems, is depicted as a function of pH.

Accordingly, all other exogenous and endogenous biomolecules present in the stomach or intestine, where the complexes are absorbed, may play a role in VO(IV) binding. Interactions with these molecules may change the charge conditions of the complex unfavorably, which will decrease their absorption efficacy. This certainly has to be taken into account in the formulation of the drug (e.g. by encapsulation techniques, these problems may well be overcome). The recent results of Sakurai et al. (5) support this prediction. In their study $VOSO_4$ was administered orally in various ways: in solution, in gelatin capsules, and in enteric-coated capsules (ECC). It was found that administration of the ECC containing vanadyl salt improved the metal ion absorption compared with that associated with the other two ways. In this case the encapsulated metal ion could reach the ileum, where its absorption was more efficient than at other gastrointestinal sites.

Transport Processes of the Insulin-Mimetic Vanadium Complexes in the Blood

After absorption, during their transport in the bloodstream, complex formation with the serum components, as active VO(IV) binders also has to be considered. The interactions of several potential IM compounds with the high molecular mass (HMM) protein constituents, e.g. albumin (HSA) and transferrin (Tf) and some of the low molecular mass (LMM) constituents (the most potent binders), e.g. lactate, phosphate, oxalate and citrate, have been studied in detail (6–14). The results of model calculations for serum conditions (14) are depicted in Figure 3.

It is seen in Figure 3 that (i) only the pyridinone derivative, dhp is a strong enough carrier to preserve a significant proportion of the VO(IV) in the original complex (10,12); in the other cases, the carrier ligands are displaced by serum components. Accordingly, at a first glance, the most important role of the carrier ligand seems to be to facilitate the absorption of VO(IV). (ii) of the two important HMM binders, Tf is much more efficient than HSA and will displace

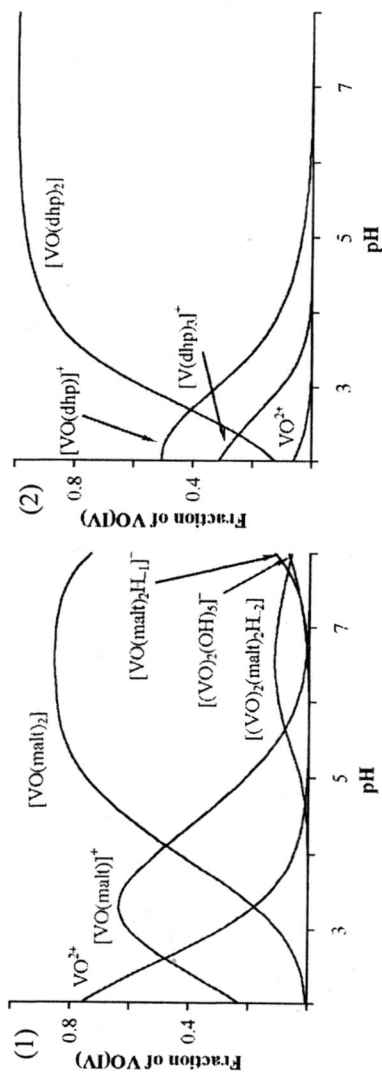

Figure 2. Species distribution diagram of (1) VO(IV)–malt 1:2, (2) VO(IV)–dhp 1:2 systems, $c(VO(IV)) = 1mM$ (based on data reported in Refs 6–10)

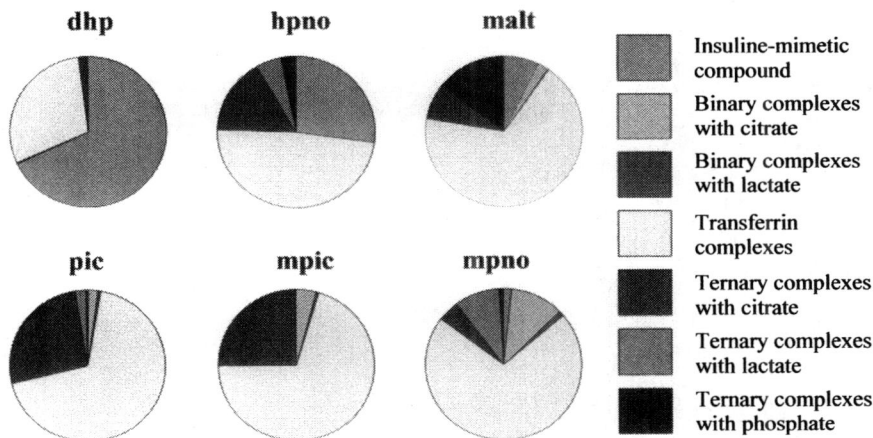

Figure 3. Speciation of various IM VO(IV) compounds (100 μM) in serum at pH 7.4 (based on data reported in Refs 6–14)

30–70% of the original carrier from the complex. (At the biologically more relevant VO(IV) concentrations (<5 mM), practically all the VO(IV) is bound to Tf.) Similarly, the predominant binding of vanadate(V) to Tf in human plasma and the negligible role of HSA in transporting vanadium was found in detailed in vitro measurements and reported by Heinemann et al. (*15*). (iii) among the LMM binders, citrate is the only "active" component, able to influence the solution state of these insulin mimetics. At physiological pH, VO(IV) exists mostly as the VO(IV)-citrate binary complex and as the VO(IV)-IM ligand-citrate mixed ligand complex, but in different proportions.

The decisive role of Tf in VO(IV) binding could be demonstrated by EPR. It has been found (*13,16,17*) that, in the presence of apoTf, all carrier complexes showed anisotropic EPR spectra even at room temperature (because of the slow tumbling motion of the VO(IV)-protein complex, the parallel and the perpendicular components could not be averaged), indicating unambiguously, that interactions between the carrier complex of VO(IV) and the protein. As seen in Figure 4, in the case of VO(pic)$_2$, the RT spectrum is practically the same as that of VO(IV)–apoTf, suggesting a complete displacement of the carrier ligand by apoTf. At the same time, in the case of the VO(dhp)$_2$ complex, signals of the anisotropic carrier complex can also be observed, suggesting only partial displacement of the carrier ligand. A similar composite spectrum of the isotropic VO(pic)$_2$ and the anisotropic VO(IV)–HSA can be observed in the VO(IV)–pic–HSA system, indicating partial displacement reaction between the protein and

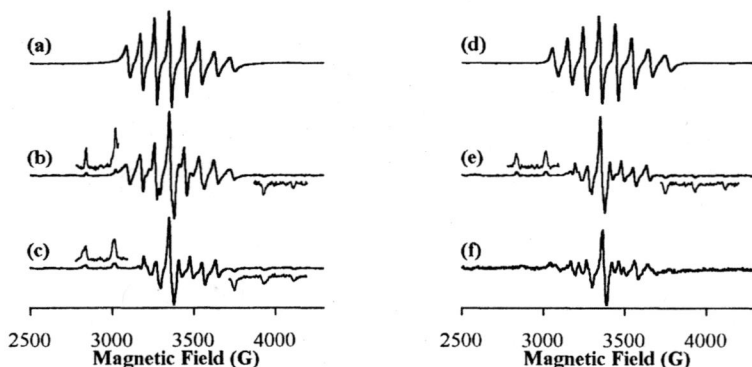

Figure 4. Room temperature EPR spectra of (a) VO(IV)–dhp 1:2,(b) VO(IV)–dhp–apoTf 1:2:1, (c) VO(IV)–apoTf 1:1, (d) VO(IV)–pic 1:2, (e) VO(IV)–pic–apoTf 1:2:1 ,(f) VO(IV)–pic–HSA 1:2:1 systems at pH 7.4, c(VO(IV))=1mM

the carrier ligand. These are in complete agreement with the species distributions depicted in Figure 3.

The results strongly indicate that vanadium must be released from the carrier compound in order for it to be pharmacologically active. Perhaps the most convincing point is the different pharmacokinetics of disappearance of vanadium and the carrier ligand (18,19). Although, in principle the ligand should be present at roughly twice the concentration of vanadium in the blood, it is found to be present at a lower concentration compared with vanadium from the 1-hour time point onwards. Thus, a number of recent studies have demonstrated fairly conclusively that IM vanadium complexes dissociate rapidly once ingested or injected. In other words, the eventual in vivo metabolic fate for chelated vanadium complexes differs little, if at all, from that of non-complexed vanadium compounds, when they are administered orally.

The partial displacement reactions of the carrier ligand of VO(dhp)$_2$ by apoTf and VO(pic)$_2$ by HSA could be used to estimate the binding constants of the VO(IV)-protein interactions. The quantitative evaluation and simulation of the RT EPR spectra provided us the conditional binding constants. In order to make calculations simple, as a first approximation, ternary complex formation between the carrier ligand and the protein was neglected. The conditional constant for binding of the first VO(IV) to apoTf was found to be, $\log K_1 = 14.3 \pm 0.6$ (25°C, pH 7.4, 0.025 M HCO$_3^-$ (14)). In the case of HSA only a limiting value could be obtained: $\log K = 10.0 \pm 1.0$ (25°C, pH 7.4, (20)). This 4 orders of magnitude difference in the VO(IV) binding constants of the two proteins means that HSA is not an efficient VO(IV) binder in the presence of apoTf. A qualitative confirmation of this result was achieved by membrane separation of the low molecular mass and the high molecular mass bound VO(IV) using a 10 kDa membrane in VO(pic)$_2$–protein samples. The results showed that apoTf could displace completely the carrier ligand from the VO(pic)$_2$ complex, while HSA was not able at all.

Based on similar ultrafiltration measurements Heinemann *et al.* obtained the same result for vanadate(V), namely the binding capacity of HSA was about 1000-fold lower than that of fresh human plasma and Tf (*15*). It has to be mentioned that this value is very much different from that reported by Chasteen *et al.* (*21*), who published a data of K(VO(IV)-Tf)/K(VO(IV)-HSA) ~ 6.

So far any ternary interactions between the proteins and the VO(IV) carrier complexes were neglected. However, NMR relaxation studies indicated that this was a simplified approach. (In these measurements the effect of the paramagnetic VO(IV) centers was measured on the spin-lattice relaxation rate of the water protons.) Namely, the ^{1}H NMR relaxation dispersion curves of the VO(pic)$_2$, the VO(IV)–protein and VO(pic)$_2$–protein samples differ from each other, indicating some ternary interaction. This may be primary interaction by partial displacement of the binding donors of the protein, or secondary through hydrogen bonding or hydrophobic interaction(s).

It is worth mentioning that the NMR relaxation measurements confirmed the results obtained for the relative VO(IV) binding strength of the two serum proteins. Namely as shown in Figure 5, when VO(IV) was added to a mixture of apoTf and HSA the relaxation curve practically coincided with that obtained in the absence of HSA. (The low relaxivity values obtained in the presence of only HSA are due to low paramagnetic VO(IV) concentration at pH 7.4, as at this pH most of the metal ion is present in EPR silent hydroxo bridged dinuclear species (*22*)).

Recently a detailed paper was published by Orvig et al. (*17*) about the interactions of VO(malt)$_2$ (BMOV) with the serum proteins. We agree with them in the main findings with one exception. Although they write that "Reaction of BMOV with apo-transferrin can be expected to completely dissociate the complex, however, it may not be the dominant effect in delivery of vanadium from vanadyl chelates to target tissues." Instead they assume that "adduct or ternary albumin complexes could be the pharmacologically active species, or at the very least the main method of vanadium delivery to cells." We also have evidence for the ternary complex formation with albumin, however in the presence of apoTf, VO(IV) binding of HSA is suppressed and thus, so is its ternary complex formation. At the same time, both RT EPR and CD spectral results support ternary complex formation with apoTf.

Namely, coordination of VO(IV) to apoTf could be followed by CD (see Figure 6) and EPR. When the strongest carrier ligand the pyridinone derivative dhp, which was able to displace the protein, at least partially, was added in increasing amount to the VO(IV)–apoTf 1:1 system, instead of a continuous decrease in the intensity of the characteristic CD signal, first a slight, but

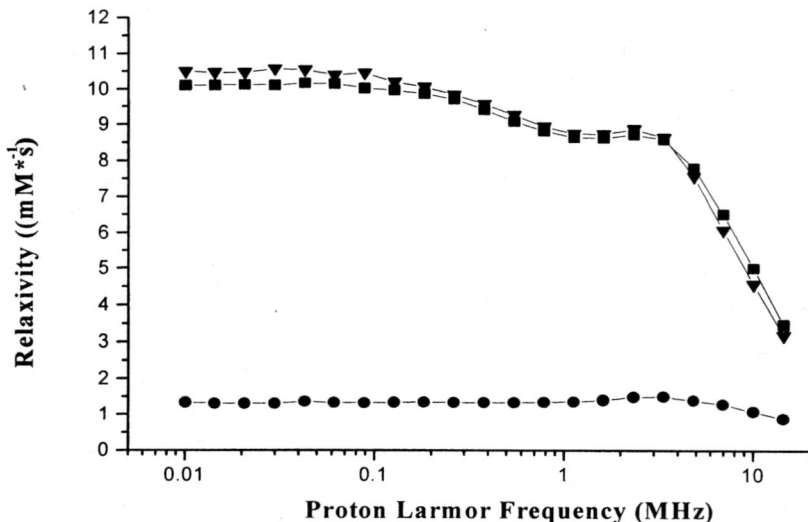

*Figure 5. NMR relaxation dispersion curves for the ●=VO(IV)–HSA 1:1,
■=VO(IV)–apoTf 1:1, ▼=VO(IV)–apoTF–HSA 1:1:1 systems
at pH 7.4, c(VO(IV)) =1 mM*

significant shift could be detected, while the RT EPR spectra changed only slightly (not shown). The spectral observations indicate the formation of a new species (with different CD spectral feature) obviously a ternary complex VO(apoTf)(dhp).

A similar behavior was observed for the corresponding maltol system. However, a higher excess of maltol was necessary for the formation of the ternary VO(apoTf)(malt) complex as compared with dhp. This difference can be explained by fact that the VO(malt)$_2$ complex is 6.5 orders of magnitude less stable than the bis complex of dhp (*9, 10*). A joint evaluation of the CD and EPR spectra provided the stability constants of all the species being in equilibrium: VO(apoTf)(malt); (VO)$_2$(apoTf)(malt); (VO)$_2$(apoTf)(malt)$_2$ and it allowed us also to refine the logβ value of the VO(apoTf) complex (*23*).

The role of albumin in vanadium binding is still an issue of contradiction (*3,17,21,23*). The CD and RT EPR spectral characteristics of the VO(IV)–apoTf–malt 1:1:1 was studied in the presence and absence of HSA. As can be seen in Figure 7 the albumin practically has no effect on the CD spectrum of the VO(IV)–apoTf–malt ternary system. Accordingly, the role of albumin in vanadium transport is probably negligible at such condition.

In the knowledge of the stability constants of all the species formed in the vanadium-IM complexes-protein systems species distribution of the vanadium for serum conditions were calculated and presented in Figure 8. As it is seen, in

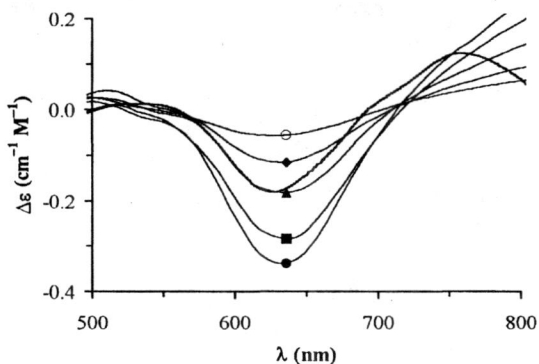

Figure 6. The effect of increasing dhp concentration on the CD spectra of the VO(IV)–apoTf 1:1 systems at pH 7.4, c(VO(IV))=1mM (dhp concentration: — = 0.0, ● = 1.2, ■ =2.4, ▲ = 4.9, ◆ = 10.2, ○ = 21.1mM)

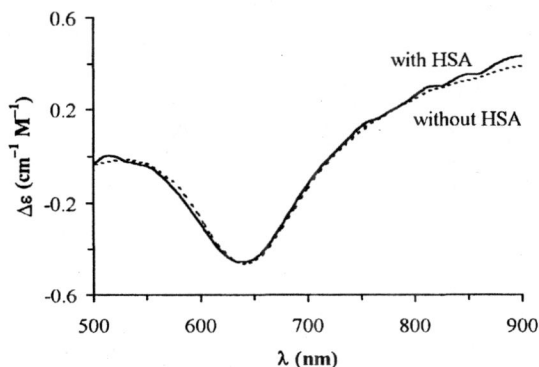

Figure 7. Effect of equimolar HSA on the CD spectra of the VO(IV)–apoTf–malt 1:1:4 system at pH 7.4, c(VO(IV))=1mM

the biologically relevant concentration range (<10μM) VO(IV) is bound mostly in the form of the 1:1 complex VO(apoTf) and formation of the ternary complexes are not significant (*24*).

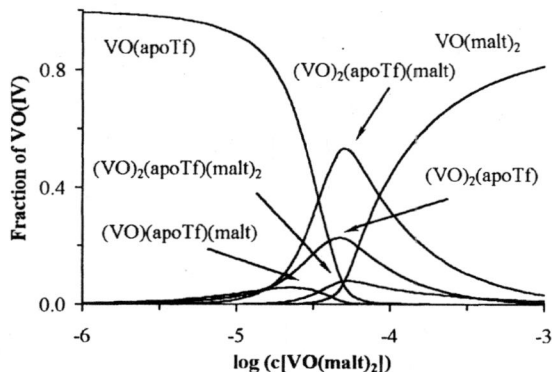

Figure 8. Species distribution diagram of the complexes formed in the VO(IV)–apoTf–malt system at pH 7.4 as a function of increasing VO(malt)₂ concentrations

All the measurements reported here were carried out with metal ion-free apoTf. The results can be applied with a good approximation to the interaction of VO(IV) with Tf too, as only ~30% of the metal-binding sites of the protein are saturated with Fe(III) under normal serum conditions (*24*), and thus there are enough free sites to bind other, mostly hard metal ions (e.g. Al(III) or VO(IV)) and to transport them in the blood stream. This partial saturation of Tf was taken into account in the speciation calculations.

Interactions of Insulin-Mimetic VO(IV) Complexes with Cell Constituents

In vivo NMR studies on rats showed (*25*) that almost independently of the oxidation state of the vanadium compound, the metal ion is transported in the blood in the IV oxidation state. Binding of VO(IV) to ligands, mostly to Tf (see above) prevents its oxidation to vanadate(V), which would otherwise occur rapidly at intracellular pH. Nevertheless, oxidation may occur resulting in the formation of a limited amount of vanadate(V). Accordingly, vanadium may be assumed to enter the cell, either in oxidation state IV through the transferrin receptor following the iron pathway, or in the oxidation state V, through the phosphate or sulfate pathway. In the intracellular medium, reducing agents can redox-interact with vanadate(V). A frequently discussed candidate for the reduction is glutathione (GSH) (26), although it is a rather ineffectual reducing agent (27,28). To what extent such redox interactions take place largely depends on the stabilization of vanadium in the oxidation state V or IV through the complexation of cell constituents such as GSH, GSSG, ATP, etc. A high intracellular excess of GSH increases the formation possibility of VO(IV) and its complexation with either GSH or GSSG. Both have been shown to be reasonably potent binders for VO(IV) (26,29,30). Other effective reducing agents, such as NADH or ascorbate, may cause the formation of V(III) species (31,32). Further hydrolytic degradation of VO(IV) may be responsible for the re-oxidation to vanadate(V). Via the redox and complexation reactions, the finely-tuned speciation of vanadium may lead to the efficiency of the metal to mimic the effects of insulin.

The above discussed results revealed that the original complexes may remain partially intact, i.e. they keep the original carrier ligand bound to vanadium(IV,V), although the endogenous binders of the biological fluids displace them partially e.g. through the formation of ternary complexes. Accordingly, they may find a way to reach the cell, where they will compete with some of the cell constituents for the central metal ion.

In order to assess the molecular form of vanadium IM complexes in cells, the interactions in the model systems of VO(IV)-malt and VO(IV)-dipicolinate with various cell components were studied by employing pH-potentiometric, and spectroscopic (EPR, CD and UV-Vis) techniques (33).

In cells, GSH is present in high excess compared with vanadium; its intracellular concentration level is millimolar, about 3 orders of magnitude more than the biologically relevant concentration of vanadium. According to speciation studies for VO(IV) concentrations above ~1 mM and more than a 10-fold excess of the ligand, the bis amino acid type binding mode with the donor set $2 \times (COO^-, NH_2)_{eq}$ is relevant in the pH range 5.0–6.5 (26,30,34,35). Participation of the thiolate donor occurs above pH ~ 7.

Concerning the competition between the carrier ligand maltol and GSH, at a 25-fold excess of GSH and 2-fold excess of carrier ligand, ternary complex formation could be detected by pH-potentiometry only with rather high uncertainty; this excess of GSH was not enough to prevent the hydrolysis of the metal ion and formation of the oligonuclear hydroxo species $\{[(VO)_2(OH)_5]^-\}_n$ (22) at pH > 8.0. At the same time the high buffer capacity of the ligand excess strongly limited the applicability of pH-potentiometry in speciation, at such high ligand concentrations. Spectroscopic measurements (EPR, UV-Vis and CD), however, provided more unambiguous results.

The new EPR signals detected (see Figure 9) could be assigned to the ternary species (VO(malt)(GSH)H$_2$, [VO(malt)$_2$(GSH)H$_2$]$^-$, [VO(malt)$_2$-

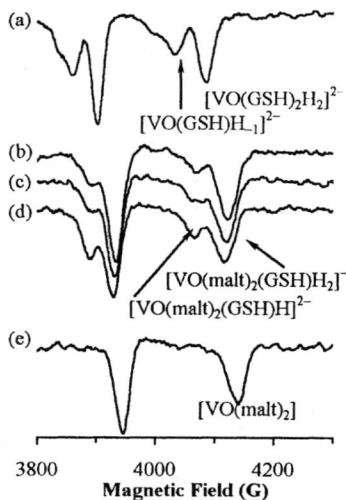

Figure 9. High field range of the EPR spectra at 77 K of frozen solutions containing:, (a) VO(IV)–GSH 1:50, VO–malt–GSH (b) 1:2:25 ,(c)1:2:50, (d) 1:2:100, and (e) VO(IV)–malt 1:2 systems, c(VO(IV)) ≈ 4mM at pH 7.0.

$(GSH)H]^{2-}$). Considering the EPR parameters (the most probably binding set of the complexes formed was estimated on the basis of the additivity rule of the EPR parameters developed by Chasteen (*36*)) and the low CD signals recorded, we concluded (*33*) that GSH most probable coordinates at the Gly end via (COO⁻, H$_2$O/O-amide) donors, while participation of the thiolate donor occurs only at pH > 7.

Among the LMM binders, the widely distributed adenosine 5'-triphosphate (ATP) may be also of importance (*37*), as it efficiently binds VO(IV) and is present in millimolar concentrations in cells. Comparing the VO(IV) complex forming properties of ATP and GSH, it can be concluded that in the whole pH range ATP is more efficient VO(IV) binder. ATP coordinates to VO(IV) through the terminal phosphate donor(s) in the weakly acidic and neutral pH range to yield $[VO(ATP)H_x]^{x-2}$ (x = 2, 1, 0) and $[VO(ATP)_2]^{6-}$ (*37*). Increasing the pH, as the proton competition for the alcoholate donors decreases, the ribose moiety becomes a more efficient binding site. In the slightly basic pH range, the complex $[VO(ATP)_2H_{-2}]^{8-}$ forms, involving a mixed binding mode, one ATP coordinating through the phosphate chain and the other through the ribose moiety. In the species $[VO(ATP)_2H_{-4}]^{10-}$, both ATP molecules coordinate to the VO(IV) via the ribose residue (*37*). The CD spectra furnish information on the species in which the ribose moiety is coordinated to the VO(IV); these species are mostly formed above the physiological pH (*33*).

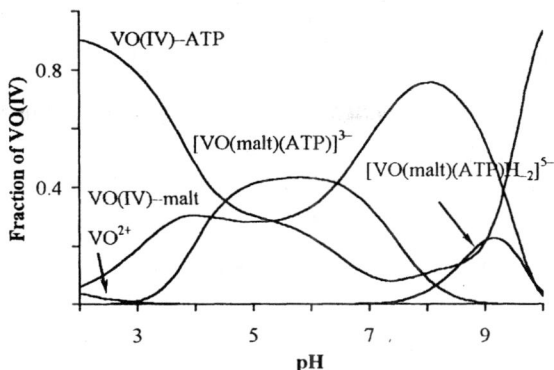

Figure 10. Species distribution diagram of the VO(IV)–malt–ATP 1:2:10 system, c(VO(IV)) = 4mM.

In the cells, ATP is present in high excess relative to VO(IV) and also to the carrier ligands. Our speciation calculations indicated that, in the ternary system with ATP as a strong VO(IV) binder, ATP might displace one of the maltols from the bis complex and/or two waters from the mono complex in the coordination sphere of VO(IV). As a result, ternary complexes will exist besides binary maltolato complexes also at physiological pH (see Figure 10). The potentiometric data could be fitted by assuming the formation of the species $[VO(malt)(ATP)]^{3-}$ and $[VO(malt)(ATP)H_{-2}]^{5-}$ in the system (*6,33*). Because of the significant difference in the type of donor groups involved in the coordination, ternary complex formation is clearly indicated by EPR spectroscopy. At pH 8–9 a significant change in the CD spectra (not shown) unambiguously indicated the coordination of the ribose residue, corresponding to the formation of $[VO(malt)(ATP)H_{-2}]^{5-}$.

When ATP and GSH are simultaneously considered as potential VO(IV) binders, GSH is not expected to be able to compete with ATP for binding to VO(IV) since ATP is a much stronger binder. The competition reaction was studied by EPR spectroscopy in the absence and in the presence of the IM complexes (see Figure 11). For the VO(IV)–ATP–GSH ternary system, when ATP was in a 10-fold excess relative to VO(IV), only the EPR signals of the VO(IV)–ATP complexes were detected. New signals appeared only at a 2.5-fold excess of ATP (EPR parameters: $A_{||}$ = $172.1\cdot10^{-4}$ cm^{-1} and $165.7\cdot10^{-4}$ cm^{-1}). Comparison of the EPR spectra of solutions containing VO(IV), maltolate, ATP and GSH at pH ~ 7 revealed significant changes only when ATP was not present in at least a 10-fold excess (see above). Below such an excess of ATP, the changes in the ratio of the predominating signals seem to indicate the participation of GSH in complex formation.

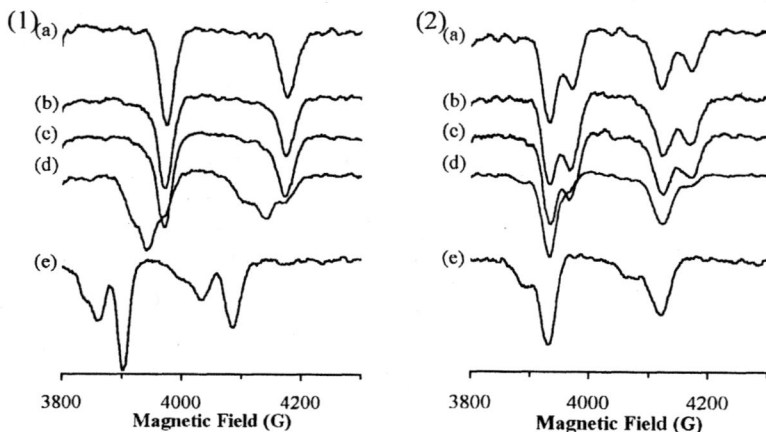

Figure 11. High field range of the EPR spectra at 77 K of frozen solutions: (1) without carrier ligand, (2) with 8mM malt, (a) VO(IV)–ATP 1:10, VO(IV)–ATP–GSH (b) 1:10:25 ,(c)1:10:50, (d) 1:2.5:50 and (e) VO(IV)–GSH 1:50; c(VO(IV)) ≈ 4mM at pH 7.0

With respect to the two important cell constituents GSH and ATP, our results indicate that, if the carrier ligands can somehow find a way to enter the cell, strong VO(IV)-binder cell constituents will partly displace the carrier ligands, and ternary complexes with relevant biomolecules of the cell will be formed. Some of the ternary species are highly anionic, which are partly neutralized and are probably in ion-paring with cations, K^+ and Mg^{2+} of the cell.

From among the important cell constituents, GSH will possibly take part in the reduction of vanadate(V) to VO(IV) and will help keep VO(IV) in this oxidation state. As a strong VO(IV) binder, ATP will chelate the metal ion, forming binary and/or ternary complexes. This strongly suggests that ATP binds relevant VO(IV) species under cell conditions, and thus might somehow be involved in the insulin-mimetic action of VO(IV) compounds. However, the time courses of these parallel redox and complexation reactions require further investigations.

Conclusion

Several examples have been given how modeling calculations can help us in characterizing the solution state of VO(IV) in the body in order to obtain information about the biologically important/active form of the metal ion in the different biofluids and tissues. We are aware of the fact that modeling is only modeling. Especially thermodynamic models can be strongly affected by kinetic factors. Accordingly, the models always need in vitro or even *in vivo* confirmation. Such measurements are not always easy to perform *e.g.* for technical reasons. In these cases modeling remains a way for the recognition, however, only with limited use.

Acknowledgment

The authors thank Professors I. Bertini and J. Costa Pessoa for the possibility of carrying out the NMRD and a part of the EPR measurements and for the valuable discussions in the evaluation of the results. This work was performed in the frame of the COST D21/009/01 project (Insulin-Mimetic Vanadium Compounds) and the Hungarian-Portuguese Intergovernmental S & T Co-operation Program for 2004–2005 and supported by the National Science Research Fund (OTKA No. T49417 and NI 61786).

References

1. Sechter, Y.; Karlish, S. J. D. *Nature* **1980**, *284*, 556.
2. Thompson, K. H.; McNeill, J. H.; Orvig, C. *Chem. Rev.* **1990**, *99*, 2885–2891.
3. Kiss, T.; Jakusch, T. In *Metallotherapeutic Drugs and Metal-based Diagnostic Agents;* Gielen, M.; Tiekink, E. R. T. Eds.; Wiley: Chichester, UK, 2005; pp 143–156 (and references therein).
4. Rehder, D.; Costa Pessoa, J.; Geraldes, C. F. G. C.; Castro, M. M. C. A.; Kabanos, T.; Kiss, T.; Meier, B.; Micera, G.; Pettersson, L.; Rangel, M.; Salifoglou, A.; Turel, I.; Wang, D. *J. Biol. Inorg. Chem.* **2002**, *7*, 384–396.
5. Sakurai, H.; Fugono, J.; Yasui, H. *Mini-Rev. Med. Chem.* **2004**, *4*, 41–48.
6. Kiss, T.; Kiss, E.; Micera, G.; Sanna, D. *Inorg. Chim. Acta* **1998**, *283*, 202–210.
7. Buglyo, P.; Kiss, E.; Fabian, I.; Kiss, T.; Sanna, D.; Garribba, E.; Micera, G. *Inorg. Chim. Acta* **2000**, *306*, 174–183.
8. Kiss, E.; Petrohan, K.; Sanna, D.; Garribba, E.; Micera, G.; Kiss, T. *J. Inorg. Biochem.* **2000**, *78*, 97–108.
9. Kiss, T.; Kiss, E.; Garribba, E.; Sakurai, H. *J. Inorg. Biochem.* **2000**, *80*, 65–73.
10. Buglyo, P.; Kiss, T.; Kiss, E.; Sanna, D.; Garribba, E.; Micera, G. *J. Chem. Soc. Dalton Trans.* **2002**, 2275–2282.
11. Harris, W.R. *Clin. Chem.* **1992**, *38*, 1809–1818.
12. Sakurai, H.; Tamura, A.; Fugano, J.; Yasui, H.; Kiss, T. *Coord. Chem. Rev.* **2003**, *245*, 31–37.

338

13. Yasui, H.; Kunori, Y.; Sakurai, H. *Chem. Lett.* **2003**, *32*, 1032–1033.
14. Kiss, T.; Jakusch, T.; Bouhsina, S.; Sakurai, H.; Enyedy, É. A. *Eur. J. Inorg. Chem.* (in press).
15. Heinemann, G.; Fichtl, B.; Mentler, M.; Vogt, W. *J. Inorg. Biochem.* **2002**, *90*, 38–42.
16. Thompson, K. H.; Liboiron, B. D.; Hanson, G. R.; Orvig, C. In *Medicinal Inorganic Chemistry;* Sessier, J. L.; Doctrow, S. R.; McMurry, T. J.; Lippard, S.J., Eds.; ACS Symposium Series 903, ACS: Washington, DC, 2005; pp 384–399.
17. Liborion, B. D.; Thompson, K. H.; Hanson, G. R.; Lam, E.; Aebischer, N.; Orvig, C. *J. Am. Chem. Soc.* **2005**, *127*, 5104–5115.
18. Thompson, K. H.; Tsukuda, Y.; Xu, Z.; Bartell, M.; McNeill, J. H.; Orvig, C. *Biol. Trace Elem. Res.* **2002**, *86*, 31–44.
19. Thompson, K. H.; Liboiron, B. D.; Sun, Y.; Bellman, K. D. D.; Setyawati, I. A.; Patrick, B. O.; Karunarante, V.; Rawji, G.; Wheeler, J.; Sutton, K.; Bhanot, S.; Cassidy, C.; McNeill, J. H.; Nguen, V. G.; Orvig, C. *J. Biol. Inorg. Chem.* **2003**, *8*, 66–74.
20. Kiss, T.; Jakusch, T.; Bouhsina, S.; Sakurai, H.; Enyedy, É. A. *Eur. J. Inorg. Chem.* **2006** (in press)
21. Chasteen, N. D.;Grady, J. K.; Holloway, C. E. *Inorg. Chem.* **1986**, *25*, 2754–2760.
22. Vilas Boas, L. F.; Costa Pessoa, J. In *Comprehensive Coordination Chemistry, Vanadium*; Wilkinson, G.; Gillard, R. D.; McCleverty, J. A., Eds.; Pergamon: Oxford, UK, 1987; Vol. 3, pp 453–583.
23. Jakusch, T.; Hollender, D.; Costa Pessoa, J.; Kiss, T. (not published).
24. Sun, H. Z.; Cox, M. C.; Li, H. Y.; Sadler, P. J. *Structure and Bonding* **1997**, *88*, 71–102.
25. Sakurai, H.; Shimomura, S.; Fukazawa, K.; Ishizu, K. *Biochem. Biophys. Res. Commun.* **1980**, *96*, 293–298.
26. Goda, T.; Sakurai, H.; Yashimura, T. *Nippon Kagaku Kaishi* **1998**, 654–661.
27. Li, J.; Elberg, G.; Crans, D. C.; Sechter, Y. *Biochemistry* **1996**, *35*, 8314–8318.
28. Macara, I. G.; Kustin, K.; Cantley, L. C. *Biochim. Biophys. Acta* **1980**, *629*, 95–106.
29. Costa Pessoa, J.; Tomaz, I.; Kiss, T.; Buglyo, P. *J. Inorg. Biochem.* **2001**, *84*, 259–270.
30. Costa Pessoa, J.; Tomaz, I.; Kiss, T.; Kiss, E.; Buglyo, P. *J. Biol. Inorg. Chem.* **2002**, *7*, 225–240.
31. Stern, A.; Davison, A. J.; Wu, Q.; Moon, J. *Arch. Biochem. Biophys.* **1992**, *299*, 125–128.
32. Kanamori, K.; Kinebuchi, Y.; Michibata, H. *Chem. Lett.* **1997**, 423–424.

33. Dörnyei, Á.; Marcão, S.; Costa Pessoa, J.; Jakusch, T.; Kiss, T. *Eur. J. Inorg. Chem.* **2006** (in press).

34. Dessi, A.; Micera, G.; Sanna, D. *J. Inorg. Biochem.* **1993**, *52*, 275–286.

35. Tasiopoulos, A. J.; Troganis, A. N.; Evangelou, A.; Raptopoulou, C. P.; Terzis, A.; Deligiannakis, Y. G.; Kabanos, T. A. *Chem. Eur. J.* **1999**, *5*, 910–921.

36. Chasteen, N.D. In *Biological Magnetic Resonance*; Lawrence, J.; Berliner, L. J.; Reuben, J., Eds.; Plenum: New York, 1981; Vol 3, pp 53–119.

37. Alberico, E.; Dewaele, D.; Kiss, T.; Micera, G. *J. Chem. Soc. Dalton Trans.* **1995**, *3*, 425–430.

Chapter 24

Vanadium Schiff Base Complexes: Chemistry, Properties, and Concerns about Possible Therapeutic Applications

João Costa Pessoa[1], I. Cavaco[1], I. Correia[1], I. Tomaz[1], P. Adão[1],
I. Vale[1], V. Ribeiro[2], M. M. C. A. Castro[3], and C. C. F. G. Geraldes[3]

[1]Instituto Superior Técnico, Centro Química Estrutural, Av. Rovisco Pais,
1049–001 Lisboa, Portugal
[2]Centro de Biomedicina Molecular e Estrutural, Universidade do Algarve,
Faro, Portugal
[3]Departmento de Bioquímica, Faculdade de Ciências e Tecnologia e Centro
de Neurociências de Coimbra, Universidade de Coimbra, P.O. Box 3126,
3001–401 Coimbra, Portugal

We discuss aspects related to the speciation of vanadium
compounds (VCs) with salen-type ligands, as well as $V^{IV}O(acac)_2$
and $V^{IV}O(phen)_2(SO_4)$, namely the stability of their V^{IV} and V^V
complexes in aqueous aerobic solutions at pH~7, and
consequences on the study of toxicity, insulin mimetic and
nuclease activity studies. We show that in these aerobic aqueous
solutions V^{IV} Schiff-base (SB) complexes of the salen-type are
normally not stable to oxidation to V^V and to hydrolysis of the
ligand. $V^{IV}O(phen)_2(SO_4)$ and $V^{IV}O(acac)_2$ and are also not
stable to oxidation, and a significant decomposition of these
complexes occurs within the first 30 minutes after their
dissolution. Therefore, when these VCs are used for *in vitro* or *in
vivo* studies the active species is not known. Reduction of the
salen SB to give amine compounds yields salan ligands which
form much more stable complexes than the parent SB. When
dissolved in non-degassed aqueous solutions at pH~7 the V^{IV}-
salan compounds oxidise to V^V-salan complexes, but no
hydrolysis is detected. At least with the cell lines tested these
VCs are not toxic possibly because they do not enter the cells

© 2007 American Chemical Society

significantly. We emphasize that to understand if the VCs enter the cells or not is an important point to sort out in insulin-mimetic studies. In fact we also report some studies of nuclease activity of several salen and salan VCs, as well as of $V^{IV}O(acac)_2$ and $V^{IV}O(phen)_2(SO_4)$ with plasmid DNA. Many of these VCs show nuclease activity even in the absence of activating agents, so toxicity resulting from this may occur. We also study several parameters relevant for the nuclease activity of VCs, namely the nature and concentration of the buffer used.

Introduction

The presence of vanadium in biological systems and its insulin-enhancing action (*1*) and anticancer activity (*2*) has driven a considerable amount of research. Particular interest has been given to the study of the potential benefits of VCs as oral insulin substitutes for the treatment of diabetes. Coordinated ligands are said to be able to improve the absorption of vanadium, reducing the dose necessary for producing equivalent effects. However, the molecular mechanisms by which the VCs exert their insulin enhancing effects have not been clarified. In fact, in *in vitro* studies the nature of the vanadium species acting is often not known, and additionally in *in vivo* studies the role of serum proteins (e.g. albumin, transferrin) and that of insulin, which may act synergistically, is also not yet understood.

Several VCs with the tetradentate SB salen-type ligands have been proposed for use as insulin enhancing agents. The ability of $V^{IV}O(salen)$ to reverse the hyperglycemic condition of alloxan-induced diabetic rats to near normal has been reported (*3*). However, the rats tended to become hypoglycemic, and withdrawal of treatment brought an immediate return to hyperglycemia.

To evaluate the use of a particular VC for oral treatment it must be understood:
1. How efficiently and in what form is the compound absorbed in the gastro-intestinal tract? The evaluation of the lipophilic-hydrophilic balance, and its speciation as a function of pH are among the important aspects to clarify.
2. How is vanadium transported in the bloodstream? The understanding of the binding of vanadium or the VC to serum proteins (e.g. serum albumin and transferrin) is important. Inorganic V^{IV} possibly binds stronger to transferrin, but other VCs may form protein-VC ternary complexes with albumin, and this preference may be changed.
3. How is vanadium delivered to cells? Protein-VC ternary or quaternary complexes may be more efficient in this respect than simple transport of inorganic V^{IV} or V^{V}.

4. Mechanism of action of the VC? For example in *in vitro* studies with 3T3-L1 adipocytes, VO(acac)$_2$ has been reported to exert its action by directly potentiating the tyrosine phosphorylation of the insulin receptor (*4*).
5. To exert its insulin-enhancing properties is it necessary or not that the VC enters into the cells? If it enters the cells then its toxic effects must be evaluated, and taken into account in order to use non-toxic effective doses. The possibility of its interaction with DNA should also be evaluated.

In the present work we mainly discuss aspects related to points 1 and 5. The complexes studied are: Cs$_2$[V^{IV}O(SO$_3$-sal)en] **1**, V^{IV}O(salan) **2**, V^{IV}O(pyren) **3**, V^{IV}O(pyran) **4**, Na[V^{IV}O(salDPA)] **5**, V^{IV}O(NEt$_3$-sal)en **6**, V^{IV}O(NEt$_3$-sal)chen **7**, V^{IV}O(pOH-sal)en **8**, V^{IV}O(acac)$_2$ **9** and V^{IV}O(phen)$_2$(SO$_4$) **10**. Figure 1 shows the structure of some of the ligands.

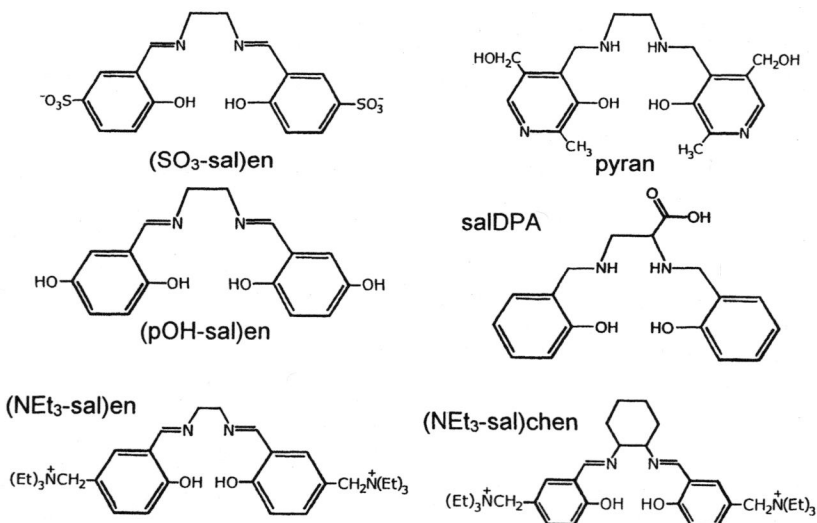

Figure 1. Molecular structures of some of the ligands of the VCs studied.

The stability of several of their V^{IV} and V^V complexes in aqueous solutions at pH ~7, and consequences on the study of toxicity, insulin mimetic and nuclease activity studies are discussed.

Results and Discussion

Salen SB V^{IV}-complexes, with general molecular formula shown in Figure 2, have the disadvantage of often not being soluble in water. Once in solution, there are often problems of oxidation to V^V and/or hydrolysis of the complex and/or of the SB ligand.

V^{IV}O(salen)

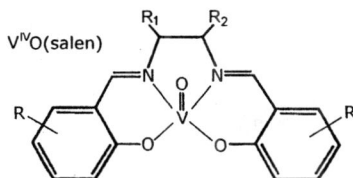

Figure 2. General molecular formula of salen-type oxovanadium(IV) complexes. One of the possible isomers is represented.

For example, the SB obtained by the condensation of pyridoxal and ethylenediamine (pyren) is moderately soluble in water and the complex formation with V^{IV}O^{2+} could be studied in anaerobic conditions (speciation in Figure 3). V^{IV}O-complexes with pyren acting as a tetradentate ligand form for pH > ca. 4, but for pH > 5 the pyren ligand hydrolyses. In aerobic conditions the V^{IV} also oxidizes. It is clear that in *in vitro* or in *in vivo* studies the VO-pyren complexes, with pyren acting as a tetradentate ligand, will never be the predominant species present. We anticipate that this conclusion may be extrapolated for most V^{IV}O-salen systems.

This instability can often be overcome by reduction of the SB to give an amine compound (hereafter designated by salan or pyran, depending on the aromatic aldehyde involved). This presents interesting possibilities, as salan ligands will be more flexible and not restrained to remain planar when coordinated. We have previously reported the preparation of several new salen- and salan-type compounds, and of their V^{IV}- and V^{V}-complexes, namely compounds derived either from pyridoxal (pyr), salicylaldehyde or from salicylaldehyde-5-sulphonate (SO$_3$-sal) with ethylenediamine (5,6). The salen-type ligands and their VCs prepared using pyr and SO$_3$-sal are moderately soluble in water; therefore they may be particularly useful for therapeutic use.

All salan ligands proved to be efficient binders of V^{IV} and V^{V}, namely pyran, salDPA and (SO$_3$-sal)an. We prepared several V^{IV}O- and V^{V}O$_2$-pyran complexes and studied their properties (5,6). The solution speciation revealed that pyran formed much more stable complexes with both V^{IV}O^{2+} and V^{V}O$_2$$^{+}$ than the corresponding SB (5). Either with pyran or with salDPA, in 1:1 M:L solutions, at pH 7, ~100% of V^{IV} or V^{V} are in the form of 1:1 complexes and no free vanadium is detected (see for example in Figure 4 speciation diagrams for the V-salDPA systems).

Stability of V-salen and V-salan Type Complexes in Aqueous Solution

From what was mentioned above about Figure 3 it is clear that at pH ~7 the SB complexes such as the V^{IV}O(salen) are not the predominant vanadium

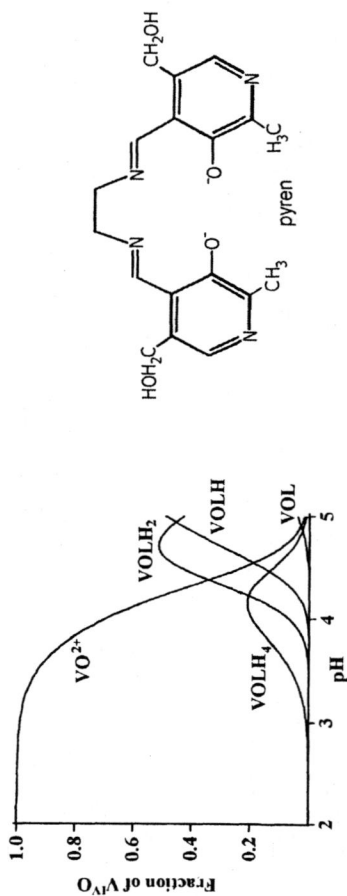

Figure 3. Speciation in the $V^{IV}O$-pyren system for $C_V=2$ mM and $L:M=2$, calculated using data from (5). The ligand is $L = pyren^{2-}$. The $VOLH_2$ species corresponds to $V^{IV}O$-complexes protonated at the pyridine N-atoms.

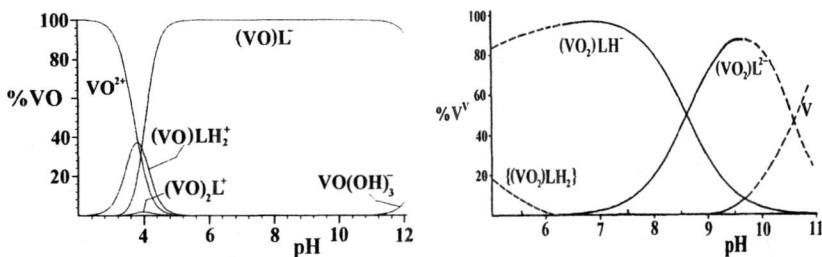

Figure 4. Speciation in the systems (A) $V^{IV}O$-salDPA (C_V = 1 mM, anaerobic conditions) and (B) $V^{V}O_2$-salDPA (C_V = 3 mM), for a L:M ratio of 1, calculated using data from (7); [L = salDPA^{3-}]. At pH = 7, ~100% of V is in the form of V-ligand complexes.

species in aqueous solution. Moreover, we measured UV-Vis spectra with time of 100 μM aqueous solutions of [$V^{IV}O$(pOH-sal)en] **8** at pH 7.0 in several buffers: phosphate, TRIS and HEPES (all 10mM, non-degassed and with stirring). Figure 5 shows the results for the phosphate buffer; those with TRIS and HEPES buffers are almost identical. It is clear that after 2 h the nature of the V species in solution changed very significantly, the V^{IV} partly oxidized to V^{V} and the SB ligand hydrolysed. After 24h the vanadium speciation is totally different from that immediately after dissolving the complex. If similar conditions are used in *in vitro* studies with cellular systems, and they often are, not much more than the overall effect observed can be stated, and any proposal for the mechanism of action of the $V^{IV}O$(salen) complexes on the cells under study will be mostly speculation.

With the V-salan complexes, their V^{IV}- and V^{V}-stability constants are much higher. The oxidation of the V^{IV}-centre may occur in aerobic conditions at pH~7, but only the corresponding V^{V}-complexes form, and in the final solution no detectable amounts of free vanadates are present, even in cell culture medium, as was shown by ^{51}V NMR studies [e.g. see refs. (5-8) and Figure 6B].

Several other types of VCs also oxidize/decompose slowly in solution. For example, in aqueous aerobic solutions at pH~7.4, the V^{IV} of $V^{IV}O$(acac)$_2$ **9** and $V^{IV}O$(phen)$_2$(SO$_4$) **10** oxidizes (**9** slower than **10**), but with **9**, the only V^{V} product detected by ^{51}V NMR after ~4h is the monovanadate (V1) - Figure 6A.

Visible spectra (400-1000 nm) of non-degassed aqueous solutions of **9** simulating cell incubation (containing 132 mM NaCl, 4 mM KCl, 1.2 mM NaH$_2$PO$_4$, 1.4 mM MgCl$_2$, 6mM glucose, 10 mM HEPES, at pH~7.4) also change with time. After a small initial increase in absorption (up to ~35 min.), the absorption decreases ca. 25% after 4 h, and ~60% after 25h. At least part of the decrease in the absorption in the visible range is due to V^{IV} oxidation, the intensity of the EPR spectra also decreases, but no V^{V} complex was detected in

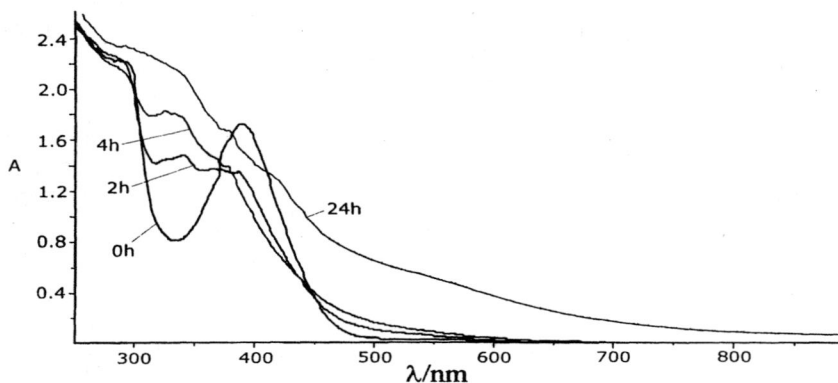

Figure 5. UV-Vis spectra measured at different time intervals of 100 µM aqueous aerobic solutions of [V^{IV}O(pOH-sal)en] 8 at pH 7.0 in 10 mM phosphate buffer. t=0 h corresponds to the moment of the addition of the complex dissolved in DMSO to the buffer (DMSO: ~4%, buffer ~96%).

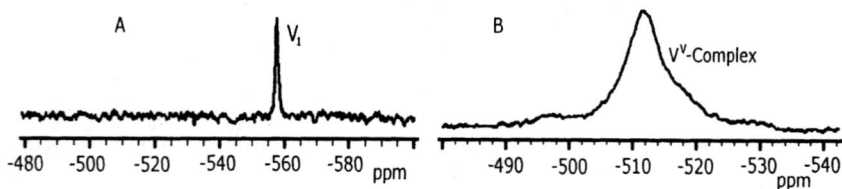

Figure 6. ^{51}V NMR spectra of the oxidation products of ~3 mM of (A) V^{IV}O(acac)$_2$ 9 and of (B) Na[V^{IV}O(salDPA)] 5 at pH~7.4 in an aqueous aerobic solution containing the Dulbecco's Modified Eagle's Medium – High Glucose.

the corresponding ^{51}V NMR spectra (Figure 6A). Similar results were obtained with $V^{IV}O(phen)_2(SO_4)$.

While no significant difference in the UV spectra of $V^{IV}O(pOHsal)en$ in different buffers were found (see above), distinct changes with time in the UV spectra of 100 μM $VO(acac)_2$ solutions were detected, at least comparing the phosphate and the TRIS buffers. The intensity of the UV spectra decreased faster in TRIS buffer aerobic solutions.

Toxicity Tests

Some of the V^{IV}-complexes synthesized, namely **1-4**, have been tested *in vitro* for their toxicity and insulin-mimetic behaviour using transformed mice fibroblasts (*1*). Most of these complexes are toxic at C_V=1mM [except VO(pyran)], and negligibly toxic or non-toxic at C_V=0.01mM and below. $V^{IV}O(pyran)$ showed no toxicity even after 36h of incubation with the fibroblasts in a 1 mM concentration. However, in aqueous aerobic solution these complexes either undergo hydrolysis and/or the V^{IV} oxidizes to V^V; therefore the studies do not really evaluate the toxicity of each of the $V^{IV}O$-complexes **1-4**, but of the mixture of V-species formed in the medium. For these four compounds, only in the case of $V^{IV}O(pyran)$ we know that the complex does not hydrolyse, but forms $V^VO_2(pyran)$ upon oxidation (*5*). The question raised here on the real composition of the incubating solutions is normally not considered or taken into account but is certainly relevant in most cases.

Viability tests were also made with Na_3VO_4 and $Na[V^{IV}O(salDPA)]$ **5** with tumoral HeLa epithelial cells (for 72h) and with 3T3 L1 fibroblasts (for 48h) in the concentration range 1-200 μM. Similar results were found for both cell types. While the $IC_{50}(Na_3VO_4)$ = 32 μM for the HeLa cells, ~100% cellular viability was found with the oxidation products of **5** in all conditions used. Some of the results obtained are shown in Figure 7. These results suggest that while vanadate enters the cells and its toxicity increases with its concentration, the V^V-complexes of salDPA do not hydrolyse, the vanadium does not enter the cells significantly and the toxicity is much lower. Similar results were obtained with the V^V-pyran complexes (*8*).

Insulin Mimetic Tests

Some tests made with complexes **1-4** with transformed mice fibroblasts were described in (*1*). The glucose intake was determined by two different methods: (a) the vitality test based on MTT (*1*), and (b) glucose consumption and lactate production by enzymatic methods (*9*). Using test (a) for complexes **1-5**, maximum activity was found in the range C_V = 0.1 to 100 μM. After 24h of

348

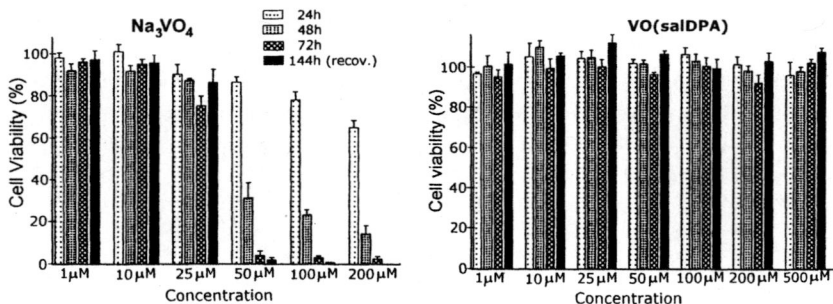

Figure 7. % cell viability of HeLa epithelial cells after exposure to sodium vanadate and to the oxidation products of Na[V^{IV}O(salDPA)], in the concentration range 1-200 μM, during different incubation periods.

incubation, in some cases some of the VCs appeared to be more effective than insulin itself. However, again the question about what was the real composition in V-species of the incubating solution may be raised. Only with pyran and with salDPA we may state that most of the vanadium is in fact coordinated by the original ligand. The effects observed could be due to free vanadates, which may enter the cells by the phosphate anion channel. With test (b), in the concentration range 1-5 μM, no significant glucose consumption and lactate production was found either with vanadate or with the oxidation products of 4 or 5. Apparently, when the complexes do not decompose and vanadium does not significantly enter the cells, no insulin-mimetic effect is detected (8).

DNA Cleavage Reactions

If vanadium enters the cells, then its possible toxic effects should be evaluated at different levels, namely if it can interact with DNA and/or promote DNA cleavage. Studies of the effect of VCs on DNA have mainly concentrated on plasmid nicking caused by the VC itself or reactivity initiated by H_2O_2 or UV radiation. Namely some (hydroxy-sal)en vanadium complexes, particularly VO(pOH-sal)en 8, have been reported to exhibit nuclease activity in the presence of an activating agent: mercaptopropionic (MPA) acid or Oxone, whereas in the absence of an activating agent no cleavage of DNA was induced. The reaction was reported to occur mainly at guanine residues (10).

DNA cleavage was analysed by monitoring the conversion of supercoiled plasmid DNA (Sc) to nicked circular DNA (Nck) and linear DNA (Lin). We made this study with several complexes in varying conditions of incubating medium, amount of complex or DNA, incubation time, temperature or under inert atmosphere or not. The complexes tested (all starting as V^{IV}-compounds) were 1-10, as well as Cu(pOHsal)en, VOSO$_4$ and NaVO$_3$.

We found that the concentration and nature of buffer could be important, e.g. cases where the use of 0.01 M TRIS buffer could induce DNA cleavage, with 0.1 M this did not occur. The use of a few-months-old solution of some reagents (e.g. bromophenol blue) appeared also to induce some different results in some cases. An incubation time of 1 h at 37°C normally gave results equivalent to ~12 h at room temperature (ca. 20°C). Incubation under N_2 in some cases yielded less DNA cleavage than in the presence of air.

Some complexes induced DNA cleavage in the absence of activating agents, namely VO(pyran), VO(acac)$_2$ and VO(phen)$_2$(SO$_4$), and more efficiently in the presence of oxone or MPA (see for example Figure 8). Others, e.g. 5 and 8, only induced DNA cleavage in the presence of activating agents.

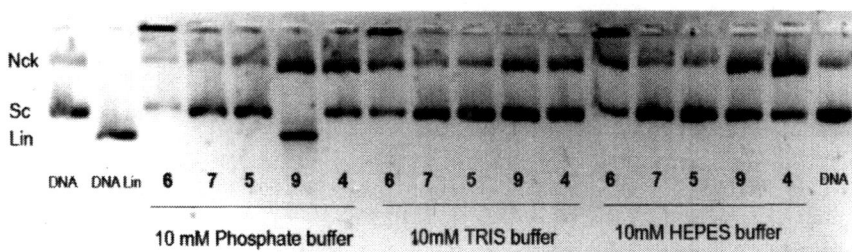

Figure 8. Cleavage of supercoiled plasmid DNA (Sc) by vanadium with no addition of activators. Nck and Lin refer to the nicked and linear DNA forms, respectively. The lanes marked DNA and DNALin refer to the plasmid DNA and to the plasmid linearized by enzymatic digestion.

Globally, for the set of complexes tested, the order of efficiency to cleave DNA is: VO(phen)$_2$(SO$_4$) > VO(acac)$_2$ > Cu((pOHsal)en) ≥ VO(pyran) VO((NEt$_3$-sal)en) > VO(salDPA) ≈ VO((NEt$_3$-sal)chen) > VO((pOHsal)en)) > [VO(SO$_3$-sal)en] > VOSO$_4$ (no cleavage). The first five complexes caused DNA cleavage without the need of additional activating agents. The results were reproducible for the two amounts of DNA tested (0.1 μg and 0.5 μg). The extent of DNA cleavage observed after incubation is slightly higher if it is carried out in aerobic conditions. For example with 5, some DNA linearization could be observed only after incubation for 1h at 37°C in aerobic conditions, but not at room temperature nor under N_2. With complex 4 in similar conditions, DNA linearization is more extensive in aerobic conditions than under N_2 atmosphere.

As may be seen in Fig. 8, VO(acac)$_2$ behaved differently in the 3 buffers tested: in phosphate buffer it could linearize DNA, while no linear form is detected in TRIS or HEPES buffers. Addition of oxone or MPA increases the extent of DNA cleavage, but MPA much less than oxone.

As mentioned above, for VO(pOHsal)en in aqueous solution no significant difference was found in the UV-Vis spectra and their change with time when the pH is set at ~7 with the phosphate, TRIS or HEPES buffers. With VO(acac)$_2$ some distinct behaviour was found. However, we believe that the different DNA cleavage ability of VO(acac)$_2$ in solutions containing distinct buffers, and the observation that some the complexes studied show DNA cleavage ability in 0.01 M TRIS buffer, but not in its 0.1 M solutions, is probably due to the TRIS molecules acting as OH radical scavengers (assuming the DNA cleavage requires hydroxyl radicals).

Conclusions

Many VCs tested *in vitro* here and elsewhere revealed their potential insulin-enhancing properties. If we envisage VCs for oral treatment of diabetes, it is important to consider that oral application normally provides an intimate contact of VCs with oxygen. Moreover, oxidation and the acidic stomach conditions will convert most complexes to a partially hydrolysed V^V species. Most *in vivo* and *in vitro* studies concerning insulin-mimetic VCs do not take this into account properly, or do not recognize that the observations made may result from a very distinct species from the originally VC used. The synergistic effect of serum proteins and that of insulin only recently started to be evaluated, but is also far form being understood.

The speciation of the V^{IV} and V^V with pyran and with salDPA is well understood (5,7), and it is known that at pH=7, ~100% of vanadium is in the form of the corresponding V^{IV}O-salan or V^{V}O$_2$-salan complexes. Both 4 and 5 were found to be non-toxic compounds. This may simply result from their low uptake from the cells, as found in human erythrocytes (8). However, while in the tests with mice fibroblasts (1) complexes 4 and 5 were found to be insulin-mimetic compounds, in the tests measuring the glucose intake rates by the hexokinase method (9) no significant stimulation was found (8). Moreover, **in vitro insulin-mimetic activity of 4 was not found in rat adipocytes (by inhibition of free fatty acids release experiments made by K. Kawabe and H. Sakurai).** The possibility of being an insulin-enhancing compound *in vivo* remains open.

As V-pyran and V-salDPA complexes were found to cause DNA cleavage, if they are significantly absorbed and keep their integrity inside the cells, then the possibility of DNA damage may be significant, and it would be toxic. At least with human erythrocytes it was found by EPR that the small amount of vanadium inside the cells is not in the form of V^{IV}O(pyran). This balance between the insulin mimetic effect of VCs and their possible DNA damage should be carefully evaluated before any compound could be considered for the treatment of diabetes. If the VC acts directly potentiating the tyrosine phosphorylation of the insulin receptor without entering the cell, problems of DNA damage could possibly be ruled out.

Acknowledgements

We thank the FEDER, Fundação para a Ciência e Tecnologia, project POCI/QUI/56949/2004, and the COST D21 Action.

References

1. Rehder, D.; Costa Pessoa, J.; Geraldes, C.F.G.C.; Castro, M.M.C.A.; Kabanos, T.; Kiss, T.; Meier, B.; Micera, G.; Pettersson, L.; Rangel, M.; Salifoglou, A.; Turel, I.; Wang, D. *J Biol Inorg Chem* **2002**, *7*, 384-396.
2. Evangelou, A.M. *Crit Rev Oncol/Hemat* **2002**, *42*, 249-265.
3. Durai, N.; Saminathan, G. *J Clin Biochem Nutr.* **1997**, *22*, 31-39.
4. Ou, H.; Yan, L.; Mustafi, D.; Makinen, M.W.; Brady, M.J. *J. Biol. Inorg. Chem.* **2005**, *10*, 874-886.
5. Correia, I.; Costa Pessoa, J.; Duarte, M.T.; Henriques, R.T.; Piedade, M.F.M.; Veiros, L.F.; Jakusch, T.; Dörnyei, A.; Kiss T.; Castro, M.M.C.A.; Geraldes, C.F.G.C.; Avecilla, F. *Chem. Eur. J.* **2004**, *10*, 2301-2317.
6. Correia, I.; Costa Pessoa, J.; Duarte, M.T.; Piedade, M.F.M.; Jackush, T.; Kiss, T.; Castro, M.M.C.A.; Geraldes, C.F.G.C.; Avecilla, F. *Eur. J. Inorg. Chem.*, 2005, 732-744.
7. Costa Pessoa, J.; Marcão, S.; Correia, I.; Gonçalves, G.; Dörnyei, A.; Kiss, T.; Jakusch, T.; Tomaz, I.; Castro, M.M.C.A.; Geraldes, C.F.G.C.; Avecilla, F. *Eur. J. Inorg. Chem.* **2006**, 3614-3621.
8. Delgado, T.C.; Tomaz, I.; Correia, I.; Costa Pessoa, J.; Jones, J.G.; Geraldes, C.F.G.C.; Castro, M.M.C.A. *J. Inorg. Biochem.*, **2005**, *99*, 2328-2339.
9. Delgado, T.C.; Castro, M.M.C.A.; Geraldes, C.F.G.C.; Jones, J.G. *Magn. Reson. Med.* **2004**, *51*, 1283-1286.
10. Verquin G.; Fontaine G.; Bria M.; Zhilinskaya E.; Abi-Aad E.; Aboukais A.; Baldeyrou B.; Bailly C.; Bernier J.L. *J. Biol. Inorg. Chem.* **2004**, *9*, 345-353.

Chapter 25

Charge Distribution in Vanadium
p-(Hydro/Semi)Quinonate Complexes

Chryssoula Drouza and Anastasios D. Keramidas[*]

Department of Chemistry, University of Cyrus, Nicosia, 1678, Cyprus
[*]Corresponding author: email: akeramid@ucy.ac.cy

The known crystal structures of co-ordination compounds containing p-dioxolene ligands in the form of hydroquinone, semiquinone or quinone have been examined. A simple method is proposed to correlate the oxidation state of these ligands with the structural distortion based on crystallographic data. The results fit well with the literature oxidation-state assignments for ligands ligated either to one or to two bridged through the ligand metal ions including the vanadium(IV/V) (hydro/semi)quinonate complexes.

o- and p- dioxolenes, catechols (Cat) and hydroquinones (Hq), and the oxidation products, o- and p- semiquinones (Sq) and quinones (Q), are important compounds in chemical and biochemical reactions such as organic electron and hydrogen transfer reactions or as strong ligators for the transfer of metal ions in biological systems (1-3). For example, electron transfer reactions between transition metal centres and p-quinone cofactors are vital for all life, occurring in key biological processes as diverse as the oxidative maintenance of biological amine levels (4), tissue (collagen and elastin) formation (5,6), photosynthesis (7,8) and aerobic (mitochondrial) respiration (9,10).

The interaction of o- and p- dioxolenes with vanadium presents additional interest, because a) it serves as a model of the possible biological function of tunichromes, compounds found in tunicate blood cells (morula cells), and b) helps in understanding the mechanism of the redox reactions of vanadium(V) in biological systems (11), such as the reduction of vanadium(V), present in sea

© 2007 American Chemical Society

water, to vanadium(III) in the blood cells of tunicates (*12, 13*). In addition, the semiquinonate complexes of vanadium are important intermediates in biochemical and chemical redox reactions, such as the oxidative C-H activation of aliphatic and aromatic hydrocarbons (*14,15*).

Dioxolenes have orbitals that can be close in energy to the transition-metal d orbitals generating the opportunity for considerable covalency between the redox-active metal centre and co-ordinated redox active ligand (*16,17*). These metal complexes which consist of two redox-active centres, are characterized by the existence of two electronic forms (valence tautomers) with different charge distribution, and consequently, different optical, electric and magnetic properties (Figure 1). These species might interconvert to each other by a reversible intramolecular electron transfer involving the metal ion and the redox active ligand and can been used for the preparation of new molecular electronic devices (*16-18*). In addition *p*-dioxolenes can link two metal centres together serving as redox active bridging ligands. Redox active bridges create the possibility of modulating the degree of electronic coupling between the various molecular components since the electronic energy of both bridge and its adjoining units will depend implicitly on the redox state of the bridge (*19*).

In marked contrast to the extensive structural chemical studies for chelate stabilized *o*-(hydro/semi)quinone metal compounds (*20-22*), examples of structurally characterized σ-bonded *p*-hydroquinone–metal compounds are less known (*23-28*) and there is only one example of σ-bonded *p*-semiquinone–metal complex reported by us (*29*). This is mainly due to the absence of a chelate co-ordination site in simple *p*-(hydro/semi)quinone and the low basicity of *p*-semiquinone ($pK_{BH}^{+} < 5$). The strategy usually applied on the preparation of such species is to synthesize substituted, in the *o*-position, *p*-hydroquinones with substituents containing one or more donor atoms, thus, enabling the metal atom to form chelate rings.

The similarity of the π-electronic levels of dioxolenes to the energy of the d-orbitals of the transition-metals characterizes these molecules as non-innocent (*30-32*). This property has often created ambiguity in assessing the oxidation state of metal ion and the dioxolene ligand. The ligands may be in *o*-Q, *o*-Sq⁻, or Cat²⁻ and *p*-Q, *p*-Sq⁻, or Hq²⁻ forms for *o*- and *p*-dioxolenes respectively, thus, the metal-ligand interaction in the complex, containing a metal of variable oxidation state, may be written as shown in Figure 1. Since stoichiometry alone can do nothing to resolve this problem, a variety of spectroscopic and spectral arguments have been used to establish the detailed structure, including X-Ray crystallography (*33,34*), infra-red (IR) and raman spectroscopy (*35*), EPR and NMR spectroscopy (*36-38*) and magnetic measurements (*35*).

Our focus in this study is on the complexes of vanadium ion that contain *p*-dioxolene ligands. In particular, we will present the X-ray crystallographic data that characterize the different oxidation states of *p*-dioxolenes, Hq²⁻, Sq⁻ and Q. The extensive literature on the *o*- analogues will be used in order to formulate the assignments of the oxidation states of *p*-dioxolenes in solid state.

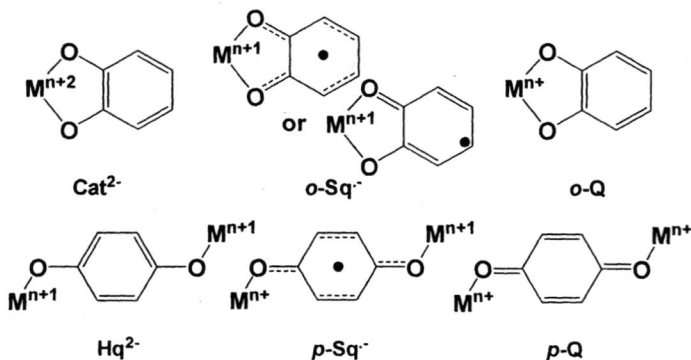

Figure 1. Valence tautomeric structures of o- and p- dioxolene complexes

Crystallographic Aspects of *o*- and *p*-Dioxolenes Ligated to Transition Metal Ions

o-Dioxolenes

The C-O and the intradiol C-C bond distances have been found to be particularly diagnostic in order to decide about the oxidation state of the quinoidal ligands of the metal complexes (20,39), with few exceptions (34). In generalized terms, the C-O bond lengths are ca. 1.23 Å, 1.29 Å and 1.35 Å, and the intradiol C-C bond lengths are ca. 1.53 Å, 1.44 Å and 1.39 Å, for o-Q, o-Sq⁻, and Cat²⁻ respectively. Carugo et.al (34) have used a statistical approach taking in consideration both the C-O and all the six intraring C/C distances, affording a diagnostic structural function Δ. They have considered the X-Ray crystallographic data of 146 ligands and they have found that this function can predict the oxidation state of the ligand for the 135 compounds. Δ is equal to 0, -1 and -2 for Q, Sq⁻ and Cat²⁻ respectively. However, the use of not suitable reference points in the analysis and sometimes the large uncertainties in the determination of the C-O and C-C bond distances, have put this statistical analysis in debate (33).

There are several vanadium catecholate complexes with vanadium in oxidation states +3, +4 and +5 characterized by X-Ray crystallography (30,20-45), but only one vanadium(V) *o*-semiquinonate complex (37). The C-O and C-C intradiol bond lengths on the catecholate complexes range from 1.33 to 1.35 Å and 1.40 to 1.42 Å respectively supporting the oxidation state of the ligand. The oxidation state of 3,5-di-butylsemiquinone (3,5-DBSQ) ligand in complex [$V^V O$(3,5-DBSQ)(3,5-DBCat)]₂ (3,5-DBCat = 3,5-di-butylcatechole) has been supported from the C-O bond lengths [1.292(8) and 1.322 (8) Å] and the longer than catecholate ligand C-C intradiol bond distance [1.44 (1) Å].

p-Dioxolenes

The bond lengths of the Hq^{2-}, p-Sq^{-} and p-Q ligands of various complexes have been collected in Table 1 from the literature. The bonds that mostly characterize the oxidation states of the p-dioxolene ligands are the C-O (ε, δ) and C-C (β) bonds (Figure 3). The mean values of these bond lengths are 1.35 ± 0.05 1.32 ± 0.03, and 1.24 ± 0.04 Å (C-O, ε, δ) and 1.39 ± 0.02, 1.370 ± 0.001, and 1.33 ± 0.02 Å (C-C, β), for Hq^{2-}, Sq^{-} and Q respectively. It is important to notice here that the different substitution on the p-dioxolene ligands, as indicated by the deviation of these values, influence the bond lengths of the ligand in lesser degree than the oxidation state. The mean values have been calculated for all complexes including ligands ligated either to one or to two metal ions. This explains the larger deviation from the mean value of the Hq^{2-} C-O bond lengths.

The structures of the anions of the hydroquinonate binuclear complex $[(VO_2)_2bicah]^{4-}$ and the tetranuclear semiquinonate complex $\{[(VO)_2O](bicas)\}_2^{6-}$ as found from the X-ray analysis of the molecules (29) are shown in Figure 2. The co-ordination environment of the vanadium atoms in both complexes is octahedral with one oxo group and the amine nitrogen of the tripod-ligating group to occupy the apical positions. Three oxygen atoms, originated from the two-carboxylate groups and the hydroquinone or semiquinone of the tripod moiety, occupy the three out of four equatorial positions of the octahedron. The fourth equatorial position belongs to a terminal oxo group for $[(VO_2)_2bicah]^{4-}$ and to the bridging oxo group for the $\{[(VO)_2O](bicas)\}_2^{6-}$ complex.

The oxidation state of the metal ions in those complexes has been unambiguously extracted after the determination of the oxidation state of the ligand and the total charge of the complex by a combination of techniques including X-Ray crystallography, elemental analysis and conductivity measurements (29). In addition, the longer V-$O_{p\text{-dioxolene}}$ bond length of the semiquinonate complex [1.887(3) Å] compared with that of the hydroquinonate complex [1.864(3) Å] confirms the +4 and +5 oxidation states of the vanadium atoms for the former and the latter complex respectively. Furthermore, these assigments are supported by the statistically significant differences in the mean values of the bond lengths around vanadium atoms between the hydroquinonate and the semiquinonate complexes. The mean value for the hydroquinonate compound [1.941(4) Å] is smaller than the one calculated for the semiquinonate complex [1.949(4) Å], due to the smaller size of vanadium ion in the higher oxidation state V^{V}.

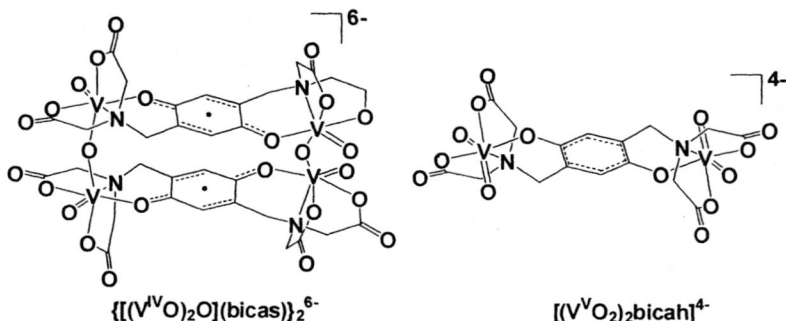

{[(V^{IV}O)_2O](bicas)}_2^{6-}   [(V^VO_2)_2bicah]^{4-}

Figure 2. Structures of anions {[(VO)_2O](bicas)}_2^{6-} and [(VO_2)_2bicah]^{4-}

The bond distances of the *p*-dioxolene ligands in these two complexes as well as the mean values of the bond distances from structures of quinone dinuclear complexes with other metal ions (*46,47*) are shown in Figure 3. The oxidation state of the ligand in [(VO_2)_2bicah]^{4-} is clearly borne out by the observation (*48,49*), that (i) the C–C bonds of the six-member ring are equidistant at 1.386 ± 0.004 Å and (ii) the C–O bond length at 1.354 (6) Å is long and typical for *p*-hydroquinonates. In contrast, the anion {[(VO)_2O](bicas)}_2^{6-} contains two *p*-semiquinonate ligands; the C–O bond length [1.322 (5) Å] is shorter than in [(VO_2)_2bicah]^{4-} and longer than the bond length expected for quinone [~ 1.26 Å] (*46,47*) and this denotes a partial double bond character. In addition, the C–C bonds of the six-member ring, 1.399 (7), 1.371 (7) and 1.427 (7) Å exhibit a long-short-long pattern expected for *p*-semiquinonates. Although the six-member ring of the ligated *p*-quinone ligands shows a similar pattern to *p*-semiquinonates, different bond lengths are expected (Figure 3).

Application of Δ Statistical Analysis to *p*-Dioxolenes

The linear relationship of the charge of *o*-dioxolenes (Δ) versus the bond lengths of the ligand is given by the equations 1 and 2 (*34*) where d_i is the experimental *i*th bond length, and d_{1i} and d_{2i} are the ith bond lengths of the pure

$$\Delta_i = -2(d_i-d_{2i})/(d_{1i}-d_{2i}) \tag{1}$$

$$\Delta = (\Sigma w_i \Delta_i)/(\Sigma w_i) \tag{2}$$

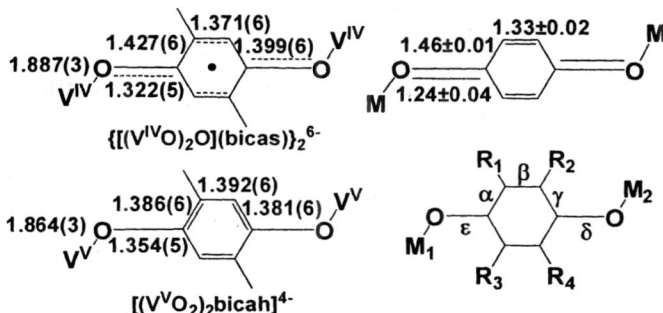

Figure 3. Bond lengths of p-dioxolene ligands in $\{[(VO)_2O](bicas)\}_2^{6-}$ and $[(VO_2)_2bicah]^{4-}$ complexes and mean bond lengths in quinone metal complexes (46,47). Numbering of bonds, M_1=metal ion, M_2=metal ion, -CH$_3$, -H, heptyl, or none for quinone. R_{1-4}=-H, -CH$_3$, t-but, -CH$_2$-, -CH=, naphthalene, porphyrin, pyridine, or -COO$^-$.

forms of catecholate and benzoquinone. The same linear relationship has also been used successfully in calculating the partial electron transfer of organic p-dioxolenes from the quinoidal distortion (50,51). Thus, it is reasonable to assume that this relationship can be applied on the calculation of the ligand charges in the p-dioxolene transition metal complexes. The values of d_{1i} and d_{2i} for hydroquinone and p-quinones respectively were considered as the experimental bond lengths of the uncomplexed organic molecules (52-58). The average values for C-O (ε, δ), C-C (α,γ) and C-C (β) are 1.377 ± 0.004 Å, 1.385 ± 0.005 Å and 1.384 ± 0.003 Å for free hydroquinone and 1.266 ± 0.003 Å, 1.347 ± 0.002 Å and 1.332 ± 0.005 for free quinone. The application of these bond lengths on equations 1 and 2 for all the complexes in Table 1 afford Δ values that are statistically significant, matching the expected hydroquinonate, semiquinonate or quinone nature of p-dioxolenes ligated either to one or to two metal ions. Complexes $[(VO_2)_2bicah]^{4-}$ and $\{[(VO)_2O](bicas)\}_2^{6-}$ gave Δ values -2.03 and -1.38 respectively, which supports the hydroquinonate and semiquinonate nature of the ligands.

The use of the crystal structures of free ligands, hydroquinone and quinone, as reference points in the analysis introduces some uncertainty, since these compounds in the crystal structure are not idealized free molecules due to hydrogen bonding and because of the use of substituted p-hydroxolene ligands for the complexation of the metal ion. In addition, the coordination of the transition metal ions causes partial charge transfer from the ligand to metal distorting the structures of p-dioxolenes. This can clearly be observed on Table 1, where co-ordination of metal ions causes shortening of C-O bond of

Table 1. Bond lengths of *p*-dioxolene ligands and calculated Δ values

Comp	α β γ δ ε (bond lengths in Å)	Δ, oxid. state
1	1.386(6) 1.392(6) 1.381(6) 1.354(5) 1.354(5)	-2.02, Hq
	1.381(7) 1.384(7) 1.383(7) 1.353(6) 1.353(6)	-1.95, Hq
2	1.38(1) 1.40(1) 1.38(1) 1.36(1) 1.36(1)	-2.15, Hq
3	1.387(2) 1.390(2) 1.387(2) 1.300(2) 1.300(2)	-1.78, Hq
4	1.3955(2) 1.3785(4) 1.3955(2) 1.3492(4) 1.3492(4)	-1.74, Hq
5	1.407(7) 1.395(8) 1.384(6) 1.380(8) 1.324(6)	-1.91, Hq
	1.416(5) 1.396(7) 1.378(6) 1.377(6) 1.316(6)	-1.88, Hq
6	1.405(6) 1.397(6) 1.406(5) 1.347(5) 1.347(5)	-1.80, Hq
7	1.39(1) 1.393(9) 1.39(1) 1.360(8) 1.360(8)	-1.98, Hq
8	1.426(4) 1.395(4) 1.378(4) 1.385(3) 1.312(2)	-1.82, Hq
9	1.397(4) 1.370(4) 1.377(4) 1.368(4) 1.339(4)	-1.77, Hq
	1.396(4) 1.373(5) 1.386(5) 1.372(5) 1.328(4)	-1.74, Hq
10	1.390(6) 1.390(5) 1.375(6) 1.373(4) 1.358(4)	-2.05, Hq
11	1.39(3) 1.38(3) 1.38(3) 1.38(2) 1.34(2)	-1.90, Hq
12	1.40(1) 1.40(1) 1.41(1) 1.36(1) 1.37(1)	-1.89, Hq
	1.41(1) 1.40(1) 1.41(1) 1.372(8) 1.372(8)	-1.86, Hq
13	1.386(3) 1.384(3) 1.386(4) 1.357(3) 1.357(3)	-1.92, Hq
14	1.38(1) 1.38(1) 1.32(1) 1.360(7) 1.369(7)	-2.30, Hq
15	1.390(3) 1.389(3) 1.403(3) 1.376(2) 1.340(2)	-1.86, Hq
16	1.35(3) 1.37(3) 1.39(2) 1.39(2) 1.32(2)	-1.96, Hq
	1.38(2) 1.44(2) 1.42(2) 1.34(2) 1.33(2)	-2.23, Hq
	1.37(2) 1.40(2) 1.35(2) 1.37(2) 1.38(2)	-2.42, Hq
17	1.404(6) 1.388(6) 1.372(6) 1.397(5) 1.320(4)	-1.94, Hq
	1.404(6) 1.392(6) 1.378(6) 1.380(5) 1.323(5)	-1.93, Hq
18	1.35(2) 1.38(2) 1.35(2) 1.353(1) 1.35(1)	-2.26, Hq
19	1.41(1) 1.41(1) 1.36(1) 1.392(9) 1.310(8)	-2.16, Hq
	1.42(1) 1.39(1) 1.39(1) 1.40(1) 1.310(8)	-1.76, Hq
20	1.404(4) 1.386(5) 1.386(4) 1.372(4) 1.334(4)	-1.83, Hq
	1.400(4) 1.390(5) 1.382(4) 1.373(4) 1.343(4)	-1.93, Hq
	1.395(4) 1.390(5) 1.383(4) 1.375(4) 1.341(4)	-1.95, Hq
	1.400(3) 1.393(5) 1.388(4) 1.370(2) 1.348(2)	-1.93, Hq
21	1.400(4) 1.391(5) 1.378(5) 1.365(4) 1.340(4)	-1.94, Hq
22	1.39(1) 1.386(8) 1.39(1) 1.350(7) 1.350(7)	-1.88, Hq
	1.386(9) 1.398(9) 1.386(9) 1.362(7) 1.362(7)	-2.07, Hq
23	1.387(4) 1.386(4) 1.387(4) 1.353(3) 1.353(3)	-1.92, Hq
24	1.413(5) 1.396(5) 1.383(5) 1.386(5) 1.341(4)	-1.93, Hq
	1.414(5) 1.404(5) 1.381(5) 1.396(4) 1.344(4)	-2.03, Hq
	1.439(5) 1.369(5) 1.402(5) 1.348(5) 1.290(4)	-1.27, Sq
25	1.427(6) 1.371(6) 1.399(6) 1.322(5) 1.322(5)	-1.38, Sq

Table 1. *Continued.*

Comp	$\alpha\,\beta\,\gamma\,\delta\,\varepsilon$ *(bond lengths in Å)*	Δ, *oxid. state*
26	1.49(2) 1.36(2) 1.49(2) 1.21(1) 1.21(1)	-0.06, Q
27	1.48(1) 1.34(1) 1.49(1) 1.233(9) 1.242(9)	-0.02, Q
28	1.46(1) 1.32(1) 1.46(1) 1.236(8) 1.236(8)	-0.09, Q
29	1.47(2) 1.32(2) 1.47(2) 1.24(2) 1.24(2)	0.00, Q
30	1.44(2) 1.35(2) 1.46(2) 1.28(2) 1.28(2)	-0.64, Q
31	1.46(1) 1.33(1) 1.51(1) 1.20(1) 1.24(1)	0.14, Q

NOTE: The numbering of the bonds is shown in Figure 3. Compounds: **1** $R_{1,3}$=-CH_2-, $R_{2,4}$=H, $M_{1,2}$=V (*29*), **2** R_{1-4}=H, $M_{1,2}$=W (*59*), **3** $R_{1,3}$=pyridine, $R_{2,4}$=H, $M_{1,2}$=Cu (*60*), **4** R_{1-4}=H, $M_{1,2}$=Fe (*61*), **5** R_4=-CH=, R_{2-4}=H, M_1=V, M_2=-CH_3 (*62*), **6** $R_{1,3}$=-COO⁻, $R_{2,4}$=H, M_1=M_2=-Zn_2 (*63*), **7** $R_{1,2}$=-CH_3, $R_{3,4}$=H, $M_{1,2}$=Ti (*64*), **8** R_1=-tBu, $R_{2,3}$=H, R_4=-CH_2-, M_1=Ni, M_2=-CH_3 (*65*), **9** R_1= -CH=, R_{2-4}=H, M_1=V, M_2=-CH_3 (*66*), **10** R_1= -CH=, R_{2-4}=H, M_1=V, M_2=-CH_3 (*66*), **11** $R_{1,4}$=H, $M_{1,2}$=V (*23*), **12** R_{1-4}=H, $M_{1,2}$=Ti (*67*), **13** R_{1-4}=H, $M_{1,2}$=Zr (*68*), **14** R_{1-4}=H, $M_{1,2}$=Ti (*67*), **15** R_1=-CH_2-, R_{2-4}=H, M_1=V, M_2=-CH_3 (*69*), **16** R_{1-4}=H, $M_{1,2}$=Mo (*70*), **17** R_1= -CH=, R_{2-4}=H, M_1=V,M_2=heptyl (*71*), **18** R_{1-4}=H,$M_{1,2}$=Mo (*72*), **19** R_1=-CH=, R_{2-4}=H, M_1=V, M_2=-CH_3 (*73*), **20** R_1= -CH=, R_{2-4}=H, M_1=V, M_2=-CH_3 (*74*), **21** R_1=-CH=, R_{2-4}=H, M_1=V, M_2=-CH_3 (*75*), **22** R_{1-4}=H, $M_{1,2}$=Zr (*25*), **23** R_{1-4}=H, $M_{1,2}$=Ti (*25*), **24** R_1=-CH_2-, R_4=-tBu $R_{2,3}$=-H, M_1=Cr, M_2=-CH_3 (*76*), **25** $R_{1,3}$=-CH_2-,$R_{2,4}$= H, $M_{1,2}$=V (*29*), **26** $R_{1,2}$=naphthalene-, $R_{3,4}$=H, $M_{1,2}$=Mo (*77*), **27** R_1=R_2=-CH_3,R_4=H, R_3==porphyrin, M_1=Zn, M_2=none (*78*), **28** $R_{1,4}$=H$M_{1,2}$=Rh (*46*), **29** $R_{1,2}$=-CH_3, $R_{3,4}$=H, $M_{1,2}$=Rh (*46*), **30** $R_{1,4}$=-$CH_3$$R_{2,3}$=H, $M_{1,2}$=Mo (*47*), **31** $R_{2,3}$=-tBu $R_{1,4}$=-H, M_1=Mo, M_2=none (*47*).

hydroquinonate ligand and elongation of the C-O bond of quinone ligand. Furthermore, ionic bonds with the counter ions in the crystal lattice will transfer some of the charge from the anions causing further structural distortion. It is reasonable to assume that using the vanadium(V) complex $[(VO_2)_2bicah]^{4-}$ and an analogous p-quinone complex instead of free hydroquinone and quinone as reference points, some of the uncertainties will be eliminated. Because, there is no structure of vanadium with p-quinone, as quinone reference point has been used the quinone complexes ligated to other metal ions (Figure 3). The result from this calculation for $\{[(VO)_2O](bicas)\}_2^{6-}$ is -1.30, which is closer to -1 expected for semiquinonates than the -1.38 found using as reference points the free p-dioxolenes. The deviation from the ideal value is attributed to the non ideal quinone reference point.

Conclusions

In this study, we have shown that the Δ statistical analysis can be applied not only to the o- but to the p-dioxolenes as well. It is interesting that this simple linear equation is working equally well to the p-dioxolenes ligated either to two or to one metal ions. The application of this equation on the p-semiquinonate tetranuclear complex $\{[(VO)_2O](bicas)\}_2^{6-}$ gave $\Delta = -1.38$ which confirms the predicted value of the oxidation state of the ligand in the literature.

Acknowledgment

This work was supported by the Cyprus Research Promotion Foundation (Grants MEDA, ENTAX/0504/08 and KAMY, ENTAX/0505/14).

References

1. Rappoport, Z., *The Chemistry of the Quinoid Compounds*. Wiley: New York, 1988; Vol. 1 and 2.
2. Buglyo, P.; Dessi, A.; Kiss, T.; Micera, G.; Sanna, D., *J. Chem. Soc., Dalton Trans.* **1993**, 5092.
3. Dertz, E. A.; Xu, J.; Stintzi, A.; Raymond, K. N., *J. Am. Chem. Soc.* **2006**, *128*, 22.
4. Dooley, D. M.; Scott, R. A.; Knowles, P. F.; Colangelo, C. M.; McGuirl, M. A.; Brown, D. E., *J. Am. Chem. Soc.* **1998**, *120*, 2599.
5. Klinman, J. P., *Chem. Rev.* **1996**, *96*, 2541.

6. McIntire, W. S., *Annu. Rev. Nutrition* **1998**, *18*, 145.
7. Calvo, R.; Abrecsch, E. C.; Bittl, R.; Feher, G.; Hofbauer, W.; Isaacson, A. R.; Lubitz, W.; Okamura, M. Y.; Paddock, M., *J. Am. Chem. Soc.* **2000**, *122*, 7327.
8. Hoganson, C. W.; Babcock, G. T., *Science* **1997**, *277*, 1953.
9. Nichols, D. G.; Ferguson, S. J., *Bioenergetics 2.* Academic Press: New York, 1992.
10. Iwata, S.; Lee, L. W.; Okada, K.; Lee, J. K.; Iwata, M.; Rasmussen, B.; Link, T. A.; Ramaswamy, S.; Jap, B. K., *Science* **1998**, *281*, 64.
11. a) Rehder, D., *Coord. Chem. Rev.* **1999**, *182*, 197. b) Crans, D. C.; Smee, J. J.; Gaidamauskas, E.; Yang, L., *Chem. Rev.* **2004**, *104*, 849.
12. Frank, P.; Hodgson, K. O., *Inorg. Chem.* **2000**, *39*, 6018.
13. Michibata, H.; Sakurai, H., Vanadium in Biological Systems; Chasteen, N. D., Ed.; Kluer Academic Publishers: Dordrecht, 1990; pp. 153-171.
14. Bonchio, M.; Conte, V.; Di Furia, F.; Modena, G., *J. Org. Chem.* **1989**, *54*, 4368.
15. Punniyamurthy, T.; Velusamy, S.; Iqbal, J., *Chem. Rev.* **2005**, *105*, 2329.
16. Pierpont, C. G., *Coord. Chem. Rev.* **2001**, *216-217*, 99.
17. Evangelio, E.; Ruiz-Molina, D., *Eur. J. Inorg. Chem.* **2005**, 2975.
18. Dei, A.; Gatteschi, D.; Sangregorio, C.; Sorace, L., *Acc. Chem. Res.* **2004**, *37*, 827.
19. Keyes, T. E.; Foster, R. J.; Jayaweera, P. M.; Coates, C. G.; McGarvey, J. J.; Vos, J. G., *Inorg. Chem.* **1998**, *37*, 5925.
20. Pierpont, C. G.; Lange, C., *Prog. Inorg. Chem.* **1994**, *41*, 331.
21. Adams, D. M.; Hendrickson, D. N., *J. Am. Chem. Soc.* **1996**, *118*, 11515.
22. Jung, O. S.; Jo, D. H.; Lee, Y. A.; Conklin, B. J.; Pierpont, C. G., *Inorg. Chem.* **1997**, *36*, 19.
23. Tanski, J. M.; Wolczanski, P. T., *Inorg. Chem.* **2001**, *40*, 346.
24. Foster, C. L.; Liu, X.; Kilner, C. A.; Thornton-Pett, M.; Halcrow, M. A., *J. Chem. Soc., Dalton Trans.* **2000**, 4563.
25. Kunzel, A.; Sokolow, M.; Liu, F.-Q.; Roesky, H. W.; Noltemeyer, M.; Schmidt, H.-G.; Uson, I., *J. Chem. Soc., Dalton Trans.* **1996**, 913.
26. McQuillan, F. S.; Berridge, T. E.; Chen, H.; Hamor, T. A.; Jones, C. J., *Inorg. Chem.* **1998**, *37*, 4959.
27. Calderazzo, F.; Englert, U.; Pampaloni, G.; Passarelli, V., *J. Chem. Soc., Dalton Trans.* **2001**, 2891.
28. Sembiring, S.; Colbran, S. B.; Craig, D. C., *J. Chem. Soc., Dalton Trans.* **1999**, 1543.
29. Drouza, C.; Tolis, V.; Gramlich, V.; Raptopoulou, C.; Terzis, A.; Sigalas, M. P.; Kabanos, T. A.; Keramidas, A. D., *Chem. Commun.* **2002**, 2786.
30. Cornman, C. R.; Colpas, G. J.; Hoeschele, J. D.; Kampf, J.; Pecoraro, V. L., *J. Am. Chem. Soc.* **1992**, *114*, 9925.

31. Justel, T.; Bendix, J.; Metzler-Nolte, N.; Weyhermuller, T.; Nuber, B.; Wieghardt, K., *Inorg. Chem.* **1998**, *37*, 35.
32. Masui, H.; Lever, A. B. P.; Auburn, P. R., *Inorg. Chem.* **1991**, *30*, 2402.
33. McGarvey, B. R.; Ozarowski, A.; Tuck, D. G., *Inorg. Chem.* **1993**, *32*, 4474.
34. Carugo, O.; Castellani, C. B.; Djinovic, K.; Rizzi, M. J., *J. Chem. Soc., Dalton Trans.* **1992**, 837.
35. Tuchagues, J.-P., M.; Hedrickson, D. N., *Inorg. Chem.* **1983**, *22*, 2545.
36. Tuck, D. G., *Chem. Soc. Rev.* **1992**, *112*, 215.
37. Cass, M. E.; Greene, D. L.; Buchanan, R. M.; Pierpont, C. G., *J. Am. Chem. Soc.* **1983**, *105*, 2680.
38. Cass, M. E.; Gordon, N. R.; Pierpont, C. G., *Inorg. Chem.* **1986**, *25*, 3962.
39. Pierpont, C. G., *Coord. Chem. Rev.* **2001**, *219-221*, 415.
40. Zanello, P.; Corsini, M., *Coord. Chem. Rev.* **2006**, *250*, 2000.
41. Manos, M. J.; Tasiopoulos, A. J.; Raptopoulou, C.; Terzis, A.; Woolins, J. D.; Slawin, A. M. Z.; Keramidas, A. D.; Kabanos, T. A., *J. Chem. Soc., Dalton Trans.* **2001**, 1556.
42. Cooper, S. R.; Koh, Y. B.; Raymond, K. N., *J. Am. Chem. Soc.* **1982**, *104*, 5092.
43. Karpishin, T. B.; Dewey, T. M.; Raymond, K. N., *J. Am. Chem. Soc.* **1993**, *115*, 1842.
44. Kabanos, T. A.; White, A. J. P.; Williams, D. J.; Woolins, J. D., *Chem. Commun.* **1992**, 17.
45. Kabanos, T. A.; Slawin, A. M. Z.; Williams, D. J.; Woolins, J. D., *Chem. Commun.* **1990**, 193.
46. Handa, M.; Mikuriya, M.; Sato, Y.; Kotera, T.; Nukada, R.; Yoshioka, D.; Kasuga, K., *Bull. Chem. Soc. Jpn.* **1996**, *69*, 3483.
47. Handa, M.; Matsumoto, H.; Namura, T.; Nagaoka, T.; Kasuga, K.; Mikuriya, M.; Kotera, T.; Nukada, R., *Chem. Lett.* **1995**, 903.
48. Chun, H.; Chaudhuri, P.; Weyhermuller, T.; Wieghardt, K., *Inorg. Chem.* **2002**, *41*, 790.
49. Chun, H.; Verani, C. N.; Chaudhuri, P.; Bothe, E.; Bill, E.; Weyhemuller, T.; Wieghardt, K., *Inorg. Chem.* **2001**, *40*, 4157.
50. Le Maqueres, P.; Lindeman, S. V.; Xochi, J. K., *J. Chem. Soc., Perkin Trans.* **2001**, 1180.
51. Sun, D.-L.; Rosokha, S. V.; Linderman, S. V.; Kochi, J. K., *J. Am. Chem. Soc.* **2003**, 125, 15950.
52. Oswald, I. D. H.; Motherwell, W. D. S.; Parsons, S., *Acta Crystallogr., Sect. E: Struct. Rep. Online* **2004**, *60*, o1967.
53. Arulsamy, N.; Bohle, D. S.; Butikofer, J. L.; Stephens, P. W.; Yee, G. T., *Chem. Commun.* **2004**, 1856.
54. Patil, A. O.; Curtin, D. Y.; Paul, I. C., *J. Am. Chem. Soc.* **1984**, *106*, 4010.
55. Pennington, W. T.; Patil, A. O.; Curtin, D. Y.; Paul, I. C., *J. Chem. Soc., Perkin Trans. 2* **1986**, 1693.

56. Wallwork, S. C.; Powell, H. M., *J. Chem. Soc., Perkin Trans. 2* **1980**, 641.
57. Bolte, M.; Margraf, G.; Lerner, W., *Cambridge Crystallographic Database* **2002**.
58. Sugiura, K.; Toyado, J.; Okamoto, H.; Okaniwa, K.; Mitani, T.; Kawamoto, A.; Tanaka, J.; Nakasuji, K., *Angew. Chem., Int. Ed. Engl.* **1992**, *31*, 852.
59. Stobie, K. M.; Bell, Z. R.; Munhoven, T. W.; Maher, J. P.; McCleverty, J. A.; Ward, M. D.; Mcinnes, E. J. L.; Totti, F.; Gatteschi, D., *J. Chem. Soc., Dalton Trans.* **2003**, 36.
60. Dinnebier, R.; Lerner, H.-W.; Ding, L.; Shankland, D., W. I. F.; Stephens, P. W.; Wagner, M., *Z. Anorg. Allg. Chem.* **2002**, *628*, 310.
61. Heistand, R. H.; Roe, A. L.; Que Jounior, L., *Inorg. Chem.* **1982**, *21*, 676.
62. Hoshina, G.; Ohba, S.; Tsuchimoto, M., *Acta Crystallogr., Sect. C: Cryst. Struct. Commun.* **1999**, *55*, 1451.
63. Rosi, N. L.; Kim, J.; Eddaoudi, M.; Chen, B.; O'Keffe, M.; Yaghi, O. M., *J. Am. Chem. Soc.* **2005**, *127*, 1504.
64. Arevalo, S.; Bonillo, M. R.; De Jesus, E.; De la Mata, E.; Flores, J. C.; Gomez-Sal, P.; P., O., *Journal Of Organometallic Chemistry* **2003**, *681*, 228.
65. Muller, J.; Kikuchi, A.; Bill, E.; Weyhermuller, T.; Hildebrandt, P.; Ould-Moussa, L.; Wieghart, K., *Inorg. Chim. Acta* **2000**, *297*, 265.
66. Sangeetha, N. R.; Pal, S., *Bull. Chem. Soc. Jpn.* **2000**, *73*, 357.
67. Tanski, J. M.; Vaid, T., P.; Lobkovsky, E. B.; Wolczanski, P. T., *Inorg. Chem.* **2000**, *39*, 4756.
68. Evans, W. J.; Ansari, M. A.; Ziller, J. W., *Polyhedron* **1998**, *17*, 299.
69. Plass, W., *Z. Anorg. Allg. Chem.* **1997**, *623*, 461.
70. McQiullan, F. S.; Berridge, T. E.; Chen, H.; Hamor, T. A.; Jones, C. J., *Inorg. Chem.* **1998**, *37*, 4959.
71. Serrette, A.; Carroll, P. J.; Swager, T. M., *J. Am. Chem. Soc.* **1992**, *114*, 1887.
72. Van An Ung; Bardwell, D. A.; Jeffrery, J. C.; Maher, J. P.; McCleverty, J. A.; Ward, M. D.; Williamson, A., *Inorg. Chem.* **1996**, *35*, 5290.
73. Tsuchimoto, M.; Kasahara, R.; Kakajima, K.; Kojima, M.; Ohba, S., *Polyhedron* **1999**, *18*, 3035.
74. Tsuchimoto, M.; Yasuda, E.; Ohba, S., *Chem. Lett.* **2000**, 562.
75. Sangeetha, N. R.; Kavita, V.; Wocadlo, S.; Powell, A. K.; Pal, S., *J.Coord. Chem.* **2000**, *51*, 55.
76. Sokolowski, A.; Bothe, E.; Bill, E.; Weyhermuller, T.; Wieghart, K., *Chem. Commun.* **1996**, 1671.
77. Handa, M.; Matsumoto, H.; Yoshioka, D.; Nukada, R.; Mikuriya, M.; Hiromitsu, I.; Kasuga, K., *Bull. Chem. Soc. Jpn.* **1998**, *71*, 1811.
78. Senge, M. O.; Speck, M.; Wiehe, A.; Diecks, H.; Aguirre, S.; Kurreck, H., *Photochem. Photobiol.* **1999**, *70*, 206.

Chapter 26

Electron Spin Lattice Relaxation of V(IV) Complexes in Glassy Solutions between 15 and 70 K

Alistair J. Fielding[1], Dong Bin Back[1], Michael Engler[1],
Bharat Baruah[2], Debbie C. Crans[2], Gareth R. Eaton[1],
and Sandra S. Eaton[1]

[1]Department of Chemistry and Biochemistry, University of Denver,
Denver, CO 80208–2436
[2]Department of Chemistry, Colorado State University,
Fort Collins, CO 80523–1872

Electron spin lattice relaxation rates for four vanadium(IV) complexes: bis(acetylacetonate)oxovanadium(IV), (VO-(acac)$_2$), **1**; bis(maltolato)oxovanadium(IV), (VO(maltol)$_2$), **2**; Cesium N,N'-ethylenebis(salicylideneiminato-5'-sulfonato)-oxovanadium(IV), Cs$_2$[VO(salen-SO$_3$)(H$_2$O)], **3**; and bis(N-hydroxyiminodiacetato)oxovanadium(IV), (Ca[V(hida)$_2$]), **4**; in 1:1 water:glycerol glasses were measured by long-pulse saturation recovery at X-band. Although these complexes have coordination spheres that vary from O$_5$ to N$_4$O and N$_2$O$_6$, the relaxation rates differ by factors of only about 2 at 15 K and 4 at 70 K. Relaxation rates for **4**, which does not contain an oxo group, are very similar to those for oxo-containing complexes. At 70 K relaxation rates decrease in the order aquo VO^{2+} > VO(acac)$_2$ ~ VO(maltol)$_2$ > [V(hida)$_2$]$^{2-}$ > [VO(salen-SO$_3$)(H$_2$O)]$^{2-}$ > vanadyl porphyrin. This order correlates with decreasing flexibility of the ligands and coordination sphere. The temperature dependence of spin lattice relaxation rates was analyzed in terms of contributions from the direct process, the Raman process, and local modes.

© 2007 American Chemical Society

Although the coordination chemistry of vanadium(IV) complexes has been studied extensively using electron paramagnetic resonance (EPR) (*1, 2*), there have been few studies of the electron spin relaxation rates of these systems. Prior studies on a limited selection of complexes include primarily low-temperature measurements in ionic lattices near 4 K (*3*) and NMR studies in fluid solution (*4*). Mechanisms of relaxation for aquo vanadyl ion (*5*) and vanadyl porphyrins (*6*) in glassy solution between about 10 and 100 K have been analyzed (*7*). Relaxation rates reflect the electronic structures of the paramagnetic center and the dynamic processes of these species and their environment. Knowledge of typical relaxation rates is useful in predicting observability of EPR signals and feasibility of electron-nuclear double resonance (ENDOR) experiments, and in selecting parameters for recording spectra. The difference between aquo vanadyl and vanadyl porphyrin suggests an impact of structure on relaxation. We report the relaxation rates of vanadium(IV) in complexes with diverse coordination geometries: five-coordinated VO(maltol)$_2$ **2** and VO(acac)$_2$ **1**, six-coordinated [VO(salen-SO$_3$)(H$_2$O)]$^{2-}$ **3**, and eight-coordinated non-oxo amavadin analog, [V(hida)$_2$]$^{2-}$ **4**. Our goal is to understand the relaxation processes that occur in glassy solutions at temperatures between about 15 and 70 K and to determine the extent to which coordination geometry impacts relaxation rates.

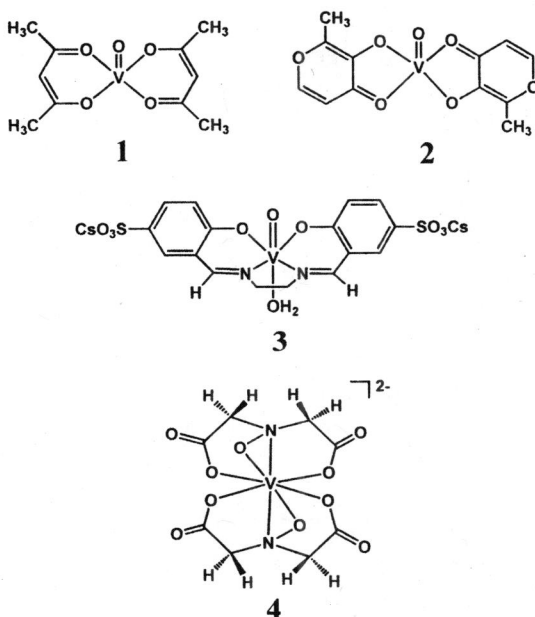

Experimental

VO(acac)$_2$, vanadyl sulfate, 3-hydroxy-2-methyl-4-pyrone, salicylaldehyde, ethylenediamine, and bromoacetic acid were purchased from Aldrich. Hydroxylamine hydrochloride was purchased from Fisher Scientific. Reagents were used as received. Deionized water was used throughout and prepared to a specific resistivity of > 18 MΩ cm (Barnstead E-pure system). VO(maltol)$_2$ (8), Cs$_2$[VO(salen-SO$_3$)(H$_2$O)] (9), and the non-oxo-vanadium(IV) complex, (Ca[V(hida)$_2$]) (10, 11) were prepared by literature methods.

Aqueous stock solutions of VO(acac)$_2$ (10 mM, pH 4.2), VO(maltol)$_2$ (5 mM, pH 4.6), Cs$_2$[VO(salen-SO$_3$)(H$_2$O)] (25 mM, pH 7.0) and Ca[V(hida)$_2$] (25 mM, pH 5.3) were prepared by dissolving in water in a volumetric flask. The pH values of the aqueous solutions were measured at 25°C using a pH meter (Orion 420A) calibrated with three buffers of pH 4, 7 and 10. The VO(salen-SO$_3$) complex is assigned as the 6-coordinate water adduct on the basis of the red color of the solution (12). The aqueous solutions were combined 1:1 v:v with glycerol to make mixtures that form a good glass when cooled quickly in liquid nitrogen.

CW spectra were recorded on a Varian E9 and computer simulated using the locally-written program, MONMER, which is based on the equations in (13).

Spin-lattice relaxation rates, $1/T_1$, as a function of temperature and position in the spectrum were measured by long-pulse saturation recovery on a locally constructed X-band spectrometer using a rectangular TE$_{102}$ cavity (14). The procedures for data acquisition and analysis are similar to those employed in a recent study of relaxation rates for Cu(II) complexes (15). Analyses of recovery curves as distributions of exponentials were performed using Brown's uniform penalty, UPEN, routines (16, 17). By minimizing the sum of the residuals on a log-log scale, a fit line for the temperature dependence of $1/T_1$ for the $m_I = -1/2$ transition was based on Eq. [1] (7).

$$\frac{1}{T_1} = A_{dir} T + A_{Ram} \left(\frac{T}{\theta_D}\right)^9 J_8\left(\frac{\theta_D}{T}\right) + A_{loc}\left[\frac{e^{\Delta_{loc}/T}}{(e^{\Delta_{loc}/T}-1)^2}\right] \qquad [1]$$

where T is temperature in Kelvin, A_{dir} is the coefficient for the contribution from the direct process, A_{Ram} is the coefficient for the contribution from the Raman process, θ_D is the Debye temperature, J_8 is the transport integral,

$$J_8\left(\frac{\theta_D}{T}\right) = \int_0^{\theta_D/T} x^8 \frac{e^x}{(e^x-1)^2} \, dx,$$

A_{loc} is the coefficient for the contribution from a local vibrational mode, and Δ_{loc} is the energy for the local mode in units of Kelvin. The expressions to describe the temperature dependence of the Raman process (18, 19) and local mode (20) were taken from the literature.

Results and Discussion

The V(IV) complexes were characterized by continuous wave EPR spectra at X-band (~ 9.2 GHz) of samples in water at ambient temperature and in glassy 1:1 water:glycerol at about 120 K (Figure 1). Solvent crystallization can cause locally high concentrations of solute that broaden CW spectra and increase relaxation rates (7, 21). Therefore glycerol was added to the solutions to ensure glass formation at low temperature. However, the presence of glycerol in the solutions for low-temperature spectroscopy raises the question whether glycerol coordinated to the V(IV). A single species was observed in fluid and frozen solution EPR spectra of VO(maltol)$_2$, [VO(salen-SO$_3$)(H$_2$O)]$^{2-}$, and [V(hida)$_2$]$^{2-}$. For each of the complexes the average of anisotropic g and A values in the glassy 1:1 water:glycerol were in good agreement with the isotropic values in water so there was no evidence for complexation of glycerol. For VO(acac)$_2$ two species were observed in water at room temperature, in agreement with prior studies (21), and in the glassy 1:1 water:glycerol. Comparison of average g and A values for the two species in fluid and glassy solution provide no evidence for glycerol complexation, but to determine whether the two species are the same would require ENDOR (22-24) or electron spin echo envelope modulation (ESEEM) experiments (25).

In the glassy solutions all orientations of the molecule with respect to the magnetic field are present so the spectra are superpositions of signals with g and A values characteristic of differing orientations. The first-derivative display emphasizes positions in the spectrum where there are abrupt changes in the amplitude of the absorption signal. These changes occur at magnetic fields that correspond to extrema in the orientation dependence of the EPR resonance. Thus "peaks" in the first derivative signal occur at magnetic fields that correspond to resonance along the principal magnetic axes, x, y, and z. The vanadium nucleus has a spin of 7/2 so for each orientation of the molecule with respect to the external field the EPR signal is split into 8 lines corresponding to m_I = -7/2, -5/2, -3/2, -1/2, 1/2, 3/2, and 7/2. At X-band the nuclear hyperfine splitting is much larger than the difference in resonance field that arises from the g anisotropy. The large hyperfine splitting along the magnetic z-axis (153x10^{-4} cm^{-1}, 171 G) determines the width of the spectrum (Figure 1). The differences between g_x and g_y and between A_x and A_y are small so these peaks overlap in the center of the spectrum. The g values and hyperfine coupling constants obtained by computer simulation of the spectra are summarized in Table 1 and are in good agreement with literature values for [V(hida)$_2$]$^{2-}$ (11), VO(maltol)$_2$ (26), and VO(acac)$_2$ (27). Despite the absence of an oxo-group the rigid-lattice EPR spectrum of [V(hida)$_2$]$^{2-}$ (Figure 1) is very similar to those for oxo-containing vanadyl complexes.

A saturation recovery curve is recorded by applying a pulse of microwaves with sufficiently large microwave magnetic field to saturate (equalize) the

Table 1. EPR g and A values at 120 K in glassy 1:1 water:glycerol[a]

Sample	g_x, A_x	g_y, A_y	g_z, A_z
Aquo-VO^{2+} [b]	1.979, 70	1.978, 70	1.933, 183
VO(acac)$_2$	1.980, 54	1.967, 56	1.950, 163
VO(maltol)$_2$	1.977, 60	1.974, 55	1.939, 169
[V(hida)$_2$]$^{2-}$	1.986, 42	1.984, 49	1.918, 153
[VO(salen-SO$_3$)H$_2$O]$^{2-}$	1.978, 46	1.974, 55	1.955, 154

[a]Values of A are in units of 10^{-4} cm^{-1}. Uncertainties in g and A values are ±0.001 and ±1, respectively. [b] EPR parameters reported in (23, 28) in water.

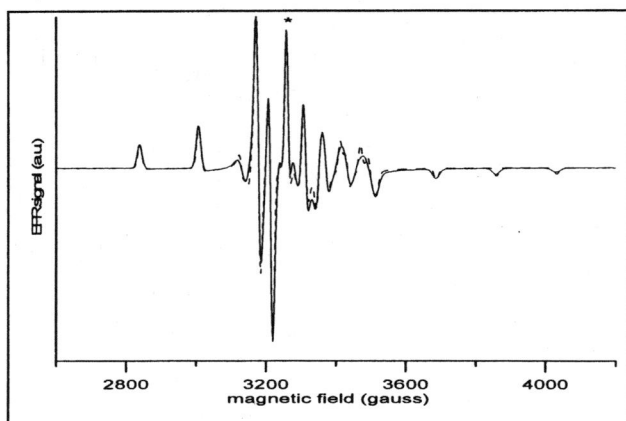

Figure 1. X-band (9.2267 GHz) spectrum of [V(hida)$_2$]$^{2-}$ in 1:1 water:glycerol at 118 K. The dashed line is a simulation using the parameters listed in Table 1. The asterisk marks the position of the perpendicular component of the m_I = -1/2 transition where most of the saturation recovery curves were recorded.

populations of the electron spin energy levels, which nulls the EPR signal. The return to equilibrium of the amplitude of the EPR signal is then recorded using very low (minimally-perturbing) microwave power (29). The time constant for return to equilibrium is the spin-lattice relaxation time, T_1. Representative saturation recovery curves for the non-oxo complex [V(hida)$_2$]$^{2-}$, recorded at the peak of the m_I = -1/2 line, are shown in Figure 2. Analogous to observations for Cu(II) complexes, better fits to the saturation recovery curves were obtained with a distribution of exponentials than with a single exponential (15). The dashed lines in Figure 2 are fit functions based on distributions of exponentials calculated using the uniform penalty method (16, 17). The widths of the distributions were 2 to 3 times the central value. Distributions in g and A values, sometimes called *g*-strain and *A*-strain, are routinely observed for Cu(II) complexes (*30, 31*) and more recently for vanadyl complexes (*32*). These distributions reflect variations in geometry that may also impact relaxation rates. The observation of distributions in relaxation rates may be characteristic of transition metal complexes.

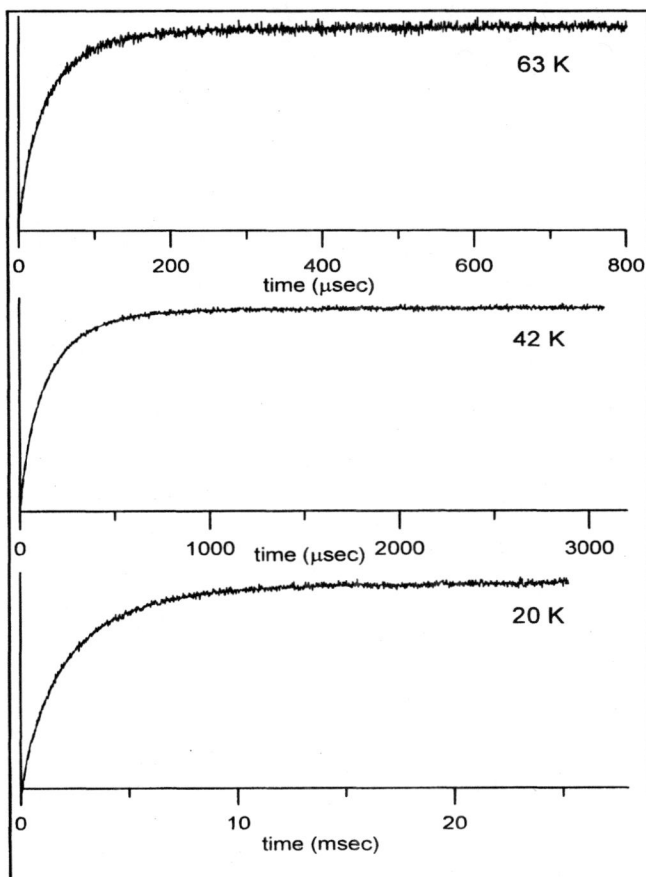

Figure 2. X-band saturation curves for 2.5 mM [V(hida)₂]²⁻ in 1:1 water:glycerol as a function of temperature recorded at the maximum in the absorption signal, which is on the mₗ = -1/2 line. Note the changes in the scales of the x axes as a function of temperature. The dashed lines are fits to the data for a distribution of exponentials.

In some systems electron spin relaxation rates are strongly dependent upon the orientation of the molecule with respect to the external field (*7*). Relaxation rates at 62 K for [V(hida)$_2$]$^{2-}$ and VO(maltol)$_2$ were measured at a series of positions in the spectrum, but the variation with orientation was less than a factor of 2. This small orientation dependence of T$_1$ is similar to what was observed previously for a vanadyl porphyrin (6).

Temperature dependence of relaxation rates

The temperature dependence of the median values from the distributions of relaxation rates measured for the m$_1$ = -1/2 transition was analyzed by fitting equation [1] to the data, as shown in Figure 3 for [V(hida)$_2$]$^{2-}$. At the lowest temperatures studied there was a small contribution from the direct process, which is characterized by a weak temperature dependence. This process involves a single photon and increases as the local concentration of spins increases. The dominant process between about 20 and 40 K is the two-photon Raman process which arises from the many motional modes with energies less than the Debye temperature. At higher temperature there is a contribution from an additional process that can be modeled as a local vibrational mode with an energy of about 205 K (140 cm^{-1}). An equally good fit to the X-band experimental relaxation rates could be obtained with a thermally-activated process instead of the local mode. However a thermally-activated process predicts relaxation rates that are dependent on microwave frequency. For a vanadyl porphyrin the relaxation rate at 80 K and 95 GHz is about the same as at 9.5 GHz which is inconsistent with a thermally-activated process (33). In view of the similarity in relaxation rates for the series of V(IV) complexes, discussed below, the additional process for [V(hida)$_2$]$^{2-}$ is assigned as local mode.

Comparison of temperature dependence of relaxation rates

The relaxation rates as a function of temperature for the four V(IV) complexes in 1:1 water:glycerol are compared (Figure 4) with previously reported data for aquo VO^{2+} in 1:1 water:glycerol and for a vanadyl porphyrin in 2:1 toluene:CHCl$_3$. Over the full temperature range there is substantial similarity in the relaxation rates for the six V(IV) complexes and in the parameters required to fit the temperature dependence (Table 2). The relaxation rates for [V(hida)$_2$]$^{2-}$, which does not have an oxo group, are similar to those for the oxo-containing complexes. The shape of the curve for VOTTP-bipy at low temperature is different than for the other complexes because of a larger contribution from the direct process, which is attributed to stacking of the planar aromatic molecules. The differences between molecules are largest at higher temperatures where local modes dominate the relaxation. At about 70 K the relaxation rates decrease in the order aquo VO^{2+} > VO(acac)$_2$ ~ VO(maltol)$_2$ >

$[V(hida)_2])^{2-} > [VO(salen-SO_3)(H_2O)]^{2-} >$ vanadyl porphyrin, which is the order in which entries are listed in Table 2.

Modulation of spin-orbit coupling is a significant mechanism of spin-lattice relaxation. Deviation of g values away from the free electron value (g = 2.0023) is a qualitative indicator of the magnitude of spin-orbit coupling. The g values for the V(IV) complexes are quite similar, which is consistent with the observation of similar spin-lattice relaxation rates. At 70 K the relaxation rates for all of the V(IV) complexes are slower than for a wide range of Cu(II) complexes (15). The g_z values for the Cu(II) complexes were in the range of 2.08 to 2.33, which is substantially further from g = 2.0023 than for V(IV), which indicates greater spin-orbit coupling and is consistent with faster relaxation for the Cu(II) complexes than for the V(IV) complexes.

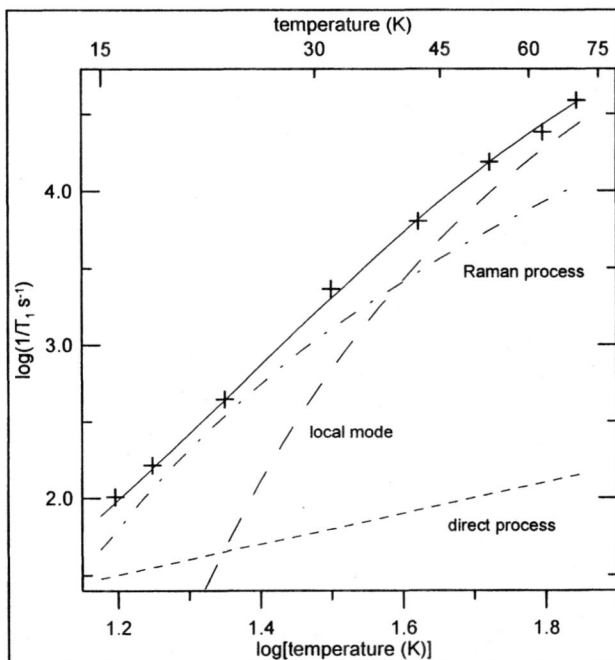

Figure 3. Temperature dependence of X-band spin lattice relaxation rates for the m_I = -1/2 line for [V(hida)$_2$]$^{2-}$ in 1:1 water glycerol. The solid lines through the data are fits obtained using Eq. [1] and the parameters in Table 2. The contributions to the relaxation from the direct process (---), Raman process (-·-·) and local mode (-··-) are shown separately.

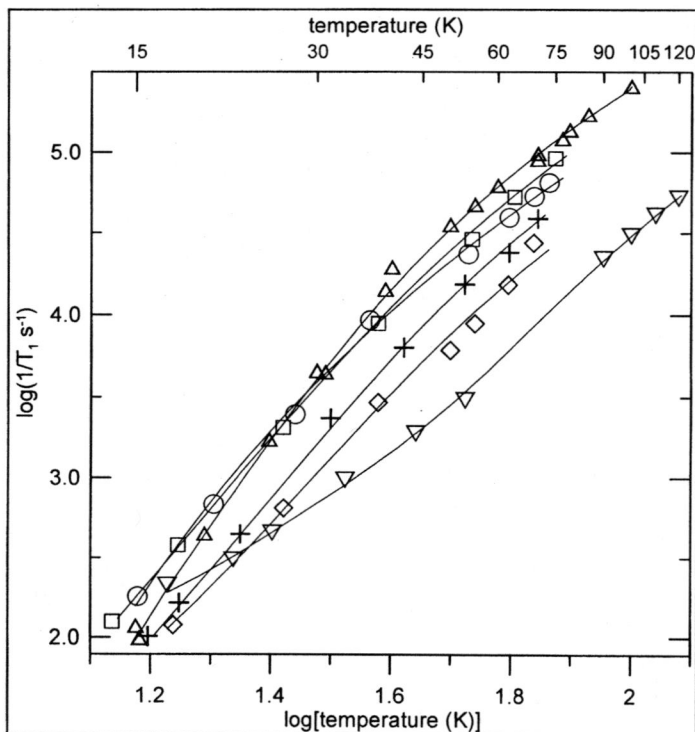

Figure 4. Temperature dependence of X-band spin lattice relaxation rates for the perpendicular lines in the spectra of aquo VO^{2+} ($\triangle$), $VO(acac)_2$ ($\square$), $VO(maltol)_2$ ($\bigcirc$) , $[V(hida)_2]^{2-}$ (+), and $[VO(salen-SO_3)H_2O]^{2-}$ ($\diamond$), in 1:1 water:glycerol and vanadylporphyrin (∇) in 2:1 toluene:CHCl_3. The solid lines through the data are fits obtained using Eq. [1] and the parameters in Table 2.

Results from prior studies (7, 15) suggest that molecular flexibility contributes to differences in relaxation rates – the less flexible the molecule, the slower the relaxation. Studies presented here allow testing of this hypothesis for V(IV) complexes with different geometries and ligands. The 5-coordinate complexes investigated are $VO(acac)_2$, $VO(maltol)_2$, and vanadyl porphyrin. In addition the 6-coordinated aquo VO^{2+} and $[VO(salen-SO_3)(H_2O)]^{2-}$ and an 8-coordinated non-oxo amavadin analog $[V(hida)_2]^{2-}$ were investigated. The more flexible metal-ligand complexes aquo VO^{2+}, $VO(acac)_2$, and $VO(maltol)_2$ are compared with the more rigid $[V(hida)_2]^{2-}$, $[VO(salen-SO_3)(H_2O)]^{2-}$ and vanadyl porphyrin. The relaxation rates decrease in the order aquo VO^{2+} > bidentate $VO(acac)_2$, $VO(maltol)_2$ > polydentate $[V(hida)_2]^{2-}$ > polydentate [VO(salen-

$SO_3)H_2O]^{2-}$ > rigid polydentate vanadyl porphyrin. The decreasing relaxation rates with increasing denticity of the ligands is consistent with slower relaxation for more rigid complexes. Within the set of polydentate ligands the slowest relaxation is observed for the porphyrin which has the rigid aromatic ligand. The similarity of relaxation rates for the non-oxo containing $[V(hida)_2]^{2-}$ and the oxo-vanadium complexes indicates that motion of the vanadyl unit is not a major determinant of relaxation rates.

Table 2. Contributions to spin-lattice relaxation at 15 to 70 K in 1:1 water:glycerol glasses[a]

Sample	coord.	g_z, g_{iso}	A_z	Direct A_{dir}	Raman A_{Ram}, θ_D	Local A_{loc}, Δ_{loc}
Aquo VO^{2+} [b]	O_6	1.933,[c] 1.969	183[c]	0	4.4×10^5, 110	9.2×10^5, 185
$VO(acac)_2$	O_5	1.950, 1.966	163	3.5	5.8×10^5, 110	7.0×10^5, 205
$VO(maltol)_2$	O_5	1.939, 1.963	169	0	7.4×10^5, 110	3.0×10^5, 205
$[V(hida)_2]^{2-}$	N_2O_6	1.918, 1.964	152	2.0	2.2×10^5, 110	4.5×10^5, 205
$[VO(salen - SO_3)H_2O]^{2-}$	N_2O_4	1.955, 1.963	154	3.0	1.5×10^5, 110	2.5×10^5, 205
VOporphyrin[b]	N_4O	1.984, 1.971	158	8.5	6.5×10^4, 100	6.5×10^5, 350

[a] Units are: A_z (10^{-4} cm^{-1}), A_{dir} (s^{-1} K^{-1}), A_{Ram} (s^{-1}), θ_D (K), A_{loc} (s^{-1}), Δ_{loc} (K).
[b] Contributions to relaxation in 2:1 toluene:chloroform reported in (*34*). [c] EPR parameters reported in (*28*).

Acknowledgments

Financial support of this work by NIH/NIBIB EB002807 (GRE and SSE) and NSF 0314719 (DCC) is gratefully acknowledged.

References

1. Chasteen, N. D. *Biol. Magn. Reson.* **1983**, *3*, 53-119.
2. Smith, T. S.; LoBrutto, R.; Pecoraro, V. L. *Coord. Chem. Rev.* **2002**, *228*, 1-18.

3. Standley, K. J.; Vaughan, R. A. *Electron Spin Relaxation Phenomena in Solids*; Plenum Press, 1969.
4. Bertini, I.; Martini, G.; Luchinat, C. In *Handbook of Electron Spin Resonance: Data Sources, Computer Technology, Relaxation, and ENDOR*; Poole, J., C. P., Farach, H., Eds.; American Institute of Physics: New York, 1994, pp 79-310.
5. Eaton, G. R.; Eaton, S. S. *J. Magn. Reson.* **1999**, *136*, 63-68.
6. Du, J.-L.; Eaton, G. R.; Eaton, S. S. *J. Magn. Reson. A* **1996**, *119*, 240-246.
7. Eaton, S. S.; Eaton, G. R. *Biol. Magn. Reson.* **2000**, *19*, 29-154.
8. Orvig, C.; Caravan, P.; Gelmini, L.; Glover, N.; Herring, F. G.; Li, H.; McNeill, J. H.; Rettig, S. J.; Setyawati, I. A.; Shuter, E.; Sun, Y.; Tracey, A. S.; Yuen, V. G. *J. Am. Chem. Soc.* **1995**, *117*, 12759-12770.
9. Evans, D. F.; Missen, P. H. *J. Chem. Soc., Dalton Trans.* **1987**, 1279-1281.
10. Bayer, E.; Kneifel, H. *Z. Naturforsch.* **1972**, *27B*, 207.
11. Smith, P. D.; Berry, R. E.; Harben, S. M.; Beddoes, R. L.; Helliwell, M.; Collison, D.; Garner, C. D. *J. Chem. Soc., Dalton Trans.* **1997**, 4509-4516.
12. Crans, D. C.; Khan, A. R.; Mahroof-Tahir, M.; Mondal, S.; Miller, S. M.; LaCour, A.; Anderson, O. P.; Jakusch, T.; Kiss, T. *Dalton Trans.* **2001**, 3337-3345.
13. Toy, A. D.; Chaston, S. H. H.; Pilbrow, J. R.; Smith, T. D. *Inorg. Chem.* **1971**, *10*, 2219-2225.
14. Quine, R. W.; Eaton, S. S.; Eaton, G. R. *Rev. Sci. Instrum.* **1992**, *63*, 4251-4262.
15. Fielding, A. J.; Fox, S.; Millhauser, G. L.; Chattopadhyay, M.; Kroneck, P. M. H.; Fritz, G.; Eaton, G. R.; Eaton, S. S. *J. Magn. Reson.* **2006**, *179*, 92-104.
16. Borgia, G. C.; Brown, R. J. S.; Fantazzini, P. *J. Magn. Reson.* **1998**, *132*, 65-77.
17. Borgia, G. C.; Brown, R. J. S.; Fantazzini, P. *J. Magn. Reson.* **2000**, *147*, 273-285.
18. Abragam, A. In *The Principles of Nuclear Magnetism*; Oxford University Press: London, 1961, pp 405-409.
19. Murphy, J. *Phys. Rev.* **1966**, *145*, 241-247.
20. Castle, J. G., Jr.; Feldman, D. W. *Phys. Rev. A* **1965**, *137*, 671-673.
21. Amin, S. S.; Cryer, K.; Zhang, B.; Dutta, S. K.; Eaton, S. S.; Anderson, O. P.; Miller, S. M.; Reul, B. A.; Brichard, S. M.; Crans, D. C. *Inorg. Chem.* **2000**, *39*, 406-416.
22. Grant, C. V.; Ball, J. A.; Hamstra, B. J.; Pecoraro, V. L.; Britt, R. D. *J. Phys. Chem. B* **1998**, *102*, 8145-8150.
23. Grant, C. V.; Cope, W.; Ball, J. A.; Maresch, G. G.; Gaffney, B. J.; Fink, W.; Britt, R. D. *J. Phys. Chem. B Condens. Matter Mater. Surf. Interfaces Biophys.* **1999**, *103*, 10627-10631.

24. Mustafi, D.; Makinen, M. W. *Inorg. Chem.* **2005**, *44*, 5580-5590.
25. Eaton, G. R.; Eaton, S. S. *Comp. Coord. Chem. II* **2004**, *2*, 49-55.
26. Hanson, G. R.; Sun, Y.; Orvig, C. *Inorg. Chem.* **1996**, *35*, 6507-6512.
27. Campbell, R. F.; Freed, J. H. *J. Phys. Chem.* **1980**, *84*, 2668-2680.
28. Albanese, N. F.; Chasteen, N. D. *J. Phys. Chem.* **1978**, *82*, 910.
29. Eaton, S. S.; Eaton, G. R. *Biol. Magn. Reson.* **2005**, *24*, 3-18.
30. Froncisz, W.; Hyde, J. S. *J. Chem. Phys.* **1980**, *73*, 3123-3131.
31. Cannistraro, S. *J. Physique (Paris)* **1990**, *51*, 131-139.
32. Mustafi, D.; Galtseva, E. V.; Krzystek, J.; Brunel, L. C.; Makinen, M. W. *J. Phys. Chem. A* **1999**, *103*, 11279-11286.
33. Eaton, S. S.; Harbridge, J.; Rinard, G. A.; Eaton, G. R.; Weber, R. T. *Appl. Magn. Reson.* **2001**, *20*, 151-157.
34. Zhou, Y.; Mitri, R.; Eaton, G. R.; Eaton, S. S. *Current Topics in Biophysics* **1999**, *23*, 63-68.

Chapter 27

Synthetic and Structural Studies of Aqueous Vanadium(IV,V) Hydroxycarboxylate Complexes

Athanasios Salifoglou

Department of Chemical Engineering, Laboratory of Inorganic Chemistry, Aristotle University of Thessaloniki, Thessaloniki 54124, Greece and Chemical Process Engineering Research Institute, Thermi, Thessaloniki 57001, Greece

Vanadium has been amply established as a competent inorganic cofactor in biological systems. Among vanadium's various functions its insulin mimetic capacity stands out and constitutes a challenge to bioinorganic chemists. Prompted by the need to comprehend the insulin mimetic action of vanadium and its interaction with biomolecular targets, extensive synthetic reactivity studies were carried out with V(IV,V) in the presence of low molecular mass ligands in aqueous solutions. Through meticulous synthetic pH-dependent efforts, new V(IV,V) soluble binary and ternary species arose that delineate the aqueous speciation and reflect the chemical reactivity of insulin mimetic vanadium involved in interactions with physiological substrates.

Introduction

Vanadium is an element widely established in abiotic and biological applications. Its employment as an inorganic cofactor in biological systems has received considerable attention in the past couple of decades due to its presence in enzymes (*1,2*), such as nitrogenase (*3*), haloperoxidases (*4,5,6*), and others. Beyond, however, its role in active sites of metalloenzymes outstanding has been its involvement in biochemical processes that affect the physiology of an organism. Characteristic to that end have been its roles in mitogenicity (*7,8*),

© 2007 American Chemical Society

antitumorigenicity (*9,10*) and more recently its insulin mimetic action (*11,12*). The latter biological function has propelled vanadium into the sphere of challenging scientific endeavors invoking multidisciplinary research in biology, medicine and chemistry.

The insulin mimetic action of vanadium has spurred considerable research activities in chemistry, in hopes of finding vanadium drugs that could be of help to patients suffering from Diabetes mellitus II. Key to this challenging feat is in depth understanding of the type of interactions involved in vanadium (bio)chemistry, once the latter element is recognized and internalized by the cell. Such interactions entail chemical reactivity of the biologically relevant oxidation states of vanadium (V(IV), V(V)) with physiological substrates of both low and high molecular mass. Consequently, the existence of well-characterized, water soluble species arises as an important goal in pursuing vanadium metallodrugs capable of insulin mimetic action. Moreover, such vanadium drugs should be further characterized by ideally no toxicity, and fully accounted for physicochemical profiles linking the bioactive vanadium species with cellular organic ligands-substrates.

Prompted by the need to understand the primary chemical interactions of vanadium initially with low molecular mass physiological and physiologically relevant ligands, we launched efforts targeting synthetic complexes of V(V) and V(IV) with the physiological ligands α-hydroxycarboxylic acids. Prominent representatives of such substrates are citric acid and malic acid. They both exist in cellular fluids, with citric acid being present at a concentration of ~0.1 mM (*13,14*). Our synthetic efforts were a) based on information on aqueous solution studies and b) guided by the need to develop pH-dependent reactivity methodologies in binary and ternary systems of V(V) and V(IV) with α-hydroxycarboxylic acids and hydrogen peroxide, all of them components of species purported to exhibit insulin mimetic properties.

Experimental Section

Materials and methods

All experiments were carried out in the open air. Nanopure quality water was used for all reactions. V_2O_5, $VOSO_4$, VCl_3, anhydrous citric acid, and H_2O_2 30% were purchased from Aldrich. Ammonia, potassium and sodium hydroxide were supplied by Fluka.

FT-Infrared spectra were recorded on a Perkin Elmer 1760X FT-infrared spectrometer. UV/Visible measurements were carried out on a Hitachi U2001 spectrophotometer in the range from 190 to 1000 nm. A ThermoFinnigan Flash EA 1112 CHNS elemental analyser was used for C,H,N analysis.

Results

Binary V(V)/V(IV)-(α-hydroxycarboxylate) systems

Both V(V) and V(IV) oxidation states were investigated in the presence of citric acid in aqueous solutions at specific pH values. The general methodology involved the pH-dependent synthesis of V(V,IV)-citrate complexes over the entire pH range. The isolated crystalline products were characterized by elemental analysis, spectroscopically (FT-IR, ^{51}V-, ^{13}C-MAS NMR), and structurally by X-ray crystallography. Chemical reactivity studies were carried out on all new complexes through acid-base chemistry, thermal and non-thermal transformations, at various pH values.

V(V)-citrate system

The chemical reactivity in the aqueous V(V)-citric acid system was studied through reactions of simple reagents at specific molar ratios and pH values. To that end, reactions between V_2O_5 and citric acid with a molar ratio 1:1 and 1:2 were employed throughout this effort and led to the successful synthesis of discrete soluble species. The appropriate base MOH ($M^+ = K^+$, Na^+, Me_4N^+, NH_4^+) was used to adjust the pH of the reaction medium and concurrently provide the necessary counterions for balancing the arising anionic complexes. Representative species of this family of well-characterized complexes include $(NH_4)_2[V_2O_4(C_6H_6O_7)_2] \cdot 2H_2O$ (1) (pH 3.0), $(NH_4)_4[V_2O_4(C_6H_5O_7)_2] \cdot 4H_2O$ (2) (pH 5.0), $(NH_4)_6[V_2O_4(C_6H_4O_7)_2] \cdot 6H_2O$ (3) (pH 8.0) (Figure 1). The employed pH values for the synthesis of 1-3 were optimally set at pH 3, 5, and 8. A representative reaction showing the synthesis of 2 is shown below:

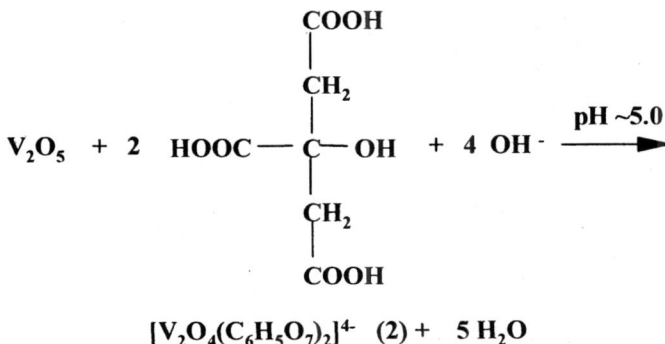

$$V_2O_5 \;+\; 2 \; HOOC-\underset{\underset{\displaystyle COOH}{\overset{\displaystyle |}{CH_2}}}{\overset{\overset{\displaystyle COOH}{\overset{\displaystyle |}{CH_2}}}{\underset{|}{C}}}-OH \;+\; 4 \; OH^- \xrightarrow{\;pH \sim 5.0\;}$$

$$[V_2O_4(C_6H_5O_7)_2]^{4-} \;\; (2) + \;\; 5 \; H_2O$$

In all three cases, the isolated materials were characterized spectroscopically and subsequently structurally by X-ray crystallography.

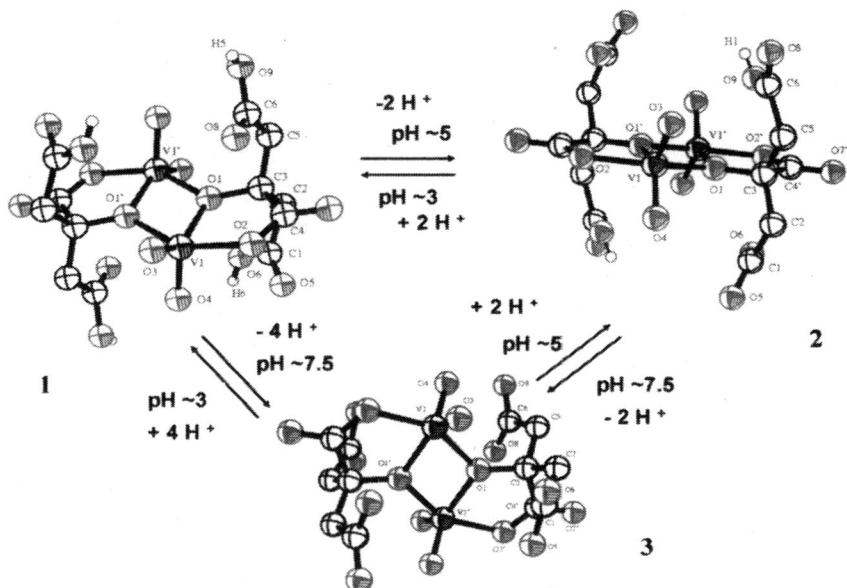

Figure 1. pH-structural variants of binary V(V)-citrate species and their pH-dependent acid-base chemistry.
(See page 3 of color inserts.)

Compounds **1-3** contain a stable planar $V^V_2O_2$ core to which attached are two citrate ligands. The latter bind to the planar core through the deprotonated central alkoxide and carboxylate anchors. The terminal carboxylate groups do not participate in the coordination to the core, dangling away from the complex.

The fundamental structural composition throughout the family of these complexes is the same, with the main difference between the various pH-structural variants being the state of protonation of the non-coordinating citrate terminal carboxylate groups. Specifically, as the pH increases progressively (from 3 to 5 to 7.5-8.0), two protons are released from the complex (one from each citrate) in each pH step, thus raising the charge of the anionic species by 2. In this regard, the acid-base chemistry among the three species establishes their involvement as competent partners in the requisite aqueous binary speciation. In so doing, it emphasizes the significance of pH as a molecular switch dictating a) the (de)protonation chemistry in the periphery of the vanadium(V)-citrate complex, and b) the chemical and geometrical features of the structural assembly of all arising complexes (*15,16*).

V(IV)-citrate system

In the binary V(IV)-citrate system, aqueous reactions between $VOSO_4$ and citric acid with molar ratios 1:1 and 1:2 were carried out in a pH-dependent fashion.   The same reactivity was observed when VCl_3 was used as a starting material.   Here as well, the employment of a base (e.g. KOH) was instrumental in a) adjusting the pH of the reaction medium, and b) concurrently providing the counterion necessary for the neutralization of the charge of the arising anionic species.   Specifically, the reaction at pH ~4.5 led to the isolation of complex $K_3[V_2O_2(C_6H_4O_7)(C_6H_5O_7)] \cdot 7H_2O$ (4) upon ethanol addition.

In a similar reaction at pH 8, the material isolated was $K_4[V_2O_2(C_6H_4O_7)_2] \cdot 6H_2O$ (5).   In this case, use of a base included MOH ($M=K^+$, Na^+, NH_4^+).   A characteristic reaction leading to the synthesis of 5 is shown below:

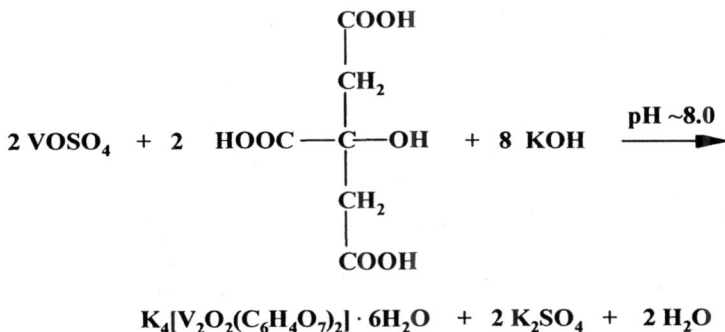

$$2\ VOSO_4\ +\ 2\ \ \text{citric acid}\ +\ 8\ KOH\ \xrightarrow{\text{pH} \sim 8.0}$$

(structure of citric acid shown: COOH–CH$_2$–C(OH)(COOH)–CH$_2$–COOH, drawn as HOOC–C(CH$_2$COOH)(OH)(CH$_2$COOH) vertical)

$$K_4[V_2O_2(C_6H_4O_7)_2] \cdot 6H_2O\ +\ 2\ K_2SO_4\ +\ 2\ H_2O$$

In all of these cases, dinuclear complexes were isolated and fully characterized.   In the high pH-structural variant 5, the citric acid employs all four coordinating anchors to bind V(IV), thus giving rise to a dinuclear $V^{IV}_2O_2$ core assembly basically formed through participation of the alkoxide group of the α-hydroxycarboxylate moiety.   In so reacting, the citrate ligand is fully deprotonated and the arisen complex contains the two vanadium(IV) ions octahedrally coordinated.   Key to the structural identity of the dinuclear V(IV)-citrate complexes is the presence of the V=O unit, contributing to the formulation of the octahedral geometry around each metal ion.   In the case of complex 4, the citrate binds to V(IV) in two different coordination modes, still creating the same $V^{IV}_2O_2$ core, albeit with variable coordination geometry and state of citrate protonation.   Both types of complexes exhibit acid-base chemistry reactivity between them, consistent with their participation in the requisite speciation scheme as competent pH-structural variants (Figure 2) (*17*).

382

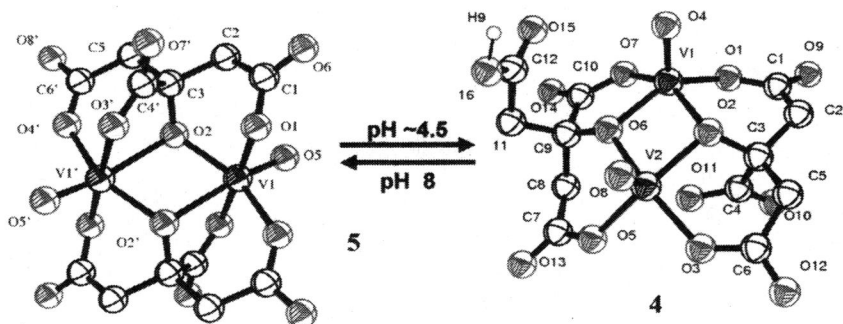

Figure 2. pH-Dependent acid-base chemistry between pH-structural
variants of binary V(IV)-citrate species.
(See page 3 of color inserts.)

Ternary V(V)-H₂O₂-hydroxycarboxylate systems

The investigated ternary systems targeted soluble complexes of the type a) V(V)-H_2O_2-malic acid, and b) V(V)-H_2O_2-citric acid. The methodology employed here was similar to the one mentioned previously in the case of the binary system of V(V)-citric acid. The isolated crystalline products were characterized by elemental analysis, spectroscopically (FT-IR, UV-Visible, NMR), and structurally by X-ray crystallography. Chemical reactivity studies were carried out through acid-base chemistry, thermal and non-thermal transformations, at pH values consistent with the nature of the original starting materials or the binary and ternary products derived synthetically.

V(V)- H_2O_2 –malate

To achieve the synthesis of ternary complexes of vanadium(V) with hydrogen peroxide and malic acid, VCl_3 was used as a convenient starting material. That reacted with malate in the presence of hydrogen peroxide at pH 4.5 and 7.5. Identical reactivity was observed with $VOSO_4$ and V_2O_5. In the case of the low pH value system, employment of a base, such as K^+, led to the isolation of complex $K_2[V_2O_2(O_2)_2(C_4H_4O_5)_2] \cdot 2H_2O$ (6). The system on the other hand, at the highest pH value, closest to the physiological value, led in the presence of ammonia or K^+ to the complexes $(Cat)_4[VO(O_2)(C_4H_3O_5)]_2 \cdot nH_2O$ (Cat = K^+, n=4 (7); Cat = NH_4^+, n=3 (8)) (18).

V(V)-H₂O₂-citrate

In the case of the ternary complexes of vanadium with hydrogen peroxide and citric acid, pH-dependent synthesis was employed successfully. The optimal pH values included pH 3.0, 5.5 and 8.0. Here as well, the presence of the base and the choice of pH were instrumental in ultimately achieving the isolation of pure soluble complexes $(Cat)_2[V_2O_4(C_6H_6O_7)_2]\cdot nH_2O$ (**9**) $(Cat^+ = Na^+, NH_4^+, n = 2; Me_4N^+, K^+, n = 4)$, $K_{10}[V_2O_2(O_2)_2(C_6H_5O_7)_2][V_2O_2(O_2)_2(C_6H_4O_7)_2]\cdot 20H_2O$ (**10**), and $(NH_4)_6[V_2O_2(O_2)_2(C_6H_4O_7)_2]\cdot 4.5H_2O$ (**11**) (*19*). A representative reaction leading to the formation of **10** is given below:

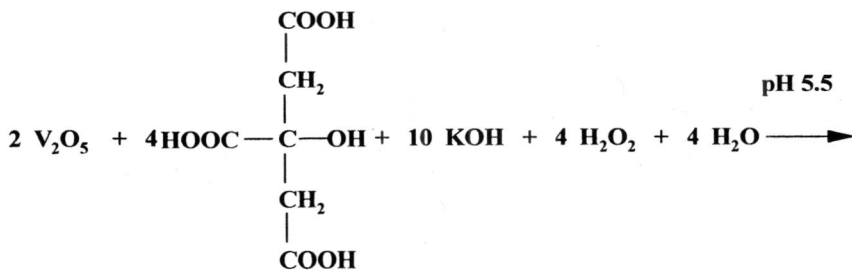

$$2\ V_2O_5\ +\ 4\,HOOC-\underset{\underset{\displaystyle COOH}{\displaystyle |\ CH_2\ |}}{\overset{\overset{\displaystyle COOH}{\displaystyle |\ CH_2\ |}}{C}}-OH\ +\ 10\ KOH\ +\ 4\ H_2O_2\ +\ 4\ H_2O\ \xrightarrow{\ pH\ 5.5\ }$$

$$K_{10}[V_2O_2(O_2)_2(C_6H_5O_7)_2][V_2O_2(O_2)_2(C_6H_4O_7)_2]\cdot 20\ H_2O$$

The basic structural motif in this family of complexes is the $[(V^VO)_2(O)_2(O_2)_2]^0$ core surrounded by two citrate ligands of variable mode of coordination and state of (de)protonation. The chemical reactivity of the isolated complexes was investigated through a) acid-base chemistry, b) thermal transformations, and c) non-thermal transformations (Figure 3).

Discussion

The aqueous synthetic chemistry of the binary V(V,IV)-(α-hydroxycarboxylate) and ternary V(V)-(α-hydroxycarboxylate)-H₂O₂ systems has proven to be quite diverse. In our attempts to delineate the nature and physicochemical properties of soluble species in the requisite speciation schemes, the synthetic structural speciation approach was used. To this end, pH-dependent synthetic reactions were carried out on both binary and ternary systems of simple V(III,IV,V) reagents with the aforementioned physiological ligands. The pH values, at which the reactions were run, were optimized to afford pure crystalline products amenable to full characterization.

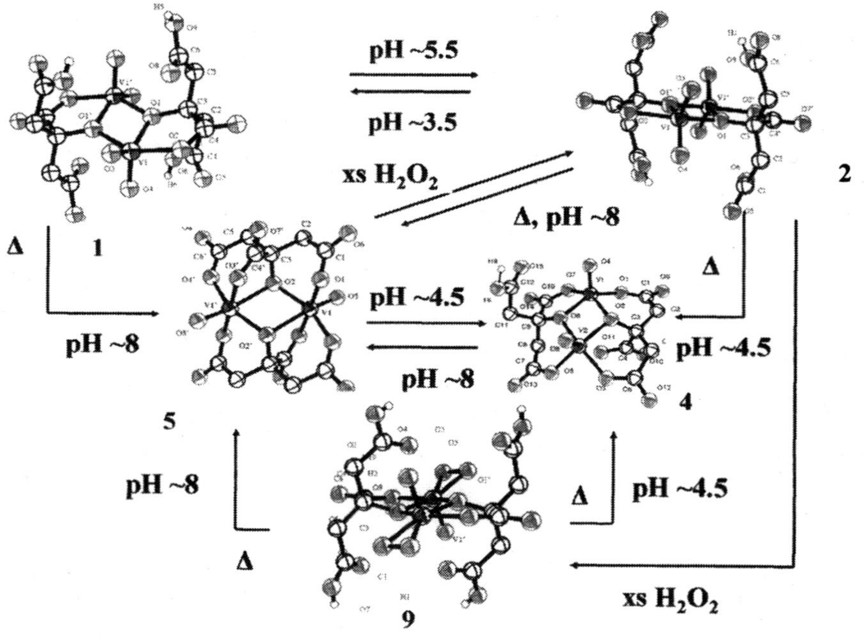

Figure 3. pH-Dependent acid-base and (non)-thermal transformations among pH-structural variants of binary and ternary V-(peroxo)-citrate species. (See page 4 of color inserts.)

In the case of the binary system species studied at the V(IV) oxidation state, the following structural features were observed: a) variable nature structures for the derived anionic complexes **4** and **5**, b) a $V^{IV}{}_2O_2$ planar rhombic core in both classes of arisen complexes; specifically both alkoxide and carboxylate binding sites in the citrate ligands were utilized in their coordination spanning over the two vanadium ions forming the $V^{IV}{}_2O_2$ core, c) a +4 oxidation state of the two vanadium ions comprising the $V^{IV}{}_2O_2$ core, d) variable coordination mode for the bound citrate ligands; specifically, an octahedral coordination geometry was observed around the vanadium ions of the planar core in the high pH ~8 structure, whereas a square pyramidal coordination geometry was observed in the low pH ~4.5 structural variant, and e) anionic charge on the complexes containing the $V^{IV}{}_2O_2$ core (albeit different for the two classes of isolated and characterized complexes **4** and **5**).

In the case of the species isolated with a V(V) oxidation state, the following structural features were observed: a) a $V^V{}_2O_2$ planar rhombic core, b) a +5 oxidation state for the two vanadium ions comprising the $V^V{}_2O_2$ core, c) a single type of coordination mode for the bound citrate ligands; specifically both central alkoxide and carboxylate binding sites in the citrate ligands were utilized in their coordination to the $V^V{}_2O_2$ core. The deprotonated α-hydroxycarboxylate moieties promoted formation of the $V^V{}_2O_2$ core in the arisen complexes , d) a coordination geometry intermediate between a square pyramid and trigonal bipyramid around the vanadium ions of the planar core, e) a variable (de)protonation state of the bound citrate ligands, with the site of (de)protonation being the terminal carboxylate groups that are not participating in coordination, and f) a variable anionic charge of the complex containing the $V^V{}_2O_2$ core, with the high pH structural variant **3** bearing the highest anionic charge 6- (due to the full pH-specific deprotonation of the citrate ligands).

In the case of the ternary systems, the following features were observed in both the malate and citrate ligands: 1) the single most important component in both classes of dinuclear peroxo complexes was the $[V^V{}_2O_2]$ unit, the integrity of which was independent of the pH, at which the complexes were synthesized and isolated. In all cases, the core was planar. 2) the peroxo moieties were bound to the vanadium ions in the same η^2-side-on fashion, and 3) the oxidation of the vanadium ions in the cores of all complexes was +5. The striking differences between the pH-structural variant species included a) the differing protonation state of the bound citrates around the planar $[V^V{}_2O_2]$ core. In particular, the terminal carboxylate groups not participating in the coordination sphere around vanadium were the molecular locus of (de)protonation in the molecule. That, in fact, was a critical factor dictating V(V) coordination geometry changes with increasing pH (vide infra), b) the mode of citrate coordination around the core, and c) the geometry around the vanadium ions. Specifically, the coordination geometry around V(V) in the low pH structural variant **9** was pentagonal pyramidal in contrast to the coordination geometry of V(V) ion in the high pH structural variant **11**, which was pentagonal bipyramidal. The aforementioned structural features emphasized the distinct nature of the two species, which was proven by solid state and solution ^{13}C-NMR spectroscopy. In all cases of pH-dependent structures, the existence of hydrogen-bonding networks was crucial in stabilizing the respective lattices. Similar hydrogen-bonding networks were previously reported in other vanadium-carboxylate complexes (*20,21,22*).

Among the outstanding discoveries in the exploration of the binary and ternary species of vanadium with hydroxycarboxylate substrates was the interconnection a) of non-peroxo binary V(V) species, b) of peroxo ternary V(V) systems, and c) between non-peroxo and peroxo V(V) species. This feat was accomplished through a) non-thermal, and b) thermal transformations of binary V(V) and V(IV) species, and ternary V(V) ternary species. These transformations established the chemical association among species of a) dinuclear cores with the same vanadium oxidation state, and b) dinuclear cores with the differing vanadium oxidation states ($V^V_2O_2$ and $V^{IV}_2O_2$). The results of reactivity studies show that the aqueous speciation schemes of the binary V(V) and V(IV)-hydroxycarboxylate systems a) are unique in the composition of vanadium species, b) contain soluble vanadium species with interdependent chemical relations, c) contain dinuclear V_2O_2 species with vanadium to hydroxycarboxylate ratio of 1:1, and d) may include V_2O_2 species with unique hydroxycarboxylate coordination modes and citrate state of (de)protonation. In the case of the ternary vanadium systems, the studies suggest that the requisite aqueous speciations a) are of unique composition in species, b) contain soluble species with bound peroxide ligands attached to the vanadium(V) ions, c) contain dinuclear V_2O_2 species with vanadium exclusively in the oxidation state V(V), d) include participant species with unique coordination modes of hydroxycarboxylate ligands bound to the vanadium ions, all depending on the state of (de)protonation of the ligands, and e) contain species linked through thermal and non-thermal transformations to non-peroxo V(V) and V(IV) binary species of corresponding aqueous distributions.

Conclusions

Understanding the role of vanadium in insulin mimetic (bio)chemistry entails in-depth perusal of the aqueous speciation of binary and ternary systems of that metal with physiological substrates of variable molecular mass. Efforts to delineate the interactions of vanadium with such biotargets were implemented through structural speciation approaches at the synthetic level. Toward this end, pH-dependent synthesis of binary and ternary soluble complexes of vanadium in aqueous media was pursued with selected physiological α-hydroxycarboxylic acid ligands (citric and malic acids). The results show that in the physiological relevant oxidation states V(V) and V(IV), the metal ion interacts with the aforementioned ligands forming dinuclear complexes containing the planar V_2O_2 core. The hydroxycarboxylate ligands bind to the central core through variable modes of coordination unique to the oxidation state and pH value, at which the synthesis of the specific species is carried out. The state of (de)protonation of the bound ligands plays an important role in the nature of the complex forming, intimately associated with the pH of synthesis and the oxidation state of the metal ion in the dinuclear V_2O_2 core. Of great significance is the establishment of the structural speciation of the requisite binary and ternary systems on the basis of knowledge of the specific structure and properties of participating species emanating from their physicochemical profile. The latter emerges as a direct consequence of their nature and chemical reactivity linking homologous species through non-thermal as well as thermal interconversions in aqueous media. The overall work projects the close connection between speciation distributions of vanadium species at variable vanadium oxidation states and among binary and ternary systems. In this respect, the present studies are in line with solution speciation studies existing for the corresponding binary and ternary systems (e.g. V(V)-citrate, V(V)-citrate-H_2O_2-citrate). Collectively, the studies reflect the complexity of soluble V(IV,V) speciation in aqueous media and potentially in biological fluids, containing low molecular mass physiological metal ion binders. On these grounds, the light shed on this area of vanadium chemistry provides the background for the exploration of the even more intricate (bio)chemistry of interaction of that metal ion with high molecular mass targets. Such efforts are expected to provide evidence for specific binary and ternary species, which in their bioavailable form influence insulin mimesis.

References

1. a) Bayer, E. *Metal Ions in Biological Systems: Amavadin, the Vanadium Compound of Amanitae*; Sigel, H.; Sigel, A., Eds.; Marcel Dekker, Inc.: New York, NY, 1995; Vol. 31, Ch. 12, pp. 407-421. b) Smith, M. J.; Ryan, D. E.; Nakanishi, K.; Frank, P.; Hodgson, K. O. *Metal Ions in Biological Systems: Vanadium in Ascidians and the Chemistry in Tunichromes;* Sigel, H.; Sigel, A., Eds.; Marcel Dekker, Inc.: New York, NY, 1995; Vol. 31, Ch. 13, pp. 423-490.
2. Fraústo da Silva, J. J. R. *Chemical Speciation and Bioavailability* **1989**, *1*, 139-150.

388

3. Liang, J.; Madden, M.; Shah, V. K.; Burris, R. H. *Biochem.* **1990**, *29*, 8577.
4. Weyand, M.; Hecht, H.; Kiess, M.; Liaud, M.; Vilter, H.; Schomburg, D. *J. Mol. Biol.* **1999**, *293*, 595-611.
5. Vilter, H. *Metal Ions in Biological Systems: Vanadium and its Role in Life*; Sigel, H.; Sigel, A., Eds.; Marcel Dekker, Inc.: New York, NY, 1995, Vol. 31, Ch. 10, pp. 325-362.
6. Butler, A. *Curr. Opin. Chem. Biol.* **1998**, *2*, 279-285.
7. Klarlund, J. K *Cell* **1985**, *41*, 707-717.
8. Smith, J. B. *Proc. Natl. Acad. Sci. USA* **1983**, *80*, 6162-6167.
9. Djordjevic, C. *Metal Ions in Biological Systems: Antitumorigenic Activity of Vanadium Compounds*; Sigel, H.; Sigel, A., Eds.; Marcel Dekker, Inc.: New York, NY, 1995; Vol. 31, Ch. 18, pp. 595-616.
10. a) Köpf-Maier, P.; Köpf, H. *Metal Compounds in Cancer Therapy*; Fricker, S. P., Ed.; Chapman and Hall: London, 1994, pp. 109-146. c) Djordjevic, C.; Wampler, G. L. *J. Inorg. Biochem.* **1985**, *25*, 51-55.
11. a) Brand, R. M.; Hamel, F. G. *Int. J. Pharm.* **1999**, *183*, 117-123. b) Drake, P. G.; Posner, B. I. *Mol. Cell Biochem.* **1998**, *182*, 79-89. c) Drake, P. G.; Bevan, A. P.; Burgess, J. W.; Bergeron, J. J.; Posner, B. I. *Endocrinology* **1996**, *137*, 4960.
12. a) Sakurai, H.; Kojima, Y.; Yoshikawa, Y.; Kawabe, K.; Yasui, H. *Coord. Chem. Rev.* **2002**, *226*, 187-198. b) Sasagawa, T.; Yoshikawa, Y.; Kawabe, K.; Sakurai, H.; Kojima, Y. *J. Inorg. Biochem.* **2002**, *88*, 108-112. c) Kanamori, K.; Nishida, K.; Miyata, N.; Okamoto, K.; Miyoshi, Y.; Tamura, A.; Sakurai, H. *J. Inorg. Biochem.* **2001**, *86*, 649-656.
13. Crans, D. C. *Metal Ions in Biological Systems: Vanadium and its Role in Life*; Sigel, H.; Sigel, A., Eds.; Marcel Dekker, Inc.: New York, NY, 1995; Vol. 31, Ch. 5, pp. 147-209.
14. Martin, R. B. *J. Inorg. Biochem.* **1986**, *28*, 181-187.
15. Kaliva, M.; Giannadaki, T.; Raptopoulou, C. P.; Terzis, A.; Salifoglou, A. *Inorg. Chem.* **2002**, *41*, 3850-3858.
16. Kaliva, M.; Raptopoulou, C. P.; Terzis, A.; Salifoglou, A. *Inorg. Biochem.* **2003**, *93*, 161-173.
17. Tsaramyrsi, M.; Kaliva, M.; Giannadaki, T.; Raptopoulou, C. P.; Tangoulis, V.; Terzis, A.; Giapintzakis, J.; Salifoglou, A. *Inorg. Chem.* **2001**, *40*, 5772.
18. Kaliva, M.; Giannadaki, T.; Raptopoulou, C. P.; Tangoulis, V.; Terzis, A. ; Salifoglou, A. *Inorg. Chem.* **2001**, *40*, 3711-3718.
19. Kaliva, M.; Raptopoulou, C. P.; Terzis, A.; Salifoglou, A. *Inorg. Chem.* **2004**, *43*, 2895-2905.

20. a) Wright, D. W.; Humiston, P. A.; Orme-Johnson, W. H.; Davis, W. M. *Inorg. Chem.* **1995**, *34*, 4194-4197. b) Velayutham, M.; Varghese, B.; Subramanian, S. *Inorg. Chem.* **1998**, *37*, 1336-1340. c) Djordjevic, C.; Lee, M.; Sinn, E. *Inorg. Chem.* **1989**, *28*, 719-723.

21. Tsaramyrsi, M.; Kavousanaki, D.; Raptopoulou, C. P.; Terzis, A.; Salifoglou, A. *Inorg. Chim. Acta*, **2001**, *320*, 47-59.

22. Burojevic, S.; Shweky, I.; Bino, A.; Summers, D. A.; Thompson, R. C. *Inorg. Chim. Acta* **1996**, *251*, 75-79.

New Materials and Processes

Chapter 28

Structural Determinants in the Oxovanadium Diphosphonate System

Wayne Ouellette and Jon Zubieta

Department of Chemistry, Syracuse University, Syracuse, NY 13244

Oxovanadium organodiphosphonates are prototypical organic-inorganic hybrid materials. While these phases are generally characterized by V-P-O layers separated by organic domains, the structural details are complex, reflecting structural determinants such as the length and identity of the organic spaces, the presence of organic or complex inorganic cationic components, and the incorporation of fluoride into the V-P-O substructure.

The widespread contemporary interest in metal oxide based solid phases is related to their significant applications to areas of chemistry and materials science as diverse as catalysis, sorption, molecular electronics, ceramics, energy storage, fuel cells and optical materials (*1-8*). This diversity of properties reflects a vast compositional range, which allows variations in covalency, geometry, and oxidation states, and a versatile crystalline architecture, which may provide diverse pore structures, coordination sites, and juxtapositions of functional groups.

The ubiquity of inorganic oxide phases in both the geosphere and the biosphere (*9-13*) suggests that naturally occurring oxides may provide useful guidelines for the preparation of synthetic phases and the modification of oxide microstructures. It is noteworthy that many of the remarkable oxide materials fashioned by nature contain mixtures of inorganic oxides coexisting with organic molecules which dramatically influence the microstructure of the inorganic and allow higher order organization of hierarchical structures (*14-16*).

© 2007 American Chemical Society

Complex structures (*17*), based on a molecular scale composite of inorganic and organic components, provide the potential for the design of novel functional materials for technological applications (*18*). An inorganic material may provide useful magnetic, dielectric or optical properties, mechanical hardness, and thermal stability, while organic compounds offer processability, structural diversity, a range of polarizabilities and luminescent properties (*19*). Consequently, the combination of the characteristics of the organic and inorganic components offers an opportunity to conflate useful properties within a single composite, providing access to a vast area of complex, multifunctional materials (*20*).

Inorganic-organic hybrid materials (*21*) are extended arrays of metal atoms or clusters bridged by polyfunctional organic molecules. An important subclass of this family of materials are the hybrid metal oxides, which contain metal-oxygen-metal (M-O-M) arrays as part of their structures. In such materials, the inorganic oxide contributes to the increased complexity, and hence functionality, through incorporation as one component in multilevel structural materials where there is a synergistic interaction between organic and inorganic components.

Metal organophosphonates are prototypical composite materials, which can exhibit a range of structures, including molecular clusters, chains, layers and three-dimensional frameworks. An important subclass of these materials are the oxovanadium organophosphonates, whose structures are often characterized by a two-dimensional network of V-P-O layers separated by hydrophobic organic domains. However, the detailed structural chemistry may be exceedingly complex, reflecting a variety of structural determinants. In an attempt to elaborate the structural systematics of these materials, we have investigated the oxovanadium organodiphosphonate system, focusing on a number of variables, specifically: (i) the length and identity of the organic tether of the diphosphonates, (ii) the introduction of organic or metal complex cations, and (iii) the incorporation of fluoride anions into the V-P-O substructure.

Variations in tether length of α, ω-alkyldiphosphonates

While the prototypical structural motif of V-P-O layers buttressed by organic linkers is retained throughout the series of neutral network materials, $[V_xO_y\{O_3P(CH_2)_nPO_3\}_z]$, significant changes evolve as the tether length increases from $n = 2$ to $n = 11$. Thus, for $n = 2$-5, the three dimensional structure $[V_2O_2(H_2O)\{O_3P(CH_2)_nPO_3\}]$ (**1**) of Figure 1a is maintained. In contrast, at $n = 6$ through $n = 8$, the two-dimensional structure of $[V_2O_2(H_2O)_4\{O_3P(CH_2)_nPO_3\}]$ (**2**), shown in Figure 1c is observed, wherein the organic chains are sandwiched between V-P-O layers. Further expansion of the tether to $n = 9$ results in the 3-D structure of $[V_2O_2(H_2O)\{HO_3P(CH_2)_9PO_3H\}_2]$

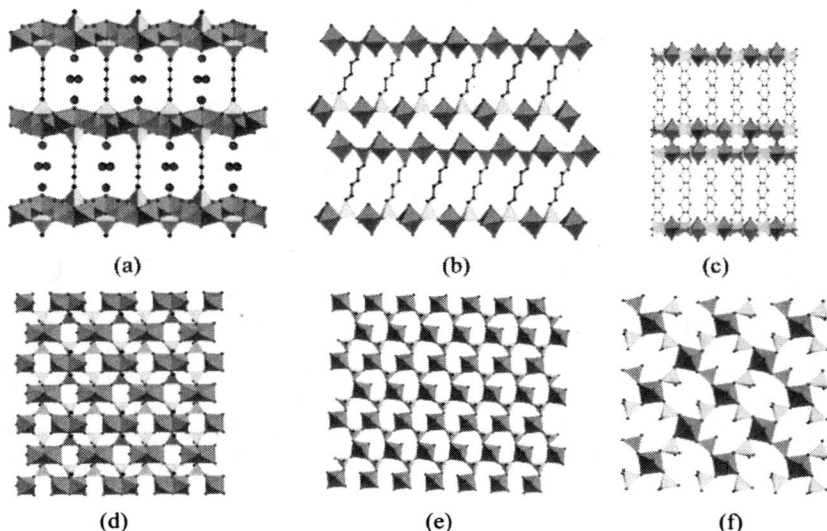

Figure 1. Polyhedral representations of the structures of compounds of the type (1), (2), and (3), viewed both parallel to the V-P-O plane and normal to the V-P-O plane: (a and b) 1; (c and d) 2; (e and f) 3.

(3) (Figure 1e), where once again slabs defined by two V-P-O networks sandwich the organic tethers. However, in this instance, the V-P-O networks of adjacent slabs link through bridging aqua ligands to produce a double layer substructure and the resultant three-dimensional connectivity.

This structural diversity is reflected in the connectivities of the component V-P-O networks of **1-3** shown in Figures 1b, d and e. Compound **1** exhibits a layer substructure constructed from face-sharing pairs of V(IV) octahedra and phosphorus tetrahedra. In contrast, the V-P-O layer of **2** does not exhibit V-O-V linkages, but rather consists of isolated $\{VO_6\}$ octahedra, each of which exhibits *cis*-oriented aqua ligands. In the case of **3**, the double layer results from aqua bridging of pairs of corner-sharing V(IV) octahedra.

Aromatic diphosphonates

As shown in Figure 2, the structures of $[VO\{HO_3P(1,4-C_6H_4)PO_3\}]$ **(4)** and $[VO\{HO_3P(C_6H_4)_2PO_3\}]$ **(5)** conform to the buttressed layer prototype. However, the V-P-O layer is quite distinct from those of the α, ω-alkyldiphosphonate materials **1-3**. The network is constructed from six coordinate V(V) sites, each bonding to oxygen donors from five phosphonate ligands and a terminal oxo-group. The V-P-O layer is three polyhedra thick

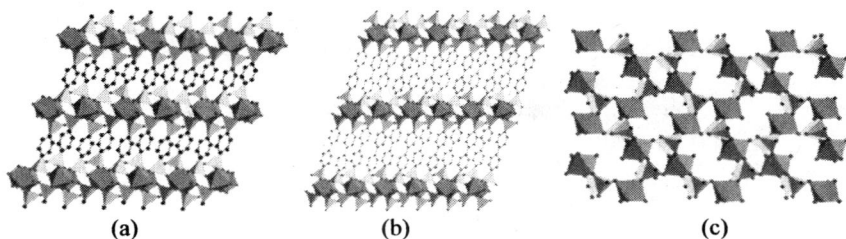

Figure 2. Polyhedral representations of the structures of (a) (4) and (b) (5); (c) a view of the V-P-O layer of 4 and 5.

with vanadium octahedra sandwiched between phosphonate tetrahedra. One phosphonate terminus bonds to three vanadium sites while the second bonds to two, leaving a pendant -OH group.

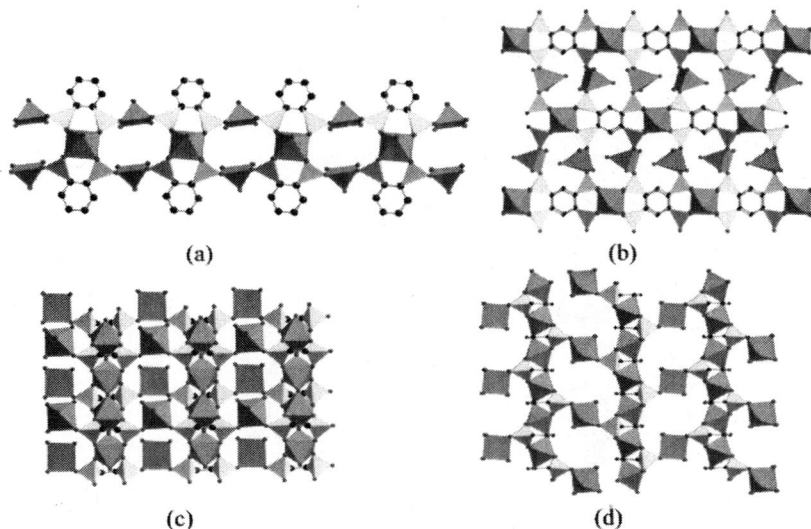

Figure 3. (a) A view of the structure of (4); (b) the V-P-O layer of 4; (c) and (d) views of the structure of (5).

Of course, the relative location of the {PO$_3$} groups can be moved about the central ring of the benzene-diphosphonate. The structural consequences are quite dramatic. Thus, [V$_2$O$_2$(H$_2$O)$_2${O$_3$P(1,2-C$_6$H$_4$)PO$_3$}] (6) (Figure 3a) is two-dimensional, with the ligand adopting a chelating role. Curiously, there are two unique V(IV) sites. The square pyramidal geometry of the first is defined by four oxygen donors from four diphosphonate groups in the basal plane and an apical oxo-group. In contrast, the second site exhibits two aqua ligands in the basal plane and two diphosphonate oxygen donors.

In contrast, the structure of $[V_2O_2(H_2O)_2\{O_3P(1,3-C_6H_4)PO_3\}]$ (7) is three dimensional (Figure 3c). Once again there are two distinct V(IV) sites. One consists of square pyramids defined by two aqua ligands and two diphosphonate oxygen donors in the basal plane and an apical oxo-group. The second consists of chains of *trans* corner-sharing octahedra with the common $\{V=O\cdots V=O\}$ short-long alternation of V-O bonds along the chain axis.

Introduction of cationic components

The introduction of organonitrogen cations results in dramatic structural perturbations. Thus, while $[H_3N(CH_2)_nNH_3][V_4O_4(OH)_2\{O_3P(CH_2)_3PO_3\}_2]$-•$xH_2O$ (8) retains the prototypical "pillared" layer architecture (Figure 4a) for n = 2-8, the V-P-O substructure is quite distinct from those of 1-3. As shown in Figure 4b, the network exhibits corner-sharing pairs of V(IV) square pyramids. It is noteworthy that the content of water of crystallization x decreases as n increases; that is, the water is "squeezed" out by the cations.

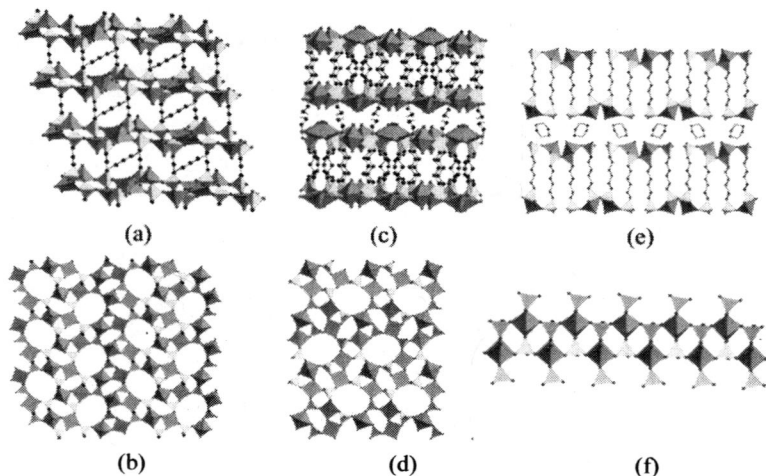

(a)   (c)   (e)

(b)   (d)   (f)

Figure 4. Polyhedral representations of the structures of materials of the types (8), (9) and (10), viewed both parallel and normal to the V-P-O planes: (a and b) 8; (c and d) 9; (e and f) 10.

When n = 5, the two-dimensional phase $[H_3N(CH_2)_2NH_3][V_4O_4(OH)_2(H_2O)$ $\{O_3P(CH_2)_5PO_3\}_2]$ (9) is obtained (Figure 4c). The tether groups of the diphosphonate ligands project from a single face of the V-P-O network, resulting in a slab-like structure with the organic tethers sandwiched between two V-P-O layers.

Substitution of piperazinium cations for the simple alkyl chain cations of **8** and **9** provides the 2-D structure of $[H_2N(CH_2CH_2)_2NH_2]$-$[V_2O_2\{O_3P(CH_2)_nPO_3H\}_2]$ (**10**) (Figure 4e). The undulating structure of **10** is a consequence of the different environments of the phosphorus termini of the diphosphonate ligand: the first bridges three vanadium sites, while the second bonds to a single site.

Curiously, when the diphosphonate tether length is reduced to n = 2, the three-dimensional $[H_3N(CH_2)_nNH_3][V_3O_3\{O_3P(CH_2)_2PO_3\}_2]$ (**11**) (Figure 5) is observed. The structure exhibits large channels of dimensions 10.4Å x 8Å defined by a 30-membered $\{V_6P_8O_{12}C_4\}$ ring and encapsulating the cation.

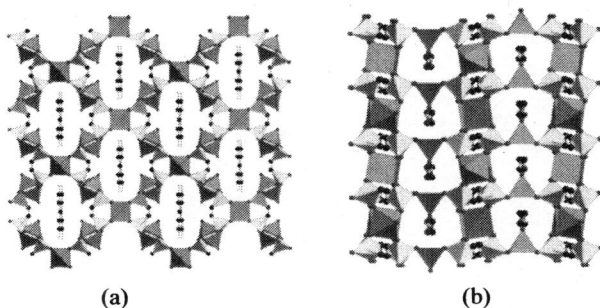

(a)   (b)

Figure 5.   Two views of the structure of (11).

When metal complex cations, such as $\{Cu_2(bisterpy)\}^{4+}$ (bisterpy = 2,2':4',4'':2'',2'''-quaterpyridyl-6',6''-di-2-pyridine), are used, the V-P-O networks common to structures **1-6** are largely disrupted and one-dimensional and cluster V-P-O substructures are observed. Coordination of the Cu(II) centers to vanadium and/or phosphorus oxygen atoms prevents aggregation of the V-P-O component into higher dimensionality substructures. Thus, $[\{Cu_2(bisterpy)(H_2O)_2\}V_4O_4\{O_3P(CH_2)_2PO_3\}\{HO_3P(CH_2)_2PO_3H\}]$ (**12**) and $[\{Cu_2(bisterpy)\}V_4O_8\{O_3P(CH_2)_3PO_3\}_2]$ (**13**) exhibit V-P-O chains, while $[\{Cu_2(bisterpy)\}VO_2\{O_3P(CH_2)_3PO_3\}\{HO_3P(CH_2)_3PO_3H_2\}]$ (**14**) is constructed from cyclic $[VO_2\{O_3P(CH_2)_3PO_3\}]_2^{6-}$ clusters and $[\{Cu_2(bisterpy)\}V_2O_4(OH)_2$-$\{O_3P(CH)_4PO_3\}]$ (**15**) contains embedded $[V_2O_4(OH)_2\ \{O_3P(CH_2)_4PO_3\}]^{4-}$ subunits (Figure 6).

Fluoride incorporation

The introduction of fluoride into the V-P-O substructure of these materials has profound structural consequences, generally related to the reduction of one

398

(a)   (b)

(c)   (d)

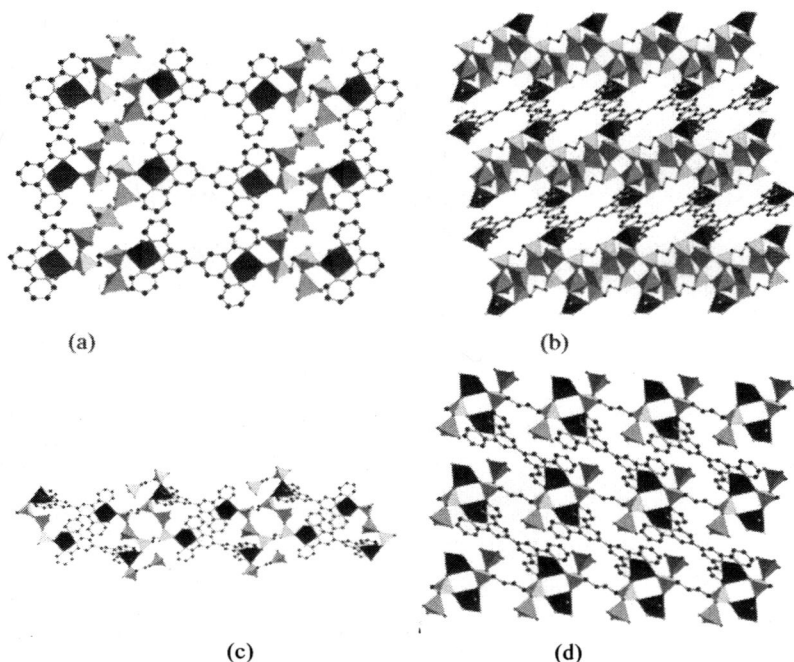

Figure 6. Polyhedral representations of the structures of (a) (12), (b) (13), (c) (14) and (d) (15).

or more vanadium sites to the V(III) oxidation state. Since vanadium(III) adopts more or less regular octahedral geometry with no short multiply-bonded {V=O} units, considerable structural variety is possible. Several structures are reminiscent of those previously discussed. Thus, $[H_2pip][V_4F_4O_2(H_2O)_2\{O_3P(CH_2)_3PO_3\}_2]$ **(16)** exhibits the prototypical "pillared" layer structure (Figure 7a). However, the V-P-O-F layer, shown in Figure 5b, contains pairs of edge-sharing V(IV) and V(III) octahedra linked through bridging fluoride ligands. In contrast, $[H_3NCH_2CH_2NH_3][V_2O_2F_2(H_2O)_2\{O_3P(CH_2)_2PO_3\}]$ **(17)** (Figure 7c) exhibits pairs of V-P-O-F layers linked through the alkyl chains of the diphosphonate ligands with the cations separating adjacent slabs, in a fashion reminiscent of compound **9**. While the theme of two-dimensional V-P-O-F slabs with cations occupying the interlamellar domain is reiterated in $[H_3NCH_2CH_2NH_3]_2[V_6F_{12}(H_2O)_2\{O_3P(CH_2)_5PO_3\}\{HO_3P(CH_2)_5PO_3\}_2]$ **(18)** (Figure 7e), the V-P-O-F network of **18** is constructed from chains of V(III) octahedra (Figure 7b).

As noted for the structure of compound **11**, shortening the diphosphonate tether length to n = 2 allows the entertainment of the organic tether within V-P-

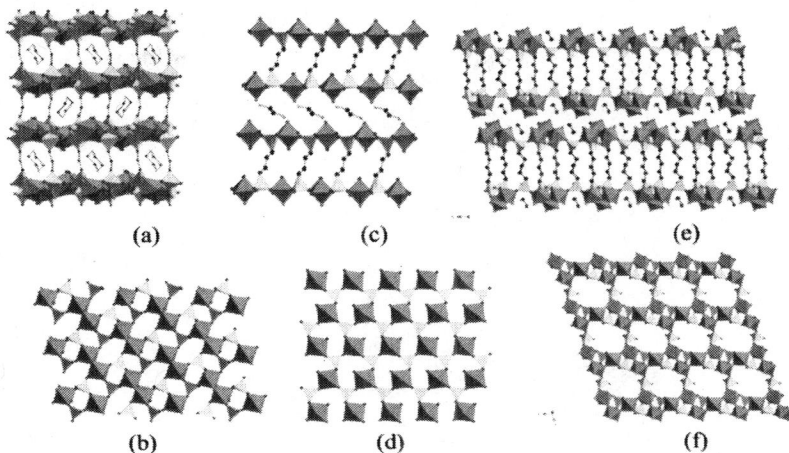

Figure 7. The structures of (16), (17) and (18) viewed both parallel and normal to the V-P-O-F planes: (a and b) 16; (c and d) 17; (e and f) 18.

O or V-P-O-F frameworks of expanded dimensionality. Thus, both $[H_3NCH_2CH_2NH_3]_2[V_7O_6F_4(H_2O)_2\{O_3P(CH_2)_2PO_3\}_4]$ (**19**) and $(H_3O)[V_3F_2(H_2O)_2\{O_3P(CH_2)_2PO_3\}_2]$ (**20**) are three-dimensional, albeit with quite distinct vanadium substructures (Figure 8).

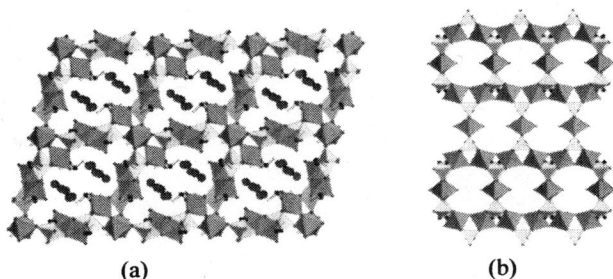

Figure 8. The three-dimensional structures of (a) (19) and (b) (H_3O) (20).

When metal complex cations are introduced to provide charge compensation, as well as an additional coordinating component, an unusually broad range of structures is observed. The tendency of Cu(II) coordination to the oxygen atoms of the V-P-O-F substructure to reduce the dimensionality of this subunit is evident in the one-dimensional structure of

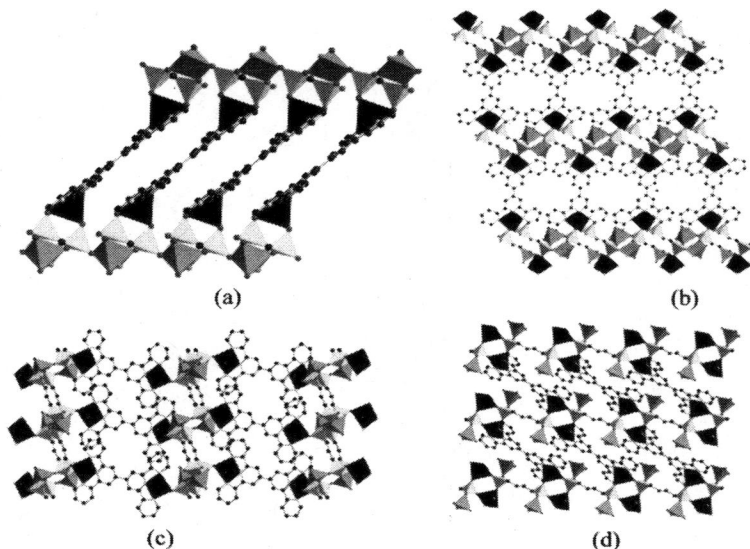

Figure 9. Views of the structures of (a) (21); (b) (22); (c) (23); and (d) (24).

$[\{Cu_2(bisterpy)\}V_2F_2O_2\{O_3PCH_2PO_3\}\{HO_3PCH_2PO_3\}]$ **(21)** and, the two-dimensional structures of $[\{Cu_2(bisterpy)\}V_2F_2O_2(H_2O)_2\}HO_3P(CH_2)_2PO_3\}_2]$ **(22)**, $[\{Cu_2(bisterpy)(H_2O)_2\}V_2F_2O_2\{O_3P(CH_2)_3PO_3\}\{HO_3P(CH_2)_3PO_3H\}]$ **(23)** and $[\{Cu_2(bisterpy)\}V_4F_4O_4(OH)(H_2O)\{O_3P(CH_2)_5PO_3\}\{HO_3P(CH_2)_5PO_3\}]$ **(24)**, all of which are constructed from V-P-O-F chains and $\{Cu_2(bisterpy)\}^{4+}$ clusters (Figure 9). However, the overall three-dimensional structure of $[\{Cu_2(bisterpy)(H_2O)\}_2V_8F_4O_8(OH)_4\{HO_3P(CH_2)_5PO_3H\}_2\{O_3P(CH_2)_5PO_3\}_3]$ **(25)** exhibits a two-dimensional V-P-O-F substructure, while $[\{Cu_2(bisterpy)\}V_4F_2O_6\{O_3P(CH_2)_4PO_3\}_2]$ **(26)** exhibits a $[V_4F_2O_6\{O_3P-(CH_2)_4PO_3\}_2]_n^{4-}$ "pillared" framework component (Figure 10).

As noted in Table 1, the introduction of $\{Cu_2(bisterpy)\}^{4+}$ rods as building units generally correlates with reduced complexity of the V-P-O or V-P-O-F substructure. Thus, for the sixteen structures reported to date, thirteen exhibit cluster of chain V-P-O(F) substructures. Only one material possesses a two-dimensional V-P-O-F subunit, which is so common in the absence of the secondary metal-ligand component. On the other hand, two of these materials are characterized by three-dimensional V-P-O(F) substructures.

Structures and Properties

The structural diversity of these vanadium-diphosphonate materials is reflected in their properties, such as magnetism and thermal stability. While the

Table I. Summary of structural characteristics for materials containing anionic V-P-O or V-O-O-F components and {Cu₂(bisterpy)}⁴⁺ cations.

Compound		V-P-O or V-P-O-F Substructure	V-X-V Linkage	V-O-Cu Linkage
[{Cu₂(bisterpy)}V₂F₂O₂{HO₃PCH₂PO₃}{O₃PCH₂PO₃}]	1-D	Chains	No	No
[{Cu₂(bisterpy)}V₂O₄{HO₃PCH₂PO₃}₂]	1-D	Clusters	No	No
[{Cu₂(bisterpy)}V₂F₂O₂(H₂O)₂(HO₃P(CH₂)₂PO₃}₂]·2H₂O	2-D	Chains	No	No
[{Cu₂(bisterpy)}V₂F₄O₄{HO₃P(CH₂)₂PO₃H}]	1-D	Clusters	No	Yes
[{Cu₂(bisterpy)(H₂O)₂}V₂O₄{HO₃P(CH₂)₂PO₃}₂]	2-D	Chains	No	No
[{Cu₂(bisterpy)(H₂O)₂}V₂F₂O₂{O₃P(CH₂)₃PO₃}{HO₃P(CH₂)₃PO₃}]	2-D	Chains	No	No
[{Cu₂(bisterpy)}₃{V₈F₆O₁₇{HO₃P(CH₂)₃PO₃}₄]·0.8H₂O	3-D	Clusters	Yes, X = O	Yes
[{Cu₂(bisterpy)}V₄O₈{O₃P(CH₂)₃PO₃}₂]·4H₂O	2-D	Chains	No	Yes
[{Cu₂(bisterpy)(H₂O)}VO₂{O₃P(CH₂)₃PO₃}{HO₃P(CH₂)₃PO₃H}]	1-D	Clusters	No	No
[{Cu₂(bisterpy)}V₂O₄{HO₃P(CH₂)₃PO₃}₂]	3-D	Chains	No	Yes
[{Cu₂(bisterpy)}₃{V₄O₈(OH)₂(HO₃P(CH₂)₃PO₃}₂{O₃P(CH₂)₃PO₃}₂]	3-D	Clusters	No	Yes
[{Cu₂(bisterpy)}V₄F₆O₆{O₃P(CH₂)₄PO₃}₂]	3-D	Framework	Yes, X = F	Yes
[{Cu₂(bisterpy)}V₂O₄(OH)₂O₃P(CH₂)₄PO₃}]·4H₂O	2-D	Clusters	No	No
[{Cu₂(bisterpy)}V₄F₄O₄(OH)(H₂O){HO₃P(CH₂)₅PO₃}{O₃P(CH₂)₅PO₃}]	2-D	Chains	Yes, X = OH	Yes
[{Cu₂(bisterpy)}₂V₈F₄O₈(OH)₄{HO₃P(CH₂)₅PO₃H}₂{O₃P(CH₂)₅PO₃}₃]	3-D	Layer	X = O and X = F	Yes
[{Cu₂(bisterpy)}V₄O₄{HO₃P(CH₂)₅PO₃}₄]·7·3H₂O	3-D	Framework	No	Yes

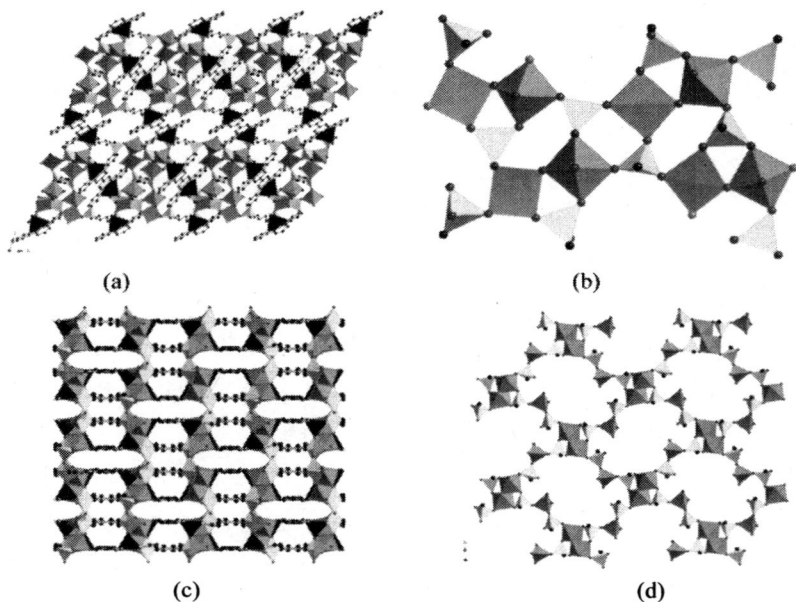

Figure 10. The three-dimensional structures of (25) and (26): (a) a view of the structure of 25 in the bc plane; (b) the vanadophosphonate substructure of 25; (c) a view of 26; (d) the V-P-O-F layers of 26.

majority of the materials conform to simple Curie-Weiss behavior, several exhibit more complicated magnetic properties. For example, the magnetic susceptibility of $[V_2O_2(H_2O)\{O_3P(CH_2)_nPO_3\}]$ (1) is best described by the Heisenberg dimer model with S = 1 (Figure 11), while that of $[H_3N(CH_2)_2NH_3][V_2O_2F_2(H_2O)_2\{O_3P(CH_2)_4PO_3\}]$ (17) is described by the Heisenberg linear antiferromagnetic chain model for V(IV) (Figure 12).

A noteworthy observation is that the oxyfluorovanadium/diphosphonate frameworks are thermally robust and are retained well past the dehydration temperature. This is evident for $[H_3NCH_2CH_2NH_3]_2[V_7O_6F_4(H_2O)_2\{O_3P(CH_2)_2PO_3\}_4]\cdot7H_2O$ (19) which loses water in the 50-250°C range and exhibits partial decomposition of the cation between 250-400°C. However, the thermodiffraction profile is unchanged to 450°C, indicating that the V-P-O-F framework is thermally robust and persistent to 450°C (Figure 13).

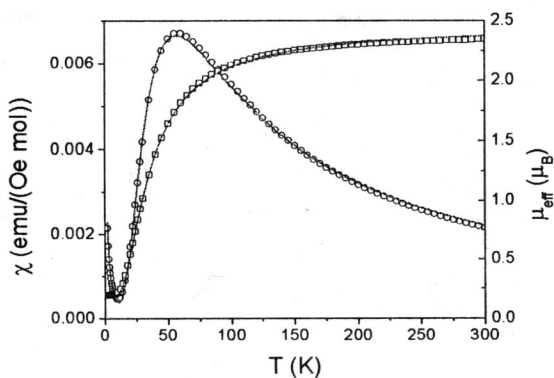

Figure 11. Dependence of the magnetic susceptibility χ(○) and of the effective moment μ_eff (□) of 1 on the temperature. The lines drawn through the data are the fits to the Heisenberg dimer model.

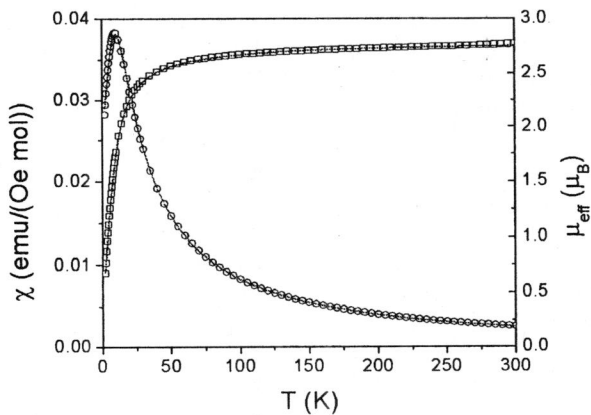

Figure 12. The dependence of the magnetic susceptibility χ(○) and the effective moment μ_eff (□) of 17 on temperature.

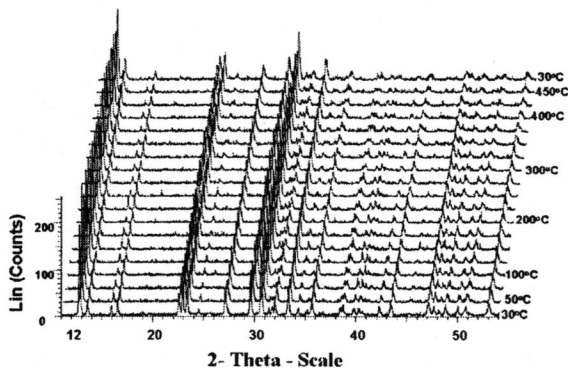

Figure 13. Thermodiffraction pattern of 19 in the temperature range 30-450°C.

Conclusions

Hydrothermal chemistry provides a facile route to the synthesis of a large number of materials of the vanadium-organodiphosphonate family. The emerging structural systematics for this class of materials reveals several significant structural determinants. These include the polyhedral and oxidation state variability of vanadium, the promiscuous possibilities for vanadium and phosphorus polyhedral connectivities, the flexibility of P-O(H) bond distances and V-O-P bond angles, variable protonation of {PO$_3$} groups and vanadium oxo-sites, coordination of aqua ligands, incorporation of varying ratios of fluoride to vanadium, organic tether lengths, the presence of cationic compounds, and the coordination of secondary metals, such as copper.

Acknowledgment

This work was funded by a grant from the National Science Foundation, CHE-0604527.

References

1. Fernandez-Garcia, M.; Martinez-Arias, A.; Hanson, J.C.; Rodriguez, J.A., Nanostructured oxides in chemistry: characterization and properties, *Chem. Rev.* **2004**, *104*, 4063-4104.
2. *Metal Oxides: Chemistry and Applications*, Fierro, J.L.G., ed., CRC Press, Boca Raton, FL 2005.

3. (a) Noguera, C., *Physics and Chemistry at Oxide Surfaces*, Cambridge University Press: Cambridge, UK 1996; (b) Kung, H.H., *Transition Metal Oxides: Surface Chemistry and Catalysis*, Elsevier: Amsterdam 1989.

4. Bruce, D.W.; O'Hare, D., eds., *Inorganic Materials*, Wiley: Chichester 1992.

5. Cheetham, A.K., Advanced inorganic materials: an open horizon, *Science* **1994**, *264*, 794-795.

6. Büchner, W.; Schliebs, R.; Winter, G.; Büchel, K.H., *Industrial Inorganic Chemistry*, VCH, New York 1989.

7. McCarroll, W.H., Oxides: solid state chemistry, *Encyclopedia of Inorganic Chemistry*. R.B. King, ed., John Wiley and Sons, New York **1994**, *vol. 6*, 2903-2946.

8. Newsam, J.M., Zeolites, *Solid State Compounds*, A.K. Cheetham and P. Day eds., Clarendon Press, Oxford 1992, 234-280.

9. Wells, A.F., *Structural Inorganic Chemistry*, 6th Ed., Oxford University Press: New York 1987.

10. Greenwood, N.N.; Earnshaw, A., *Chemistry of the Elements*, 2nd Ed., Butterworth-Heinemann, Oxford, England 1997.

11. Hench, L.L., *Inorganic Biomaterials, Materials Chemistry, an Emerging Discipline*, L.V. Interrante, L.A. Casper, A.B. Ellis, eds., ACS Series *245*, chapter 21, pp. 523-547, 1995.

12. Smyth, J.R.; Jacobsen, S.D.; Hazen, R.M., Comparative crystal chemistry of dense oxide minerals, *Rev. Mineral. Geochem.* **2001**, *41*, 157-186.

13. Mason, B., *Principles of Geochemistry*, 3rd ed., Wiley, New York 1966.

14. Lowenstan, H.A.; Weiner, S., *On Biomineralization*, Oxford University Press, New York, 1989.

15. Cölfen, H.; Mann, S., High order organization by mesoscale self-assembly and transformation of hybrid nanostructures, *Angew. Chem., Int. Ed. Engl.* **2003**, *42*, 2350-2365.

16. Soler-Illia, G.J. de A.A.; Sanchez, C.; Lebean, B.; Patarin, J., Chemical strategies to design textured materials: from microporous and mesoporous oxides to nanonetworks and hierarchical structures, *Chem. Rev.* **2002**, *102*, 4093-4138.

17. Complexity is a subject of significant and general scientific interest. Complexity in chemistry refers to the description and manipulation of systems of molecules, as in living cells and materials. In the latter context, organic-inorganic hybrid structures partake of the chemical complexity of materials, with the attendant complications of predictability and rational design. See, for example: Whitesides, G.M.; Ismagilov, R.F., Complexity in chemistry, *Science* **1999**, *284*, 89-92. The relationship between complexity and functionality is abundantly evident in biological systems. Chemists may learn from biology and make the creative leap to the design of inorganic materials whose structures are influenced by organic

molecules. See, for example: Kiss, I.Z.; Hudson, J.L., Chemical complexity: spontaneous and engineered structures, *AICLE Journal* **2003**, *49*, 2234-2241; Lehn J.M., Toward complex matter: supramolecular chemistry and self-organization, *Proc. Natl. Acad. Sci.* **2002**, *99*, 4763-4768; Lehn, J.M., Toward self-organization and complex matter, *Science*, **2002**, *295*, 2400-2403; Förster, S.; Plantenberg, T., From self-organizing polymers to nanohybrid and biomaterials, *Angew. Chem., Int. Ed. Engl.* **2002**, *41*, 688-714; Miller, A.D., Order for free: molecular diversity and complexity promote self-organization, *Chem. Biochem.* **2002**, *3*, 45-46.

18. Janiak, C., Engineering coordination polymers towards applications, *Dalton Trans.* **2003**, 2781-2804, and references therein.
19. Mitzi, D.B.. Templating and structural engineering in organic-inorganic perovskites, *Dalton Trans.* **2001**, 1-12.
20. The properties of inorganic-organic hybrids, specifically metal-organic frameworks (MOF) have been extensively elaborated in recent years. For example, porosity: (a) Chen, B.; Ockwig, N.W.; Millward, A.R.; Contreras, S.D.; Yaghi, O.M., High H_2 adsorption in a microporous metal-organic framework with open-metal sites, *Angew. Chem. Int. Ed.. Eng.* **2005**, *44*, 4745-4749; (b) Bradshaw, D.; Claridge, J.B.; Cussen, E.J.; Prior, T.J.; Rosseinsky, M.J., Design, chirality, and flexibility in nanoporous molecule-based materials, *Chem. Res.* **2005**, *38*, 273-282; (c) Ohmori, O.; Kawano, M.; Fujita, M., A two-in-one crystal: uptake of two different guests into two distinct channels of a biporous coordination network, *Angew. Chem., Int. Ed.* **2005**, *44*, 1962-1964; (d) Wu, C-D.; Lin, W., Highly porous, homochiral metal-organic frameworks: solvent-exchange-induced single-crystal to single-crystal transformations, *Angew. Chem., Int. Ed.* **2005**, *44*, 1958-1961. Gas sorption: (a) Sudik, A.C.; Millward, A.R.; Ockwig, N.W.; Cote, A.P.; Kim, J.; Yaghi, O.M., Design, synthesis, structure, and gas (N_2, Ar, CO_2, CH_4, and H_2) sorption properties of porous metal-organic tetrahedral and heterocuboidal polyhedra, *J. Am. Chem. Soc.* **2005**, *127*, 7110-7118; (b) Kitaura, R.; Kitagawa, S.; Kubtoa, Y.; Kobayashi, T.C.; Kindo, K.; Mita Y.; Matsuo, A.; Kobayashi, M.; Chang H-C.; Ozawa, T.C.; Suzuki, M.; Sakata, M.; Takata, M., Formation of a one-dimensional array of oxygen in a microporous metal-organic solid, *Science* **2002**, *298*, 2358-2361.
a. Chiral separations: Bradshaw, D.; Prior, T.J.; Cussen, E.J.; Claridge, J.B.; Rosseinsky, M.J., Permanent microporosity and enantioselective sorption in a chiral open framework, *J. Am. Chem. Soc.* **2004**, *126*, 6106-6114. Molecular sensing: Halder, G.J.; Kepert, C.J.; Moubaraki, B.; Murray, K.S.; Cashion, J.D., Guest-dependent spin crossover in a nanoporous molecular framework material, *Science (Washington, DC, US)* **2002**, *298*, 1762-1765.

21. The literature on organic-inorganic hybrid materials is voluminous. Some recent reviews and representative articles include: (a) Kitagawa, S.; Noro, S., Coordination polymers: infinite systems, *Comprehensive Coordination Chemistry II* **2004**, *7*, 231-261; (b) Rao, C.N.R.; Natarajan, S.; Vaidhyanathan, R., Metal carboxylates with open architectures, *Angew. Chem., Int. Ed.* **2004**, *43*, 1466-1496; (c) Yaghi, O.M.; O'Keeffe, M.; Ockwig, N.W.; Chae, H.K.; Eddaoudi, M.; Kim, J., Reticular synthesis and the design of new materials, *Nature* **2003**, *423*, 705-714.

Chapter 29

Stepwise Synthesis of Disk- and Ball-Shaped Polyoxovanadates: All-Inorganic Coordination Chemistry of Polyoxovanadates

Yoshihito Hayashi, Takayuki Shinguchi, Taisei Kurata, and Kiyoshi Isobe

Department of Chemistry, Graduate School of Natural Science, Kanazawa University, Kanazawa 920–1192, Japan

The new macrocyclic chemistry of polyoxovanadate that is reminiscent of crown ethers chemistry has been investigated. The chemistry between the polyanions as the *ligand* and polycations as *metals* defines a new class of *hetero*-polyoxovanadates and provides an example of all-inorganic coordination chemistry of a well-defined soluble species. The synthesis of hexavanadate-palladium, octavanadate-dicopper, and decavanadate-di(μ-hydroxo)tetra-nickel *complexes* have been demonstrated. To design polyoxovanadate cages, the reductive coupling method by *inorganic* catalytic systems has also been investigated with tetrabutylammonium salts. All the *iso*-polyoxovanadates(V) available were tested for the growth reaction for the synthesis of reduced polyoxovanadates. The coupling results were confirmed by the isolation of reduced complexes, such as a heptadecavanadate, $[V_{17}O_{42}]^{4-}$.

© 2007 American Chemical Society

Introduction

Polyoxovanadates have unique characteristics among the polyoxo-anions inherited from the uniqueness of the vanadium element, such as a preference of square pyramidal coordination, variety of valencies, unique structures, and of course its beautiful colors. In an alkaline medium, metavanadate species, $[VO_3]^-$ are dominant species. In the solid state, $K[VO_3]$ has a linear structure with the tetrahedral corner sharing of VO_4 units, and $K[VO_3]\cdot H_2O$ has a chain of distorted trigonal pyramidal structure. These tendencies to form a chain through the corner-sharing of tetrahedral $[MO_3]_m^{n-}$ units are common feature for the *meta*-polyoxo species such as polysilicate $(SiO_3)_m^{n-}$. In addition to those linear forms, polyoxoanions are known to have a wide variety of cyclic structures as well as metal adducts of those *meta*-species, such as metasilicates and metaphosphates which constitute a large part of mineral frameworks. The metal complex adducts of the polyoxovanadates were also studied extensively by Zubieta's group (*1*).

Present methods for the synthesis of polyoxovanadate have to control a precise experimental condition to extract one of the chemical species involved in the intricate equilibrium. The polyoxovanadates form polyanions of various condensation degrees, in which the coordination number of vanadium varies from 4 to 6. In contrast, simpler PO_4^{3-} species stay only as a monomer due to the stronger P–O bonds which have a higher covalent character. The phosphorous species may condense only under rigorous conditions, such as a thermal dehydration. For the synthesis of versatile polyoxovanadates, artistic techniques are required to fine tune the multidimensional concentration/pH diagram, and even the change of the reaction scale or dropping rate of the reactants may affect the isolation and the spontaneous crystallization of the product out of the equilibrium mixture.

So we are taking a new look at an experimentally successful technique —the synthesis of polyoxovanadates in organic solvents—that exploits tetrabutylammonium salts to provide a solubility in organic media, while reducing or eliminating the formation of complicated equilibrium species in an aqueous solution (*2*). The currently known example of the metavanadate species in acetonitrile involved a cyclic tetravanadate, $[V_4O_{12}]^{4-}$ and cyclic trivanadate, $[V_3O_9]^{3-}$ which were isolated and its structure were reported recently (*3, 4*). Mechanistic details on the formation of these condensed species, however, are difficult to elucidate. The presence of cyclic pentamer $[V_5O_{15}]^{5-}$ and hexamer $[V_6O_{18}]^{6-}$ has been indicated by solution studies (*5, 6*), but cyclic vanadates larger than the V_4 unit have yet to be isolated and structurally characterized.

In this chapter, we show the designing principles of a polyoxovanadate molecules toward the synthesis of a host molecule. The example of a host molecule come in two broad classes: crown ether type as a cation host and cryptand as an anion host, both of which have alternating donor sites through carbon chains. The inorganic counterparts of those organic hosts may be an

Anderson type as a metal cation host and *Keggin* type (7) as a T_d oxo-anion host (Figure 1), which are often observed in molybdate and tungstate chemistry. The versatile flexibility of the polyoxovanadate may be utilized to design such an inorganic host molecule. To achieve that goal, we approach the problems by developing the two methods: (*a*) the stepwise growth reaction of polyoxovanadates by reductive coupling methods, and (*b*) the synthesis of the all-inorganic *complexes* of polyoxovanadates as a precursor for a host molecule.

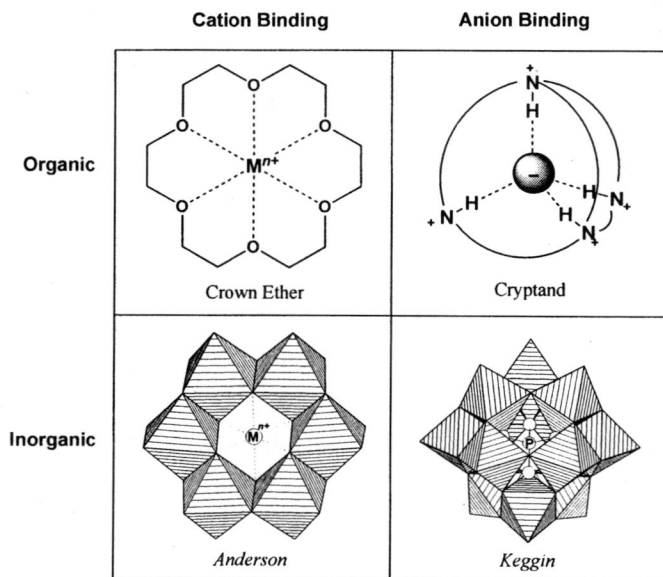

Figure 1. Comparison between macrocyclic host molecules and representative polyoxometalates.

The Reductive Coupling of Polyoxovanadates

Inorganic chemists have always dreamed of building an inorganic molecule that could form a specific geometry for transporting molecule like molecular sieves or for transforming molecule like a tailor made catalysis. But one thing has stood in their way, the synthetic methodology for metal-oxygen bond formation has been limited, and the weak inorganic M–O bonds does not always have enough bonding energy to hold their geometry in a protic media. The synthesis of inorganic metal-oxygen compounds also requires crystallization of the compounds. One answer could come from a reductive coupling synthesis in organic media (*8*). A polyoxovanadate with the quaternary tetraalkylammonium cation is an anionic molecule that is soluble in organic media. These soluble anionic molecules are used as starting materials for non-aqueous polyoxometalate synthesis.

The decavanadate, $[V_{10}O_{28}]^{6-}$ is a most well known species for isopolyvanadates. In organic media there are a few additional species such as pentavanadate. Nevertheless, the representative isopolyvanadates are known to be a few kinetically stable molecules. The available iso-V(V) are trivanadate $[V_3O_9]^{3-}$, tetravanadate $[V_4O_{12}]^{4-}$, pentavanadate $[V_5O_{14}]^{3-}$, decavanadate $[V_{10}O_{28}]^{6-}$, dodecavanadate $[V_{12}O_{32}(CH_3CN)]^{4-}$, and tridecavanadate $[V_{13}O_{34}]^{3-}$.

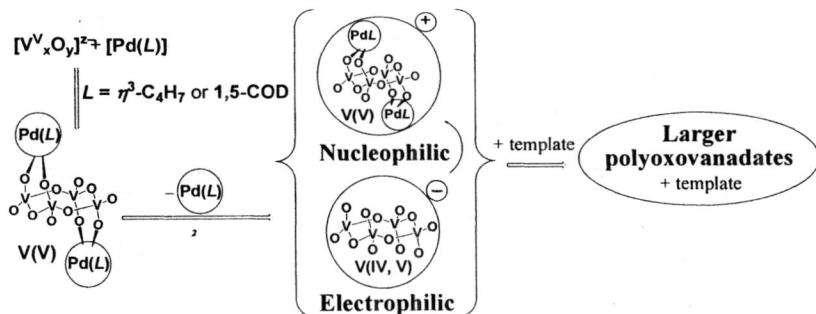

Figure 2. Reductive coupling scheme for the polyoxovanadates.

By using all the available iso-V(V) species mentioned above, we have pursued the framework growth reactions of polyoxovanadates. We developed the reductive coupling methods (Figure 2) to produce a reduced polyoxovanadate by using organometallic palladium complex through the following three steps: (1) reaction with the organometallic palladium complex to produce the supported complex with or without an isolation of this intermediate, (2) refluxing the acetonitrile solution to promote the oxidation reaction of the organometallic group by vanadates(V) resulting in a formation of reduced polyoxovanadate species, and (3) the spontaneous coupling between the reduced species as an electrophile and the starting complex as a nucleophile, prompts a coupling toward a larger polyoxovanadate. To complete the coupling reaction, addition of a stoichiometric amount of proton such as *p*-toluene sulfonic acid is required, and oxygen free environments are necessary to prevent reoxidation of the reduced species.

Condensation of Decavanadate to Heptadecavanadate

The reductive coupling reaction was performed by using the representative polyoxovanadate, decavanadate. The decavanadate, $(n\text{-}Bu_4N)_3[H_3V_{10}O_{28}]$ has a structure that fuses octahedral hexavanadate cores. The reductive coupling was carried out with $[Pd(1,5\text{-}COD)Cl_2]$ in acetonitrile, and the reduced heptadecavanadate $(n\text{-}Bu_4N)_4[V_{17}O_{42}]$ (9) was isolated after solvent extraction procedure. The similar conditions without organometallic reagents produce V(V) species such as $[CH_3CN \cdot (V_{12}O_{32})]^{4-}$ or $(n\text{-}Bu_4N)_3[H_3V_{13}O_{34}]$ depending on the reaction conditions (10, 11).

The anion is composed of seventeen {VO_6} units sharing the edge of the octahedra forming the NaCl type close-packed cluster as shown in Figure 3. The structure may be also regarded as fused decavanadate (Figure 4). In an alternative view, the V–O (3×3×3) cubic closest packed distorted-supercubane (octuple cubane) core, $V_{13}O_{14}$, is observed in this cluster with four V=O capping groups. Each V=O group is bounded to the four faces of supercubane with approximate D_{4h} symmetry. The structurally related hetero-tetradecavanadate, $K_7[AsV_{14}O_{40}]$ has been reported (12). The differences of the structure is its iso-polyanion framework which V^{4+} sits on the center, while the hetero-$[AsV_{14}O_{40}]^{7-}$ has As^{5+} on the center with removal of two V=O capping group from the iso-polyanion framework of our case.

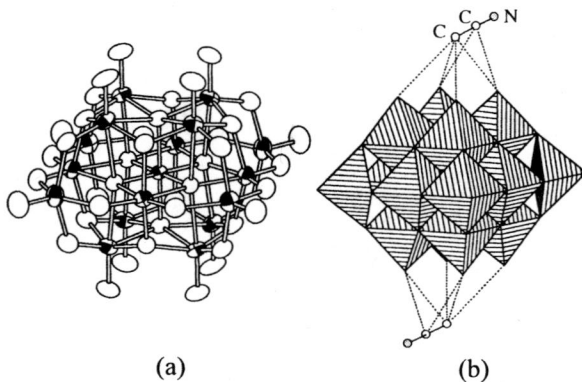

(a)   (b)

Figure 3. Molecular structure of the heptadecavanadate, (a) thermal ellipsoid view, (b) polyhedral view with two acetonitrile molecules.

The valencies of $V^V_{12}V^{IV}_5$ anion are estimated from the redox titration and bond valence calculations. The vanadium atom at the center and four capped V=O units are regarded as the V(IV) site. The magnetic moment at room temperature shows the value of 3.41 B.M. (per molecule) indicating the antiferromagnetic interaction, and corresponding to the arrangement of V(IV) sites which locate as far as possible each other.

Interesting interactions with acetonitrile molecules are found that all the four terminal oxygen atoms have close contacts with methyl goup (C_{methyl}–O = 3.298(8)~3.341(8)) at the pseudo-four-fold corner of supercubane core. Similar orientation of methyl groups heading toward the oxide-mimic surfaces have been observed in inorganic-organic hybrid polyoxoanion supported organometallic compounds (*13*, *14*). The mean bond valences for the oxygen atoms of the cluster show no sign of protonation.

Framework Splitting Reaction of Heptadecavanadate to Reduced Tri-decavanadate

The highly symmetrical $[V^V_{12}V^{IV}_5O_{42}]^{4-}$ framework was split by the reaction with *Lewis* acid, $[Cu(CH_3CN)_4](BF_4)$. The deep blue crystal of $[(n-C_4H_9)_4N]_4[H_4V^V_9V^{IV}_4O_{34}]$ (*15*) was isolated by recrystallization (Figure 5). The structure of the anion is the same as the oxidized form reported by Hill (*11*). The *Lewis* acid is a key reagent and the use of protic acids such as *p*-TsOH or Cl₃CCOOH gives no isolable product. The crystal of $[(n-C_4H_9)_4N]_4[H_4V^V_9V^{IV}_4O_{34}]$ is stable under ambient conditions. The intense blue color (680 nm, $\varepsilon = 2300$) is characteristic of the reduced vanadates, while the color of the oxidized form is yellow. The comparison of the bond lengths between reduced $[H_4V^V_9V^{IV}_4O_{34}]^{3-}$ and $[V^V_{13}O_{34}]^{3-}$ suggests the volume expansion of the cluster framework due to the four-electron reduction.

The redox titration also suggests that the four V(IV) centers exist in the cluster, and the reduced centers sit on the farthest sites from one another to minimize the repulsion of negative charges, occupying the vertices of pseudotetrahedral geometry in the cluster.

The dimeric structure of the cluster through eight hydrogen bonds is shown in Figure 5. The protonation sites are estimated from the bond valence calculation. The similar dimeric structure is found in $[H_3V_{10}O_{28}]^{3-}$ (*16*) with six hydrogen bonds.

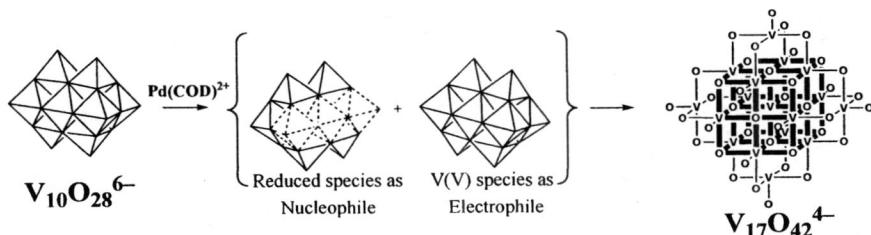

Figure 4. Proposed coupling scheme of the decavanadate. The heptadecavanadate framework is shown with the emphasis of the super-cubane framework.

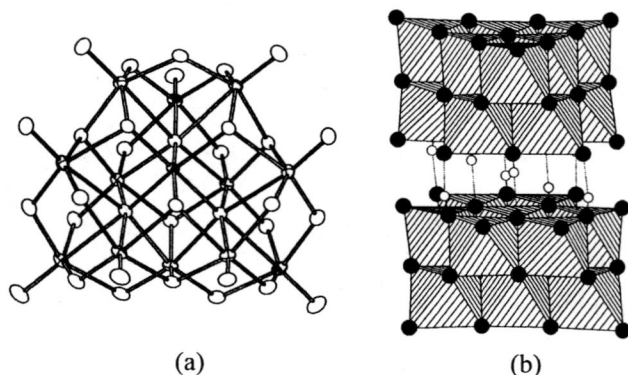

(a)   (b)

Figure 5. Molecular structure of reduced tridecavanadate, $[V_{13}O_{34}]^{7-}$, (a) thermal ellipsoid view; V^V and V^{IV} are represented by equatorial and equatorial octant ellipsoids, respectively, and (b) polyhedral view of $(H_4V_{13}O_{34})_2^{6-}$ dimer. The dotted lines indicate the hydrogen bonds.

The cyclicvoltammetry of the oxidized form shows only two reversible one-electron reduction waves (*11*), not enough to give our reduced form. When the four-electron oxidation of the reduced form was performed, the color of the solution faded to brown, and $[V_{12}O_{32}(CH_3CN)]^{4-}$ was isolated. The result suggests the conversion between the oxidized tridecavanadate and four-electron reduced tridecavanadate is irreversible as suggested from the electrochemistry.

Stepwise Growth Reaction of Cyclic Tetravanadate to Spherical Pentadecavanadate

The multistep growth reaction of the polyoxovanadate can be also performed (Figure 6). In the first step, the organometallic group supported complex we proposed in the coupling scheme (Figure 2) was isolated. The supported complex $[\{(\eta^3\text{-}C_4H_7)Pd\}_2V_4O_{12}]^{2-}$ was obtained by the reaction of $[Pd(\eta^3\text{-}C_4H_7)Cl]_2$ with $(n\text{-}Bu_4N)[VO_3]$ at room temperature. The two $(\eta^3\text{-}C_4H_7)Pd$ groups are coordinated on both side of the tetravanadate ring. The geometrical data are in agreement with the reported structures of the related tetravanadate supported compounds (*17, 18*). The four vanadium atoms and the bridging oxygens are found to be planar which is a part of the V_4O_4 eight membered ring.

The second step is the thermal coupling of the supporting complex. When the solution of $[\{(\eta^3\text{-}C_4H_7)Pd\}_2V_4O_{12}]^{2-}$ was refluxed to prompt the reduction of V(V) to V(IV), the solution turned deep blue to give a condensed cluster $(n\text{-}Bu_4N)_4[V_{10}O_{26}]\cdot H_2O$ (*19*).

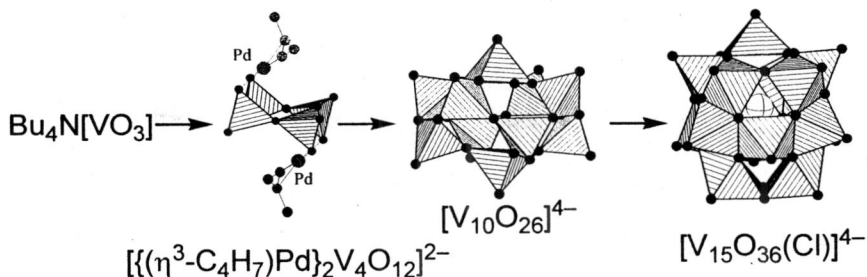

Figure 6. The reductive coupling with the isolation of intermediates of each step.

From the further reaction of $[V_{10}O_{26}]^{4-}$ with $[Pd(COD)Cl_2]$, the recrystallization gives deep blue crystals of $(n\text{-}Bu_4N)_4[V_{15}O_{36}(Cl)]$. The $[V^{IV}_6V^V_9O_{36}(Cl)]^{4-}$ anion is a spherical cluster (D_{3h}) by linkage of fifteen tetragonal VO_5 pyramids encapsulating chloride anion and contains six V^{IV} and nine V^V centers. The V–V distances (2.83–3.03 Å) are significantly shorter than the values of reported V_{15} clusters (3.2–3.6 Å) which have higher negative charges of –6 to –7 (20, 21, 22). The shorter metal–metal bond lengths are due to the lower negative charge on $[V_{15}O_{36}(Cl)]^{4-}$ which decreases the cluster shell size.

Catalytic Condensation

To make the structure growth reaction more practical, we carry out the catalytic reaction of the reductive coupling to save precious palladium sources (Figure 7). In contrast to a catalytic system seen in organic synthesis, the *inorganic* vanadates are synthesized with oxidized olefine species as byproducts.

The proposed catalytic cycle involves the coordination of the organometallic Pd(COD) group on the polyoxovanadate, (a) the formation of the polyoxometalate supported organometallic compounds, (b) the redox reaction between V(V) centers and COD of the palladium complex affords a reduced polyoxovanadate species (as nucleophile), then spontaneously coupled with the polyoxovanadate(V) (as electrophile) resulting a condensation of the polyoxovanadate: the growth reaction, (c) the elimination of the oxidized organic species from the palladium to give off $[Pd_2Cl_6]^{2-}$ species in an acetonitrile, and (d) the olefin coordination reaction between the excess COD and the palladium source in (c) to reproduce $Pd(COD)Cl_2$, which is set to repeat the cycle. The yields of these catalytic synthesis are comparable to the original synthesis as shown in Table 1.

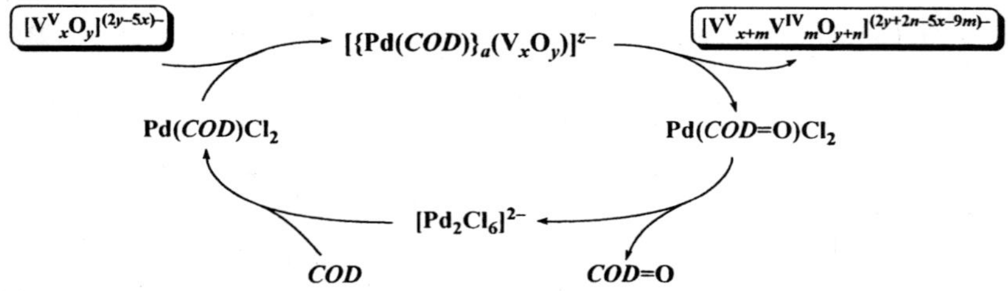

Figure 7. Proposed mechanism for the "inorganic" catalytic condensation.

Table 1. The yields(%) of reductive coupling reactions[a]

Starting material	Coupling product	Yield(%)[b]	Yield(catalytic)[c]
$[V_4O_{12}]^{4-}$ or $[V_5O_{14}]^{3-}$	$[V_{10}O_{26}]^{4-}$	87	65
$[V_5O_{14}]^{3-}$	$[V_{10}O_{26}]^{4-}$	73	73
$[H_3V_{10}O_{28}]^{3-}$	$[V_{17}O_{42}]^{4-}$	78	86
$[V_{12}O_{32}(CH_3CN)]^{4-}$	$[V_{15}O_{36}(Cl)]^{4-}$	80	86
$[V_{10}O_{26}]^{4-}$	$[V_{15}O_{36}(Cl)]^{4-}$	86	79
$[V_{13}O_{34}]^{3-}$	$[V_{17}O_{42}]^{4-}$	81	81

[a] The yields are a typical yield for the crystalline powder based on V. After recrystallization, the total yields of the X-ray quarity crystals decrease around 25%.

[b] 0.2 mmol of V and Pd compound in 5 cm^3 of acetonitrile.

[c] 0.2 mmol of V, 0.04 mmol of Pd complex, and 0.2 cm^3 of COD in 5 cm^3 of acetonitrile.

The synthetic pathways of the reductive coupling scheme are summarized in Figure 8.

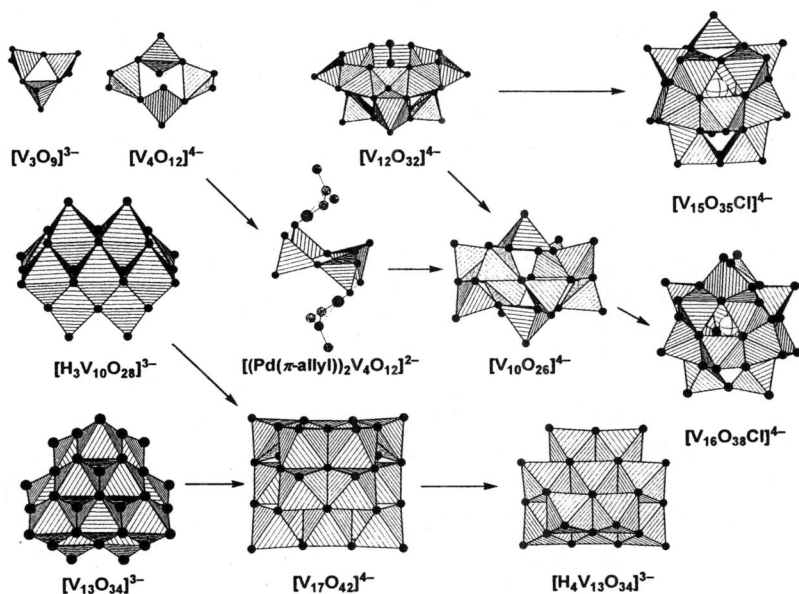

$[V_3O_9]^{3-}$ $[V_4O_{12}]^{4-}$ $[V_{12}O_{32}]^{4-}$

$[V_{15}O_{35}Cl]^{4-}$

$[H_3V_{10}O_{28}]^{3-}$ $[(Pd(\pi\text{-allyl}))_2V_4O_{12}]^{2-}$ $[V_{10}O_{26}]^{4-}$

$[V_{16}O_{38}Cl]^{4-}$

$[V_{13}O_{34}]^{3-}$ $[V_{17}O_{42}]^{4-}$ $[H_4V_{13}O_{34}]^{3-}$

Figure 8. Reductive coupling reactions of polyoxovanadates with VO_4, VO_5, VO_6 groups represented by tetrahedra, square pyramids, octahedra, respectively. The T_d vanadates produce open framework clusters such as cyclic or spherical molecules, while the O_h vanadates give condensed close-packed molecules.

Coordination Chemistry of Cyclic Metavanadate Species

The chemical species of polyoxovanadates are greatly influenced by its experimental conditions. In acetonitrile, metavanadate species exist as an equilibrium mixture of cyclic tetravanadates that differ in their chain orientation, number of protonation and may be a number of cyclization. Upon the addition of multivalent transition metal cationic salts, there are shifts in the equilibrium to the formation of the larger cyclic species, which are investigated in this study through synthetic works. With varying transition metal cations, both the polyoxovanadate ring sizes and the number of transition metals are studied *(23)*. In this reaction between a *hetero*-cation and a polyoxoanion, and then if we consider an anion as a ligand, this is a coordination chemistry of polyoxoanion. Here, we developed a new coordination chemistry which is a chemistry between all-inorganic polyoxovanadate ligands and metal cationic species.

Many polyoxovanadates involve reversible reactions, so it is important that we understand how varying the conditions will affect the composition of an equilibrium mixture. This knowledge allows us to manipulate conditions to make the cyclic vanadate species larger, which resemble a structure of a crown ether. The overall stoichiometry of the reactions is summarized in the following equation.

$$a[VO_3]^- + bM^{n+} \rightarrow [M_b(VO_3)_a]^{n \times b - a}$$

Palladium Complex of Cyclic Hexavanadate

The hexavanadate palladium complex was formed when a mixture of $\{(C_2H_5)_4N\}[VO_3]$ and bis(benzonitrile)dichloropalladium(II) was reacted with a molar ratio of $\{(C_2H_5)_4N\}[VO_3]$ to $Pd(C_6H_5CN)_2Cl_2$ as 6:1.

$$6[(C_2H_5)_4N][VO_3] + Pd(C_6H_5CN)_2Cl_2 \rightarrow$$
$$[(C_2H_5)_4N]_4[PdV_6O_{18}] + 2[(C_2H_5)_4N]Cl + 2C_6H_5CN$$

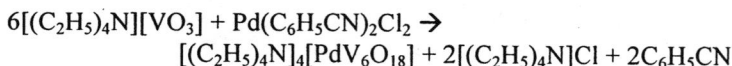

The structure determination reveals the formation of a *complex* between Pd^{2+} and cyclic hexavanadate, $[VO_3]_6^{6-}$, composed of six VO_4 units bridged together through corner-sharing of the oxygen atoms (Figure 9). The hexavanadate ring exists in a boat conformation; two of the six VO_4 units at the flag positions are not coordinated to Pd^{2+} leaving two terminal oxygen atoms per unit, whereas the remaining four VO_4 units are coordinated to Pd^{2+} through an oxygen atom leaving one terminal oxygen per unit.

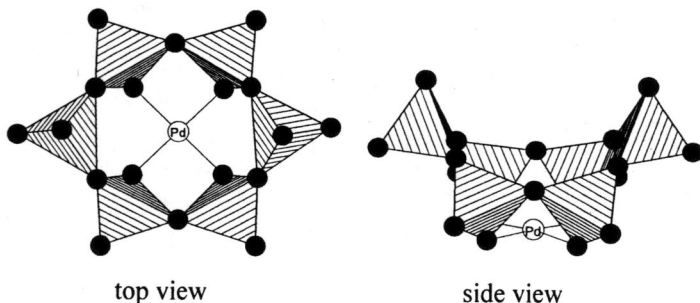

top view   side view

Figure 9. Structure of $[PdV_6O_{18}]^{4-}$ anion.

For this reaction, the use of a precursor with labile benzonitrile ligands was essential. When the dichloro(cyclooctadiene)palladium(II) was applied, it resulted in a formation of a reduced decavanadate, $[V_{10}O_{26}]^{4-}$.

The reaction of $[(C_4H_9)_4N][VO_3]$ with di-μ-chlorobis(2-methylallyl)dipalladium(II) is known to afford the mixed valence decavanadate through the tetravanadate-supported organometallic complex (24).

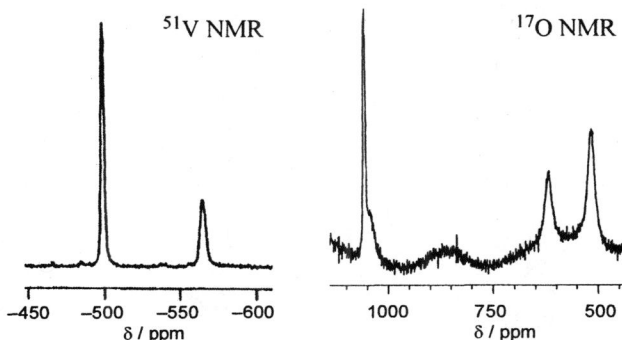

Figure 10. ^{51}V NMR and ^{17}O NMR of $[PdV_6O_{18}]^{4-}$.

The ^{51}V NMR spectrum in acetonitrile exhibits resonances at –499 and –565 ppm with the 2:1 intensity ratio (Figure 10). The downfield peak is assigned to the vanadiums that are connected to Pd^{2+} through an oxygen bridge, whereas the upfield peak is assigned to the vanadiums at the flag position of the boat-type conformation. In contrast, the chemical shift of the tetravanadate in acetonitrile is reported at –570 and –574 ppm (25). The conformational changes of the ring between boat and chair are not observed from the variable temperature ^{51}V NMR up to 75 °C. The ^{17}O NMR spectrum exhibits three resonances at 517, 618, and 1059 ppm, corresponding to the two types of bridging and terminal oxygens, respectively.

Dicopper Complex of Cyclic Decavanadate

The decavanadate dicopper complex was prepared similarly, but a different molar ratio according to the following equation.

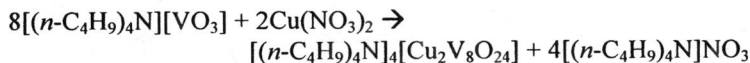

$$8[(n\text{-}C_4H_9)_4N][VO_3] + 2Cu(NO_3)_2 \rightarrow$$
$$[(n\text{-}C_4H_9)_4N]_4[Cu_2V_8O_{24}] + 4[(n\text{-}C_4H_9)_4N]NO_3$$

The structure of $[Cu_2V_8O_{24}]^{4-}$ is similar to the structure of $[V_{10}O_{26}]^{4-}$ (19) and both compounds have a crown ring of eight edge-shared tetrahedral VO_4 units. The interior of both rings also contain same charged groups, that is, two Cu^{2+} in our compound or two $[V^{IV}O]^{2+}$ in the mixed valence decavanadate; the total charge of −4 in both compounds is attained by the inclusion of two cations with charge of +2 within the $(VO_3)^{8-}$ ring. However, the symmetry of copper complex(C_s) is lower than that of the mixed valence decavanadate (S_8). The distortion of the complex is due to the weak interaction at the fifth coordination sites of the copper with the bridging oxygen (Figure 11).

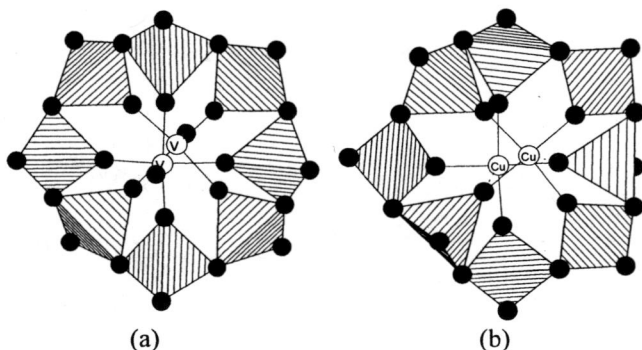

(a)   (b)

Figure 11. Molecular structure of (a) $[V_{10}O_{26}]^{4-}$ and (b) $[Cu_2V_8O_{24}]^{4-}$.

Tetra-nickel Complex of Cyclic Decavanadate

In the case of decavanadate nickel complex, the complex formation was accomplished by the following equation.

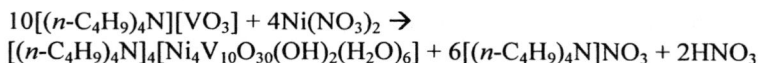

$$10[(n\text{-}C_4H_9)_4N][VO_3] + 4Ni(NO_3)_2 \rightarrow$$
$$[(n\text{-}C_4H_9)_4N]_4[Ni_4V_{10}O_{30}(OH)_2(H_2O)_6] + 6[(n\text{-}C_4H_9)_4N]NO_3 + 2HNO_3$$

The slow evaporation of the light yellow acetonitrile solution gave the largest metavanadate ring complex, $[Ni_4V_{10}O_{30}(OH)_2(H_2O)_6]^{4-}$ as yellow green crystals (Figure 12). Occasionally, the crude solid included a small impurity, which was identified as Ni^{2+} hydrolysis products, may produce. But, the addition of an excess of the $[VO_3]^-$ salt resulted in the conversion of the hydrolysis species to the expected cyclic compound. The decavanadate ring holds a nickel tetramer that is composed with a di-μ-hydroxo core with two capping groups of nickel aqua complex. In this decavanadate ring, two VO_4

units at the flag positions are not coordinated to Ni^{2+}, and one of those terminal oxygens oriented inward to the center are interacted with the μ-hydroxo ligands through the hydrogen bonds ($O\cdots O = 2.813$ Å).

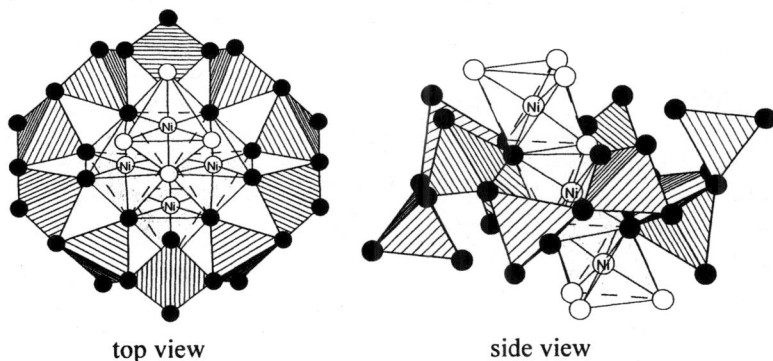

top view   side view

Figure 12. Molecular structure of $[\{Ni_4 (OH)_2(H_2O)_6\} \{V_{10}O_{30}\}]^{+}$. Oxygen atoms, hydroxo groups, and water ligands are represented by dark, gray, and white spheres, respectively.

In these *all-inorganic complex*, the size of the vanadate rings correlates with the total charge of the central metal ions. The higher the total charge of the central unit, the larger the vanadate rings are formed. The positive charges of cationic *hetero*-metals at the center conpensates the high negative charges of larger cyclic polyoxovanadates $[VO_3]_n^{n-}$ that are sufficiently large to include metal cations within the ring.

Thus, the addition of Lewis acidic cations to the $[VO_3]^-$ solution may drive the reaction towards the direction to the formation of larger ring species. The metal cations act as a template in gathering the vanadate *ligand,* and the cyclic vanadate $[VO_3]_n^{n-}$ species can function as the macrocyclic oxo ligands. These oxide based macrocyclic *complexes* dissolves in acetonitrile well, and can be handled on the bench top without a care, but the control of molar ratio, concentration, and the quality of the $[VO_3]^-$ solution have to maintain to prevent the precipitation of the hydrolysis products, when those complex are synthesized.

Conclusions

Synthetic studies of heteropolyoxovanadates and reduced polyoxovanadates have allowed us to gain insight into the metal template effect on the formation of the cyclic polyoxovanadate species in acetonitrile. Novel cyclic polyoxovanadates with three vanadate ring sizes were prepared by the reaction between the alkylammonium metavanadate and transition metal ions in acetonitrile. The structure growth reactions were also performed by the reductive couping reaction which preserved the coordination environments of vanadium atoms. While the T_d or square pyramidal vanadate gave the cage cluster with square pyramidal sites, the O_h vanadate gave the NaCl style condensed cluster. Our simple synthetic methodology for these new classes of inorganic complexes can be extended for the synthesis of other polyoxovanadate complexes with different cationic groups and perhaps different ring sizes. Furthermore, it may prove to be interesting to explore reactions with larger cationic hydroxide species, and these disk shaped complexes can be utilized as a scaffold for a synthesis of a spherical molecule.

Acknowledgements

This work was partly supported by a Grants-in-Aid for Scientific Research, No. 16550051 from the Ministry of Education, Culture, Sports, Science and Technology of Japan.

References

1. Finn, R. C.; Zubieta, J.; Haushalter, R. C. *Prog. Inorg. Chem.* **2003**, *51*, 421–601.
2. Errington, R. J. In *Polyoxometalate Molecular Science*; Borrás-Almenar, J. J., Coronado, E., Müller, A., Pope, M. Eds.; NATO Science Series, II: Mathematics, Physics and Chemistry, **2003**, *98*, 55–78.
3. Roman, P.; San Jose, A.; Luque, A.; Gutierrez-Zorrilla, J. M. *Inorg. Chem.* **1993**, *32*, 775–776.
4. Hamilton, E. E.; Fanwick, P. E.; Wilker, J. J. *J. Am. Chem. Soc.* **2002**, *124*, 78–82.
5. Andersson, I.; Pettersson, L.; Hastings, J. J.; Howarth, O. W. *J. Chem. Soc., Dalton Trans.* **1996**, 3357–3361.
6. Tracey, A. S.; Jaswal, J. S.; Angus-Dunne, S. J. *Inorg. Chem.* **1995**, *34*, 5680–5685.
7. Pope, M. T. *Heteropoly and Isopoly Oxometalates*; Springer-Verlag: Berlin, 1983.
8. Müller, A.; Meyer, J.; Krickemeyer, E.; Beugholt, C.; Bogge, H.; Peters, F.; Schmidtmann, M.; Kogerler, P.; Koop, M. J. *Chem.-A Eur. J.* **1998**, *4*, 1000–1006.
9. Hayashi, Y.; Fukuyama, K.; Takatera, T.; Uehara, A. *Chem. Lett.* **2000**, *(7)*, 770-771.

10. Day, V. W.; Klemperer, W. G.; Yaghi, O. M. *J. Am. Chem. Soc.* **1989**, *111*, 4518.
11. Hou, D.; Hagen, K. S.; Hill, C. L. *J. Am. Chem. Soc.* **1992**, *114*, 5864.
12. Müller, A.; Döring, J.; Khan, I.; Wittneben, V. *Angew. Chem., Int. Ed. Engl.* **1991**, *30*, 210.
13. Hayashi, Y.; Ozawa, Y.; Isobe, K. *Inorg. Chem.* 1991, *30*, 1025–33.
14. Hayashi, Y.; Mueller, F.; Lin, Y.; Miller, S. M.; Anderson, O. P.; Finke, R. G. *J. Am. Chem. Soc.* **1997**, *119*, 11401–11407.
15. Kurata, T.; Hayashi, Y.; Uehara, A.; Isobe, K. *Chem. Lett.* **2003**, *32*, 1040–1041.
16. Day, V. W.; Klemperer, W. G.; Maltbie, D. J. *J. Am. Chem. Soc.* **1987**, *109*, 2991–3002.
17. Day, V. W.; Klemperer, W. G.; Yagasaki, A. *Chem. Lett.* **1990**, *(8)*, 1267–70.
18. Abe, M.; Isobe, K.; Kida, K.; Yagasaki, A. *Inorg. Chem.* **1996**, *35*, 5114–5115.
19. Bino, A.; Cohen, S.; Heitner-Wirguin, C. *Inorg. Chem.* **1982**, 21, 429–31.
20. Müller, A.; Krickemeyer, E.; Penk, M.; Walberg, H-J.; Bögge, H. *Angew. Chem., Int. Ed. Engl.* **1987**, *26*, 1045.
21. Müller, A.; Penk, M.; Rohlfing, R; Krickemeyer, E.; Döring, J. *Angew. Chem., Int. Ed. Engl.* **1990**, *29*, 926.
22. Müller, A.; Rohlfing, R; Krickemeyer; Bögge, H. *Angew. Chem., Int. Ed. Engl.* **1993**, *32*, 909.; Yamase, T.; Ohtaka, K. *J. Chem. Soc., Dalton Trans.* **1994**, 2599.
23. Kurata, T.; Uehara, A.; Hayashi, Y.; Isobe, K. *Inorg. Chem.* **2005**, *44*, 2524–2530.
24. Hayashi, Y.; Miyakoshi, N.; Shinguchi, T.; Uehara, A. *Chem. Lett.* **2001**, 170–171.
25. Nakano, H.; Ozeki, T.; Yagasaki, A. *Inorg. Chem.* **2001**, *40*, 1816–1819.

Chapter 30

A Vanadium-Based Homogeneous Chemical Oscillator

Kan Kanamori and Yuya Shirosaka

Department of Chemistry, Faculty of Science, University of Toyama, Gofuku 3190, Toyama 930–8555, Japan

It has been found that a dichloromethane solution of [V(IV)OCl$_2$(bpy)] or [V(III)Cl$_3$(CH$_3$CN)(bpy)] in the presence of air exhibits a new oscillating reaction. The initial pale green color turned to dark orange after an induction period. The color of the solution changed back to pale green, and this pattern repeated. The dark orange species was revealed to be [{V(V)OCl$_2$(bpy)}$_2$(μ-O)] by X-ray crystallography. Thus, a redox reaction between vanadium(V) and vanadium(IV) species is responsible for the oscillatory reaction. Dissolved dioxygen may work as an oxidizing agent and formaldehyde included in dichloromethane as a contaminant would be a reducing agent. Addition of chloride to a reaction solution considerably increased the induction period.

© 2007 American Chemical Society

Manos et al. found that a spontaneous reduction of vanadium(IV) to vanadium(III) occurred when they added 2,2′-bipyridine (bpy) to a solution of a bare vanadium(IV) complex, [V(IV)Cl$_2$(acac)$_2$], in dry organic solvent (*1*). Reduction of vanadium(IV) to vanadium(III) is an important subject with regard to the accumulation and reduction of vanadium by ascidians (tunicates). Since a strictly dry environment is not likely to exist in living organisms, we investigated whether a similar spontaneous reduction would occur in non-dry solvents. We used non-dry ethanol as a solvent and a common vanadyl complex, [V(IV)O(acac)$_2$], instead of [V(IV)Cl$_2$(acac)$_2$]. We found that reduction of vanadium(IV) to (III) also occurred when a large excess (up to 60 equivalents) of HCl gas was introduced into a reaction mixture containing [V(IV)O(acac)$_2$] and bpy (unpublished work). In this experiment, a pale green complex was obtained when more than 80 equivalents of HCl gas was introduced into an ethanolic solution of [V(IV)O(acac)$_2$]. The pale green complex was found to be [V(IV)OCl$_2$(bpy)] where the acac ligands in the starting material were substituted by bpy and chloride. The above compositional assignment is supported by elemental analysis and IR spectroscopy. We dissolved the pale green complex in dichloromethane in order to observe the absorption spectrum. After measurement of a UV-vis spectrum, the remaining solution was kept in a volumetric flask with a stopper. A few days later the color of the solution suddenly turned to dark orange. After that, the color of the solution changed back to pale green, and this pattern repeated, indicating the discovery of a new, vanadium-based oscillator (*2*).

Experimental

Preparation of [V(IV)OCl$_2$(bpy)]

[V(IV)O(acac)$_2$] was dissolved in ethanol. An ethanolic solution of HCl (more than 80 equivalents) was added to the above solution and the resulting solution was stirred for 15 min. Then, an ethanolic solution containing an equivalent amount of bpy was added. The reaction mixture was stirred for one day. The solution was evaporated to dryness. An appropriate amount of acetonitrile was added to the residue. The mixture was again evaporated to dryness. This procedure was repeated three times. Pale green powder was collected by filtration.

Preparation of [V(III)Cl$_3$(CH$_3$CN)(bpy)]

[V(III)Cl$_3$(THF)$_3$] was dissolved in acetonitrile. An acetonitrile solution of bpy was added to the solution to yield a green solution. The resulting solution was evaporated to some extent, and a green precipitate was deposited. The green precipitate was collected by filtration.

Preparation of [{V(V)OCl$_2$(bpy)}$_2$(μ-O)]

[V(III)Cl$_3$(CH$_3$CN)(bpy)] was dissolved in dichloromethane by stirring under aerobic conditions. After the color of the solution changed to dark orange, the solution was evaporated to some extent, and kept in a freezer at −30 °C. Dark orange crystals were deposited after a few days. Although the dark orange color sometimes disappeared, it generally continues for a long period when the solution is allowed to stand at low temperature without stirring.

Observation of Oscillations

An appropriate amount of [V(IV)OCl$_2$(bpy)] or [V(III)Cl$_3$(CH$_3$CN)(bpy)] (typically 0.34 mM) was added to dichloromethane under aerobic conditions using a 20-ml volumetric flask with a stopper. The resulting mixture was stirred continuously. Initially the mixture was turbid but became clear, usually after several hours. Since the oscillating reaction we found is very slow, and it takes a long time to observe oscillations for solutions under various conditions, we followed the oscillations by capturing frames of several different reaction solutions simultaneously with a digital video camera. For a selected system, we also followed the oscillations with a UV-vis spectrophotometer.

Results and Discussion

Observation of Oscillations with a Digital Video Camera

We captured video frames of several dichloromethane solutions of [V(IV)OCl$_2$(bpy)] under various conditions every hour. An example is shown in Fig. 1. To make a graph showing the oscillations, we set a y-value to 0 when the color of the solution was pale green (almost colorless) and set it to 1 when the color was orange regardless of its depth. The graph thus obtained is shown in Fig. 2(A). As can be seen in Fig. 2(A), the solution exhibited oscillations in color, though the periods corresponding to the orange color and their intervals were irregular. The induction period before the first color change occurred was also irregular. Stirring velocity seems to be one of the factors affecting the induction and oscillating periods. We will discuss other factors below. Videos of an aerobic solution of [V(III)Cl$_3$(CH$_3$CN)(bpy)] also exhibited oscillations as shown in Fig. 3(A).

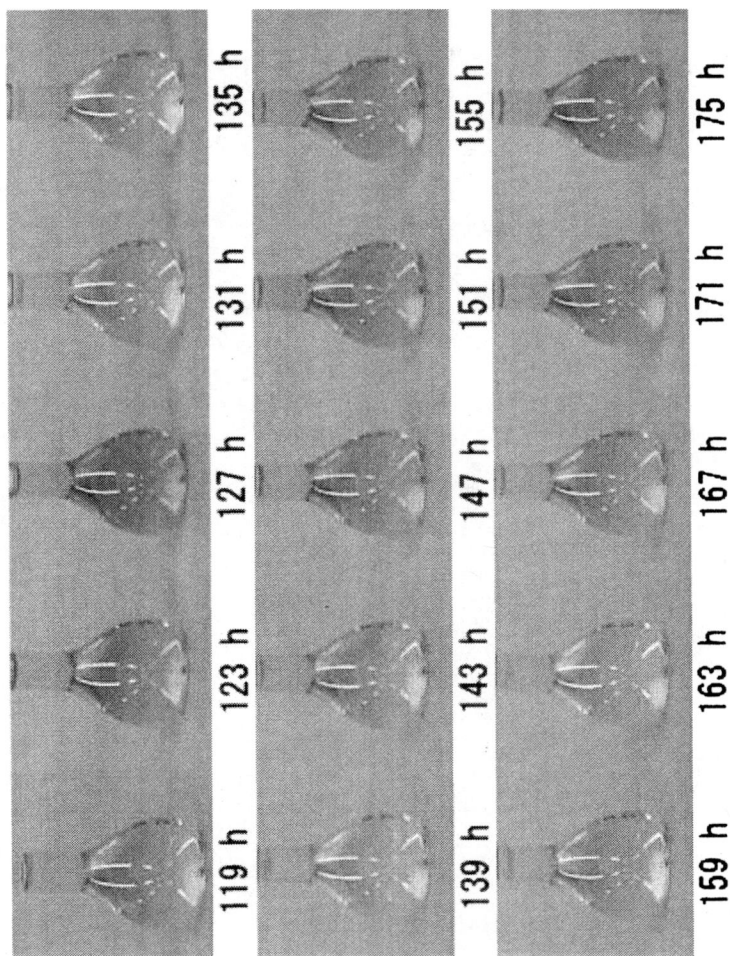

Figure 1. Video frames of the oscillating reaction: $[V(IV)OCl_2(bpy)]$ (0.34 mM) in 20 ml of dichloromethane; stirring rate = 900 rpm. (See page 4 of color inserts.)

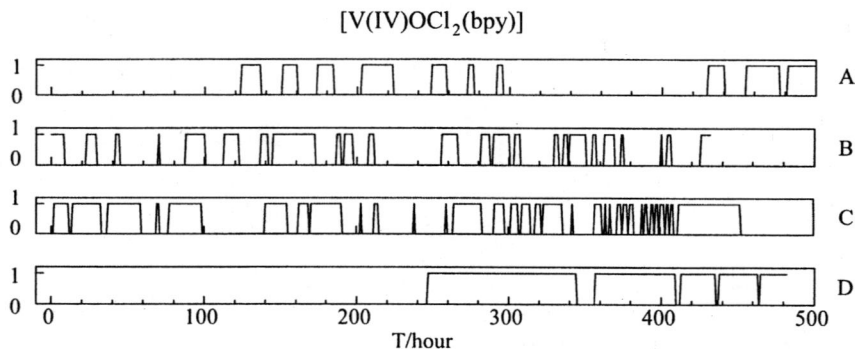

$[V(IV)OCl_2(bpy)]$

Figure 2. Oscillating pattern observed by video camera for $[V(IV)OCl_2(bpy)]$: "break down" signifies instrumental failure.
A: $[V(IV)] = 0.34$ mM; B: $[V(IV)] = 0.34$ mM, $[Bza] = 12.5$ mM;
C: $[V(IV)] = 0.34$ mM, $[Bza] = 12.5$ mM, Volume of Soln = Half;
D: $[V(IV)] = 0.34$ mM, $[Bza] = 12.5$ mM, $[TEAC] = 5.0$ mM.

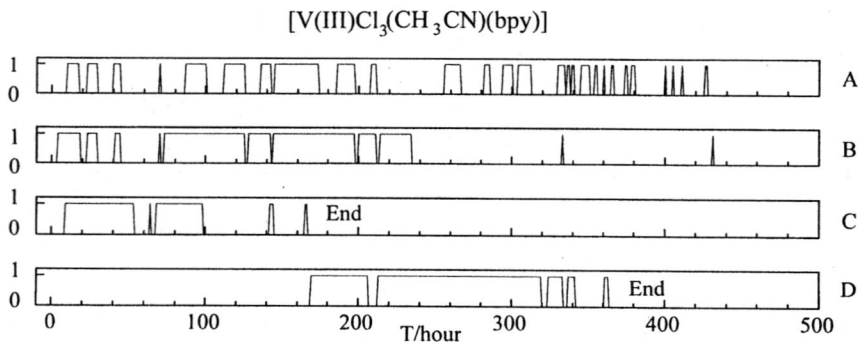

$[V(III)Cl_3(CH_3CN)(bpy)]$

Figure 3. Oscillating pattern observed by video camera for
$[V(III)Cl_3(CH_3CN)(bpy)]$: "End" signifies experiment terminated.
A: $[V(III)] = 0.34$ mM; B: $[V(III)] = 0.34$ mM, $[Bza] = 12.5$ mM;
C: $[V(III)] = 0.34$ mM, $[Bza] = 12.5$ mM, Volume of Soln = Half;
D: $[V(III)] = 0.34$ mM, $[Bza] = 12.5$ mM, $[TEAC] = 5.0$ mM.

Oscillations in the [V(IV)OCl$_2$(bpy)] system were also recorded by measuring UV-vis spectra. In this case, we used a larger vessel (about 50 ml, rather than a 20-ml volumetric) mounted in a temperature-controlled spectrophotometer. The induction period was very long. The reason is not clear at present. During the induction period, a spectral change was observed in the UV region (data not shown), indicating that a pre-reaction occurred before the first color change. We added benzaldehyde after 9 days in order to enhance the reaction. The color of the solution turned immediately to light orange, and the intensity of the orange color changed periodically as shown in Fig. 4. The period of the intensity change was about 24 hours.

Characterization of the Dark Orange Species

The absorption spectrum of the dark orange species formed in a dichloromethane solution of [V(IV)OCl$_2$(bpy)] is shown in Fig. 5(A). The observed spectral features (the band positions as well as the band shape) resemble those of a V(III)-O-V(III) dimer. The absorption spectrum of [V(III)$_2$(μ-O)(L-his)$_4$] (3) is shown in Fig. 5(B) for comparison. The dark orange species was found to be EPR silent, whereas the initial pale green complex exhibited an 8-line EPR spectrum typical of vanadium(IV) species. Therefore, we first assumed that the dark orange complex would be an oxo-bridged dinuclear vanadium(III) complex, and thus the oscillations were presumed to occur between vanadium(IV) and vanadium(III). This assumption was mistaken, since the X-ray crystal structure analysis shown below revealed that the dark orange species is a vanadium(V) complex, and thus oscillations occurred between vanadium(V) and vanadium(IV) species. An Ortep perspective view of the dark orange complex's molecular structure is shown in Fig. 6. The complex has an oxo-bridged dinuclear structure as expected. However, the V1-O1 and V2-O2 distances indicate that the oxidation state of the vanadium in not +3 but +5, because these distances are 1.586(4) and 1.582(4) Å, respectively. Such short V-O distances clearly indicate a double bond character for V1-O1 and V2-O2 bonds. Therefore, O1 and O2 atoms should not be water oxygen atoms but oxo ligands, and the oxidation state of the vanadium must be +5. It is surprising that the present oxo-bridged dinuclear vanadium(V) complex exhibited an optical absorption spectrum very similar to oxo-bridged dinuclear vanadium(III) complexes.

Reducing and Oxidizing Agents

Since it was found that a redox reaction between vanadium(V) and vanadium(IV) species is responsible for the oscillatory reactions, some reducing

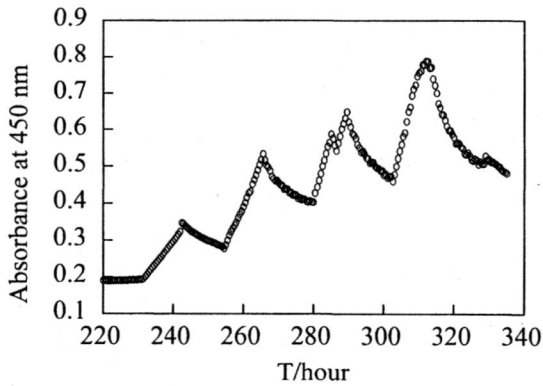

Figure 4. Time-dependent Absorbance at 450 nm for V(IV)/Bza System.

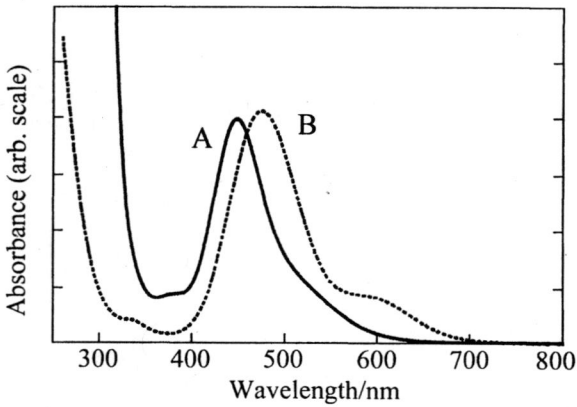

Figure 5. Abospriton spectra of the dark orange species (A) and
[V(III)$_2$(μ-O)(L-his)$_4$] (B).

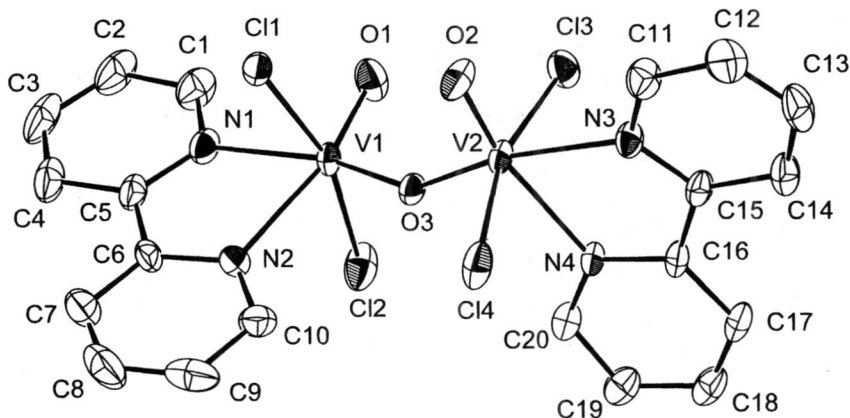

Figure 6. Perspective view of [{V(V)OCl₂(bpy)}₂(μ-O)].

agent as well as some oxidizing agent must participate in the oscillation. Dissolved dioxygen probably functions as an oxidizing agent, since deaeration of a reaction solution strongly inhibits the oscillation. What then is the reducing agent? We simply dissolved [V(IV)OCl₂(bpy)] or [V(III)Cl₃(CH₃CN)(bpy)] in dichloromethane and did not add any reducing agent. The purity report, which was obtained from Wako Chemical, Japan, indicates that the dichloromethane we employed is contaminated by a very small amount of formaldehyde and chlorine. Since formaldehyde has reducing ability, we think at present that it is the reducing agent. In order to examine the effect induced by an addition of aldehyde to a reaction solution, we observed oscillations for a solution containing an excess amount of aldehyde. In this experiment, we used benzaldehyde instead of formaldehyde, because gaseous formaldehyde is difficult to handle in organic solvents. The observed oscillations are shown in Fig. 2(B) and Fig. 3(B) for [V(IV)OCl₂(bpy)] and [V(III)Cl₃(CH₃CN)(bpy)], respectively. As can be seen in these figures, the induction period before the first development of the orange color decreased, which is contrary to our expectation. The reason is not clear at present. There may be two orange species in the reaction solution as suggested by the UV-vis spectra shown in Fig. 4.

Other dependencies were investigated. Oscillations were observed for the systems in which the volume of the solution was reduced by one- half, and thus the volume of air was increased. The results are shown in Fig. 2(C) and Fig 3(C). The effect induced by air content is not clear. Addition of chloride (as tetraethylammonium chloride (TEAC)) considerably increases the induction period before the first development of the dark orange color (Fig. 2(D) and Fig. 3(D)). This effect may indicate that direct coordination of dioxygen to vanadium would be requisite for production of the oxo-bridged vanadium(V) dimer that is responsible for the dark orange color, and free chloride suppress this reaction.

Further studies are required before a mechanism for this fascinating new chemical oscillator can be developed.

Acknowledgement

We wish to thank Prof. Kenneth Kustin, Brandeis University Emeritus, for his valuable suggestions and for encouraging us to embark on a study of vanadium oscillations.

References

1. Manos, M. J.; Tasiopoulos, A. J.; Raptopoulou, C.; Terzis, A.; Woollins, J. D.; Slawin, A, M, Z.; Keramidas, A. D.; Kabanos, T. A. *J. Chem. Soc., Dalton Trans.* **2001**, 1556.
2. Epstein, I. R.; Pojman, J. A. *An Introduction to Nonlinear Chemical Dynamics: Oscillations, Waves, Patterns, and Chaos;* Oxford University Press: New York, 1998.
3. Kanamori, K.; Teraoka, M.; Maeda, H.; Okamoto, K. *Chem. Lett.* **1993**, 1731.

Indexes

Author Index

Subject Index

A

Absorption processes in gastrointestinal tract, insulin-mimetic vanadium complexes, 325–326*f*

Actin cytoskeleton, decavanadate interaction, 253–254*f*

Acute intraperitoneal administration, vanadium dipicolinic acid complexes, rats with STZ-induced diabetes, 98–100, 101*t*, 105

Administration routes, vanadium dipicolinic acid complexes, to rats with streptozocin-induced diabetes, 93–109

Air levels, vanadium, 218–219

Aldimines, reductive coupling, 8

Alkane functionalization under mild conditions, vanadium catalyzed, 51–60

Alkane metal-catalyzed carboxylation, 54*f*, 56–57

Alkane metal-catalyzed hydroxylation and halogenation, 53–56

Alkene oxybromination in two-phase system using 1-methyl-3-butylimidazolium [bmim$^+$], 31–33

Alkenols, vanadium(V)-catalyzed oxidation, 38–50

Alkylating agents and vanadates, reactivity, 301–302, 303*f*

Alkylation prevention by vanadium and selenium compounds, 303–305

Alkylation reactivity, selenium oxo species, 302–303

α, ω-Alkyldiphosphonates, tether length variations, 393–394

Alveolar macrophage function, vanadium effects, 220–224

See also Macrophage iron homeostasis

Amavadine, catalyst for alkane functionalization, 52–53*f*

Amino acid sequences, vanabins from *Ascidia sydneiensis*, 270–272*f*

Aqueous speciation, selenium and vanadium inorganic oxo species, 298, 300

Aqueous stability, vanadium(IV)-salen and salan type complexes, 343, 345–347

Aqueous vanadium(IV,V) hydroxycarboxylate complexes, 377–389

Aromatic diphosphonates, structures, 394–396

Ascidia ahodori, 266

Ascidia ceratodes, 282, 288–292

Ascidia gemmata, 265–266*t*

Ascidia sydneiensis samea, 268, 270, 271*f*, 274, 276

Ascidians, genes and proteins involved in vanadium accumulation, 264–280

See also Tunicates

Ascophyllum nodosum, 61, 149

B

Benzene, hydroxylation, 35–36

Benzylphenylsulfide sulfoxygenation, vanadium catalyst, 66–67, 68*t*

Binary vanadium(IV,V)-(α-hydroxycarboxylate) aqueous systems, 379–385

Bioinorganic vanadium coordination complexes mimicking active site, vanadium haloperoxidases, 187–195